STUDENT'S SOLUTIONS MANUAL

CARRIE GREEN

MATHEMATICAL IDEAS
TWELFTH EDITION

Charles D. Miller

Vern E. Heeren
American River College

John Hornsby
University of New Orleans

Addison-Wesley
is an imprint of

PEARSON

Addison-Wesley
is an imprint of

www.pearsonhighered.com

Contents

CONTENTS

CONTENTS

Chapter 1

1.1 Exercises

1. This is an example of a deductive argument because a specific conclusion, "you can expect it to be ready in ten days," is drawn from the two given premises.

3. This is an example of inductive reasoning because you are reasoning from a specific pattern to the conclusion that "It will also rain tomorrow."

5. This represents deductive reasoning since you are moving from a general rule (addition) to a specific result (the sum of 85 and 20).

7. This is a deductive argument where you are reasoning from the two given premises. The first, "If you build it, they will come" is a general statement and the conclusion, "they will come" is specific.

9. This is an example of inductive reasoning because you are reasoning from a specific pattern of all previous attendees to a conclusion that the next one "I" will also be accepted into graduate school.

11. This is an example of inductive reasoning because you are reasoning from a specific pattern to a generalization as to what is the next element in the sequence.

13. Writing exercise; answers will vary.

15. Each number in the list is obtained by adding 3 to the previous number. The most probable next term is $18 + 3 = 21$.

17. Each number in the list is obtained by multiplying the previous number by 4. The most probable next term is $4 \times 768 = 3072$.

19. Beginning with the third term, each number in the sequence is the sum of the two previous terms.
$$9 = (3 + 6)$$
$$15 = (6 + 9)$$
$$24 = (9 + 15)$$
$$39 = (15 + 24)$$
The most probable next term is
$$63 = (24 + 39).$$

21. The numerators and denominators are consecutive counting numbers. The probable next term is $\dfrac{11}{12}$.

23. The most probable next term is $6^3 = 216$. Observe the sequence:
$$1 = 1^3$$
$$8 = 2^3$$
$$27 = 3^3$$
$$64 = 4^3$$
$$125 = 5^3$$
This sequence is made up of the cubes of each counting number.

25. The probable next term is 52. Note that each term (after the first) may be computed by adding successively 5, 7, 9, and 11 to each preceding term. Thus, it follows that a probable next term would be $39 + 13 = 52$.

27. The probable next term is 5 since the sequence of numbers seems to add one more 5 each time the 5's precede the number 3.

29. There are many possibilities. One such list is 10, 20, 30, 40, 50, … .

31.
$$(9 \times 9) + 7 = 88$$
$$(98 \times 9) + 6 = 888$$
$$(987 \times 9) + 5 = 8888$$
$$(9876 \times 9) + 4 = 88,888$$
Observe that on the left, the pattern suggests that the digit 5 will be appended to the first number. Thus, we get (98,765 × 9) which is added to 3. On the right, the pattern suggests appending another digit 8 to obtain 888,888. Therefore,
$$(98,765 \times 9) + 3 = 888,885 + 3 = 888,888$$
By computation, the conjecture is verified.

33.
$$3367 \times 3 = 10,101$$
$$3367 \times 6 = 20,202$$
$$3367 \times 9 = 30,303$$
$$3367 \times 12 = 40,404$$
Observe that on the left, the pattern suggests that 3367 will be multiplied by the next multiple of 3, which is 15. On the right, the pattern suggests the result 50,505. The pattern suggests the following equation:
$$3367 \times 15 = 50,505.$$
Multiply 3367×15 to verify the conjecture.

35.
$$34 \times 34 = 1156$$
$$334 \times 334 = 111,556$$
$$3334 \times 3334 = 11,115,556$$
The pattern suggests the following equation:
$$33,334 \times 33,334 = 1,111,155,556.$$
Multiply $33,334 \times 33,334$ to verify the conjecture.

37.
$$3 = \frac{3(2)}{2}$$
$$3 + 6 = \frac{6(3)}{2}$$
$$3 + 6 + 9 = \frac{9(4)}{2}$$
$$3 + 6 + 9 + 12 = \frac{12(5)}{2}$$
The pattern suggests the following equation:
$$3 + 6 + 9 + 12 + 15 = \frac{15(6)}{2}.$$
Since both the left and right sides equal 45, the conjecture is verified.

39.
$$5(6) = 6(6 - 1)$$
$$5(6) + 5(36) = 6(36 - 1)$$
$$5(6) + 5(36) + 5(216) = 6(216 - 1)$$
$$5(6) + 5(36) + 5(216) + 5(1296) = 6(1296 - 1)$$
Observe that the last equation may be written as:
$$5(6^1) + 5(6^2) + 5(6^3) + 5(6^4) = 6(6^4 - 1).$$
Thus, the next equation would likely be:
$$5(6) + 5(36) + 5(216) + 5(1296) + 5(6^5)$$
$$= 6(6^5 - 1)$$
or,
$$5(6) + 5(36) + 5(216) + 5(1296) + 5(7776).$$
$$= 6(7776 - 1)$$

41.
$$\frac{1}{2} = 1 - \frac{1}{2}$$
$$\frac{1}{2} + \frac{1}{4} = 1 - \frac{1}{4}$$
$$\frac{1}{2} + \frac{1}{4} + \frac{1}{8} = 1 - \frac{1}{8}$$
$$\frac{1}{2} + \frac{1}{4} + \frac{1}{8} + \frac{1}{16} = 1 - \frac{1}{16}$$
Observe that the last equation may be written as
$$\frac{1}{2^1} + \frac{1}{2^2} + \frac{1}{2^3} + \frac{1}{2^4} = 1 - \frac{1}{2^4}.$$
The next equation would be
$$\frac{1}{2^1} + \frac{1}{2^2} + \frac{1}{2^3} + \frac{1}{2^4} + \frac{1}{2^5} = 1 - \frac{1}{2^5}, \text{ or}$$

$$\frac{1}{2} + \frac{1}{4} + \frac{1}{8} + \frac{1}{16} + \frac{1}{32} = 1 - \frac{1}{32}.$$
Using the common denominator 32 for each fraction, the left and right side add (in each case) to $\frac{31}{32}$. The conjecture is, therefore, verified.

43. $1 + 2 + 3 + \ldots + 200$
Pairing and adjoining the first term to the last term, the second term to the second-to-last term, etc., we have:
$1 + 200 = 201$, $2 + 199 = 201$,
$3 + 198 = 201, \ldots$
There are 100 of these sums. Therefore,
$100 \times 201 = 20,100.$

45. $1 + 2 + 3 + \ldots + 800$
Pairing and adjoining the first term to the last term, the second term to the second-to-last term, etc., we have:
$1 + 800 = 801$, $2 + 799 = 801$,
$3 + 798 = 801, \ldots$
There are 400 of these sums. Therefore,
$400 \times 801 = 320,400.$

47. $1 + 2 + 3 + \ldots + 175$
Note that there are an odd number of terms. So consider omitting, for the moment, the last term and take $1 + 174 = 175$,
$2 + 173 = 175$, $3 + 172 = 175$, etc. There are $\frac{174}{2} = 87$ of these pairs in addition to the last term. Thus, $(87 \times 175) + 175$, or
$88 \times 175 = 15,400.$

49.
$$2 + 4 + 6 + \ldots + 100 = 2(1 + 2 + 3 + \cdots + 50)$$
$$= 2[25(1 + 50)]$$
$$= 2(1275)$$
$$= 2550$$

51. These are the number of chimes a clock rings, starting with 12 o'clock, if the clock rings the number of hours on the hour and 1 chime on the half-hour. The next most probable number is the number of chimes at 3:30, which is 1.

53. **(a)** Here are three examples.

623	841	584
−326	−148	−485
297	693	99

In each result, the middle digit is always 9, and the sum of the first and third digits is always 9 (considering 0 as the first digit if the difference has only two digits).

(b) Writing exercise; answers will vary.

55. $142,857 \times 1 = 142,857$
$142,857 \times 2 = 285,714$
$142,857 \times 3 = 428,571$
$142,857 \times 4 = 571,428$
$142,857 \times 5 = 714,285$
$142,857 \times 6 = 857,142$
Each result consists of the same six digits, but in a different order. But $142,857 \times 7 = 999,999$. Thus, the pattern doesn't continue.

1.2 Exercises

1. If we choose any term after the first term, and subtract the preceding term, the common difference is 10. Therefore, this is an arithmetic sequence. The next term in the sequence is $46 + 10 = 56$.

3. If any term after the first is multiplied by 3, the following term is obtained. Therefore, this is a geometric sequence. The next term in the sequence is $405 \cdot 3 = 1215$.

5. There is neither a common difference nor a common ratio. This is neither an arithmetic nor a geometric sequence.

7. If any term after the first is multiplied by $\frac{1}{2}$, the following term is obtained. Therefore, this is a geometric sequence. The next term in the sequence is $16 \cdot \frac{1}{2} = 8$.

9. There is neither a common difference nor a common ratio. This is neither an arithmetic nor a geometric sequence.

11. If we choose any term after the first term, and subtract the preceding term, the common difference is 2. Therefore, this is an arithmetic sequence. The next term in the sequence is $20 + 2 = 22$.

13. 1   4   11   22   37   56   <u>79</u>
    3   7   11   15   19   <u>23</u>
      4   4   4   4   (4)
Each line represents the difference of the two numbers above it. The number 23 is found from adding the predicted difference, (4), in line three to 19 in line 2. And 79 is found by adding 23, in line two, to 56 in line one. Thus, our next term in the sequence is 79.

15. 6   20   50   102   182   296   <u>450</u>
   14   30   52   80   114   <u>154</u>
     16   22   28   34   <u>40</u>
       6   6   6   (6)
Thus, our next term in the sequence is $154 = 296 = 450$.

17. 0   12   72   240   600   1260   2352   <u>4032</u>
   12   60   168   360   660   1092   <u>1680</u>
     48   108   192   300   432   <u>588</u>
       60   84   108   132   <u>156</u>
         24   24   24   (24)
Thus, our next term in the sequence is $1680 + 2352 = 4032$.

19. 5   34   243   1022   3121   7770   16799   <u>32758</u>
   29   209   779   2099   4649   9029   <u>15959</u>
     180   570   1320   2550   4380   <u>6930</u>
       390   750   1230   1830   <u>2550</u>
         360   480   600   <u>720</u>
           120   120   (120)
Thus, our next term in the sequence is $15959 + 16799 = 32,758$.

21. 1   2   4   8   16   31   (57)   99
     1   2   4   8   15   <u>26</u>   42
       1   2   4   7   11   <u>16</u>
         1   2   3   4   <u>5</u>
           1   1   1   (1)
The next term of the sequence is 57. Following this pattern, we predict that the number of regions determined by 8 points is 99. Use $n = 8$ in the formula
$$\frac{n^4 - 6n^3 + 23n^2 - 18n + 24}{24}.$$
$$\frac{8^4 - 6 \times 8^3 + 23 \times 8^2 - 18 \times 8 + 24}{24}$$
$$= \frac{4096 - 3072 + 1472 - 144 + 24}{24}$$
$$= \frac{2376}{24}$$
$$= 99$$
Thus, the result agrees with our prediction.

23. By the pattern, the next equation is $(4321 \times 9) - 1 = 38,888$.
To verify, calculate left side and compare, $38,889 - 1 = 38,888$.

25. $999,999 \times 2 = 1,999,998$
$999,999 \times 3 = 2,999,997$
By the pattern, the next equation is $999,999 \times 4 = 3,999,996$.
To verify, multiply left side to get $3,999,996 = 3,999,996$.

27.
$$3^2 - 1^2 = 2^3$$
$$6^2 - 3^2 = 3^3$$
$$10^2 - 6^2 = 4^3$$
$$15^2 - 10^2 = 5^3$$

Following this pattern, we see that the next equation will start with 21^2 since $15 + 6 = 21$. This equation will be $21^2 - 15^2 = 6^3$. The left side is $441 - 225 = 216$. The right side also equals 216.

29.
$$2^2 - 1^2 = 2 + 1$$
$$3^2 - 2^2 = 3 + 2$$
$$4^2 - 3^2 = 4 + 3$$

Following this pattern, we see that the next equation will be $5^2 - 4^2 = 5 + 4$.
To verify, the left side is $25 - 16 = 9$. The right side also equals 9.

31.
$$1 = 1 \times 1$$
$$1 + 5 = 2 \times 3$$
$$1 + 5 + 9 = 3 \times 5$$

The last term on the left side is 4 more than the previous last term. The first factor on the right side is the next counting number; the second factor is the next odd number. Thus, the probable next equation is
$1 + 5 + 9 + 13 = 4 \times 7$.
To verify, calculate both sides to arrive at $28 = 28$.

33. $1 + 2 + 3 + \ldots + 300$
$$S = \frac{300(300 + 1)}{2} = \frac{90300}{2} = 45,150$$

35. $1 + 2 + 3 + \ldots + 675$
$$S = \frac{675(675 + 1)}{2} = \frac{456300}{2} = 228,150$$

37. $1 + 3 + 5 + 7 + \ldots + 101$
Note that
$$n = \frac{1 + 101}{2} = 51 \text{ terms, so that } S = 51^2 = 2601.$$

39. $1 + 3 + 5 + \ldots + 999$
Observe that
$$n = \frac{1 + 999}{2} = 500 \text{ terms, so that}$$
$$S = 500^2 = 250,500.$$

41. Since each term in the second series is twice that of the first series, we might expect the sum to be twice as large or
$$S = 2 \times \frac{n(n + 1)}{2} = n(n + 1).$$

43. Writing exercise; answers will vary.

45. Figurate

Number	1st	2nd	3rd	4th	5th	6th	7th	8th
Triangular	1	3	6	10	15	21	28	36
Square	1	4	9	16	25	36	49	64
Pentagonal	1	5	12	22	35	51	70	92
Hexagonal	1	6	15	28	45	66	91	120
Heptagonal	1	7	18	34	55	81	112	148
Octagonal	1	8	21	40	65	96	133	176

47. $8(1)+1 = 9 = 3^2$; $8(3)+1 = 25 = 5^2$; $8(6)+1 = 49 = 7^2$; $8(10)+1 = 81 = 9^2$

49. The square numbers are 1, 4, 9, 25, 36, … .
  $1 \div 4 = 0$, remainder 1
  $4 \div 4 = 1$, remainder 0
  $9 \div 4 = 2$, remainder 1
  $16 \div 4 = 4$, remainder 0
  $25 \div 4 = 6$, remainder 1
  $36 \div 4 = 9$, remainder 0
  The pattern of remainders is 1, 0, 1, 0, 1, 0, … .

51. The square number 25 may be represented by the sum of the two triangular numbers 10 and 15. The square number 36 may be represented by the sum of the two triangular numbers 15 and 21.

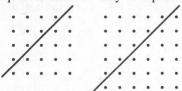

53. To find the sixteenth square number, use
  $S_n = n^2$ with $n = 16$.
  $S_{16} = 16^2 = 256$

55. To find the ninth pentagonal number, use
  $P_n = \dfrac{n(3n-1)}{2}$ with $n = 9$.
  $P_9 = \dfrac{9(26)}{2} = 117$.

57. To find the tenth heptagonal number, use
  $Hp_n = \dfrac{n(5n-3)}{2}$ with $n = 10$.
  $Hp_{10} = \dfrac{10(47)}{2} = 235$

59. Since each coefficient in parentheses appears to step up by 1, we would predict:

$$N_n = \frac{n(7n-5)}{2}.$$

$$N_6 = \frac{6(37)}{2} = 111$$

This verifies our prediction for $n = 6$.

61. The triangular numbers are
1, 3, 6, 10, 15, 21, 28, 36, 45, ...
Adding consecutive triangular numbers, for example,
$1 + 3 = 4, 3 + 6 = 9, 6 + 10 = 16, ...$, will give square numbers.

63. In each case, you get a perfect cube number. That is, if we take the 2nd and 3rd triangular numbers 3 and 6, $6^2 - 3^2 = 36 - 9 = 27$

which is the perfect cube number 3^3.

65. This sequence has a common difference of 4, so it is an arithmetic sequence with
$n = 11$
$a_1 = 2$
$d = 4.$
Using the formula,
$a_n = a_1 + (n-1)d$
$a_{11} = 2 + (11-1) \cdot 4$
$a_{11} = 2 + 10 \cdot 4$
$a_{11} = 42$
The eleventh term in the sequence is 42.

67. This sequence has a common difference of 20, so it is an arithmetic sequence with
$n = 21$
$a_1 = 19$
$d = 20.$
Using the formula,
$a_n = a_1 + (n-1)d$
$a_{21} = 19 + (21-1) \cdot 20$
$a_{21} = 19 + 20 \cdot 20$
$a_{21} = 419$
The 21st term in the sequence is 419.

69. This sequence has a common difference of $\frac{1}{2}$, so it is an arithmetic sequence with
$n = 101$
$a_1 = \frac{1}{2}$
$d = \frac{1}{2}.$
Using the formula,

$$a_n = a_1 + (n-1)d$$

$$a_{101} = \frac{1}{2} + (101-1) \cdot \frac{1}{2}$$

$$a_{101} = \frac{1}{2} + 100 \cdot \frac{1}{2}$$

$$a_{101} = \frac{101}{2}$$

The 101st term in the sequence is $\frac{101}{2}$.

71. This sequence has a common ration of 2, so it is a geometric sequence with
$n = 11$
$a_1 = 2$
$r = 2.$
Using the formula,
$a_n = a_1 \cdot r^{n-1}$
$a_{11} = 2 \cdot 2^{11-1}$
$a_{11} = 2 \cdot 2^{10}$
$a_{11} = 2048$
The eleventh term in the sequence is 2048.

73. This sequence has a common ration of $\frac{1}{2}$, so it is a geometric sequence with
$n = 12$
$a_1 = 1$
$r = \frac{1}{2}.$
Using the formula,

$$a_n = a_1 \cdot r^{n-1}$$

$$a_{12} = 1 \cdot \left(\frac{1}{2}\right)^{12-1}$$

$$a_{12} = 1 \cdot \left(\frac{1}{2}\right)^{11}$$

$$a_{12} = \frac{1}{2048}$$

The 12th term in the sequence is $\frac{1}{2048}$.

75. This sequence has a common ration of $\frac{1}{4}$, so it is a geometric sequence with
$n = 8$
$a_1 = 40$
$r = \frac{1}{4}.$
Using the formula,

$$a_n = a_1 \cdot r^{n-1}$$

$$a_8 = 40 \cdot \left(\frac{1}{4}\right)^{8-1}$$

$$a_8 = 40 \cdot \left(\frac{1}{4}\right)^{7}$$

$$a_8 = \frac{40}{16384}$$

$$a_8 = \frac{5}{2048}$$

The 8th term in the sequence is $\frac{5}{2048}$.

1.3 Exercises

1. When 8 girls leave, twice as many boys as girls remain. Because we started with an equal number of boys and girls, 8 is half the number of boys in the classroom. Thus, there are 16 boys, so there are 16 girls, for a total of 32 students.

3. In each row, multiply the first digit by the last digit. The result is the two-digit number between them.
$3 \times 8 = 24$
$7 \times 3 = 21$
$8 \times 5 = 40$
$4 \times 9 = 36$
The x is the digit in the ones place in the result $8 \times 4 = 40$, so x is 0.

5. Let X = the number of schools that took part in the race. Then there were $3x$ participants in the race. Erin finished in the middle position, so $3x$ is an odd number that is divisible by 3. Also, $3x > 28$, since Iliana finished 28th. The smallest odd number that is divisible by 3 and greater than 28 is 33. This must be the number of participants in the race because if there were 39 participants, then the middle position would have been 20, and Erin would have finished after Katelyn. So $3x = 33$, and $x = 11$. 11 schools sent participants to the cross-country meet.

7. Draw several squares, and for each square draw a different line that goes through the center of the square. Any line through the center of the square will divide the square into two halves that have the same size and shape. There are infinitely many lines that can be drawn through the center of a square, so there are infinitely many ways to cut a square in half.

9. The key to this problem is to think about the way the books are placed on the shelf. When they are placed in alphabetical order from left to right, that means volume A is on the far left and volume Z is on the far right. But think about where the covers for these volumes are. The front cover for volume A is on its right side, touching the back cover of volume B. The back cover for volume Z is on its left side, touching the front cover for volume Y. So the bookworm starts with the front cover of volume A $\left(\frac{1}{4} \text{ inch}\right)$ and eats through the entire volumes B through Y (24 books, 2 inches each, 48 inches total), then finishes by eating the back cover of volume Z $\left(\frac{1}{4} \text{ inch}\right)$. So the total amount eaten is 48.5 inches.

11. If we let n = the number of cats she has, then we can interpret her response as
$n = \left(\frac{5}{6}n + 7\right)$ and solve for n.

$$n = \left(\frac{5}{6}n + 7\right)$$
$$6n = 5n + 42$$
$$n = 42$$

13. Let f = the number of gallons in a full tank. Then

$$\frac{1}{8}f + 15 = \frac{3}{4}f$$
$$8\left(\frac{1}{8}f + 15\right) = 8\left(\frac{3}{4}f\right)$$
$$f + 120 = 6f$$
$$120 = 5f$$
$$f = 24 \text{ gallons}$$

Thus a full tank is 24 gallons. The van started with $\frac{1}{8}$ of a full tank or

$\frac{1}{8} \cdot 24 = 3$ gallons. Then 15 gallons were added for a total of $(3 + 15 = 18)$ gallons. So $(24 - 18) = 6$ gallons are needed to fill the tank.

15. 18, 38, 24, 46, 42
    8, 24, 8, 24, 8
By trial and error we might notice that if we multiply the two digits of each of the numbers in the first row, we get the corresponding number in the second row of numbers (8 and 24, which repeat).

17. I put the ring in the box and put my lock on the box. I send you the box. You put your lock on, as well, and send it back to me. I then remove my lock with my key and send you the box (with your lock still on) back to you, so you can remove your lock with your key and get the ring.

19. Given the sequence 16, 80, 48, 64, A, B, C, D, where each term is the arithmetic mean of the previous two terms: e.g. $48 = \dfrac{16+64}{2}$. We therefore know that $A = \dfrac{48+64}{2} = 56$; $B = \dfrac{64+56}{2} = 60$; $C = \dfrac{56+60}{2} = 58$; $D = \dfrac{60+58}{2} = 59$.

21. Choose a sock from the box labeled *red and green socks*. Since it is mislabeled, it contains only *red* socks or only *green* socks, determined by the sock you choose. If the sock is green, relabel this box *green socks*. Since the other two boxes were mislabeled, switch the remaining label to the other box and place the label that says *red and green socks* on the unlabeled box. No other choice guarantees a correct relabeling, since you can remove only one sock.

23. The total number of dots on each die is $1 + 2 + 3 + 4 + 5 + 6 = 21$. Thus the top die has (21 – dots showing), unseen dots, or $21 - (1 + 2 + 3) = 21 - 6 = 15$. The middle die has $21 - (4 + 6) = 21 - 10 = 11$. The bottom die has $21 - (5 + 1) = 21 - 6 = 15$ dots not shown. The total is $15 + 11 + 15 = 41$ dots not shown. This is option D.
Alternatively, since each die has 21 dots, there are $21 \times 3 = 63$ total dots. Thus, there are $63 - 22 = 41$ unseen dots.

25. Visualize (or create unfolded box strip) with "1" on top, and folding "2," "3," and "4" around the middle. Option A satisfies this result.

27. One example of a solution follows.

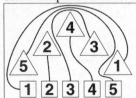

29. By trial and error, the following arrangement will work:

9 7 2 14   11 5 4 12 13 3 6 10   15   1 8

31. Use trial and error. One possible solution is as follows.

33. For units column assume that 1 is borrowed from the *a* digit. This suggests that $b = 9$ since $12 - 9 = 3$. To arrive at 7 in the tens column, we know that 8 must be subtracted from 15. Thus, $a = 6$ (remember that we borrowed one from that column, also). We borrowed one from the 7, as well, so that $c = 6 - 4 = 2$ in the hundreds column. Thus, $a + b + c = 6 + 9 + 2 = 17$.
This is represented by option D.

35. This exercise can be solved algebraically. If we let D = the total distance of the trip, and x = the distance traveled while asleep, then

$$x + \frac{1}{2}x = \frac{1}{2}D$$
$$2x + x = D$$
$$3x = D$$
$$x = \frac{1}{3}D$$

Thus, the distance traveled while asleep is $\frac{1}{3}$ of the total distance traveled.

37. Fill the big bucket. Pour into the small bucket. This leaves 4 gallons in the larger bucket. Empty the small bucket. Pour from the big bucket to fill up the small bucket. This leaves 1 gallon in the big bucket. Empty the small bucket. Pour 1 gallon from the big bucket to the small bucket. Fill up the big bucket. Pour into the small bucket. This leaves 5 gallons in the big bucket. Pour out the small bucket. This leaves exactly 5 gallons in the big bucket to take home. The above sequence is indicated by the following table.

Big bucket	7	4	4	1	1	0	7	5	5
Small bucket	0	3	0	3	0	1	1	3	0

39. Count systematically.

	No. of rows × No. of columns
15	1×1 rectangles
12	1×2 rectangles
9	1×3 rectangles
10	2×1 rectangles
8	2×2 rectangles
6	2×3 rectangles
5	3×1 rectangles
4	3×2 rectangles
3	3×3 rectangles
6	1×4 rectangles
4	2×4 rectangles
2	3×4 rectangles
3	1×5 rectangles
2	2×5 rectangles
<u>1</u>	3×5 rectangles
90	total rectangles

This gives a total of 90 rectangles.

41. One strategy is to assume the car was driving near usual highway speed limits (55-75 mph). We begin by trying 55 mph. In two hours the car would have traveled 110 miles. Adding 110 miles to the odometer reading, 15951, we get $110 + 15951 = 16061$ miles, which is palindromic. Thus, the speed of the car was 55 miles per hour.

43. Similar to Example 5 in the text, we might examine the units place and tens place for repetitive powers of 7 in order to explore possible patterns.

$$7^1 = \quad 07 \qquad 7^5 = \quad 16,807$$
$$7^2 = \quad 49 \qquad 7^6 = \quad 117,649$$
$$7^3 = \quad 343 \qquad 7^7 = \quad 823,543$$
$$7^4 = 2401 \qquad 7^8 = 5,764,801$$

Since the final two digits cycle over four values, we might consider dividing the successive exponents by 4 and examining their remainders. (Note: We are using inductive reasoning when we assume that this pattern will continue and will apply when the exponent is 1997.) Dividing the exponent 1997 by 4, we get a remainder of 1. This is the same remainder we get when dividing the exponent 1 (on 7^1) and 5 (on 7^5). Thus, we expect that the last two digits for 7^{1997} would be 07 as well.

45. Start with a smaller problem.

$$\begin{array}{rrrr}
1000 & 10,000 & 100,000 & 1,000,000 \\
-50 & -50 & -50 & -50 \\
\hline
950 & 9950 & 99,950 & 999,950
\end{array}$$

Following this pattern, the number that is the result of $10^{50}-50$ has one zero, one 5, and $(50-2)=48$ 9's. Thus, $(48\times9)+5+0=432+5+0=437$.

47. Similar to Example 5 in the text (and Exercise 46 above), we might examine the units place for repetitive powers of 7 in order to explore possible patterns.

$$\begin{array}{ll}
7^1 = 7 & 7^5 = 16,807 \\
7^2 = 49 & 7^6 = 117,649 \\
7^3 = 343 & 7^7 = 823,543 \\
7^4 = 2401 & 7^8 = 5,764,801
\end{array}$$

Since the units digit cycles over four values, we might consider dividing the successive exponents by 4 and examining their remainders. Divide the exponent 491 by 4 to get a quotient of 122 and a remainder of 3. Reasoning inductively, the units digit would be the same as that of 7^3 and 7^7, which is 3.

49. Joanie will want to use as many eight-cent stamps as possible. Since no multiple of 8 has 3 as its last digit, she will need an odd number of five-cent stamps. Working backward, find the largest multiple of 8 that is less than 153 and has 8 as its last digit. The number is 128, or $8 \cdot 16$. So Joanie should use 16 eight-cent stamps and 5 five-cent stamps, for a total of 21 stamps.

51. To find the minimum number of socks to pull out, guess and check. There are two colors of socks. If you pull out 2 socks, you could have 2 of one color or 1 of each color.

You must pull out more than 2 socks. If you pull out 3 socks, you might have 3 of one color or 1 of one color and 2 of the other. In either case, you have a matching pair, so 3 is the minimum number of socks to pull out.

53. To count the triangles, it helps to draw sketches of the figure several times. There are 5 triangles formed by two sides of the pentagon and a diagonal. There are 4 triangles formed with each side of the pentagon as a base, so there are $4\times5=20$ triangles formed in this way. Each point of the star forms a small triangle, so there are 5 of these. Finally, there are 5 triangles formed with a diagonal as a base. In each, the other two sides are inside the pentagon. (None of these triangles has a side common to the pentagon.) Thus, the total number of triangles in the figure is $5+20+5+5=35$.

55. Use trial and error to find the smallest perfect number. Try making a chart such as the following one.

Number	Divisors other than itself	Sum
1	None	
2	1	1
3	1	1
4	1, 2	3
5	1	1
6	1, 2, 3	6

Six is the smallest perfect number.

57. Working backward, we see that if the lily pad doubles its size each day so that it completely covers the pond on the twentieth day, the pond was half-covered on the previous (or nineteenth) day.

59. From condition (2), we can figure that since the author is living now, the year must be 196_, since $9-3=6$. Then, from condition (1), $23-(1+9+6)=7$, so the year is 1967.

61. By Eve's statement, Adam must have $2 more than Eve. But according to Adam, a loss of $1 from Eve to Adam gives Adam twice the amount that Eve has. By trial and error, the counting numbers 5 and 7 are the first to satisfy both conditions. Thus Eve has $5, and Adam has $7.

63. The first digit in the answer cannot be 0, 2, 3, or 5, since these digits have already been used. It cannot be more than 3, since one of the factors is a number in the 30's, making it impossible to get a product over 45,000. Thus, the first digit of the answer must be 1. To find the first digit in the 3-digit factor, use estimation. Dividing a number between 15,000 and 16,000 by a number between 30 and 40 could give a result with a first digit of 3, 4, or 5. Since 3 and 5 have already been used, this first digit must be 4. Thus, the 3-digit factor is 402. We now have the following.

$$
\begin{array}{r}
\underline{4}\ 0\ 2 \\
\times\ \ \ \ \ \underline{3} \\
\hline
\underline{1}\ \ 5,\ \ \ \ \
\end{array}
$$

To find the units digit of the 2-digit factor, use trial and error with the digits that have not yet been used: 6, 7, 8, and 9.

$36 \times 402 = 14,472$ (too small and reuses 2 and 4)
$37 \times 402 = 14,874$ (too small and reuses 4)
$38 \times 402 = 15,276$ (reuses 2)
$39 \times 402 = 15,678$ (correct)

The correct problem is as follows.

$$
\begin{array}{r}
4\ 0\ 2 \\
\times\ \ \ \ 3\ 9 \\
\hline
1\ \ 5,\ 6\ 7\ 8
\end{array}
$$

Notice that a combination of strategies was used to solve this problem.

65. Notice that the first column has three given numbers. Thus, $34 - (6 + 11 + 16) = 1$
is the first number in the second row. (Note: You could use the diagonal to solve for missing number in the same manner.) Then, $34 - (1 + 15 + 14) = 4$ is in the second row, third column. The diagonal from upper left to lower right has three given numbers. Therefore, $34 - (6 + 15 + 10) = 3$ is in the fourth row, fourth column. Continue filling in the missing numbers until the magic square is completed.

6	12	7	9
1	15	4	14
11	5	10	8
16	2	13	3

67. 25 pitches: Game tied 0 to 0 going into the 9th inning. Each pitcher has pitched a minimum of 24 pitches (three per inning). The winning pitcher pitches 3 more (fly ball/out) pitches for a total of 27. The losing (visiting team) pitcher pitches 1 more (for a total of 25) which happens to be a home run, thus, losing the game by a score of 1-0. (Note: the same result occurs if the losing pitcher gives up one homerun in any inning.)

69. Draw a sketch, visualize, or cut a piece of paper to build the cube. The cube may be folded with Z on the front.

Then, E is on top and M is on the left face. This places Q opposite the face marked Z. (D is on the bottom and X is on the right face.)

71. A solution, found by trial and error, is shown here.

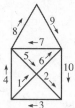

73. Solve this problem by making a list. First, find the ways he can use pennies to make 15 cents.
15 pennies
10 pennies, 1 nickel
5 pennies, 1 dime
5 pennies, 2 nickels
Find additional ways he can use nickels.
3 nickels
1 nickel, 1 dime
There are 6 ways to make 15 cents, so there are 6 ways he can pay 15 cents for a chocolate mint.

75. The triangle is a right triangle because $10^2 + 24^2 = 26^2$. The area of the triangle is $\frac{1}{2}(24)(10) = 120$ sq cm. The rectangle has the same area as the triangle, with width 3 cm, so the length of the rectangle is $\frac{120}{3} = 40$ cm. Then the perimeter of the rectangle is $2(40) + 2(3) = 80 + 6 = 86$ cm.

77. Jessica is married to James or Dan. Since Jessica is married to the oldest person in the group, she is not married to James, who is younger than Cathy. So Jessica is married to Dan, and Cathy is married to James. Since Jessica is married to the oldest person, we know that Dan is 36. Since James is older than Jessica but younger than Cathy, we conclude that Cathy is 31, James is 30, and Jessica is 29.

79. This is a problem with a "catch." The obvious answer is that only one month, February, has 28 days. However, the problem does not specify exactly 28 days, so any month with at least 28 days qualifies. All 12 months have 28 days.

81. Find the pattern in the last digit by repeatedly multiplying 7 by itself:

$$7^1 = 7$$
$$7^2 = 49$$
$$7^3 = 343$$
$$7^4 = 2401$$
$$7^5 = 16,807$$
$$7^6 = 117,649$$

The pattern in the last digit of each power of 7 is 7, 9, 3, 1 and then the pattern repeats. Divide the exponent 1783 by 4, the number of digits in the pattern: $1783 \div 4 = 445$ with a remainder of 3. The third digit in the pattern is 3, so the last digit of $49,327^{1783}$ is 3.

83. The maximum number of squares is 6. One possible array is as follows.

	X	X
X		X
X	X	

85. The sequence of numbers is
1, 1, 2, 3, 5, 8, 13, 21, 34, 55, 89, 144, ...
Look at several examples of four successive terms.

Terms	Product of first and fourth	Product of middle terms
1, 1, 2, 3	$1 \cdot 3 = 3$	$1 \cdot 2 = 2$
5, 8, 13, 21	$5 \cdot 21 = 105$	$8 \cdot 13 = 104$
2, 3, 5, 8	$2 \cdot 8 = 16$	$3 \cdot 5 = 15$
8, 13, 21, 34	$8 \cdot 34 = 272$	$13 \cdot 21 = 273$

The products in each row differ by 1.

1.4 Exercises

Using a graphing calculator, such as the TI-84, we would enter the expressions as indicated on the left side of the equality then push [Enter] to arrive at the answer. When using scientific or other types of calculators some adjustments will have to be made. See observations related to the solutions for Exercise 13 below. It is a good idea to review your calculator handbook for related examples.

1. $39.7 + (8.2 - 4.1) = 43.8$

3. $\sqrt{5.56440921} = 2.3589$

5. $\sqrt[3]{418.508992} = 7.48$

7. $2.67^2 = 7.1289$

9. $5.76^5 \approx 6340.338097$

Observe that when using a calculator, the numerator must be grouped in parentheses as must the denominator. This will make the last operation (the indicated) division.

11. $\dfrac{(14.32 - 8.1)}{(2 \times 3.11)} = 1$

13. $\sqrt[5]{1.35} \approx 1.061858759$. Observe that many scientific calculators have only the $\sqrt[2]{}$ function built into the calculator. For an index larger than 2, you might want to think of the nth root of a number b as equivalent to the exponential expression $b^{1/n}$. For example, $\sqrt[5]{1.35} = (1.35)^{1/5}$. Then use your exponentiation function (button) to calculate the 5th root of 1.35. Note that you will enter the exponent $\frac{1}{5}$ on the calculator in parentheses as $(1 \div 5)$.

15. $\frac{\pi}{\sqrt{2}} \approx 2.221441469$

17. $\sqrt[4]{\frac{2143}{22}} \approx 3.141592653$

19. $\frac{\sqrt{2}}{\sqrt[3]{6}} \approx 0.7782717162$

21. Choose a five-digit number such as 73,468.
$$73468 \times 9 = 661212$$
$$6+6+1+2+1+2 = 18$$
$$1+8 = 9$$
Choose a six-digit number such as 739,216.
$$739216 \times 9 = 6652944$$
$$6+6+5+2+9+4+4 = 36$$
$$3+6 = 9$$
Yes, the same result holds.

23. $$(-3) \div (-8) = 0.375$$
$$(-5) \div (-4) = 1.25$$
$$(-2.7) \div (-4.3) \approx 0.6279069767$$
Dividing a negative number by another negative number gives a <u>positive</u> number.

25. $5.6^0 = 1;\ \pi^0 = 1;\ 2^0 = 1;\ 120^0 = 1;$
Raising a nonzero number to the power 0 gives a result of <u>1</u>.

27. $\frac{1}{7} \approx .1428571$

$\frac{1}{(-9)} \approx -.1111111$

$\frac{1}{3} \approx .3333333$

$\frac{1}{(-8)} = -.125$

The sign of the reciprocal of a number is <u>the same as</u> the sign of the number.

29. $(0/8) = 0;\ (0/-2) = 0;\ (0/\pi) = 0$
Zero divided by a nonzero number gives a quotient of <u>0</u>.

31. $(-3) \times (-4) \times (-5) = -60$
$(-3) \times (-4) \times (-5) \times (-6) \times (-7) = -2520$
$(-3) \times (-4) \times (-5) \times (-6) \times (-7) \times (-8) \times (-9)$
$= -181440$
Multiplying an *odd* number of negative numbers gives a <u>negative</u> product.

33. Writing exercise; answers will vary.

35. Writing exercise; answers will vary.

37. $563 \div 9 \approx 62.555556$. Since more than 62 are needed, we require 63 pages.

39. $800 \div 60 \approx 13.333$. Since more than 13 are needed, we require 14 containers.

41. $140,000 \div 80 \approx 160,000 \div 80 = \$2000$; option B

43. $40,249 \div 968 \approx \left(\frac{40,000}{1000}\right) = 40$; option A

45. Approximating the numbers for ease of calculation we have
1100 yards $\div$ 100 passes = 11 yards/catch; option D

47. Add the given percentages of known countries (52% + 13% + 30%) to get 95%. Subtract from 100% to get remaining area of pie chart:
$100\% - 95\% = 5\%$

49. Thirteen percent of the 2,000,000 immigrants we would expect to arrive from Europe. Thus,
$13\% \times 2,000,000 = 0.13 \times 2,000,000$
$$= 260,000$$

51. U.S. milk production was greater than 175 billion pounds in 2005, 2006, and 2007.

53. 2001: about 165 billion pounds;
2007: about 185 billion pounds

55. The greatest increase in price occurred from 2000 to 2005. The increase was about $\$2.25 - \$1.40 = \$0.85$.

57. The price of a gallon of gas was decreasing.

Chapter 1 Test

1. This is an example of inductive reasoning, since you are reasoning from a specific pattern to the general conclusion that she will again exceed her annual sales goal.

2. This is a deductive argument because you are reasoning from the stated general property to the specific result, 176^2 is a natural number.

3. Add the elements in the center column together to get 38. Each column and diagonal must also add to 38. This yields the following magic hexagon.

4. The specific pattern seems to indicate that the second factor in the product is a multiple of 17 and the digits on the right side of the equation increase by 1. If this pattern is correct, then the next term in the sequence would be
$65,359,477,124,183 \times 68$
$= 4,444,444,444,444,444$
since $4 \times 17 = 68$. This can be verified by multiplying $65,359,477,124,183 \times 68$ on your calculator.

5.
```
3   11   31   69   131   223   351
  8   20   38   62   92   128
    12   18   24   30   36
       6    6    6   (6)
```
Thus, our next term in the sequence is $128 + 223 = 351$.

6. Using the method of Gauss, we have
$1 + 250 = 251, 2 + 249 = 251$, etc.
There are $\dfrac{250}{2} = 125$ such pairs, so the sum can be calculated as $125 \times 251 = 31,375$.

7. The next predicted octagonal number is 65, since the next equation on the list would be $65 = 1 + 7 + 13 + 19 + 25$, where $25 = 19 + 6$.

8. Beginning with the first five octagonal numbers and applying the method of successive differences, we get
```
1   8   21   40   65   96   133   176
  7   13   19   25   31   37   43
    6   6   6   6   (6)   (6)
```
Dividing each octagonal number by 4 we get the following pattern of remainders: 1, 0, 1, 0, 1, 0, 1, 0, ...

9. After the first two terms (both of which are 1), we can find the next by adding the two proceeding terms. That is, to get the 3rd term, add $1 + 1 = 2$; the 4th term, $1 + 2 = 3$; the 5th term, $2 + 3 = 5$; and so forth.

10. To make the fraction as small as possible we want the smallest possible numerator (24) and the largest possible denominator (96). Thus, we get the fraction $\dfrac{24}{96}$, which reduces to $\dfrac{1}{4}$.

11. Examine the units place for repetitive powers of 9 in order to explore possible patterns.
$9^1 = 9 \qquad 9^3 = 729 \qquad 9^5 = 59049$
$9^2 = 81 \qquad 9^4 = 6561 \qquad 9^6 = 531441$
If we divide the exponent 1997 by 2 (since the pattern of the units digit cycles after every 2nd power), we get a remainder of 1. Noting that in the line of 9^1, where each exponent when divided by 2 yields a remainder of 1, there is a units digit of 9, we reason inductively that 9^{1997} has the same units digit, 9.

12. There are 5 smaller triangles representing the extremities of the inside star. There are 5 triangles outside (between) the extremities of the star. Each (outside) line segment forms the base of (5) isosceles triangles that have their apex at each point of the star. Using the line segment connecting two points of the star as a base, two triangles can be formed; one (outside) with a point on the star as an opposite vertex; and one (inside) with opposite vertex at the intersection of any two lines forming the star. There is a total of 10 of these isosceles triangles. This gives a complete total of 35 triangles.

13. Answers will vary. One possible solution is $1 + 2 + 3 - 4 + 5 + 6 + 78 + 9 + 0 = 100$.

14. Dr. Small is 36 inches tall, and he shrinks 2 inches per year. So he will disappear after $36/2 = 18$ years. In 18 years, Ms. Tall will grow $18 \cdot \frac{2}{3} = 12$ inches, so she will be $96 + 12 = 108$ inches tall, or 9 ft.

15. Observe the following patterns on successive powers of 11, 14, and 16 in order to determine the units value of each term in the sum $11^{11} + 14^{14} + 16^{16}$.

$11^1 = 11$ $14^1 = 14$ $16^1 = 16$

$11^2 = 121$ $14^2 = 196$ $16^2 = 256$

$11^3 = 1331$ $14^3 = 2744$ $16^3 = 4096$

                        $14^4 = 38146$

Thus, we would expect 11^{11} to have the same unit digit value of 1. Since powers of 14 have units digits which cycles between 4 and 6, we observe that division of the exponents by 2 yield remainders of 1 or 0. We might expect the same pattern to continue to 14^{14}. Division of the exponent by 2 gives a remainder of 0. We get the same remainder, 0, for all even powers on 14, and each of these numbers has a units digit of 6. The powers of 16 seem to all have the same unit value of 6. Thus, if we add the units digits $1 + 6 + 6 = 13$, we see that the units digit of this sum is 3.

16. Making the following observations
$9 \times 1 = 9$
$9 \times 2 = 18 \quad (1 + 8 = 9)$
$9 \times 3 = 27 \quad (2 + 7 = 9)$
$9 \times 4 = 36 \quad (3 + 6 = 9)$
$9 \times 5 = 45 \quad (4 + 5 = 9)$
suggests that the sum of the digits in the product will always be 9.

17. $\sqrt{98.16} \approx 9.907572861$ But answers may vary depending upon what calculator you are using.

18. $3.25^3 = 34.328125$

19. The ratio of made shots to those attempted is approximately $\frac{30}{100} = \frac{3}{10}$. So in 10 attempts, we would expect her to make about 3 shots; option B

20. (a) The unemployment rate decreased between 1998 and 1999, between 199 and 2000, and between 2003 and 2004.

    (b) Between 2000 and 2003, the unemployment rate was increasing.

    (c) 2003: 6.0%; 2004: 5.5%; The unemployment rate declined by 0.5%.

Chapter 2

2.1 Exercises

1. {1, 3, 5, 7, 9} matches F, the set of odd positive integer less than 10.

3. {..., −4, −3, −2, −1} matches E, the set of all negative integers.

5. {2, 4, 8, 16, 32} matches B, the set of the five least positive integer powers of 2, since each element represents a successive power of 2 beginning with 2^1.

7. {2, 4, 6, 8, 10} matches H, the set of the five least positive integer multiples of 2, since this set represents the first five positive even integers. Remember that all even numbers are multiples of 2.

9. The set of all counting numbers less than or equal to 6 can be expressed as {1, 2, 3, 4, 5, 6}.

11. The set of all whole numbers not greater than 4 can be expressed as {0, 1, 2, 3, 4}.

13. In the set {6, 7, 8, ..., 14}, the ellipsis (three dots) indicates a continuation of the pattern. A complete listing of this set is {6, 7, 8, 9, 10, 11, 12, 13, 14}.

15. The set {−15, −13, −11, ..., −1} contains all integers from −15 to −1 inclusive. Each member is two larger than its predecessor. A complete listing of this set is {−15, −13, −11, −9, −7, −5, −3, −1}.

17. The set {2, 4, 8, ..., 256} contains all powers of two from 2 to 256 inclusive. A complete listing of this set is {2, 4, 8, 16, 32, 64, 128, 256}.

19. A complete listing of the set {x|x is an even whole number less than 11} is {0, 2, 4, 6, 8, 10}. Remember that 0 is the first whole number.

21. The set of all counting numbers greater than 20 is represented by the listing {21, 22, 23, ...}.

23. The set of Great Lakes is represented by {Lake Erie, Lake Huron, Lake Michigan, Lake Ontario, Lake Superior}.

25. The set {x|x is a positive multiple of 5} is represented by the listing {5, 10, 15, 20, ...}.

27. The set {x|x is the reciprocal of a natural number} is represented by the listing $$\left\{1, \frac{1}{2}, \frac{1}{3}, \frac{1}{4}, \frac{1}{5}, \ldots\right\}.$$

Note that in Exercises 29–31, there are other ways to describe the sets.

29. The set of all rational numbers may be represented using set-builder notation as {x|x is a rational number}.

31. The set {1, 3, 5, ..., 75} may be represented using set-builder notation as {x|x is an odd natural number less than 76}.

33. {−9, −8, −7, ..., 7, 8, 9} is the set of single-digit integers.

35. {Alabama, Alaska, Arizona, ..., Wisconsin, Wyoming} is the set of states of the United States.

37. The set {2, 4, 6, ..., 932} is finite since the cardinal number associated with this set is a whole number.

39. The set $\left\{\frac{1}{2}, \frac{2}{3}, \frac{3}{4}, \ldots\right\}$ is infinite since there is no last element, and we would be unable to count all of the elements.

41. The set {x|x is a natural number greater than 50} is infinite since there is no last element, and therefore its cardinal number is not a whole number.

43. The set {x|x is a rational number} is infinite since there is no last element, and therefore its cardinal number is not a whole number.

45. For any set A, n(A) represents the cardinal number of the set, that is, the number of elements in the set. The set A = {0, 1, 2, 3, 4, 5, 6, 7} contains 8 elements. Thus, n(A) = 8.

47. The set A = {2, 4, 6, ..., 1000) contains 500 elements. Thus, n(A) = 500.

49. The set A = {a, b, c, ..., z} has 26 elements (letters of the alphabet). Thus, $n(A) = 26$.

51. The set A = the set of integers between −20 and 20 has 39 members. The set can be indicated as {−19, −18, ..., 18, 19}, or 19 negative integers, 19 positive integers, and 0. Thus, $n(A) = 39$.

53. The set $A = \left\{\dfrac{1}{3}, \dfrac{2}{4}, \dfrac{3}{5}, \dfrac{4}{6}, ..., \dfrac{27}{29}, \dfrac{28}{30}\right\}$ has 28 elements. Thus, $n(A) = 28$.

55. Writing exercise; answers will vary.

57. The set {$x|x$ is a real number} is well defined since we can always tell if a number is real and belongs to this set.

59. The set {$x|x$ is a difficult course} is not well defined since membership is a value judgment, and there is no clear-cut way to determine whether a particular course is "difficult."

61. $5 \in$ {2, 4, 5, 7} since 5 is a member of the set.

63. $-12 \notin$ {3, 8, 12, 18} because −12 is not a member of the set.

65. {3} $\notin$ {2, 3, 4, 6} since the elements are not sets themselves.

67. $8 \in$ {11 − 2, 10 − 2, 9 − 2, 8 − 2} since 8 = 10 − 2.

69. The statement $3 \in$ {2, 5, 6, 8} is false since the element 3 is not a member of the set.

71. The statement b ∈ {h, c, d, a, b} is true since b is contained in the set.

73. The statement $9 \notin$ {6, 3, 4, 8} is true since 9 is not a member of the set.

75. The statement {k, c, r, a} = {k, c, a, r} is true since both sets contain exactly the same elements.

77. The statement {5, 8, 9} = {5, 8, 9, 0} is false because the second set contains a different element from the first set, 0.

79. The statement {4} ∈ {{3}, {4}, {5}} is true since the element, {4}, is a member of the set.

81. The statement {$x|x$ is a natural number less than 3} = {1, 2} is true since both represent sets with exactly the same elements.

83. The statement $4 \in A$ is true since 4 is a member of set A.

85. The statement $4 \notin C$ is false since 4 is a member of the set C.

87. Every element of C is also an element of A is true since the members, 4, 10, and 12 of set C, are also members of set A.

89. Writing exercise; answers will vary.

91. An example of two sets that are not equivalent and not equal would be {3} and {c, f}. Other examples are possible.

93. An example of two sets that are equivalent but not equal would be {a, b} and {a, c}. Other examples are possible.

95. (a) The actors with a return of at least $7.40 are those listed in the set {Drew Barrymore, Leonardo DiCaprio, Samuel L. Jackson, Jim Carrey}.

(b) The actors with a return of at most $3.75 are those listed in the set {Will Ferrell, Ewan McGregor}.

2.2 Exercises

1. {p}, {q}, {p, q}, $\varnothing$ matches D, the subsets of {p, q}.

3. {a, b} matches B, the complement of {c, d}, if U = {a, b, c, d}.

5. {−2, 0, 2} $\not\subseteq$ {−2, −1, 1, 2}

7. {2, 5} $\subseteq$ {0, 1, 5, 3, 7, 2}

9. $\varnothing \subseteq$ {a, b, c, d, e}, since the empty set is considered a subset of any given set.

11. {−5, 2, 9} $\not\subseteq$ {$x|x$ is an odd integer} since the element "2" is not an element of the second set.

13. {P, Q, R} $\subseteq$ {P, Q, R, S} and {P, Q, R} $\subseteq$ {P, Q, R, S}, i.e. both.

15. {9, 1, 7, 3, 5} $\subseteq$ {1, 3, 5, 7, 9}

17. $\varnothing \subseteq \{0\}$ or $\varnothing \subsetneq \{0\}$, i.e., both.

19. $\{0, 1, 2, 3\} \not\subseteq \{1, 2, 3, 4\}$; therefore, neither. Note that if a set is not a subset of another set, it cannot be a proper subset either.

21. $A \subset U$ is true since all sets must be subsets of the universal set by definition, and U contains at least one more element than A.

23. $D \subseteq B$ is false since the element "d" in set D is not also a member of set B.

25. $A \subset B$ is true. All members of A are also members of B, and there are elements in set B not contained in set A.

27. $\varnothing \not\subset A$ is false since $\varnothing$ is a subset of all sets.

29. $\varnothing \subseteq \varnothing$ is true since the empty set, $\varnothing$, is considered a subset of all sets including itself. Note that all sets are subsets of themselves.

31. $D \not\subseteq B$ is true. Set D is not a subset of B because the element "d," though a member of set D, is not also a member of set B.

33. There are exactly 6 subsets of C is false. Since there are 3 elements in set C, there are $2^3 = 8$ subsets.

35. There are exactly 3 proper subsets of A is true. Since there are 2 elements in set A, there are $2^2 = 4$ subsets, and one of those is the set A itself, so there are 3 proper subsets of A.

37. There is exactly one subset of $\varnothing$ is true. The only subset of $\varnothing$ is $\varnothing$ itself.

39. The Venn diagram does not represent the correct relationships among the sets since D is not a subset of A. Thus, the answer is false.

41. Since the given set has 6 elements, there are
(a) $2^6 = 64$ subsets, and
(b) $2^6 - 1 = 63$ proper subsets.

43. The set
$\{x \mid x$ is an odd integer between -4 and $6\}$
$= \{-3, -1, 1, 3, 5\}$. Since the set contains 5 elements, there are (a) $2^5 = 32$ subsets and (b) $2^5 - 1 = 32 - 1 = 31$ proper subsets.

45. The complement of $\{2, 3, 4, 6, 8\}$ is $\{5, 7, 9, 10\}$, that is, all of the elements in U not also in the given set.

47. The complement of $\{1, 3, 4, 5, 6, 7, 8, 9, 10\}$ is $\{2\}$.

49. The complement of the universal set, U, is the empty set, $\varnothing$.

51. In order to contain all of the indicated characteristics, the universal set $U =$ {Higher cost, Lower cost, Educational, More time to see the sights, Less time to see the sights, Cannot visit relatives along the way, Can visit relatives along the way}.

53. Since D contains the set of characteristics of the driving option, $D' =$ {Higher cost, More time to see the sights, Cannot visit relatives along the way}.

55. The set of element(s) common to F' and D' is $\varnothing$, the empty set, since there are no common elements.

57. The only possible set is $\{A, B, C, D, E\}$. (All are present.)

59. The possible subsets of three people would include $\{A, B, C\}$, $\{A, B, D\}$, $\{A, B, E\}$, $\{A, C, D\}$, $\{A, C, E\}$, $\{A, D, E\}$, $\{B, C, D\}$, $\{B, C, E\}$, $\{B, D, E\}$, and $\{C, D, E\}$.

61. The possible subsets consisting of one person would include $\{A\}$, $\{B\}$, $\{C\}$, $\{D\}$, and $\{E\}$.

63. Adding the number of subsets in Exercises 57–62, we have
$1 + 5 + 10 + 10 + 5 + 1 = 32$ ways that the group can gather.

65. Because at least one member must attend, sending no members (the empty set) is not a possible subset of the members that can be sent. So the total number of different delegations that can possibly be sent is
$2^{25} - 1 = 33,554,431$.

67. **(a)** Consider all possible subsets of a set with four elements (the number of bills). The number of subsets would be $2^4 = 16$. Since 16 includes also the empty set (and we must choose one bill), we will subtract one from this or $16 - 1 = 15$ possible sums of money.

(b) Removing the condition says, in effect, that we may also choose no bills. Thus, there are $2^4 = 16$ subsets or possible sums of money; it is now possible to select no bills.

69. (a) There are s subsets of B that do not contain e. These are the subsets of the original set A.

(b) There is one subset of B for each of the original subsets of set A, which is formed by including e as the element of that subset of A. Thus, B has s subsets which do contain e.

(c) The total number of subsets of B is the sum of the numbers of subsets containing e and of those not containing e. This number is $s + s$ or $2s$.

(d) Adding one more element will always double the number of subsets, so we conclude that the formula 2^n is true in general.

2.3 Exercises

1. The intersection of A and B, $A \cap B$, matches B, the set of elements common to both A and B.

3. The difference of A and B, $A - B$, matches A, the set of elements in A that are not in B.

5. The Cartesian product of A and B, $A \times B$, matches E, the set of ordered pairs such that each first element is from A and each second element is from B, with every element of A paired with every element of B.

7. $X \cap Y = \{a, c\}$ since these are the elements that are common to both X and Y.

9. $Y \cup Z = \{a, b, c, d, e, f\}$ since these are the elements that are contained in Y or Z (or both).

11. $X \cup U = \{a, b, c, d, e, f, g\} = U$. Observe that any set union with the universal set will give the universal set.

13. $X' = \{b, d, f\}$ since these are the only elements in U not contained in X.

15. $X' \cap Y' = \{b, d, f\} \cap \{d, e, f, g\} = \{d, f\}$

17. $X \cup (Y \cap Z) = \{a, c, e, g\} \cup \{b, c\}$
$ = \{a, b, c, e, g\}$
Observe that the intersection must be done first.

19. $(Y \cap Z') \cup X$
$= (\{a, b, c\} \cap \{a, g\}) \cup \{a, c, e, g\}$
$= \{a\} \cup \{a, c, e, g\}$
$= \{a, c, e, g\}$
$= X$

21. $(Z \cup X')' \cap Y$
$= (\{b, c, d, e, f\} \cup \{b, d, f\})' \cap \{a, b, c\}$
$= \{b, c, d, e, f\}' \cap \{a, b, c\}$
$= \{a, g\} \cap \{a, b, c\}$
$= \{a\}$

23. $X - Y = \{e, g\}$
Since these are the only two elements that belong to X and not to Y.

25. $X \cap (X - Y) = \{a, c, e, g\} \cap \{e, g\} = \{e, g\}$
Observe that we must find $X - Y$ first.

27. $X' - Y = \{b, d, f\} - \{a, b, c\} = \{d, f\}$
Observe that we must find X' first.

29. $(X \cap Y') \cup (Y \cap X')$
$= (\{a, c, e, g\} \cap \{d, e, f, g\}) \cup (\{a, b, c\}$
$\phantom{= (\{a, c, e, g\} \cap \{d, e, f, g\})} \cap \{b, d, f\})$
$= \{e, g\} \cup \{b\}$
$= \{b, e, g\}$

31. $A \cup (B' \cap C')$ is the set of all elements that are in A, or are not in B and not in C.

33. $(C - B) \cup A$ is the set of all elements that are in C but not in B, or they are in A.

35. $(A - C) \cup (B - C)$ is the set of all elements that are in A but not C, or are in B but not in C.

37. The smallest set representing the universal set U is $\{e, h, c, l, b\}$.

39. T', the complement of T, is the set of effects in U that are not adverse effects of tobacco use: $T' = \{l, b\}$.

41. $T \cup A$ is the set of all adverse effects that are either tobacco related or alcohol related: $T \cup A = \{e, h, c, l, b\} = U$.

43. $B \cup C$ is the set of all tax returns showing business income or filed in 2009.

45. $C - A$ is the set of all tax returns filed in 2009 without itemized deductions.

47. $(A \cup B) - D$ is the set of all tax returns with itemized deductions or showing business income, but not selected for audit.

49. $A \subseteq (A \cup B)$ is always true since $A \cup B$ will contain all of the elements of A.

51. $(A \cap B) \subseteq A$ is always true since the elements of $A \cap B$ must be in A.

53. $n(A \cup B) = n(A) + n(B)$ is not always true. If there are any common elements to A and B, they will be counted twice.

55. (a) $X \cup Y = \{1, 2, 3, 5\}$

(b) $Y \cup X = \{1, 2, 3, 5\}$

(c) For any sets X and Y, $X \cup Y = Y \cup X$. This conjecture indicates that set union is a commutative operation.

57. (a) $X \cup (Y \cup Z)$
$= \{1, 3, 5\} \cup (\{1, 2, 3\} \cup \{3, 4, 5\})$
$= \{1, 3, 5\} \cup \{1, 2, 3, 4, 5\}$
$= \{1, 3, 5, 2, 4\}$

(b) $(X \cup Y) \cup Z$
$= (\{1, 3, 5\} \cup \{1, 2, 3\}) \cup \{3, 4, 5\}$
$= \{1, 3, 5, 2\} \cup \{3, 4, 5\}$
$= \{1, 3, 5, 2, 4\}$

(c) For any sets X, Y, and Z,
$X \cup (Y \cup Z) = (X \cup Y) \cup Z$.

This conjecture indicates that set union is an associative operation.

59. (a) $(X \cup Y)' = \{1, 3, 5, 2\}' = \{4\}$

(b) $X' \cap Y' = \{2, 4\} \cap \{4, 5\} = \{4\}$

(c) For any sets X and Y,
$(X \cup Y)' = X' \cap Y'$.

Observe that this conjecture is one form of De Morgan's Laws.

61. For example,
$X \cup \varnothing = \{N, A, M, E\} \cup \varnothing$
$= \{N, A, M, E\}$
$= X$;
for any set X, $X \cup \varnothing = X$.

63. The statement $(3, 2) = (5 - 2, 1 + 1)$ is true.

65. The statement $(6, 3) = (3, 6)$ is false. The parentheses indicate an ordered pair (where order is important) and corresponding elements in the ordered pairs must be equal.

67. The statement $\{6, 3\} = \{3, 6\}$ is true since order is not important when listing elements in sets.

69. The statement $\{(1, 2), (3, 4)\} = \{(3, 4), (1, 2)\}$ is true. Each set contains the same two elements, the order of which is unimportant.

71. To form the Cartesian product $A \times B$, list all ordered pairs in which the first element belongs to A and the second element belongs to B:
With $A = \{2, 8, 12\}$ and $B = \{4, 9\}$,
$A \times B = \{(2, 4), (2, 9), (8, 4), (8, 9), (12, 4), (12, 9)\}$.
To form the Cartesian product $B \times A$, list all ordered pairs in which the first element belongs to B and the second element belongs to A:
$B \times A = \{(4, 2), (4, 8), (4, 12), (9, 2), (9, 8), (9, 12)\}$.

73. For $A = \{d, o, g\}$ and $B = \{p, i, g\}$,
$A \times B = \{(d, p), (d, i), (d, g), (o, p), (o, i), (o, g), (g, p), (g, i), (g, g)\}$;
$B \times A = \{(p, d), (p, o), (p, g), (i, d), (i, o), (i, g), (g, d), (g, o), (g, g)\}$.

75. For $A = \{2, 8, 12\}$ and $B = \{4, 9\}$,
$n(A \times B) = n(A) \times n(B) = 3 \times 2 = 6$, or by counting the generated elements in Exercise 71 we also arrive at 6. In the same manner, $n(B \times A) = 2 \times 3 = 6$.

77. For $n(A) = 35$ and $n(B) = 6$,
$n(A \times B) = n(A) \times n(B) = 35 \times 6 = 210$
$n(B \times A) = n(B) \times n(A) = 6 \times 35 = 210$

79. To find $n(B)$ when $n(A \times B) = 72$ and $n(A) = 12$, we have:
$n(A \times B) = n(A) \times n(B)$
$72 = 12 \times n(B)$
$6 = n(B)$

81. Let $U = \{a, b, c, d, e, f, g\}$, $A = \{b, d, f, g\}$, and $B = \{a, b, d, e, g\}$.

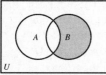

83. The set operations for $B \cap A'$ indicate those elements in B and not in A.

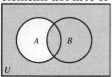

85. The set operations for $A' \cup B$ indicate those elements not in A or in B.

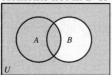

87. The set operations for $B' \cup A$ indicate those elements not in B or in A.

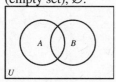

89. The set operations $B' \cap B$ indicate those elements not in B and in B at the same time, and since there are no elements that can satisfy both conditions, we get the null set (empty set), $\varnothing$.

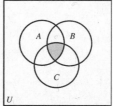

91. The indicated set operations mean those elements not in B or those not in A as long as they are also not in B. It is a help to shade the region representing "not in A" first, then that region representing "not in B." Identify the intersection of these regions (covered by both shadings). As in algebra, the general strategy when deciding which order to do operations is to begin inside parentheses and work out.

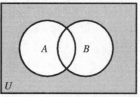

$A' \cap B'$

Finally, the region of interest will be that "not in B" along with (union of) the above intersection—$(A' \cap B')$. That is, the final region of interest is given by

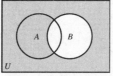

93. The complement of U, U', is the set of all elements not in U. But by definition, there can be no elements outside the universal set. Thus, we get the null (or empty) set, $\varnothing$, when we complement U.

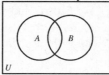

95. Let $U = \{m, n, o, p, q, r, s, t, u, v, w\}$, $A = \{m, n, p, q, r, t\}$, $B = \{m, o, p, q, s, u\}$, and $C = \{m, o, p, r, s, t, u, v\}$.
Placing the elements of these sets in the proper location on a Venn diagram will yield the following diagram.

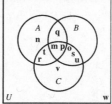

It helps to identify those elements in the intersection of A, B, and C first, then those elements not in this intersection but in each of the two set intersections (e.g., $A \cap B$, etc.), next, followed by elements that lie in only one set, etc.

97. The set operations $(A \cap B) \cap C$ indicate those elements common to all three sets.

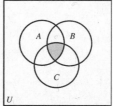

99. The set operations $(A \cap B) \cup C'$ indicate those elements in A and B at the same time along with those outside of C.

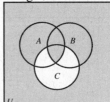

101. The set operations $(A' \cap B') \cap C$ indicate those elements that are in C while simultaneously outside of both A and B.

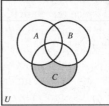

103. The set operations $(A \cap B') \cup C$ indicate those elements that are in A and at the same time outside of B, along with those in C.

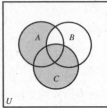

105. The set operations $(A \cap B') \cap C'$ indicate the region in A and outside B and at the same time outside C.

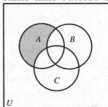

107. The set operations $(A' \cap B') \cup C'$ indicate the region that is both outside A and at the same time outside B, along with the region outside C.

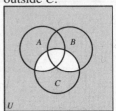

109. The shaded area indicates the region $(A \cup B)'$ or $A' \cap B'$.

111. Since this is the region in A or in B but, at the same time, outside of A and B, we have the set $(A \cup B) \cap (A \cap B)'$ or $(A \cup B) - (A \cap B)$.

113. The shaded area may be represented by the set $(A \cap B) \cup (A \cap C)$; that is, the region in the intersection of A and B along with the region in the intersection of A and C or, by the distributive property, $A \cap (B \cup C)$.

115. The region is represented by the set $(A \cap B) \cap C'$, that is, the region outside of C but inside both A and B, or $(A \cap B) - C$.

117. If $A = A - B$, then A and B must not have any common elements, or $A \cap B = \varnothing$.

119. $A = A - \varnothing$ is true for any set A.

121. $A \cup \varnothing = \varnothing$ is true only if A has no elements, or $A = \varnothing$.

123. $A\varnothing = A$ is true only if A has no elements, or $A = \varnothing$.

125. $A \cup A = \varnothing$ is true only if A has no elements, or $A = \varnothing$.

127. $A \cup B = A$ only if B is a subset of A, or $B \subseteq A$.

129. $A \cap A' = \varnothing$

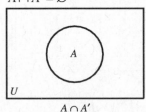

$A \cap A'$

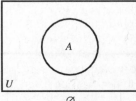

$\varnothing$

Thus, by the Venn diagrams, the statement is always true.

131. $(A \cap B) \subseteq A$

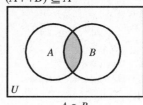

$$A \cap B$$

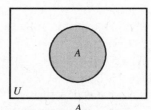

$$A$$

Thus, by the Venn diagrams, the shaded region is in A; therefore, the statement is always true.

133. If $A \subseteq B$, then $A \cup B = A$.

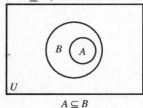

$$A \subseteq B$$

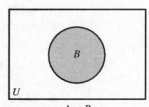

$$A \cup B$$

Thus, the statement is not always true.

135. $(A \cup B)' = A' \cap B'$ (De Morgan's second law).

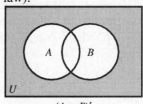

$$(A \cup B)'$$

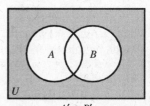

$$A' \cap B'$$

Thus, by the Venn diagrams, the statement is always true.

137. (a) $Q \cup H = \{x | x$ is a real number$)$, since the real numbers are made up of all rational and all irrational numbers.

(b) $Q \cap H = \varnothing$, since there are no common elements.

2.4 Exercises

1. (a) $n(A \cap B) = 5$ since A and B have 5 elements in common.

(b) $n(A \cup B) = 7$ since there are a total of 7 elements in A or in B.

(c) $n(A \cap B') = 0$ since there are 0 elements which are in A and, at the same time, outside B.

(d) $n(A' \cap B) = 2$ since there are 2 elements which are in B and, at the same time, outside A.

(e) $n(A' \cap B') = 8$ since there are 8 elements which are outside of A and, at the same time, outside of B.

3. (a) $n(A \cap B \cap C) = 1$ since there is only one element shared by all three sets.

(b) $n(A \cap B \cap C') = 3$ since there are 3 elements in A and B while, at the same time, outside of C.

(c) $n(A \cap B' \cap C) = 4$ since there are 4 elements in A and C while, at the same time, outside of B.

(d) $n(A' \cap B \cap C) = 0$ since there are 0 elements which are outside of A while, at the same time, in B and C.

(e) $n(A' \cap B' \cap C) = 2$ since there are 2 elements outside of A and outside of B while, at the same time, in C.

(f) $n(A \cap B' \cap C') = 8$ since there are 8 elements in A which at the same time, are outside of B and outside of C.

(g) $n(A' \cap B \cap C') = 2$ since there are 2 elements outside of A and, at the same time, outside of C but inside of B.

(h) $n(A' \cap B' \cap C') = 6$ since there are 6 elements which are outside all three sets at the same time.

5. Using the Cardinal Number Formula, $n(A \cup B) = n(A) + n(B) - n(A \cap B)$, we have $n(A \cup B) = 12 + 14 - 5 = 21$.

7. Using the Cardinal Number Formula, $n(A \cup B) = n(A) + n(B) - n(A \cap B)$, we have $25 = 20 + 12 - n(A \cap B)$. Solving for $n(A \cap B)$, we get $n(A \cap B) = 7$.

9. Using the Cardinal Number Formula, $n(A \cup B) = n(A) + n(B) - n(A \cap B)$, we have $55 = n(A) + 35 - 15$. Solving for $n(A)$, we get $n(A) = 35$.

11. Using the Cardinal Number Formula, we find that
$$n(A \cup B) = n(A) + n(B) - n(A \cap B)$$
$$25 = 19 + 13 - n(A \cap B)$$
$$n(A \cap B) = 19 + 13 - 25$$
$$= 7.$$
Then $n(A \cap B') = 19 - 17 = 12$ and $n(B \cap A') = 13 - 7 = 6$.
Since $n(A') = 11$ and $n(B \cap A') = 6$, we have $n(A \cup B)' = n(A' \cap B') = 11 - 6 = 5$.
Completing the cardinalities for each region, we arrive at the following Venn diagram.

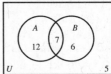

13. Since $n(B) = 28$ and $n(A \cap B) = 10$, we have $n(B \cap A') = 28 - 10 = 18$. Since $n(A') = 25$ and $n(B \cap A') = 18$, it follows that $n(A' \cap B') = 7$.
By De Morgan's laws, $A' \cup B' = (A \cap B)'$.
So, $n[(A \cap B)'] = 40$.
Thus, $n(A \cap B') = 40 - (18 + 7) = 15$.
Completing the cardinalities for each region, we arrive at the following Venn diagram.

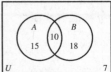

15. Fill in the cardinal numbers of the regions, beginning with $(A \cap B \cap C)$. Since $n(A \cap B \cap C) = 15$ and $n(A \cap B) = 35$, we have $n(A \cap B \cap C') = 35 - 15 = 20$. Since $n(A \cap C) = 21$, we have $n(A \cap C \cap B') = 21 - 15 = 6$. Since $n(B \cap C) = 25$, we have $n(B \cap C \cap A') = 25 - 15 = 10$. Since $n(C) = 49$, we have $n(C \cap A' \cap B') = 49 - (6 + 15 + 10) = 18$. Since $n(A) = 57$, we have $n(A \cap B' \cap C') = 57 - (20 + 15 + 6) = 16$. Since $n(B') = 52$, we have
$$n(A \cup B \cup C)' = n(A' \cap B' \cap C')$$
$$= 52 - (16 + 6 + 18)$$
$$= 12.$$
Completing the cardinalities for each region, we arrive at the following Venn diagram.

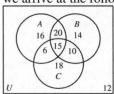

17. Fill in the cardinal numbers of the regions, beginning with the $(A \cap B \cap C)$.
$n(A \cap B \cap C) = 5$ and $n(A \cap C) = 13$, so $n(A \cap C \cap B') = 13 - 5 = 8$. $n(B \cap C) = 8$, so $n(B \cap C \cap A') = 8 - 5 = 3$. $n(A \cap B') = 9$, so $n(A \cap B' \cap C') = 9 - 8 = 1$. $n(A) = 15$, so $n(A \cap B \cap C') = 15 - (1 + 8 + 5) = 1$.
$n(B \cap C') = 3$, so $n(B \cap A' \cap C') = 3 - 1 = 2$.
$n(A' \cap B' \cap C') = n(A \cup B \cup C)' = 21$.
$n(B \cup C) = 32$, so $n(A' \cap B' \cap C) = 32 - (1 + 2 + 5 + 3) = 21$
Completing the cardinalities for each region, we arrive at the following Venn diagram.

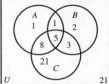

19. Complete a Venn diagram showing the cardinality for each region. Let W = set of projects Joe Long writes. Let P = set of projects Joe long produces.
Begin with $(W \cap P)$. $n(W \cap P) = 3$. Since $n(W) = 5$, $n(W \cap P') = 5 - 3 = 2$. Since $n(P) = 7$, $n(P \cap W') = 7 - 3 = 4$.

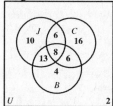

Interpreting the resulting cardinalities we see that:

(a) He wrote but did not produce $n(W \cap P') = 2$ projects.

(b) He produced but did not write $n(P \cap W') = 4$ projects.

21. Construct a Venn diagram and label the number of elements in each region. Let J = set of fans who like Jazmine Sullivan, C = set of fans who like Carrie Underwood, and B = set of fans who like Brad Paisley. Begin with the region indicating the intersection of all three sets, $n(J \cap C \cap B) = 8$.
Since $n(J \cap C) = 14$,
$n(J \cap C \cap B') = 14 - 8 = 6$.
Since $n(C \cap B) = 14$,
$n(C \cap B \cap J') = 14 - 8 = 6$.
Since $n(J \cap B) = 21$,
$n(J \cap B \cap C') = 21 - 8 = 13$.
Since $n(J) = 37$, the number of elements inside J and not in C or B is $37 - (13 + 8 + 6) = 10$. Since $n(C) = 36$, the number of elements inside C and not in J or B is $36 - (6 + 8 + 6) = 16$. Since $n(B) = 31$, the number of elements inside B and not in J or C is $31 - (13 + 8 + 6) = 4$. Since $n(U) = 65$, there are
$65 - (10 + 6 + 16 + 13 + 8 + 6 + 4)$
$= 2$ elements outside the three sets. That is, $n(J \cup C \cup B)' = 2$. The completed Venn diagram is as follows:

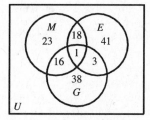

(a) There are $6 + 13 + 6 = 25$ fans that like exactly two of these singers.

(b) There are $10 + 16 + 4 = 30$ fans that like exactly one of these singers.

(c) There are 2 fans that like none of these singers.

(d) There are 10 fans that like Jazmine, but neither Carrie nor Brad.

(e) There are $13 + 6 = 19$ fans that like Brad and exactly one of the other two.

23. Let U be the set of people interviewed, and let M, E, and G represent the sets of people using microwave ovens, electric ranges, and gas ranges, respectively.
Construct a Venn diagram and label the cardinal number of each region, beginning with the region $(M \cap E \cap G)$.
$n(M \cap E \cap G) = 1$. Since $n(M \cap E) = 19$,
$n(M \cap E \cap G') = 19 - 1 = 18$. Since $n(M \cap G) = 17$,
$n(M \cap G \cap E') = 17 - 1 = 16$. Since $n(G \cap E) = 4$, $n(G \cap E \cap M') = 4 - 1 = 3$.
Since $n(M) = 58$,
$$n(M \cap G' \cap E') = n(M \cap (G \cup E)')$$
$$= 58 - (18 + 16 + 1)$$
$$= 23.$$
Since $n(E) = 63$,
$n(E \cap M' \cap G') = 63 - (18 + 3 + 1) = 41$.
Since $n(G) = 58$,
$$n(G \cap M' \cap E') = n(G \cap (M \cup E)')$$
$$= 58 - (16 + 3 + 1)$$
$$= 38.$$

(a) The number of respondents that use exactly two of these appliances is $16 + 18 + 3 = 37$.

(b) The number of respondents that use at least two of these appliances is $37 + 1 = 38$.

25. Construct a Venn diagram and label the cardinal number of each region.
Let H = the set of respondents who think Hollywood is unfriendly toward religion, M = the set of respondents who think the media are unfriendly toward religion, S = the set of respondents who think scientists are unfriendly toward religion. Then we are given the following information.

$n(H) = 240$ $n(H \cap M) = 145$
$n(M) = 160$ $n(S \cap M) = 122$
$n(S) = 181$ $n(H \cap M \cap S) = 110$
 $n(H \cup M \cup S)' = 219$

Since $n(H \cap M) = 145$,
$n(H \cap M \cap S') = 145 - 110 = 35$.
Since $n(S \cap M) = 122$,
$n(S \cap M \cap H') = 122 - 110 = 12$.
The total number of respondents who think
exactly two of these groups are unfriendly
toward religion is 80, so
$n(H \cap S \cap M') = 80 - (35 + 12) = 33$.
Since $n(H) = 240$,
$n(H \cap S' \cap M') = 240 - (33 + 110 + 35)$
$ = 62$.
Since $n(M) = 160$,
$n(M \cap H' \cap S') = 160 - (35 + 110 + 12) = 3$.
Since $n(S) = 181$,
$n(S \cap H' \cap M') = 181 - (33 + 110 + 12) = 26$.

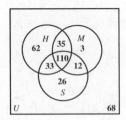

(a) The total number of respondents
surveyed is
$62 + 33 + 110 + 35 + 3 + 12 + 26 + 219$
$= 500$.

(b) The number of respondents who think
exactly one of these three groups is
unfriendly toward religion is
$62 + 3 + 26 = 91$.

27. Construct a Venn diagram to represent the
survey data beginning with the region
representing the intersection of S, B, and C.
Rather than representing each region as a
combination of sets and set operations, we
will label the regions a–h. There are
21 patients in Nadine's survey that are in the
intersection of all three sets, i.e., in region d.
Since there are 31 patients in $B \cap C$, we can
deduce that there must be 10 patients in
region c. Similarly since there are
33 patients in $B \cap S$, there must be
12 patients in region e. From the given
information, there is a total of 51 patients in
regions c, d, e, and g. Thus, there are
$51 - (10 + 21 + 12) = 8$ patients in region g.
Since there are 52 patients in S, we can

deduce that there are
$52 - (12 + 21 + 8) = 11$ patients in region f.
Similarly, there are 4 patients in region b,
and 7 patients in region h. There is a total of
73 patients found in regions b–h. Thus,
there must be 2 patients in region a.

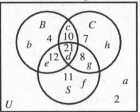

(a) The number of these patients who had
either high blood pressure or high
cholesterol levels, but not both is
represented by regions b, e, g, and h for
a total of 31 patients.

(b) The number of these patients who had
fewer than two of the indications listed
are found in regions a, b, f, and h for a
total of 24 patients.

(c) The number of these patients who were
smokers but had neither high blood
pressure nor high cholesterol levels are
found in region f, which has
11 members.

(d) The number of these patients who did
not have exactly two of the indications
listed would be those excluded from
regions c, e, and g (representing
patients with exactly two of the
indications). We arrive at a total of 45
patients.

29. (a) The set $A \cap B \cap C \cap D$ is region 1 (in
text).

(b) The set $A \cup B \cup C \cup D$ includes the
regions 1, 2, 3, 4, 5, 6, 7, 8, 9, 10, 11,
12, 13, 14, and 15.

(c) The set $(A \cap B) \cup (C \cap D)$ includes
the set of regions
$\{1, 3, 9, 11\} \cup \{1, 2, 4, 5\}$ or the
regions 1, 2, 3, 4, 5, 9, and 11.

(d) The set $(A' \cap B') \cap (C \cup D)$ includes
the set of regions
$\{5, 13, 8, 16\} \cap \{1, 2, 3, 4, 5, 6, 7, 8, 9,$
$10, 12, 13\}$, which is represented by
regions 5, 8, and 13.

31. (a) $n(J \cap G) = 9$, coming from the intersection of the first row with the first column.

(b) $n(S \cap N) = 9$, coming from the intersection of second row and the third column.

(c) $n(N \cup (S \cap F)) = 20$ since there are 20 players who are in either N(total of 15) or in S and F (just 5), at the same time.

(d) $n(S' \cap (G \cup N)) = 20$ since there are $9 + 4 + 5 + 2 = 20$ players who are not in S, but are in G or in N.

(e) $n((S \cap N') \cup (C \cap G')) = 27$
There are 27 players who are in S but not in $N(12 + 5)$, or who are in C but not in G $(8 + 2)$.

(f) $n(N' \cap (S' \cap C')) = 15$
There are 15 $(9 + 6)$ players who are not in N and at the same time are not in S and not in C.

33. Writing exercise; answers will vary

EXTENSION: INFINITE SETS AND THEIR CARDINALITIES

1. The set {6} has the same cardinality as B, {26}. The cardinal number is 1.

3. The set $\{x|x$ is a natural number$\}$ has the same cardinality as A, $\aleph_0$.

5. The set $\{x|x$ is an integer between 5 and 6$\}$ has the same cardinality, 0, as F since there are no members in either set.

7. One correspondence is:
{I,   II,   III}
$\updownarrow$ $\updownarrow$ $\updownarrow$
{x,   y,   z}
Other correspondences are possible.

9. One correspondence is:
{a,   d,   i,   t,   o,   n}
$\updownarrow$ $\updownarrow$ $\updownarrow$ $\updownarrow$ $\updownarrow$ $\updownarrow$
{a,   n,   s,   w,   e,   r}
Other correspondences are possible.

11. $n(\{a, b, c, d, ..., k\}) = 11$
By counting the number of letters a through k, we establish the cardinality to be 11.

13. $n(\varnothing) = 0$ since there are no members.

15. $n(\{300, 400, 500, ...\}) = \aleph_0$ since this set can be placed in a one-to-one correspondence with the counting numbers (i.e., is a countable infinite set).

17. $n\left(\left\{-\dfrac{1}{4}, -\dfrac{1}{8}, -\dfrac{1}{12}, ...\right\}\right) = \aleph_0$ since this set can be placed in a one-to-one correspondence with the counting numbers.

19. $n(\{x|x$ is an odd counting number$\}) = \aleph_0$ since this set can be placed in a one-to-one correspondence with the counting numbers.

21. $n(\{Jan, Feb, Mar, ..., Dec\}) = 12$ since there are twelve months indicated in the set.

23. "$\aleph_0$ bottles of beer on the wall, $\aleph_0$ bottles of beer, take one down and pass it around, $\boxed{\aleph_0}$ bottles of beer on the wall." This is true because $\aleph_0 - 1 = \aleph_0$.

25. The answer is both. Since the sets {u, v, w} and {v, u, w} are equal sets (same elements), they must then have the same number of elements and thus are equivalent.

27. The sets {X, Y, Z} and {x, y, z} are equivalent because they contain the same number of elements (same cardinality) but not the same elements.

29. The sets $\{x|x$ is a positive real number$\}$ and $\{x|x$ is a negative real number$\}$ are equivalent because they have the same cardinality, c. They are not equal since they contain different elements.

Note that each of the following answers shows only one possible correspondence.

31. {2, 4, 6, 8, 10, 12, ..., 2n, ...}
$\updownarrow$ $\updownarrow$ $\updownarrow$ $\updownarrow$ $\updownarrow$ $\updownarrow$ $\updownarrow$
{1, 2, 3, 4, 5, 6, ...,   n, ...}

33. $\{1,000,000,\ 2,000,000,\ 3,000,000,\ ...,\ 1,000,000n,\ ...\}$

$\quad\quad\updownarrow\quad\quad\quad\updownarrow\quad\quad\quad\updownarrow\quad\quad\quad\quad\updownarrow$

$\quad\ \{1,\quad\quad\quad 2,\quad\quad\quad 3,\ ...,\quad\quad\quad n,\ ...\}$

35. $\{2,\ 4,\ 8,\ 16,\ 32,\ ...,\ 2^n,\ ...\}$

$\quad\ \updownarrow\ \updownarrow\ \updownarrow\ \updownarrow\quad\ \updownarrow$

$\quad\ \{1,\ 2,\ 3,\ 4,\ 5,\ ...,\ n,\ ...\}$

37. The statement "If A and B are infinite sets, then A is equivalent to B" is not always true. For example, let $A =$ the set of counting numbers and $B =$ the set of real numbers. Each has a different cardinality.

39. The statement "If set A is an infinite set and A is not equivalent to the set of counting numbers, then $n(A) = c$" is not always true. For example, A could be the set of all subsets of the set of real numbers. Then, $n(A)$ would be an infinite number greater than c.

41. **(a)** Use the figure (in the text), where the line segment between 0 and 1 has been bent into a semicircle and positioned above the line, to prove that $\{x|x$ is a real number between 0 and 1$\}$ is equivalent to $\{x|x$ is a real number$\}$.

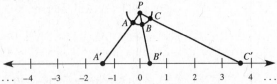

Rays emanating from point P will establish a geometric pairing for the points on the semicircle with the points on the line.

(b) The fact part (a) establishes about the set of real numbers is that the set of real numbers is infinite, having been placed in a one-to-one correspondence with a proper subset of itself.

43. $\{3,\ 6,\ 9,\ 12,\ ...,\quad 3n,\ ...\}$

$\quad\ \updownarrow\ \updownarrow\ \updownarrow\ \updownarrow\quad\quad \updownarrow$

$\quad\ \{6,\ 9,\ 12,\ 15,\ ...,\ 3n+3,\ ...\}$

45. $\left\{\dfrac{3}{4},\ \dfrac{3}{8},\ \dfrac{3}{12},\ \dfrac{3}{16},\ ...,\ \dfrac{3}{4n},\ ...\right\}$

$\quad\ \updownarrow\ \ \updownarrow\ \ \updownarrow\ \ \updownarrow\quad\quad \updownarrow$

$\quad\ \left\{\dfrac{3}{8},\ \dfrac{3}{12},\ \dfrac{3}{16},\ \dfrac{3}{20},\ ...,\ \dfrac{3}{4n+4},\ ...\right\}$

47. $\left\{\dfrac{1}{9},\ \dfrac{1}{18},\ \dfrac{1}{27},\ ...,\ \dfrac{1}{9n},\ ...\right\}$

$\quad\ \updownarrow\ \ \updownarrow\ \ \updownarrow\quad\quad \updownarrow$

$\quad\ \left\{\dfrac{1}{18},\ \dfrac{1}{27},\ \dfrac{1}{36},\ ...,\ \dfrac{1}{9n+9},\ ...\right\}$

49. Writing exercise; answers will vary.

51. Writing exercise; answers will vary.

Chapter 2 Test

1. $A \cup C = \{a, b, c, d\} \cup \{a, e\}$

$\quad\quad\quad\ = \{a, b, c, d, e\}$

2. $B \cap A = \{b, e, a, d\} \cap \{a, b, c, d\}$
    $\qquad = \{a, b, d\}$

3. $B' = \{b, e, a, d\}' = \{c, f, g, h\}$

4. $A - (B \cap C')$
    $\qquad = A - (\{b, e, a, d\} \cap \{b, c, d, f, g, h\})$
    $\qquad = \{a, b, c, d\} - \{b, d\}$
    $\qquad = \{a, c\}$

5. $b \in A$ is true since b is a member of set A.

6. $C \subseteq A$ is false since the element e, which is a member of set C, is not also a member of set A.

7. $B \subset (A \cup C)$ is true since all members of set B are also members of $A \cup C$.

8. $c \notin C$ is true because c is not a member of set C.

9. $n[(A \cup B) - C] = 4$ is false. Because,
    $n[(a \cup B) - C] = n[\{a, b, c, d, e\} - \{a, e\}]$
    $\qquad = n(\{b, c, d\})$
    $\qquad = 3$

10. $\varnothing \subset C$ is true. The empty set is considered a subset of any set. C has more elements than $\varnothing$ which makes $\varnothing$ a proper subset of C.

11. $(A \cap B')$ is equivalent to $(B \cap A')$ is true. Because, $n(A \cap B') = n(\{c\}) = 1$, $n(B \cap A') = n(\{e\}) = 1$.

12. $(A \cup B)' = A' \cap B'$ is true by one of De Morgan's laws.

13. $n(A \times C) = n(A) \times n(C) = 4 \times 2 = 8$

14. The number of proper subsets of A is $2^4 - 1 = 16 - 1 = 15$.

Answers may vary for Exercises 15–18.

15. A word description for $\{-3, -1, 1, 3, 5, 7, 9\}$ is the set of all odd integers between -4 and 10.

16. A word description for $\{$January, February, March, ..., December$\}$ is the set of months of the year.

17. Set-builder notation for $\{-1, -2, -3, -4, ...\}$ would be $\{x | x$ is a negative integer$\}$.

18. Set-builder notation for $\{24, 32, 40, 48, ..., 88\}$ would be $\{x | x$ is a multiple of 8 between 20 and 90$\}$.

19. $\varnothing \subseteq \{x | x$ is a counting number between 20 and 21$\}$ since the empty set is a subset of any set.

20. $\{4, 9, 16\}$ neither $\{4, 5, 6, 7, 8, 9, 10\}$ since the element 16 is not a member of the second set.

21. $X \cup Y'$

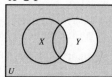

22. $X' \cap Y'$

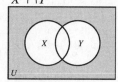

23. $(X \cup Y) - Z$

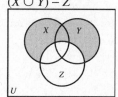

24. $[(X \cap Y) \cup (Y \cap Z) \cup (X \cap Z)]$
    $\qquad - (X \cap Y \cap Z)$

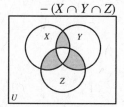

25. $A \cap T = \{$Electric razor, Telegraph, Zipper$\}$ $\cap$ $\{$Electric razor, Fiber optics, Geiger counter, Radar$\} = \{$Electric Razor$\}$

26. $(A \cup T)'$
    $= \{$Electric razor, Telegraph, Zipper$\} \cup$ $\{$Electric razor, Fiber optics, Geiger counter, Radar$\})'$
    $= \{$Electric razor, Fiber optics, Geiger counter, Radar, Telegraph, Zipper$\}'$
    $= \{$Adding machine, Barometer, Pendulum clock, Thermometer$\}$

27. $A - T'$
= {Electric razor, Telegraph, Zipper}
   − {Electric razor, Fiber optics, Geiger
      counter, Radar}′
= {Electric razor, Telegraph, Zipper}
   − {Adding machine, Barometer,
      Pendulum clock, Telegraph,
      Thermometer, Zipper}
= {Electric razor}

28. Writing exercise; answers will vary.

29. (a) $n(A \cup B) = 12 + 3 + 7 = 22$

(b) $n(A \cap B') = n(A - B) = 12$
These are the elements in A but outside
of B.

30. Let G = set of students who are receiving
government grants. Let S = set of students
who are receiving private scholarships. Let
A = set of students who are receiving aid
from the college.
Complete a Venn diagram by inserting the
appropriate cardinal number for each region
in the diagram. Begin with the intersection
of all three sets: $n(G \cap S \cap A) = 8$. Since
$n(S \cap A) = 28$, $n(S \cap A \cap G') = 28 - 8 = 20$.
Since $n(G \cap A) = 18$,
$n(G \cap A \cap S') = 18 - 8 = 10$. Since
$n(G \cap S) = 23$, $n(G \cap S \cap A') = 23 - 8 = 15$.
Since $n(A) = 43$,
$n(A \cap (G \cup S)') = 43 - (10 + 8 + 20)$
$\qquad\qquad = 43 - 38$
$\qquad\qquad = 5.$
Since $n(S) = 55$,
$n(S \cap (G \cup A)') = 55 - (15 + 8 + 20)$
$\qquad\qquad = 55 - 43$
$\qquad\qquad = 12.$
Since $n(G) = 49$,
$n(G \cap (S \cup A)') = 49 - (10 + 8 + 15)$
$\qquad\qquad = 49 - 33$
$\qquad\qquad = 16.$

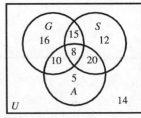

Thus,

(a) $n(G \cap (S \cup A)') = 16$ have a
government grant only.

(b) $n(S \cap G') = 32$ have a private
scholarship but not a government grant.

(c) $16 + 12 + 5 = 33$ receive financial aid
from only one of these sources.

(d) $10 + 15 + 20 = 45$ receive aid from
exactly two of these sources.

(e) $n(G \cup S \cup A)' = 14$ receive no financial
aid from any of these sources.

(f) $n(S \cap (A \cup G)') + n(A \cup G \cup S)'$
$= 12 + 14$
$= 26$
received private scholarships or no aid
at all.

Chapter 3

1. "February 2, 2009, was a Monday" is a declarative sentence that is true and, therefore, is considered a statement.

3. "Listen my children and you shall hear of the midnight ride of Paul Revere" is not a declarative sentence and does not have the property of being true or false. Hence, it is not considered a statement.

5. "$5 + 9 \neq 14$ and $4 - 1 = 12$" is a declarative sentence that is false and, therefore, is considered a statement.

7. "Some numbers are positive" is a declarative sentence that is true and, therefore, is a statement.

9. "Accidents are the main cause of deaths of children under the age of 7" is a declarative sentence that has the property of being true or false and, therefore, is considered to be a statement.

11. "Where are you going tomorrow?" is a question, not a declarative sentence and, therefore, is not considered a statement.

13. "Kevin 'Catfish' McCarthy once took a prolonged continuous shower for 340 hours, 40 minutes" is a declarative sentence that has the property of being either true or false and, therefore, is considered to be a statement.

15. "I read the Detroit Free Press, and I read the Sacramento Bee" is a compound statement because it consists of two simple statements combined by the connective "and."

17. "Tomorrow is Saturday" is a simple statement because only one assertion is being made.

19. "Jay Beckenstein's wife loves Ben and Jerry's ice cream" is not compound because only one assertion is being made.

21. "If Lorri Morgan sells her quota, then Michelle Cook will be happy" is a compound statement because it consists of two simple statements combined by the connective "if...then."

23. The negation of "Her aunt's name is Hermione" is "Her aunt's name is not Hermione."

25. A negation of "Every dog has its day" is "At least one dog does not have its day."

27. A negation of "Some books are longer than this book" is "No book is longer than this book."

29. A negation of "No computer repairman can play blackjack" is "At least one computer repairman can play blackjack."

31. A negation of "Everybody loves somebody sometime" is "Someone does not love somebody sometime."

33. A negation of "$x > 12$" (without using a slash sign) would be "$x \leq 12$."

35. A negation for "$x \geq 5$" would be "$x < 5$."

37. Writing exercise; answers will vary.

Let p represent the statement "She has green eyes," and let q represent "He is 60 years old." Translate each symbolic compound statement into words.

39. A translation for "$\sim p$" is "She does not have green eyes."

41. A translation for "$p \wedge q$" is "She has green eyes and he is 60 years old."

43. A translation for "$\sim p \vee q$" is "She does not have green eyes or he is 60 years old."

45. A translation for "$\sim p \vee \sim q$" is "She does not have green eyes or he is not 60 years old."

47. A translation for "$\sim(\sim p \wedge q)$" is "It is not the case that she does not have green eyes and he is 60 years old."

49. "Chris collects DVDs and Josh is not an art major" may be symbolized as $p \wedge \sim q$.

51. "Chris does not collect DVDs or Josh is an art major" may be symbolized as $\sim p \vee q$.

53. "Neither Chris collects DVDs nor Josh is an art major" may be symbolized as ~$(p \lor q)$ or equivalently, ~$p \land$ ~q.

55. Writing exercise; answers will vary.

Refer to the sketches labeled A, B, and C in the text, and identify the sketch (or sketches) that is (are) satisfied by the given statement involving a quantifier.

57. The condition that "all pictures have frames" is satisfied by group C.

59. The condition that "At least one picture does not have a frame" is met by groups A and B.

61. The condition that "At least one picture has a frame" is satisfied by groups A and C.

63. The condition that "all pictures do not have frames" is satisfied by group B. Observe that this statement is equivalent to "No pictures have a frame."

65. Since all whole numbers are integers, the statement "Every whole number is an integer" is true.

67. Since $\frac{1}{2}$ is a rational number but not an integer, the statement "There exists a rational number that is not an integer" is true.

69. Since rational numbers are real numbers, the statement "All rational numbers are real numbers" is true.

71. Since $\frac{1}{2}$ is a rational number but not an integer, the statement "Some rational numbers are not integers" is true.

73. The number 0 is a whole number but not positive. Thus, the statement "Each whole number is a positive number" is false.

75. Writing exercise; answers will vary.

77. We might write the statement "There is no one here who has not done that at one time or another" using the word "every" as "Every person here has done that at one time or another."

3.2 Exercises

1. If q is false, then $(p \land$ ~$q) \land q$ must be false, since both conjuncts (parts of the conjunction) must be true for the compound statement to be true.

3. If $p \land q$ is true, and p is true, then q must also be true in order for the conjunctive statement to be true. Observe that both conjuncts must be true for a conjunctive statement to be true.

5. If ~$(p \lor q)$ is true, both components (disjuncts) must be false. Thus, the disjunction itself is false making its negation true.

In exercises 7–17, p represents a false statement and q represents a true statement.

7. Since $p = $ F, ~$p = $ ~F = T. That is, replace p by F and determine the truth of ~F.

9. Since p is false and q is true, we may consider the "or" statement as
F $\lor$ T
    T,
by logical definition of an "or" statement. That is $p \lor q$ is true.

11. With the given truth values for p and q, we may consider $p \lor$ ~q as
F $\lor$ ~T
F $\lor$ F
    F,
by the logical definition of "$\lor$."

13. With the given truth values for p and q, we may consider ~$p \lor$ ~q as
~F $\lor$ ~T
T $\lor$ F
    T.
Thus, the compound statement is true.

15. Replacing p and q with the given truth values, we have
~(F $\land$ ~T)
~(F $\land$ F)
    ~F
    T.
Thus, the compound statement ~$(p \land$ ~$q)$ is true.

17. Replacing p and q with the given truth values, we have
$$\sim[\sim F \wedge (\sim T \vee F)]$$
$$\sim[T \wedge (F \vee F)]$$
$$\sim[T \wedge F]$$
$$\sim F$$
$$T.$$
Thus, the compound statement $\sim[\sim p \wedge (\sim q \vee p)]$ is true.

19. The statement $6 \geq 2$ is a disjunction since it means "$6 > 2$" or "$6 = 2$."

In exercises 21–27, p represents a true statement, and q and r represent false statements.

21. Replacing p, q and r with the given truth values, we have
$$(T \wedge F) \vee \sim F$$
$$F \vee T$$
$$T.$$
Thus, the compound statement $(p \wedge r) \vee \sim q$ is true.

23. Replacing p, q and r with the given truth values, we have
$$T \wedge (F \vee F)$$
$$T \wedge F$$
$$F.$$
Thus, the compound statement $p \wedge (q \vee r)$ is false.

25. Replacing p, q and r with the given truth values, we have
$$\sim(T \wedge F) \wedge (F \vee \sim F)$$
$$\sim F \wedge (F \vee T)$$
$$T \wedge T$$
$$T.$$
Thus, the compound statement $\sim(p \wedge q) \wedge (r \vee \sim q)$ is true.

27. Replacing p, q and r with the given truth values, we have
$$\sim[(\sim T \wedge F) \vee F]$$
$$\sim[(F \wedge F) \vee F]$$
$$\sim[F \vee F]$$
$$\sim F$$
$$T.$$
Thus, the compound statement $\sim[(\sim p \wedge q) \vee r]$ is true.

29. Replacing p, q and r with the given truth values, we have
$$\sim[\sim q \vee (r \wedge \sim p)]$$
$$\sim[\sim F \vee (F \wedge \sim T)]$$
$$\sim[T \vee F]$$
$$\sim T$$
$$F.$$
Thus, the compound statement $\sim[\sim q \vee (r \wedge \sim p)]$ is false.

Let p represent the statement "16 < 8," which is false, let q represent "5 $\not>$ 4," which is false and let r represent "17 $\leq$ 17," which is true. [E.g. p = F, q = F and r = T.]

31. Replacing p and r with the given truth values, we have
$$F \wedge T$$
$$F.$$
The compound statement $p \wedge r$ is false.

33. Replacing q adn r with the observed truth values, we have
$$\sim F \vee \sim T$$
$$T \vee F$$
$$T.$$
The compound statement $\sim q \vee \sim r$ is true.

35. Replacing p, q and r with the observed truth values, we have
$$(F \wedge F) \vee T$$
$$F \vee T$$
$$T.$$
The compound statement $(p \wedge q) \vee r$ is true.

37. Replacing p, q and r with the observed truth values, we have
$$(\sim T \wedge F) \vee \sim F$$
$$(F \wedge F) \vee T$$
$$F \vee T$$
$$T.$$
The compound statement $(\sim r \wedge q) \vee \sim p$ is true.

39. Since there are two simple statements (p and r), we have $2^2 = 4$ combinations of truth values, or rows in the truth table, to examine.

41. Since there are four simple statements (p, q, r, and s), we have $2^4 = 16$ combinations of truth values, or rows in the truth table, to examine.

43. Since there are seven simple statements (p, q, r, s, t, u, and v), we have $2^7 = 128$ combinations of truth values, or rows in the truth table, to examine.

45. If the truth table for a certain compound statement has 128 rows, then there must be seven distinct component statements ($2^7 = 128$).

47. $\sim p \wedge q$

p	q	$\sim p$	$\sim p \wedge q$
T	T	F	F
T	F	F	F
F	T	T	T
F	F	T	F

49. $\sim(p \wedge q)$

p	q	$p \wedge q$	$\sim(p \wedge q)$
T	T	T	F
T	F	F	T
F	T	F	T
F	F	F	T

51. $(q \vee \sim p) \vee \sim q$

p	q	$\sim p$	$\sim q$	$(q \vee \sim p)$	$(q \vee \sim p) \vee \sim q$
T	T	F	F	T	T
T	F	F	T	F	T
F	T	T	F	T	T
F	F	T	T	T	T

53. $\sim q \wedge (\sim p \vee q)$

p	q	$\sim q$	$\wedge$	$(\sim p$	$\vee$	$q)$
T	T	F	F	F	T	T
T	F	T	F	F	F	F
F	T	F	F	T	T	T
F	F	T	T	T	T	F
		1	3	1	2	1

55. $(p \vee \sim q) \wedge (p \wedge q)$

p	q	$(p$	$\vee$	$\sim q)$	$\wedge$	$(p$	$\wedge$	$q)$
T	T	T	T	F	T	T	T	T
T	F	T	T	T	F	T	F	F
F	T	F	F	F	F	F	F	T
F	F	F	T	T	F	F	F	F
		1	2	1	3	1	2	1

57. $(\sim p \wedge q) \wedge r$

p	q	r	$(\sim p$	$\wedge$	$q)$	$\wedge$	r
T	T	T	F	F	T	F	T
T	T	F	F	F	T	F	F
T	F	T	F	F	F	F	T
T	F	F	F	F	F	F	F
F	T	T	T	T	T	T	T
F	T	F	T	T	T	F	F
F	F	T	T	F	F	F	T
F	F	F	T	F	F	F	F
			1	2	1	3	1

59. $(\sim p \wedge \sim q) \vee (\sim r \vee \sim p)$

p	q	r	$(\sim p$	$\wedge$	$\sim q)$	$\vee$	$(\sim r$	$\vee$	$\sim p)$
T	T	T	F	F	F	F	F	F	F
T	T	F	F	F	F	T	T	T	F
T	F	T	F	F	T	F	F	F	F
T	F	F	F	F	T	T	T	T	F
F	T	T	T	F	F	T	F	T	T
F	T	F	T	F	F	T	T	T	T
F	F	T	T	T	T	T	F	T	T
F	F	F	T	T	T	T	T	T	T
			1	2	1	3	1	2	1

61. $\sim(\sim p \wedge \sim q) \vee (\sim r \vee \sim s)$

p	q	r	s	~	(~p	∧	~q)	∨	(~r	∨	~s)
T	T	T	T	T	F	F	F	T	F	F	F
T	T	T	F	T	F	F	F	T	F	T	T
T	T	F	T	T	F	F	F	T	T	T	F
T	T	F	F	T	F	F	F	T	T	T	T
T	F	T	T	T	F	F	T	T	F	F	F
T	F	T	F	T	F	F	T	T	F	T	T
T	F	F	T	T	F	F	T	T	T	T	F
T	F	F	F	T	F	F	T	T	T	T	T
F	T	T	T	T	T	F	F	T	F	F	F
F	T	T	F	T	T	F	F	T	F	T	T
F	T	F	T	T	T	F	F	T	T	T	F
F	T	F	F	T	T	F	F	T	T	T	T
F	F	T	T	F	T	T	T	F	F	F	F
F	F	T	F	F	T	T	T	T	F	T	T
F	F	F	T	F	T	T	T	T	T	T	F
F	F	F	F	F	T	T	T	T	T	T	T
				3	1	2	1	4	2	3	2

63. "You can pay me now or you can pay me later" has the symbolic form $(p \vee q)$. The negation, $\sim(p \vee q)$, is equivalent, by one of De Morgan's laws, to $(\sim p \wedge \sim q)$. The corresponding word statement is "You can't pay me now and you can't pay me later."

65. "It is summer and there is no snow" has the symbolic form $p \wedge \sim q$. The negation, $\sim(p \wedge \sim q)$, is equivalent by De Morgan's to $\sim p \vee q$. The word translation for the negation is "It is not summer or there is snow."

67. "I said yes but she said no" is of the form $p \wedge q$. The negation, $\sim(p \wedge q)$, is equivalent, by De Morgan's, to $\sim p \vee \sim q$. The word translation for the negation is "I did not say yes or she did not say no." (Note: The connective "but" is equivalent to that of "and.")

69. "$6 - 1 = 5$ and $9 + 3 \neq 7$" is of the form $p \wedge \sim q$. The negation, $\sim(p \wedge \sim q)$, is equivalent, by De Morgan's, to $\sim p \vee q$. The translation for the negation is "$6 - 1 \neq 5$ or $9 + 3 = 7$."

71. "Prancer or Vixen will lead Santa's sleigh next Christmas" is of the form $p \vee q$. The negation, $\sim(p \vee q)$, is equivalent, by De Morgan's, to $\sim p \wedge \sim q$. A translation for the negation is "Neither Prancer nor Vixen will lead Santa's sleigh next Christmas."

73. "For every real number x, $x < 14$ or $x > 6$" is <u>true</u> since for any real number at least one of the component statements is true.

75. "There exists an integer n such that $n > 0$ and $n < 0$" is <u>false</u> since any integer which is true for one of the component statements will be false for the other.

77. $p \veebar q$

p	q	$p \veebar q$
T	T	F
T	F	T
F	T	T
F	F	F

Observe that it is only the first line in the truth table that changes for "exclusive disjunction" since the component statements cannot both be true at the same time.

79. "$3 + 1 = 4 \veebar 2 + 5 = 7$" is <u>false</u> since both component statements are true.

81. "$3 + 1 = 6 \veebar 2 + 5 = 7$" is <u>true</u> since the first component statement is false and the second is true.

83. The lady is behind Door 2. *Reasoning:* Suppose that the sign on Door 1 is true. Then the sign on Door 2 would also be true, but this is impossible. So the sign on door 2 must be true, and the sign on door 1 must be false. Because the sign on Door 1 says the lady is in Room 1 and this is false, the lady must be behind Door 2.

3.3 Exercises

1. The statement "You can believe it if you see it on the Internet" becomes "If you see it on the Internet, then you can believe it."

3. The statement "Every integer divisible by 10 is divisible by 5" becomes "If an integer is divisible by 10, then it is divisible by 5."

5. The statement "All marines love boot camp" becomes "If the soldier is a marine, then the soldier loves boot camp."

7. The statement "No pandas live in Idaho" becomes "If it is a panda, then it does not live in Idaho."

9. The statement "An opium-eater cannot have self-command" becomes "If it is an opium eater, then it has no self-command."

11. The statement "If the antecedent of a conditional statement is false, the conditional statement is true" is <u>true</u>, since a false antecedent will always yield a true conditional statement.

13. The statement "If q is true, then $(p \wedge q) \to q$ is true" is <u>true</u>, since with a true consequent the conditional statement is always true (even though the antecedent may be false).

15. The negation of "If pigs fly, I'll believe it" is "If pigs don't fly, I won't believe it." This statement is <u>false</u>. The negation is "Pigs fly and I don't believe it."

17. "Given that ~p is true and q is false, the conditional $p \to q$ is true" is a <u>true</u> statement since the antecedent, p, must be false.

19. Writing exercise; answers will vary.

21. "$T \to (7 < 3)$" is a <u>false</u> statement, since the antecedent is true and the consequent is false.

23. "$F \to (5 \neq 5)$" is a <u>true</u> statement, since a false antecedent always yields a true conditional statement.

25. "$(5^2 \neq 25) \to (9 > 0)$" is a <u>true</u> statement, since a false antecedent always yields a true conditional statement.

Let s represent the statement "She has a bird for a pet," let p represent the statement "he trains dogs," and let m represent "they raise alpacas."

27. "~$m \to p$" expressed in words, becomes "If they do not raise alpacas, then he trains dogs."

29. "$s \to (m \wedge p)$" expressed in words, becomes "If she has a bird for a pet, then they raise alpacas and he trains dogs."

31. "~$p \to (\sim m \vee s)$" expressed in words, becomes "If he does not train dogs, then they do not raise alpacas or she has a bird for a pet."

Let b represent the statement "I ride my bike," let s represent the statement "it snows" and let p represent "the play is canceled."

33. The statement "If I ride my bike, then the play is canceled," can be symbolized as "$b \to p$."

35. The statement "If the play is canceled, then it does not snow" can be symbolized as "$p \to \sim s$."

37. The statement "The play is canceled, and if it snows then I do not ride my bike" can be symbolized as "$p \wedge (s \to {\sim}b)$."

39. The statement "It snows if the play is canceled" can be symbolized as "$p \to s$."

Assume that p and r are false, and q is true.

41. Replacing r and q with the given truth values, we have
${\sim}F \to T$
$\quad T \to T$
$\qquad T.$
Thus, the compound statement ${\sim}r \to q$ is true.

43. Replacing p and q with the given truth values, we have
$T \to F$
$\quad F.$
Thus, the compound statement $q \to p$ is false.

45. Replacing p and q with the given truth values, we have
$F \to T$
$\quad T.$
Thus, the compound statement $p \to q$ is true.

47. Replacing p, r and q with the given truth values, we have
${\sim}F \to (T \wedge F)$
$\quad T \to F$
$\qquad F.$
Thus, the compound statement ${\sim}p \to (q \wedge r)$ is false.

49. Replacing p, r and q with the given truth values, we have
${\sim}T \to (F \wedge F)$
$\quad F \to F$
$\qquad T.$
Thus, the compound statement ${\sim}q \to (p \wedge r)$ is true.

51. Replacing p, r and q with the given truth values, we have
$(F \to {\sim}T) \to ({\sim}F \wedge {\sim}F)$
$\quad (F \to F) \to (T \wedge T)$
$\qquad\quad T \to T$
$\qquad\qquad T.$
Thus, the compound statement $(p \to {\sim}q) \to ({\sim}p \wedge {\sim}r)$ is true.

53. Writing exercise; answers will vary.

55. ${\sim}q \to p$

p	q	${\sim}q$	$\to$	p
T	T	F	T	T
T	F	T	T	T
F	T	F	T	F
F	F	T	F	F
		1	2	1

57. $({\sim}p \to q) \to p$

p	q	$({\sim}p$	$\to$	$q)$	$\to$	p
T	T	F	T	T	T	T
T	F	F	T	F	T	T
F	T	T	T	T	F	F
F	F	T	F	F	T	F
		1	2	1	3	2

59. $(p \vee q) \to (q \vee p)$

p	q	$(p$	$\vee$	$q)$	$\to$	$(q$	$\vee$	$p)$
T	T	T	T	T	T	T	T	T
T	F	T	T	F	T	F	T	T
F	T	F	T	T	T	T	T	F
F	F	F	F	F	T	F	F	F
		1	2	1	3	1	2	1

Since this statement is always true (column 3), it is a tautology.

61. $({\sim}p \to {\sim}q) \to (p \wedge q)$

p	q	$({\sim}p$	$\to$	${\sim}q)$	$\to$	$(p$	$\wedge$	$q)$
T	T	F	T	F	T	T	T	T
T	F	F	T	T	F	T	F	F
F	T	T	F	F	T	F	F	T
F	F	T	T	T	F	F	F	F
		1	2	1	3	1	2	1

63. $[(r \lor p) \land \sim q] \to p$

p	q	r	$[(r$	$\lor$	$p)$	$\land$	$\sim q]$	$\to$	p
T	T	T	T	T	T	F	F	T	T
T	T	F	F	T	T	F	F	T	T
T	F	T	T	T	T	T	T	T	T
T	F	F	F	T	T	T	T	T	T
F	T	T	T	T	F	F	F	T	F
F	T	F	F	F	F	F	F	T	F
F	F	T	T	T	F	T	T	F	F
F	F	F	F	F	F	F	T	T	F
			1	2	1	3	2	4	3

65. $(\sim r \to s) \lor (p \to \sim q)$

p	q	r	s	$(\sim r$	$\to$	$s)$	$\lor$	$(p$	$\to$	$\sim q)$
T	T	T	T	F	T	T	T	T	F	F
T	T	T	F	F	T	F	T	T	F	F
T	T	F	T	T	T	T	T	T	F	F
T	T	F	F	T	F	F	F	T	F	F
T	F	T	T	F	T	T	T	T	T	T
T	F	T	F	F	T	F	T	T	T	T
T	F	F	T	T	T	T	T	T	T	T
T	F	F	F	T	F	F	T	T	T	T
F	T	T	T	F	T	T	T	F	T	F
F	T	T	F	F	T	F	T	F	T	F
F	T	F	T	T	T	T	T	F	T	F
F	T	F	F	T	F	F	T	F	T	F
F	F	T	T	F	T	T	T	F	T	T
F	F	T	F	F	T	F	T	F	T	T
F	F	F	T	T	T	T	T	F	T	T
F	F	F	F	T	F	F	T	F	T	T
				1	2	1	3	1	2	1

67. The statement is not a tautology if only <u>one</u> F appears in the final column of a truth table, since a tautology requires all T's in the final column.

69. The negation of "If that is an authentic Rolex watch, I'll be surprised" is "That is an authentic Rolex watch and I am not surprised."

71. The negation of "If the English measures are not converted to metric measures, then the spacecraft will crash on the surface of Saturn" is "The English measures are not converted to metric measures and the spacecraft does not crash on the surface of Saturn."

73. The negation of "If you want to be happy for the rest of your life, never make a pretty woman your wife" is "You want to be happy for the rest of your life and make a pretty woman your wife."

75. An equivalent statement to "If you give your plants tender, loving care, they will flourish" is "You do not give your plants tender, loving care or they flourish."

77. An equivalent statement to "If she doesn't, he will" is "She does or he will."

79. An equivalent conditional statement to "All residents of Pensacola are residents of Florida" is "If you are a resident of Pensacola, then you are a resident of Florida." An equivalent statement would be "The person is not a resident of Pensacola or is a resident of Florida."

81. The statements $p \rightarrow q$ and $\sim p \vee q$ are equivalent if they have the same truth tables.

p	q	p	$\rightarrow$	q	$\sim p$	$\vee$	q
T	T	T	T	T	F	T	T
T	F	T	F	F	F	F	F
F	T	F	T	T	T	T	T
F	F	F	T	F	T	T	F
		1	2	1	1	2	1

Since the truth values in the final columns for each statement are the same, the statements are equivalent.

83.

p	q	p	$\rightarrow$	q	$\sim q$	$\rightarrow$	$\sim p$
T	T	T	T	T	F	T	F
T	F	T	F	F	T	F	F
F	T	F	T	T	F	T	T
F	F	F	T	F	T	T	T
		1	2	1	1	2	1

Since the truth values in the final columns for each statement are the same, the statements are equivalent.

85.

p	q	p	$\wedge$	$\sim q$	$\sim q$	$\rightarrow$	$\sim p$
T	T	T	F	F	F	T	F
T	F	T	T	T	T	F	F
F	T	F	F	F	F	T	T
F	F	F	F	T	T	T	T
		1	2	1	1	2	1

Since the truth values in the final columns for each statement are not the same, the statements are not equivalent. Observe that

since they have opposite truth values, each statement is the negation of the other.

87.

p	q	p	$\rightarrow$	$\sim q$	$\sim p$	$\vee$	$\sim q$
T	T	T	F	F	F	F	F
T	F	T	T	T	F	T	T
F	T	F	T	F	T	T	F
F	F	F	T	T	T	T	T
		1	2	1	1	2	1

Since the truth values in the final columns for each statement are the same, the statements are equivalent.

89.

p	q	q	$\rightarrow$	$\sim p$	p	$\rightarrow$	$\sim q$
T	T	T	F	F	T	F	F
T	F	F	T	F	T	T	T
F	T	T	T	T	F	T	F
F	F	F	T	T	F	T	T
		1	3	2	1	3	2

Since the truth values in the final columns for each statement are the same, the statements are equivalent.

91. In the diagram, two series circuits are shown, which correspond to $p \wedge q$ and $p \wedge \sim q$. These circuits, in turn, form a parallel circuit. Thus, the logical statement is $(p \wedge q) \vee (p \wedge \sim q)$.
One pair of equivalent statements listed in the text includes
$(p \wedge q) \vee (p \wedge \sim q) \equiv p \wedge (q \vee \sim q)$.
Since $(q \vee \sim q)$ is always true, $p \wedge (q \vee \sim q)$ simplifies to $p \wedge T \equiv p$.

93. In the diagram, a series circuit is shown, which corresponds to $\sim q \wedge r$. This circuit, in turn, forms a parallel circuit with p. Thus, the logical statement is $p \vee (\sim q \wedge r)$.

95. In the diagram, a parallel circuit corresponds to $p \vee q$. This circuit is parallel to $\sim p$. Thus, the total circuit corresponds to the logical statement $\sim p \vee (p \vee q)$.
This statement in turn, is equivalent to $(\sim p \vee p) \vee (\sim p \vee q)$.
Since $\sim p \vee p$ is always true, we have
$T \vee (\sim p \vee q) \equiv T$.

97. The logical statement, $p \wedge (q \vee {\sim}p)$, can be represented by the following circuit.

The statement, $p \wedge (q \vee {\sim}p)$, simplifies to $p \wedge q$ as follows:
$$p \wedge (q \vee {\sim}p) \equiv (p \wedge q) \vee (p \wedge {\sim}p)$$
$$\equiv (p \wedge q) \vee F$$
$$\equiv p \wedge q.$$

99. The logical statement, $(p \vee q) \wedge ({\sim}p \wedge {\sim}q)$, can be represented by the following circuit.

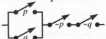

The statement, $(p \vee q) \wedge ({\sim}p \wedge {\sim}q)$, simplifies to F as follows:
$$(p \vee q) \wedge ({\sim}p \wedge {\sim}q)$$
$$\equiv [p \wedge ({\sim}p \wedge {\sim}q)] \vee [q \wedge ({\sim}p \wedge {\sim}q)]$$
$$\equiv [p \wedge {\sim}p \wedge {\sim}q] \vee [q \wedge {\sim}q \wedge {\sim}p]$$
$$\equiv [F \wedge {\sim}q] \vee [F \wedge {\sim}p]$$
$$\equiv F \vee F$$
$$\equiv F.$$

101. The logical statement, $[(p \vee q) \wedge r] \wedge {\sim}p$, can be represented by the following circuit.

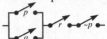

The statement, $[(p \vee q) \wedge r] \wedge {\sim}p$, simplifies to $(r \wedge {\sim}p) \wedge q$ as follows:
$$[(p \vee q) \wedge r] \wedge {\sim}p$$
$$\equiv [(p \wedge r) \vee (q \wedge r)] \wedge {\sim}p$$
$$\equiv [(p \wedge r) \wedge {\sim}p] \vee [(q \wedge r) \wedge {\sim}p]$$
$$\equiv [p \wedge r \wedge {\sim}p] \vee [(q \wedge r) \wedge {\sim}p]$$
$$\equiv [(p \wedge {\sim}p) \wedge r] \vee [(r \wedge {\sim}p) \wedge q]$$
$$\equiv (F \wedge r) \vee [(r \wedge {\sim}p) \wedge q]$$
$$\equiv F \vee [(r \wedge {\sim}p) \wedge q]$$
$$\equiv (r \wedge {\sim}p) \wedge q \text{ or } q \wedge (r \wedge {\sim}p).$$

103. The logical statement, ${\sim}q \rightarrow ({\sim}p \rightarrow q)$, can be represented by the following circuit.

The statement, ${\sim}q \rightarrow ({\sim}p \rightarrow q)$, simplifies to $p \vee q$ as follows:
$${\sim}q \rightarrow ({\sim}p \rightarrow q) \equiv {\sim}q \rightarrow (p \vee q)$$
$$\equiv q \vee (p \vee q)$$
$$\equiv q \vee p \vee q$$
$$\equiv p \vee q \vee q$$
$$\equiv p \vee (q \vee q)$$
$$\equiv p \vee q.$$

105. Referring to Figures 5 and 6 of Example 6 in the text:
Cost per year of the circuit in Figure 5
= number of switches × $.06 × 24 hr
$$\times 365 \text{ days}$$
= (4) × (.06) × 24 × 365
= $2102.40.

Cost per year of the circuit in Figure 6
= number of switches × $.06 × 24 hr
$$\times 365 \text{ days}$$
= (3) × (.06) × 24 × 365
= $1576.80.
Thus, the savings is $2102.40 − $1576.80 = $525.60.

3.4 Exercises

For each given conditional statement (symbolically as $p \rightarrow q$), write (a) the converse ($q \rightarrow p$), (b) the inverse (${\sim}p \rightarrow {\sim}q$), and (c) the contrapositive (${\sim}q \rightarrow {\sim}p$) in if...then forms. Wording may vary in the answers to Exercises 1–9.

1. The conditional statement: If beauty were a minute, then you would be an hour.

  (a) *Converse*: If you were an hour, then beauty would be a minute.

  (b) *Inverse*: If beauty were not a minute, then you would not be an hour.

  (c) *Contrapositive*: If you were not an hour, then beauty would not be a minute.

3. The conditional statement: If it ain't broke, don't fix it.

  (a) *Converse*: If you don't fix it, then it ain't broke.

  (b) *Inverse*: If it's broke, then fix it.

  (c) *Contrapositive*: If you fix it, then it's broke.

It is helpful to restate the conditional statement in an "if...then" form for the exercises 5–8 and 10.

5. The conditional statement: If you walk in front of a moving car, then it is dangerous to your health.

  (a) *Converse*: If it is dangerous to your health, then you walk in front of a moving car.

(b) *Inverse*: If you do not walk in front of a moving car, then it is not dangerous to your health.

(c) *Contrapositive*: If it is not dangerous to your health, then you do not walk in front of a moving car.

7. The conditional statement: If they are birds of a feather, then they flock together.

  (a) *Converse*: If they flock together, then they are birds of a feather.

  (b) *Inverse*: If they are not birds of a feather, then they do not flock together.

  (c) *Contrapositive*: If they do not flock together, then they are not birds of a feather.

9. The conditional statement: If you build it, then he will come.

  (a) *Converse*: If he comes, then you built it.

  (b) *Inverse*: If you don't build it, then he won't come.

  (c) *Contrapositive*: If he doesn't come, then you didn't build it.

11. The conditional statement: $p \rightarrow \sim q$.

  (a) *Converse*: $\sim q \rightarrow p$

  (b) *Inverse*: $\sim p \rightarrow q$.

  (c) *Contrapositive*: $q \rightarrow \sim p$.

13. The conditional statement: $\sim p \rightarrow \sim q$.

  (a) *Converse*: $\sim q \rightarrow \sim p$.

  (b) *Inverse*: $p \rightarrow q$.

  (c) *Contrapositive*: $q \rightarrow p$.

15. The conditional statement: $p \rightarrow (q \lor r)$.

  (a) *Converse*: $(q \lor r) \rightarrow p$.

  (b) *Inverse*: $\sim p \rightarrow \sim(q \lor r)$ or $\sim p \rightarrow (\sim q \land \sim r)$.

  (c) *Contrapositive*: $(\sim q \land \sim r) \rightarrow \sim p$.

17. Writing exercise, answers will vary.

Writing the statements, Exercises 19–39, in the form "if p, then q" we arrive at the following results.

19. The statement "If it is muddy, I'll wear my galoshes" becomes "If it is muddy, then I'll wear my galoshes."

21. The statement "'19 is positive' implies that $19 + 1$ is positive" becomes "If 19 is positive, then $19 + 1$ is positive."

23. The statement "All integers are rational numbers" becomes "If a number is an integer, then it is a rational number."

25. The statement "Doing logic puzzles is sufficient for driving me crazy" becomes "If I do logic puzzles, then I am driven crazy."

27. "A day's growth of beard is necessary for Jeff Marsalis to shave" becomes "If Jeff Marsalis is to shave, then he must have a day's growth of beard."

29. The statement "I can go from Boardwalk to Baltic Avenue only if I pass GO" becomes "If I go from Boardwalk to Baltic, then I pass GO."

31. The statement "No whole numbers are not integers" becomes "If a number is a whole number, then it is an integer."

33. The statement "The Nationals will win the pennant when their pitching improves" becomes "If their pitching improves, then the Nationals will win the pennant."

35. The statement "A rectangle is a parallelogram with a right angle" becomes "If the figure is a rectangle, then it is a parallelogram with a right angle."

37. The statement "A triangle with two perpendicular sides is a right triangle" becomes "If a triangle has two perpendicular sides, then it is a right triangle."

39. The statement "The square of a two-digit number whose units digit is 5 will end in 25" becomes "If a two-digit number whose units digit is 5 is squared, then it will end in 25."

41. Option D is the answer since "r is necessary for s" represents the converse, $s \rightarrow r$, of all of the other statements.

43. Writing exercise; answers will vary.

45. The statement "$6 = 9 - 3$ if and only if $8 + 2 = 10$" is <u>true</u>, since this is a biconditional composed of two true statements.

47. The statement "$8 + 7 \neq 15$ if and only if $3 \times 5 \neq 8$" is <u>false</u>, since this is a biconditional consisting of a false and a true statement.

49. The statement "George H.W. Bush was president if and only if George W. Bush was not president" is <u>false</u>, since this is a biconditional consisting of a true and a false statement.

51. The statements "Michael Jackson is alive" and "Michael Jackson is dead" are <u>contrary</u>, since both cannot be true at the same time.

53. The statements "That animal has four legs" and "That same animal is a cat" are <u>consistent</u>, since both statements can be true.

55. The statements "This number is a whole number" and "This number is irrational" are <u>contrary</u>, since both cannot be true at the same time.

57. The statements "This number is an integer" and "This same number is a rational number" are <u>consistent</u>, since both statements can be true.

59. Answers will vary. One example is: That man is Otis Taylor; that man sells books.

3.5 Exercises

1. Draw an Euler diagram where the region representing "amusement parks" must be inside the region representing "locations that have thrill rides" so that the first premise is true.

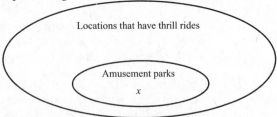

Locations that have thrill rides

Amusement parks

x

x represents *Universal Orlando*

Let x represent the amusement park "*Universal Orlando*." By the second premise, x must lie in the "amusement parks" region. Since this forces the conclusion to be true, the argument is <u>valid</u>.

3. Draw an Euler diagram where the region representing "politicians" must be inside the region representing "those who lie, cheat, and steal" so that the first premise is true.

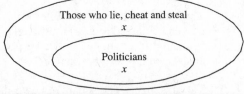

Those who lie, cheat and steal

x

Politicians

x

x represents *that man*

Let x represent "that man." By the second premise, x must lie in the "those who lie, cheat and steal" region. Thus, he could be inside or outside the inner region. Since this allows for a false conclusion (he doesn't have to be in the "politicians" region for both premises to be true), the argument is <u>invalid</u>.

5. Draw an Euler diagram where the region representing "dogs" must be inside the region representing "creatures that love to bury bones" so that the first premise is true.

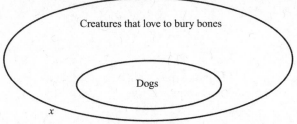

x represents Puddles

Let x represent "Puddles." By the second premise, x must lie outside the region representing "creatures that love to bury bones." Since this forces the conclusion to be true, the argument is <u>valid</u>.

7. Draw an Euler diagram where the region representing "residents of Minnesota" must be inside the region representing "those who know how to live in freezing temperatures" so that the first premise is true.

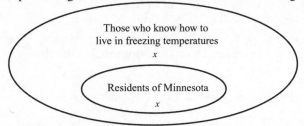

x represents Jessica Rockswold

Let x represent "Jessica Rockswold." By the second premise, x must lie in the "those who know how to live in freezing temperatures" region. Thus, she could be inside or outside the inner region. Since this allows for a false conclusion (she doesn't have to be in the "residents of Minnesota" region for both premises to be true), the argument is <u>invalid</u>.

9. Draw an Euler diagram where the region representing "dinosaurs" intersects the region representing "plant-eaters." This keeps the first premise true.

x represents Danny

Let x represent "Danny." By the second premise, x must lie in the region representing "plant-eaters." Thus, he could be inside or outside the region "dinosaurs." Since this allows for a false conclusion, the argument is <u>invalid</u>.

11. Draw an Euler diagram where the region representing "nurses" intersects the region representing "those who wear blue uniforms." This keeps the first premise true.

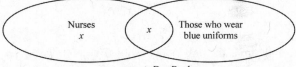

x represents Dee Boyle

Let x represent "Dee Boyle." By the second premise, x must lie in the region representing "nurses." Thus, she could be inside or outside the region "those who wear blue uniforms." Since this allows for a false conclusion, the argument is <u>invalid</u>.

13. Interchanging the second premise and the conclusion of Example 3 (in the text) yields the following argument,

All magnolia trees have green leaves.
That plant is a magnolia tree.
That plant has green leaves.

Draw an Euler diagram where the region representing "Magnolia trees" must be inside the region representing "Things that have green leaves" so that the first premise is true.

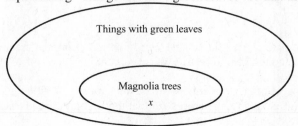

x represents that plant

Let *x* represent "That plant." By the second premise, *x* must lie inside the region representing "Magnolia trees." Since this forces the conclusion to be true, the argument is valid, which makes the answer to the question <u>yes</u>.

15. The following is a valid argument which can be constructed from the given Euler diagram.

All people with blue eyes have blond hair.
Natalie Graham does not have blond hair.
Natalie Graham does not have blue eyes.

17. The following represents one way to diagram the premises so that they are true; however, the argument is <u>invalid</u> since, according to the diagram, all birds are planes, which is false even though the stated conclusion is true.

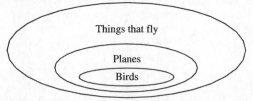

19. The following Euler diagram yields true premises. It also forces the conclusion to be true.

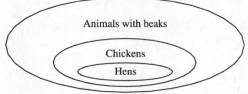

Thus, the argument is <u>valid</u>. Observe that the diagram is the only way to show true premises.

21. The following Euler diagram represents true premises.

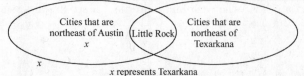

x represents Texarkana

But *x* can reside inside or outside of the "Cities that are northeast of Austin" diagram. In the one case (*x* inside) the conclusion is true. In the other case (*x* outside) the conclusion is false. Since true premises must always give a true conclusion, the argument is <u>invalid</u>.

23. The following Euler diagram represents the two premises as being true and we are forced into a true conclusion.

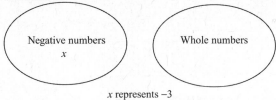

x represents −3

Thus, the argument is <u>valid</u>.

The premises marked A, B, and C are followed by several possible conclusions (Exercises 25–29). Take each conclusion in turn, and check whether the resulting argument is valid or invalid.

    A. All people who drive contribute to air pollution.

    B. All people who contribute to air pollution make life a little worse.

    C. Some people who live in a suburb make life a little worse.

Diagram the three premises to be true.

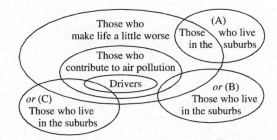

25. We are not forced into the conclusion, "Some people who live in a suburb contribute to air pollution" since option (*A*) represents true premises and a false conclusion. Thus, the argument is <u>invalid</u>.

27. We are not forced into the conclusion, "Suburban residents never drive" since diagram (C) represents true premises where this conclusion is false. Thus, the argument is <u>invalid</u>.

29. The conclusion, "Some people who make life a little worse live in a suburb" yields a <u>valid</u> argument since all three options (*A–C*) represent true premises and force this conclusion to be true.

EXTENSION: **LOGIC PROBLEMS AND SUDOKU**

    1. Because the orange mints are not Fresh Air mints, they must be Inti-Mints. Then the Fresh Air mints are spearmint. The person who ate the tuna-salad sandwich brought the wintergreen TKO mints. Neither Uma nor Xerxes had the Fresh Air ints, so neither of them had the spanakopita. And Nash didn't have the spearmint mints, so he didn't have spanakopita either. So Nash must have had the French onion soup, and the vanilla Liplickers. Draw must have had the spanakopita, so he also had the Fresh Air spearmint mints. Xerxes must have had the garlic shrimp, followed by orange Inti-mints. Uma had the tuna-salad sandwich, followed by wintergreen TKO mints. Ilse had the buffalo-chicken sandwich and the cinnamon Deltoids. Here are the final combinations:
Drew, spanakopita, Fresh Air, spearmint;
Ilse, buffalo-chicken sandwich, Deltoids, cinnamon;
Nash, French onion soup, Liplickers, vanilla;
Uma, tuna-salad sandwich, TKO, wintergreen;
Xerxes, garlic shrimp, Inti-mints, orange

3. Toni's lucky element is not wood. Earl was not the fifth person and not born in the Year of the Rooster. The person born in the Year of the Ox was not fifth either. The third person was not born in the Year of the Horse because Ivana was not the third person. the person whose element is earth was not first or second. Toni was not born in the Year of the Dragon. The fourth person's element is not water. Philip must have been either first or second and his lucky element is not earth. The first person's element is not metal, so the first person's element must be fire. This person was not born in the Year of the Dragon, so the person whose element is earth was born in the Year of the Dragon, and that is the fourth person. Then Philip is the second person. The person born in the Year of the Ox had his or her fortune told before the person whose element is metal, so this person was not third. Therefore the third person was born in the Year of the Cow, and this person's element is water so the second person's element is metal. Now the first person must have been born in the Year of the Ox. Ivana was born in the Year of the Horse, so she must have been fifth. Then Lucy's element is earth, and she was the fourth person and born in the Year of the Dragon. That means Earl was first and born in the Year of the Fox, Philip was born in the Year of the Rooster, and Toni's element was water.

Here are the final combinations:
1st, Earl, Ox, fire; 2nd, Philip, Rooster, metal; 3rd, Toni, Cow, water; 4th, Lucy, Dragon, earth; 5th, Ivana, Horse, wood

5.

7	4	2	1	6	8	9	3	5
3	1	9	4	5	7	6	2	8
8	6	5	9	3	2	7	1	4
6	2	4	7	8	9	1	5	3
1	3	8	5	4	6	2	9	7
5	9	7	2	1	3	8	4	6
2	5	3	8	7	1	4	6	9
9	7	6	3	2	4	5	8	1
4	8	1	6	9	5	3	7	2

7.

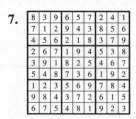

9.

6	1	7	9	2	3	5	8	4
4	5	9	8	6	1	2	7	3
3	2	8	4	5	7	9	6	1
8	3	5	6	4	9	1	2	7
7	6	1	5	3	2	8	4	9
2	9	4	7	1	8	3	5	6
5	7	2	3	9	6	4	1	8
1	8	3	2	7	4	6	9	5
9	4	6	1	8	5	7	3	2

3.6 Exercises

1. Let p represent "James Taylor comes to town," q represent "I will go to the concert," and r represent "I'll call in sick for work." The argument is then represented symbolically by:

$p \rightarrow q$
$q \rightarrow r$
$p \rightarrow r$.
This is the <u>valid</u> argument form "reasoning by transitivity."

3. Let p represent "Julie Nhem works hard enough" and q represent "she will get a promotion." The argument is then represented symbolically by:

$p \rightarrow q$
p
q.
This is the <u>valid</u> argument form "modus ponens."

5. Let p represent "He doesn't have to get up at 3:00 A.M." and q represent "he is ecstatic." The argument is then represented symbolically by:

$p \rightarrow q$
q
p.
Since this is the form "fallacy of the converse," it is invalid and considered a <u>fallacy</u>.

7. Let p represent "Mariano Rivera pitches" and q represent "the Yankees win." The argument is then represented symbolically by:

$p \rightarrow q$
$\sim q$
$\sim p$.
This is the <u>valid</u> argument form "modus tollens."

9. Let p represent "we evolved a race of Isaac Newtons" and q represent "that would not be progress." The argument is then represented symbolically by:

$p \rightarrow q$

$\sim p$

$\sim q$.

Note: that since we let q represent "that <u>would not</u> be progress," then $\sim q$ represents "that <u>is</u> progress." Since this is the form "fallacy of the inverse," it is invalid and considered a <u>fallacy</u>.

11. Let p represent "She uses e-commerce" and q represent "she pays by credit card." The argument is then represented symbolically by:

$p \vee q$ (or $q \vee p$)

$\sim q$

p.

Since this is the form "disjunctive syllogism," it is a <u>valid</u> argument.

To show validity for the arguments in the following exercises, we must show that the conjunction of the premises implies the conclusion. That is, the conditional statement $[P_1 \wedge P_2 \wedge ... \wedge P_n] \rightarrow C$ must be a tautology. For exercises 13 and 14 we will use the standard (long format) to develop the corresponding truth tables. For the remainder of the exercises we will use the alternate (short format) to create the truth tables.

13. Form the conditional statement

$[(p \vee q) \wedge p] \rightarrow \sim q$

from the argument. Complete a truth table.

p	q	$p \vee q$	$(p \vee q) \wedge p$	$\sim q$	$[(p \vee q) \wedge p] \rightarrow \sim q$
T	T	T	T	F	F
T	F	T	T	T	T
F	T	T	F	F	T
F	F	F	F	T	T

Since the conditional, formed by the conjunction of premises implying the conclusion, is not a tautology, the argument is <u>invalid</u>.

15. Form the conditional statement

$[(\sim p \rightarrow \sim q) \wedge q] \rightarrow p$

from the argument. Complete a truth table.

p	q	$[(\sim p$	$\rightarrow$	$\sim q)$	$\wedge$	$q]$	$\rightarrow$	p
T	T	F	T	F	T	T	T	T
T	F	F	T	T	F	F	T	T
F	T	T	F	F	F	T	T	F
F	F	T	T	T	F	F	T	F
		1	2	1	3	2	4	3

Since the conditional, formed by the conjunction of premises implying the conclusion, is a tautology, the argument is <u>valid</u>.

17. Form the conditional statement
$[(p \to q) \land (q \to p)] \to (p \land q)$
from the argument. Complete a truth table.

p	q	$[(p \to q)$	$\land$	$(q \to p)]$	$\to$	$(p \land q)$
T	T	T	T	T	T	T
T	F	F	F	T	T	F
F	T	T	F	F	T	F
F	F	T	T	T	F	F
		1	3	2	4	3

Since the conditional,, formed by the conjunction of premises implying the conclusion, is not a tautology, the argument is <u>invalid</u>.

19. Form the conditional statement
$[(p \to \sim q) \land q] \to \sim p$
from the argument. Complete a truth table.

p	q	$[(p$	$\to$	$\sim q)$	$\land$	$q]$	$\to$	$\sim p$
T	T	T	F	F	F	T	T	F
T	F	T	T	T	F	F	T	F
F	T	F	T	F	T	T	T	T
F	F	F	T	T	F	F	T	T
		1	2	1	3	2	4	3

Since the conditional, formed by the conjunction of premises implying the conclusion, is a tautology, the argument is <u>valid</u>.

21. Form the conditional statement $\{[(p \land q) \lor (p \lor q)] \land q\} \to \sim q$ from the argument. Complete a truth table.

p	q	$\{[(p$	$\land$	$q)$	$\lor$	$(p$	$\lor$	$q)$	$\land$	$q\}$	$\to$	$\sim q$
T	T	T	T	T	T	T	T	T	T	T	F	F
T	F	T	F	F	T	T	T	F	F	F	T	T
F	T	F	F	T	T	F	T	T	T	T	F	F
F	F	F	F	F	F	F	F	F	F	F	T	T
		1	2	1	3	1	2	1	4		5	

Since the conditional formed by the conjunction of premises implying the conclusion, is not a tautology, the argument is <u>invalid</u>.

23. Form the conditional statement
$[(\sim p \vee q) \wedge (\sim p \rightarrow q) \wedge p] \rightarrow \sim q$
from the argument. Complete a truth table.

p	q	[(~p	∨	q)	∧	(~p	→	q)]	∧	p]	→	~q
T	T	F	T	T	T	F	T	T	T	T	F	F
T	F	F	F	F	F	F	T	F	F	T	T	T
F	T	T	T	T	T	T	T	T	F	F	T	F
F	F	T	T	F	F	T	F	F	F	F	T	T
		1	2	1	3		2		4	3	5	4

Since the conditional, formed by the conjunction of premises implying the conclusion, is not a tautology, the argument is <u>invalid</u>.

25. Form the conditional statement
$\{[(\sim p \wedge r) \rightarrow (p \vee q)] \wedge (\sim r \rightarrow p)\} \rightarrow (q \rightarrow r)$
from the argument.

p	q	r	{[(~p	∧	r)	→	(p ∨ q)]	∧	(~r	→	p)}	→	(q → r)
T	T	T	F	F	T	T	T	T	F	T	T	T	T
T	T	F	F	F	F	T	T	T	T	T	T	F	F
T	F	T	F	F	T	T	T	T	F	T	T	T	T
T	F	F	F	F	F	T	T	T	T	T	T	T	T
F	T	T	T	T	T	T	T	T	F	T	F	T	T
F	T	F	T	F	F	T	T	F	T	F	F	T	F
F	F	T	T	T	T	F	F	F	F	T	F	T	T
F	F	F	T	F	F	T	F	F	T	F	F	T	T
			1	2	1	3	2	4	2	3	2	5	4

The F in the final column 5 shows us that the statement is not a tautology and hence, the argument is <u>invalid</u>.

27. Every time something squeaks, I use WD-40.
<u>Every time I use WD-40, I must go to the hardware store.</u>
Every time something squeaks, I go to the hardware store.

29. Let p represent "Joey loves to watch movies," q represent "Terry likes to jog," and r represent "Carrie drives a school bus." The argument is then represented symbolically by:
p
$q \rightarrow \sim p$
<u>$\sim q \rightarrow r$</u>
r.
Construct the truth table for $[p \wedge (q \rightarrow \sim p) \wedge (\sim q \rightarrow r)] \rightarrow r$.

p	q	r	[p	∧	(q	→	~p)	∧	(~q	→	r)]	→	r
T	T	T	T	F	T	F	F	F	F	T	T	T	T
T	T	F	T	F	T	F	F	F	F	T	F	T	F
T	F	T	T	T	F	T	F	T	T	T	T	T	T
T	F	F	T	T	F	T	F	F	T	F	F	T	F
F	T	T	F	F	T	T	T	F	F	T	T	T	T
F	T	F	F	F	T	T	T	F	F	T	F	T	F
F	F	T	F	F	F	T	T	F	T	T	T	T	T
F	F	F	F	F	F	T	T	F	T	F	F	T	F
			2	3	1	2	1	4	2	3	2	5	4

Since the conditional, formed by the conjunction of premises implying the conclusion, is a tautology, the argument is <u>valid</u>.

31. Let *p* represent "the social networking craze continues," *q* represent "downloading music will remain popular," and *r* represent "American girl dolls are favorites." The argument is then represented symbolically by:

$p \rightarrow q$

$r \vee q$

$\underline{\sim r}$

$\sim p.$

Construct the truth table for $[(p \rightarrow q) \wedge (r \vee q) \wedge (\sim r)] \rightarrow \sim p$. (Note: we do not have to complete a column under each simple statement *p*, *q*, and *r*, as we did in exercises above, since it is easy to compare the appropriate index columns to create the truth value for each connective.)

p	q	r	[(p → q)	∧	(r ∨ q)	∧	(~r)]	→	~p
T	T	T	T	T	T	F	F	T	F
T	T	F	T	T	T	T	T	F	F
T	F	T	F	F	T	F	F	T	F
T	F	F	F	F	F	F	T	T	F
F	T	T	T	T	T	F	F	T	T
F	T	F	T	T	T	T	T	T	T
F	F	T	T	T	T	F	F	T	T
F	F	F	T	F	F	F	T	T	T
			1	2	1	3	2	4	3

Since the conditional, formed by the conjunction of premises implying the conclusion, is not a tautology, the argument is <u>invalid</u>. Note: If you are completing the truth table along rows (rather than down columns), you could stop after completing the second row, knowing that with a false conditional, the statement will not be a tautology.

33. Let *p* represent "The Dolphins will be in the playoffs," *q* represent "Chad leads the league in passing," and *r* represent "Tony coaches the Dolphins." The argument is then represented symbolically by

$p \leftrightarrow q$

$r \vee q$

$\underline{\sim r}$

$\sim p.$

Construct the truth table for $[(p \leftrightarrow q) \wedge (r \vee q) \wedge (\sim r)] \rightarrow \sim p$.

p	q	r	[(p ↔ q)	∧	(r ∨ q)	∧	(~r)]	→	~p
T	T	T	T	T	T	F	F	T	F
T	T	F	T	T	T	T	T	F	F
T	F	T	F	F	T	F	F	T	F
T	F	F	F	F	F	F	T	T	F
F	T	T	F	F	T	F	F	T	T
F	T	F	F	F	T	F	T	T	T
F	F	T	T	T	T	F	F	T	T
F	F	F	T	F	F	F	T	T	T
			1	2	1	3	2	4	3

Since the conditional, formed by the conjunction of premises implying the conclusion, is not a tautology, the argument is <u>invalid</u>. (Note: If you are completing the truth table along rows, rather than down columns, you could stop after completing the second row, knowing that with a false conditional, the statement will not be a tautology.)

35. Let p represent "Dr. Hardy is a department chairman," q represent "he lives in Atlanta," and r represent "his first name is Larry." The argument is then represented symbolically by

$p \rightarrow q$

$q \wedge r$

$\sim r \rightarrow \sim p$.

Construct the truth table for $[(p \rightarrow q) \wedge (q \wedge r)] \rightarrow (\sim r \rightarrow \sim p)$.

p	q	r	[(p → q)	∧	(q ∧ r)]	→	(~r	→	~p)
T	T	T	T	T	T	T	F	T	F
T	T	F	T	F	F	T	T	F	F
T	F	T	F	F	F	T	F	T	F
T	F	F	F	F	F	T	T	F	F
F	T	T	T	T	T	T	F	T	T
F	T	F	T	F	F	T	T	T	T
F	F	T	T	F	F	T	F	T	T
F	F	F	T	F	F	T	T	T	T
			1	2	1	3	1	2	1

Since the conditional, formed by the conjunction of premises implying the conclusion, is a tautology, the argument is <u>valid</u>.

The following exercises involve Quantified arguments and can be analyzed, as such, by Euler diagrams. However, the quantified statements can be represented as conditional statements as well. This allows us to use a truth table–or recognize a valid argument form–to analyze the validity of the argument.

37. Let p represent "you are a man," q represent "you are created equal," and r represent "you are a woman." The argument is then represented symbolically by:

$p \rightarrow q$

$q \rightarrow r$

$p \rightarrow r$.

This is a "reasoning by Transitivity" argument form and, hence, is <u>valid</u>.

39. We apply reasoning by repeated transitivity to the six premises. A conclusion from this reasoning, which makes the argument valid, is reached by linking the first antecedent to the last consequent. This conclusion is "If I tell you the time, then my life will be miserable."

Answers in Exercises 39–46 may be replaced by their contrapositives.

41. The statement "all my poultry are ducks" becomes "if it is my poultry, then it is a duck."

43. The statement "guinea pigs are hopelessly ignorant of music" becomes "if it is a guinea pig, then it is hopelessly ignorant of music."

45. The statement "no teachable kitten has green eyes" becomes "if it is a teachable kitten, then it does not have green eyes."

47. The statement "I have not filed any of them that I can read" becomes "if I can read it, then I have not filed it."

49. (a) "No ducks are willing to waltz" becomes "if it is a duck, then it is not willing to waltz" or $p \rightarrow \sim s$.

(b) "No officers ever decline to waltz" becomes "if one is an officer, then one is willing to waltz" or $r \rightarrow s$.

(c) "All my poultry are ducks" becomes "if it is my poultry, then it is a duck" or $q \rightarrow p$.

(d) The three symbolic premises are
$$p \rightarrow \sim s$$
$$r \rightarrow s$$
$$q \rightarrow p.$$
Begin with q, which only appears once. Replacing $r \rightarrow s$ with its contrapositive, $\sim s \rightarrow \sim r$, rearrange the three premises.
$$q \rightarrow p$$
$$p \rightarrow \sim s$$
$$\sim s \rightarrow \sim r$$
By repeated use of reasoning by transitivity, the conclusion which provides a valid argument is $q \rightarrow \sim r$. In words, "if it is my poultry, then it is not an officer," or "none of my poultry are officers."

51. (a) "Promise-breakers are untrustworthy" becomes "if one is a promise-breaker, then one is not trustworthy" or $r \rightarrow \sim s$.

(b) "Wine-drinkers are very communicative" becomes "if one is a wine-drinker, then one is very communicative" or $u \rightarrow t$.

(c) "A person who keeps a promise is honest" becomes "if one is not a promise-breaker, then one is honest" or $\sim r \rightarrow p$.

(d) "No teetotalers are pawnbrokers" becomes "if one is not a wine-drinker, then one is not a pawnbroker" or $\sim u \rightarrow \sim q$.

(e) "One can always trust a very communicative person" becomes "if one is very communicative, then one is trustworthy" or $t \rightarrow s$.

(f) The symbolic premise statements are
$$r \rightarrow \sim s$$
$$u \rightarrow t$$
$$\sim r \rightarrow p$$
$$\sim u \rightarrow \sim q$$
$$t \rightarrow s.$$
Begin with q, which only appears once. Using the contrapositive of $\sim u \rightarrow \sim q$, ($q \rightarrow u$), and $r \rightarrow \sim s$, ($s \rightarrow \sim r$), rearrange the five premises as follows:
$$q \rightarrow u$$
$$u \rightarrow t$$
$$t \rightarrow s$$
$$s \rightarrow \sim r$$
$$\sim r \rightarrow p.$$
By repeated use of reasoning by transitivity, the conclusion which provides a valid argument is $q \rightarrow p$. In words, this conclusion can be stated as "if one is a pawnbroker, then one is honest," or "all pawnbrokers are honest."

53. Begin by changing each quantified premise to a conditional statement.

(a) The statement "all the dated letters in this room are written on blue paper" becomes "if it is dated, then it is on blue paper" or $r \rightarrow w$.

(b) The statement "none of them are in black ink, except those that are written in the third person" becomes "if it is not in the third person, then it is not in black ink" or ~u → ~t.

(c) The statement "I have not filed any of them that I can read" becomes "if I can read it, then it is not filed" or v → ~s.

(d) The statement "none of them that are written on one sheet are undated" becomes "if it is on one sheet, then it is dated" or x → r.

(e) The statement "all of them that are not crossed are in black ink" becomes "if it is not crossed, then it is in black ink" or ~q → t.

(f) The statement "all of them written by Brown begin with 'Dear Sir'" becomes "if it is written by Brown, then it begins with 'Dear Sir'" or y → p.

(g) The statement "all of them written on blue paper are filed" becomes "if it is on blue paper, then it is filed" or w → s.

(h) The statement "none of them written on more than one sheet are crossed" becomes "if it is not on more than one sheet, then it is not crossed" or ~x → ~q.

(i) The statement "none of them that begin with 'Dear Sir' are written in the third person" becomes "if it begins with 'Dear Sir,' then it is not written in the third person" or p → ~u.

(j) The symbolic premise statements are
(a) r → w
(b) ~u → ~t
(c) v → ~s
(d) x → r
(e) ~q → t
(f) y → p
(g) w → s
(h) ~x → ~q
(i) p → ~u.
Begin with y, which appears only once. Using contrapositives of v → ~s (s → ~v), ~q → t (~t → q), and ~x → ~q (q → x), rearrange the nine statements.

y → p
p → ~u
~u → ~t
~t → q
q → x
x → r
r → w
w → s
s → ~v.
By repeated use of reasoning by transitivity, the conclusion that makes the argument valid is y → ~v.
In words, the conclusion can be stated as "if it is written by Brown, then I can't read it," or equivalently "I can't read any of Brown's letters."

Chapter 3 Test

1. The negation of "6 − 3 = 3" is "6 − 3 ≠ 3."

2. The negation of "all men are created equal" is "some men are not created equal."

3. The negation of "some members of the class went on the field trip" is "no members of the class went on the field trip." An equivalent answer would be "all members of the class did not go on the field trip."

4. The negation of "if that's the way you feel, then I will accept it" is "that's the way you feel and I won't accept it." Remember that ~(p → q) ≡ (p ∧ ~q).

5. The negation of "she applied and got a student loan" is "she did not apply or did not get a student loan." Remember that ~(p ∧ q) ≡ (~p ∨ ~q).

Let p represent "you will love me" and let q represent "I will love you."

6. The symbolic form of "If you won't love me, then I will love you" is "~p → q."

7. The symbolic form of "I will love you if you will love me." is "p → q."

8. The symbolic form of "I won't love you if and only if you won't love me" is "~q ↔ ~p."

9. Writing the symbolic form "~p ∧ q" in words, we get "you won't love me and I will love you."

10. Writing the symbolic form "~$(p \lor \sim q)$" in words, we get "it is not the case that you will love me or I won't love you" (or equivalently, by De Morgan's, "you won't love me and I will love you").

Assume that p is true and that q and r are false for Exercises 11–14.

11. Replacing q and r with the given truth values, we have

$\sim$F $\land$ $\sim$F

  T $\land$ T

    T.

The compound statement $\sim q \land \sim r$ is true.

12. Replacing p, q and r with the given truth values, we have

F $\lor$ (T $\land$ $\sim$F)

F $\lor$ (T $\land$ T)

F $\lor$ T

   T.

The compound statement $r \lor (p \land \sim q)$ is true.

13. Replacing r with the given truth value (s not known), we have

F $\rightarrow$ ($s \lor$ F)

F $\rightarrow$ not known

   T.

The compound statement $r \rightarrow (s \lor r)$ is true.

14. Replacing p and q with the given truth values, we have

T $\leftrightarrow$ (T $\rightarrow$ F)

T $\leftrightarrow$ (F)

   F.

The compound statement $p \leftrightarrow (p \rightarrow q)$ is false.

15. Writing exercise; answers will vary.

16. The necessary condition for

  (a) a conditional statement to be false is that the antecedent must be true and the consequent must be false.

  (b) a conjunction to be true is that both component statements true.

  (c) a disjunction to be false is that both component statements must be false.

17.

p	q	p	$\land$	($\sim p$	$\lor$	q)
T	T	T	T	F	T	T
T	F	T	F	F	F	F
F	T	F	F	T	T	T
F	F	F	F	T	T	F
		2	3	1	2	1

18.

p	q	$\sim$	($p \land q$)	$\rightarrow$	($\sim p$	$\lor$	$\sim q$)
T	T	F	T	T	F	F	F
T	F	T	F	T	F	T	T
F	T	T	F	T	T	T	F
F	F	T	F	T	T	T	T
		2	1	3	1	2	1

Since the last completed column (3) is all true, the conditional is a tautology.

19. The statement "some negative integers are whole numbers" is <u>false</u>, since all whole numbers are non-negative.

20. The statement "all irrational numbers are real numbers" is <u>true</u>, because the real numbers are made up of both the rational and irrational numbers.

The wording may vary in the answer in Exercises 21–26.

21. "All integers are rational numbers" can be stated as "if the number is an integer, then it is a rational number."

22. "Being a rhombus is sufficient for a polygon to be a quadrilateral" can be stated as "if a polygon is a rhombus, then it is a quadrilateral."

23. "Being divisible by 2 is necessary for a number to be divisible by 4" can be stated as "if a number is divisible by 4, then it is divisible by 2." Remember that the "necessary" part of the statement becomes the consequent.

24. "She digs dinosaur bones only if she is a paleontologist" can be stated as "if she digs dinosaur bones, then she is a paleontologist." Remember that the "only if" part of the statement becomes the consequent.

25. The conditional statement: If a picture paints a thousand words, then the graph will help me understand it.

   (a) *Converse*: If the graph will help me understand it, then a picture paints a thousand words.

   (b) *Inverse*: If a picture doesn't paint a thousand words, then the graph won't help me understand it.

   (c) *Contrapositive*: If the graph doesn't help me understand it, then a picture doesn't paint a thousand words.

26. The conditional statement: $\sim p \rightarrow (q \wedge r)$.

   (a) Converse: $(q \wedge r) \rightarrow \sim p$.

   (b) Inverse: $p \rightarrow \sim(q \wedge r)$, or $p \rightarrow (\sim q \vee \sim r)$.

   (c) Contrapositive: $\sim(q \wedge r) \rightarrow p$, or $(\sim q \vee \sim r) \rightarrow p$.

27. Complete an Euler diagram as:

Those who save money

Members of athletic club
x

x represents Don O'Neal

Since, when the premises are diagrammed as being true, and we are forced into a true conclusion, the argument is <u>valid</u>.

28. (a) Let p represent "he eats liver" and q represent "he will eat anything." The argument is then represented symbolically by:

$p \rightarrow q$

$\underline{p}$

q.

This is the valid argument form "modus ponens," hence the answer is A.

   (b) Let p represent "you use your seat belt" and q represent "you will be safer." The argument is then represented symbolically by:

$p \rightarrow q$

$\underline{\sim p}$

$\sim q$.

The answer is F, a fallacy of the inverse.

   (c) Let p represent "I hear *Mr. Bojangles*," q represent "I think of her," and r represent "I smile." The argument is then represented symbolically by:

$p \rightarrow q$

$\underline{q \rightarrow r}$

$p \rightarrow r$.

This is the valid argument form "reasoning by transitivity," hence the answer is C.

   (d) Let p represent "she sings" and q represent "she dances."
The argument is then represented symbolically by:

$p \vee q$

$\underline{\sim p}$

q.

This is the valid argument form "disjunctive syllogism," hence the answer is D.

29. Let p represent "I write a check," q represent "it will bounce," and "the bank guarantees it." The argument is then represented symbolically by:

$p \rightarrow q$

$r \rightarrow \sim q$

$\underline{r}$

$\sim p.$

Construct the truth table for $\{[(p \rightarrow q) \wedge (r \rightarrow \sim q)] \wedge r\} \rightarrow (\sim p)$.

p	q	r	$\{[(p \rightarrow q)$	$\wedge$	$(r$	$\rightarrow$	$\sim q)]$	$\wedge$	$r\}$	$\rightarrow$	$(\sim p)$
T	T	T	T	F	T	F	F	F	T	T	F
T	T	F	T	T	F	T	F	F	F	T	F
T	F	T	F	F	T	T	T	F	T	T	F
T	F	F	F	F	F	T	T	F	F	T	F
F	T	T	T	F	T	F	F	F	T	T	T
F	T	F	T	T	F	T	F	F	F	T	T
F	F	T	T	T	T	T	T	T	T	T	T
F	F	F	T	T	F	T	T	F	F	T	T
			2	3	1	2	1	4	3	5	4

Since the conditional, formed by the conjunction of premises implying the conclusion, is a tautology, the argument is <u>valid</u>.

30. Construct the truth table for $[(\sim p \rightarrow \sim q) \wedge (q \rightarrow p)] \rightarrow (p \vee q)$.

p	q	$[(\sim p$	$\rightarrow$	$\sim q)$	$\wedge$	$(q$	$\rightarrow$	$p)]$	$\rightarrow$	$(p$	$\vee$	$q)$
T	T	F	T	F	T	T	T	T	T	T	T	T
T	F	F	T	T	T	F	T	T	T	T	T	F
F	T	T	F	F	F	T	F	F	T	F	T	T
F	F	T	T	T	T	F	T	F	F	F	F	F
		1	2	1	3	1	2	1	4	2	3	2

Since the conditional, formed by the conjunction of premises implying the conclusion, is not a tautology, the argument is <u>invalid</u>.

Chapter 4

4.1 Exercises

For Reference:

EGYPTIAN

Number	Symbol	Description
1			Stroke
10	∩	Heel Bone
100	9	Scroll
1000	↑	Lotus Flower
10,000	↑	Pointing Finger
100,000	∽	Burbot Fish
1,000,000	☥	Astonished Person

CHINESE

Number	Symbol
0	零
1	一
2	二
3	三
4	田
5	五
6	六
7	七
8	八
9	九
10	十
100	百
1000	千

1. $(1 \times 10,000) + (3 \times 1000) + (0 \times 100)$
   $+ (3 \times 10) + (6 \times 1) = 13,036$

3. $(7 \times 1,000,000) + (6 \times 100,000)$
   $+ (3 \times 10,000) + (0 \times 1000) + (7 \times 100)$
   $+ (2 \times 10) + (9 \times 1) = 7,630,729$

5. [Egyptian symbols]

7. [Egyptian symbols]

9. [Egyptian symbols]

11. [Egyptian symbols]

13. [Egyptian symbols]

15. $(1 \times 100) + (1 \times 50) + (3 \times 10) + (2 \times 1)$
   $= 182$

17. $1000^2 \times [(1 \times 10) + (5 - 1)] = 1,000,000 \times (14)$
   $= 14,000,000$

19. MMDCCCLXI

21. $\overline{\text{XXV}}$DCXIX

23. $(9 \times 100) + (3 \times 10) + (5 \times 1) = 935$

25. $(3 \times 1000) + (7 \times 1) = 3007$

27. [Chinese numeral symbols]

29. [Chinese numeral symbols]

31. [Chinese numeral symbols] to [Chinese numeral symbols]

33. [Chinese numeral symbols] to [Chinese numeral symbols]

35. There is a total of one scroll, eleven heelbones, and six strokes. Group ten heelbones to create a second scroll.
   $(2 \times 100) + (1 \times 10) + (6 \times 1) = 200 + 10 + 6$
   $= 216$

37. There is a total of five pointing fingers, three lotus flowers, five scrolls, nine heelbones, leaving eleven strokes. Group ten strokes to create another heelbone. The ten heelbones then create another scroll.

$(5 \times 10{,}000) + (3 \times 1000) + (6 \times 100)$
$\qquad + (0 \times 10) + (1 \times 1)$
$= 50{,}000 + 3000 + 600 + 0 + 1$
$= 53{,}601$

39. After subtracting, there is one scroll, one heelbone, and three strokes.

$(1 \times 100) + (1 \times 10) + (3 \times 1) = 100 + 10 + 3$
$\qquad\qquad\qquad\qquad\qquad = 113$

41. Regroup the pointing finger to make ten lotus flowers for a total of eleven. Then one lotus flower must be regrouped to make ten scrolls for a total of twelve, and one scroll must be regrouped to make ten heelbones. Regroup one heelbone to make ten strokes. Then ten lotus flowers less three yields seven; eleven scrolls less six yields five; nine heelbones remain; fourteen strokes less six yields eight.

$(7 \times 1000) + (5 \times 100) + (9 \times 10) + (8 \times 1)$
$= 7000 + 500 + 90 + 8$
$= 7598$

43.

1	47
2	94
4	188
8	376
16	752
32	1504

$32 \cdot 47 = 1504$

45.

1	127
2	254
4	508
8	1016
16	2032
32	4064
64	8128

$64 \cdot 127 = 8128$

47.

thirty golden basins	∩∩∩
a thousand silver basins	ℓ
four hundred ten silver bowls	9999 ∩
thirty golden bowls	∩∩∩
3000 shekels	ℓℓℓ
500 shekels	99999
50 shekels	∩∩∩∩∩
400 shekels	9999

30×3000

	1	3000
→	2	6000 ←
→	4	12,000 ←
→	8	24,000 ←
→	16	48,000 ←

$30 \times 3000 = 6000 + 12{,}000 + 24{,}000 + 48{,}000$
$\qquad\qquad = 90{,}000$

500×1000

	1	1000
	2	2000
→	4	4000 ←
	8	8000
→	16	16,000 ←
→	32	32,000 ←
→	64	64,000 ←
→	128	128,000 ←
→	256	256,000 ←

500×1000
$= 4000 + 16000 + 32{,}000 + 64{,}000 + 128{,}000$
$\qquad + 256{,}000$
$= 500{,}000$

50×410

	1	410
→	2	820 ←
	4	1640
	8	3280
→	16	6560 ←
→	32	13,120 ←

$50 \times 410 = 820 + 6560 + 13{,}120 = 20{,}500$

30×40

	1	400
→	2	800 ←
→	4	1600 ←
→	8	3200 ←
→	16	6400 ←

$30 \times 400 = 800 + 1600 + 3200 + 6400$
$= 12,000$

Now add 90,000, 500,000, 20,500, and 12,000 using Egyptian symbols.

〳〳〳〳〳〳〳〳〳 + ᐇᐇᐇᐇᐇ
+ 〳〳 𓆛𓆛𓆛𓆛𓆛 + 〳 𓏤𓏤
= ᐇᐇᐇᐇᐇ

〳〳〳〳〳〳〳〳〳〳〳〳 𓏤𓏤 𓆛𓆛𓆛𓆛𓆛

Regroup ten pointing fingers to one burbot fish.

ᐇᐇᐇᐇᐇᐇ 〳〳 𓏤𓏤 𓆛𓆛𓆛𓆛𓆛

The total value of the treasure is
$(6 \times 100,000) + (2 \times 10,000) + (2 \times 1000) + (5 \times 100) = 622,500$ shekels.

49. Writing exercise; answers will vary.

51. Writing exercise; answers will vary.

53. 99,999; five distinct symbols allows only five positions.

55. The largest number is $44,444_{\text{five}}$, which is equivalent to 3124_{ten}.

57. $10^d - 1$; examine Exercise 53 to see that $10^5 - 1 = 100,000 - 1 = 99,999$.

59. $7^d - 1$

61. The greatest number is 9999.

63. DCLXVI is equivalent to
$500 + 100 + 50 + 10 + 5 + 1 = 666$.

4.2 Exercises

1. Mayan; $2 \cdot 5 + 2 \cdot 1 = 12$

3. Babylonian; $3 \cdot 10 + 2 \cdot 1 = 32$

5. Greek; $200 + 30 + 4 = 234$

7. Mayan; $12 \cdot 20 + 2 = 242$

9. Babylonian; $21 \cdot 60 + 22 \cdot 1 = 1282$

11. Babylonian; $43 \cdot 60 + 21 \cdot 1 = 2601$

13. Mayan; $7 \cdot 360 + 6 \cdot 20 + 0 \cdot 1 = 2640$

15. Mayan;
$8 \cdot 7200 + 6 \cdot 360 + 9 \cdot 20 + 14 \cdot 1 = 59,954$

17. Babylonian;
$22 \cdot 3600 + 21 \cdot 60 + 14 \cdot 1 = 80,474$

19. Greek;
$1 \cdot 10,000 + 5 \cdot 1000 + 100 + 40 + 9$
$= 15,149$

21. ‹‹▼

23. $293 = 4 \cdot 60 + 53$
▼▼‹‹‹ ▼
▼▼ ‹‹ ▼▼

25. $1514 = 25 \cdot 60 + 14$
‹‹▼▼▼ ‹▼▼
▼▼ ‹▼▼

27. $5190 = 3600 + 26 \cdot 60 + 30$
▼‹‹▼▼▼ ‹‹
▼▼▼ ‹

29. $43,205 = 12 \cdot 3600 + 5$
‹▼▼ ▼▼▼
▼▼

31. ═

33. $151 = 7 \cdot 20 + 11$
••
⸺

35. $4694 = 13 \cdot 360 + 0 \cdot 20 + 14$
•••
⬭
••••

37. $64,712 = 8 \cdot 7200 + 19 \cdot 360 + 13 \cdot 20 + 12$
•••
••••
•••
••

39. λθ

41. ϙβ

43. υιβ

45. ,βψξθ

47. Μ̄,δψκϛ

4.3 Exercises

1. $84 = (8 \times 10) + (4 \times 1) = (8 \times 10^1) + (4 \times 10^0)$

3. 9446
$= (9 \times 1000) + (4 \times 100) + (4 \times 10) + (6 \times 1)$
$= (9 \times 10^3) + (4 \times 10^2) + (4 \times 10^1) + (6 \times 10^0)$

5. 4924
$$= (4\times1000)+(9\times100)+(2\times10)+(4\times1)$$
$$= (4\times10^3)+(9\times10^2)+(2\times10^1)+(4\times10^0)$$

7. 14,206,040
$$= (1\times10,000,000)+(4\times1,000,000)$$
$$\quad + (2\times100,000)+(0\times10,000)$$
$$\quad + (6\times1000)+(0\times100)+(4\times10)+(0\times1)$$
$$= (1\times10^7)+(4\times10^6)+(2\times10^5)+(0\times10^4)$$
$$\quad + (6\times10^3)+(0\times10^2)+(4\times10^1)$$
$$\quad + (0\times10^0)$$

9. $(7 \times 10) + (5 \times 1) = 75$

11. $(4 \times 1000) + (3 \times 100) + (8 \times 10) = 4380$

13. $(7 \times 10,000,000) + (4 \times 100,000)$
$$\quad + (1 \times 1000) + (9 \times 1) = 70,401,009$$

15.
$$37 = (3\times10^1)+(7\times10^0)$$
$$\underline{+\,42 = (4\times10^1)+(2\times10^0)}$$
$$= (7\times10^1)+(9\times10^0)$$
$$= 70+9$$
$$= 79$$

17.
$$85 = (8\times10^1)+(5\times10^0)$$
$$\underline{+\,32 = (3\times10^1)+(2\times10^0)}$$
$$= (5\times10^1)+(3\times10^0)$$
$$= 50+3$$
$$= 53$$

19.
$$75 = (7\times10^1)+(5\times10^0)$$
$$\underline{+\,34 = (3\times10^1)+(4\times10^0)}$$
$$= (10\times10^1)+(9\times10^0)$$
$$= (1\times10^2)+(9\times10^0)$$
$$= 100+9$$
$$= 109$$

21.
$$434 = (4\times10^2)+(3\times10^1)+(4\times10^0)$$
$$\underline{+\,299 = (2\times10^2)+(9\times10^1)+(9\times10^0)}$$
$$= (6\times10^2)+(12\times10^1)+(13\times10^0)$$
$$= (6\times10^2)+(12\times10^1)+(1\times10^1)$$
$$\quad + (3\times10^0)$$
$$= (6\times10^2)+(13\times10^1)+(3\times10^0)$$
$$= (6\times10^2)+(1\times10^2)+(3\times10^1)$$
$$\quad + (3\times10^0)$$
$$= (7\times10^2)+(3\times10^1)+(3\times10^0)$$
$$= 700+30+3$$
$$= 733$$

23.
$$54 = (5\times10^1)+(4\times10^0)$$
$$\underline{-\,48 = (4\times10^1)+(8\times10^0)}$$

Since, in the units position, we cannot subtract 8 from 4, we use the distributive property to modify the top expansion as follows.

$$(4\times10^1)+(1\times10^1)+(4\times10^0)$$
$$(4\times10^1)+(10\times10^0)+(4\times10^0)$$

$$54 = (4\times10^1)+(14\times10^0)$$
$$\underline{-\,48 = (4\times10^1)+(8\times10^0)}$$
$$= (0\times10^1)+(6\times10^0)$$
$$= 6$$

25.
$$645 = (6\times10^2)+(4\times10^1)+(5\times10^0)$$
$$\underline{-\,439 = (4\times10^2)+(3\times10^1)+(9\times10^0)}$$

Since, in the units position, we cannot subtract 9 from 5, we use the distributive property to modify the top expansion as follows.

$$(6\times10^2)+(3\times10^1)+(1\times10^1)+(5\times10^0)$$
$$(6\times10^2)+(3\times10^1)+(10\times10^0)+(5\times10^0)$$

$$645 = (6\times10^2)+(3\times10^1)+(15\times10^0)$$
$$\underline{-\,439 = (4\times10^2)+(3\times10^1)+(9\times10^0)}$$
$$= (2\times10^2)+(0\times10^1)+(6\times10^0)$$
$$= 200+6$$
$$= 206$$

27. Reading the abacus from the right, the number represented by this abacus is
$[(1 \times 5) + (1 \times 1)] + (1 \times 50) + (2 \times 100)$
$= 6 + 50 + 200$
$= 256.$

29. The number represented by this abacus is
$[(1 \times 5) + (4 \times 1)] + (1 \times 50) + (2 \times 100)$
$\quad + (3 \times 1000) + [(1 \times 50,000)$
$\quad + (1 \times 10,000)]$
$= 4 + 5 + 50 + 200 + 3000 + 50,000$
$\quad + 10,000$
$= 63,259$

31. $38 = (3 \times 10) + [(1 \times 5) + (3 \times 1)]$

33. 2547
$= (2 \times 1000) + (1 \times 500) + (4 \times 10)$
$\quad + [(1 \times 5) + (2 \times 1)]$

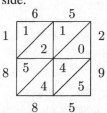

35. 65×29 is written around the top and right side.

$$
\begin{array}{c|cc|}
 & 6 & 5 \\
\hline
1 & 1/2 & 1/0 & 2 \\
8 & 5/4 & 4/5 & 9 \\
\hline
 & 8 & 5 \\
\end{array}
$$

Obtain the numbers inside each box by finding the products of all the pairs of digits on the top and side: $6 \times 2 = 12$, $5 \times 2 = 10$, etc. Then add diagonally starting from the bottom right, placing the sums outside. For example, $0 + 4 + 4 = 8$. Finally, read the answer around the left side and the bottom as 1885.

37. 525×73 is written around the top and right side.

$$
\begin{array}{c|ccc|}
 & 5 & 2 & 5 \\
\hline
3 & 3/5 & 1/4 & 3/5 & 7 \\
8 & 1/5 & 0/6 & 1/5 & 3 \\
\hline
 & 3 & 2 & 5 \\
\end{array}
$$

Find each number inside the boxes by finding the product of all the pairs of digits on the top and side. Then add diagonally

beginning from the bottom right. For example, $5 + 1 + 6 = 12$. Write the 2 outside the box and carry the 1. Now add $1 + 3 + 4 + 5 = 13$. Again carry to the next diagonal above. Reading the answer around the left side and the bottom as 38,325.

39. 723×4198 is written around the top and right side.

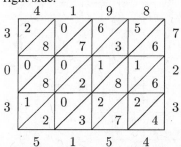

Find each number inside the boxes by finding the product of all the pairs of digits on the top and side. Then add diagonally beginning from the bottom right. For example, $6 + 2 + 7 = 15$. Write the 5 outside the box and carry the 1. Now add $1 + 6 + 1 + 8 + 2 + 3 = 21$. Again carry the 2 to the next diagonal above. Read the answer around the left side and the bottom as 3,035,154.

41. Select the rods for 6 and 2 and place them side by side. Use the index to locate the row or level for a multiplier of 8.

Index	6	2
1	0 / 6	0 / 2
2	1 / 2	0 / 4
3	1 / 8	0 / 6
4	2 / 4	0 / 8
5	3 / 0	1 / 0
6	3 / 6	1 / 2
7	4 / 2	1 / 4
→ 8	4 / 8	1 / 6
9	5 / 4	1 / 8

Te resulting lattice is shown below.

The product of 8 and 62 is 496.

43. Select the rods for 8, 3, 5, and 4 and place them side by side. use the index to first locate the row or level for multipliers of 2 and 6.

Index	8	3	5	4
1	0 / 8	0 / 3	0 / 5	0 / 4
→ 2	1 / 6	0 / 6	1 / 0	0 / 8
3	2 / 4	0 / 9	1 / 5	1 / 2
4	3 / 2	1 / 2	2 / 0	1 / 6
5	4 / 0	1 / 5	2 / 5	2 / 0
→ 6	4 / 8	1 / 8	3 / 0	2 / 4
7	5 / 6	2 / 1	3 / 5	2 / 8
8	6 / 4	2 / 4	4 / 0	3 / 2
9	7 / 2	2 / 7	4 / 5	3 / 6

5 | 4 / 8 | 1 / 8 | 3 / 0 | 2 / 4 |

    0   1   2   4

The product 6 × 8354 = 50,124.
Find the product of 2 × 8354 in a similar way, but using the index for a multiplier of 2.

1 | 1 / 6 | 0 / 6 | 1 / 0 | 0 / 8 |

    6   7   0   8

To create the table below, write the multiplicand on the top row and the multiplier in the right hand column as shown. Insert the product of 6 and 8354 as the first entry. Insert the product of 2 and 8354 as the second entry, shifted one column to the left because it is actually 20 × 8354. The final answer is found by addition; 26 × 8354 = 217,204.

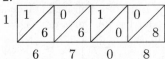

	8354	
	50124	6
	16708	2
	217204	

45. Complete missing place value with 0.
  283
− 041

Replace digits in subtrahend (041) with the nines complement of each and add.

   283
+ 958
——
1241
Delete the first digit on left and add that 1 to the remaining part of the sum:
241 + 1 = 242.

47. Complete missing place values with 0.
  50000
−00199

Replace digits in subtrahend (00199) with the nines complement of each and add.
  50000
+ 99800
——
149800
Delete the first digit on left and add that 1 to the remaining part of the sum:
49,800 + 1 = 49,801.

49. To multiply 5 and 92 using the Russian peasant method, write each number at the top of a column.
→ 5   92 ←
   2 184
→ 1 368 ←

Divide the first column by 2 and double the second column until 1 is obtained in the first column. Ignore the remainders when dividing. Add the numbers in the second column that correspond to the odd numbers in the first: 92 + 368 = 460.

51. To multiply 62 and 529 using the Russian peasant method, write each number at the top of a column.
   62 529
→ 31 1058   ←
→ 15 2116   ←
→ 7 4232   ←
→ 3 8464   ←
→ 1 16,928 ←

Divide the first column by 2 and double the second column until 1 is obtained in the first column. Ignore the remainders when dividing. Add the numbers in the second column that correspond to the odd numbers in the first column.
1058 + 2116 + 4232 + 8464 + 16,928
= 32,798

53. Writing exercise; answers will vary.

4.4 Exercises

1. 1, 2, 3, 4, 5, and 6 are the first six digits. To represent the number seven, 10 is used meaning $(1 \times 7^1) + (0 \times 7^0)$. The next six numbers would be 11, 12, 13, 14, 15, and 16. To express the number fourteen, 20 is used meaning $(2 \times 7^1) + (0 \times 7^0)$. Continue in this pattern: 21, 22, 23, 24, 25, 26.

3. 1, 2, 3, 4, 5, 6, 7, and 8 are the first eight digits. To represent the number nine, 10 is used which means $(1 \times 9^1) + (0 \times 9^0)$. The next eight numbers are 11, 12, 13, 14, 15, 16, 17, 18. To express the number eighteen, 20 is used which means $(2 \times 9^1) + (0 \times 9^0)$. Continue in this pattern: 21, 22.

5. 13_{five} is the number just before, and 20_{five} is the number just after the given number.

7. $\text{B6E}_{\text{sixteen}}$ is the number just before, and $\text{B70}_{\text{sixteen}}$ is the number just after the given number.

9. Three distinct symbols are needed.

11. Eleven distinct symbols are needed.

13. The smallest four-digit number in base three is 1000, which means $(1 \times 3^3) = 27$. The largest four-digit number in base three is 2222, which means $(2 \times 3^3) + (2 \times 3^2) + (2 \times 3^1) + (2 \times 3^0)$. This is equivalent to 54 + 18 + 6 + 2 = 80.

15. $(2 \times 5^1) + (4 \times 5^0) = 10 + 4 = 14$
Using the calculator shortcut:
$(2 \times 5) + 4 = 14$.

17. $2^3 + 2^1 + 2^0 = 8 + 2 + 1 = 11$
Using the calculator shortcut:
$[(1 \times 2 + 0) \times 2 + 1] \times 2 + 1 = [5] \times 2 + 1$
$$= 11.$$

19. $(3 \times 16^2) + (11 \times 16^1) + (12 \times 16^0)$
$= 3 \times 256 + 11 \times 16 + 12$
$= 956$
Using the calculator shortcut:
$(3 \times 16 + 11) \times 16 + 12 = 956$.

21. $(2 \times 7^3) + (3 \times 7^2) + (6 \times 7^1) + (6 \times 7^0)$
$= 686 + 147 + 42 + 6$
$= 881$
Using the calculator shortcut:
$[(2 \times 7 + 3) \times 7 + 6] \times 7 + 6 = [125] \times 7 + 6$
$$= 881.$$

23. $(7 \times 8^4) + (0 \times 8^3) + (2 \times 8^2) + (6 \times 8^1)$
$\quad + (6 \times 8^0)$
$= 28{,}672 + 128 + 48 + 6$
$= 28{,}854$
Using the calculator shortcut:
$\{[(7 \times 8 + 0) \times 8 + 2] \times 8 + 6\} \times 8 + 6$
$= 28{,}854.$

25. $(2 \times 4^3) + (0 \times 4^2) + (2 \times 4^1) + (3 \times 4^0)$
$= 128 + 8 + 3$
$= 139$
Using the calculator shortcut:
$[(2 \times 4 + 0) \times 4 + 2] \times 4 + 3 = 139.$

27. $(4 \times 6^4) + (1 \times 6^3) + (5 \times 6^2) + (3 \times 6^1)$
$\quad + (3 \times 6^0)$
$= 5184 + 216 + 180 + 18 + 3$
$= 5601$
Using the calculator method:
$\{[(4 \times 6 + 1) \times 6 + 5] \times 6 + 3\} \times 6 + 3$
$= 5601.$

29. The base five place values, starting from the right, are 1, 5, 25, 125, and so on. Since 86 is between 25 and 125, we will need some 25's but no 125's. Begin by dividing 86 by 25; then divide the remainder obtained from this division by 5. Finally, divide the remainder obtained from the previous division by 1, giving a remainder of 0.
$86 \div 25 = 3$, remainder 11
$11 \div 5 = 2$, remainder 1
$1 \div 1 = 1$, remainder 0
The digits of the answer are found by reading quotients from the top down.
$86 = 321_{\text{five}}$
Shortcut:

5	86		Rem
5	17	←	1
5	3	←	2
	0	←	3

Read the answer from the remainder column, reading from the bottom up.
$86 = 321_{\text{five}}$

31.

2	19		Rem
2	9	←	1
2	4	←	1
2	2	←	0
2	1	←	0
	0	←	1

$$19 = 10011_{two}$$

33.

16	147		Rem
16	9	←	3
	0	←	9

$$147 = 93_{sixteen}$$

35.

5	36401		Rem
5	7280	←	1
5	1456	←	0
5	291	←	1
5	58	←	1
5	11	←	3
5	2	←	1
	0	←	2

$$36401 = 2131101_{five}$$

37.

2	586		Rem
2	293	←	0
2	146	←	1
2	73	←	0
2	36	←	1
2	18	←	0
2	9	←	0
2	4	←	1
2	2	←	0
2	1	←	0
	0	←	1

$$586 = 1001001010_{two}$$

39.

3	8407		Rem
3	2802	←	1
3	934	←	0
3	311	←	1
3	103	←	2
3	34	←	1
3	11	←	1
3	3	←	2
3	1	←	0
	0	←	1

$$8407 = 102112101_{three}$$

41.

6	9346		Rem
6	1557	←	4
6	259	←	3
6	43	←	1
6	7	←	1
6	1	←	1
	0	←	1

$$9346 = 111134_{six}$$

43. First convert 43_{five} to base ten.

$$(4 \times 5) + 3 = 23$$

Then convert 23 to base seven.

7	23		Rem
7	3	←	2
	0	←	3

$$43_{five} = 32_{seven}$$

45. First convert 6748_{nine} to base ten.

$$(6 \times 9^3) + (7 \times 9^2) + (4 \times 9) + 8 = 4985$$

Then convert 4985 to base four.

4	4985		Rem
4	1246	←	1
4	311	←	2
4	77	←	3
4	19	←	1
4	4	←	3
4	1	←	0
	0	←	1

$$6748_{nine} = 1031321_{four}$$

47. Replace each octal digit with its 3-digit binary equivalent. Then combine all the binary equivalents into a single binary numeral.

$$
\begin{array}{ccc}
3 & 6 & 7 \\
\downarrow & \downarrow & \downarrow \\
011 & 110 & 111
\end{array}
$$

$$367_{\text{eight}} = 11110111_{\text{two}}$$

49. Starting at the right, break the digits into groups of three. Then convert the groups to their octal equivalents. (Refer to Table 7.)

$$
\begin{array}{ccc}
100 & 110 & 111 \\
\downarrow & \downarrow & \downarrow \\
4 & 6 & 7
\end{array}
$$

$$100110111_{\text{two}} = 467_{\text{eight}}$$

51. Each hexadecimal digit yields a 4-digit binary equivalent. (See Table 8.)

$$
\begin{array}{cc}
D & C \\
\downarrow & \downarrow \\
1101 & 1100
\end{array}
$$

$$DC_{\text{sixteen}} = 11011100_{\text{two}}$$

53. Starting at the right, break the digits into groups of four. Then convert the groups to their hexadecimal equivalent. (Refer to Table 8.)

$$
\begin{array}{cc}
10 & 1101 \\
\downarrow & \downarrow \\
2 & D
\end{array}
$$

$$101101_{\text{two}} = 2D_{\text{sixteen}}$$

55. In order to compare these numbers, we need to write them in the same base. Convert each of them to decimal form (base ten).

$$42_{\text{seven}} = (4 \times 7^1) + (2 \times 7^0) = 28 + 2 \text{ or } 30$$

$$37_{\text{eight}} = (3 \times 8^1) + (7 \times 8^0) = 24 + 7 \text{ or } 31$$

$$1D_{\text{sixteen}} = (1 \times 16^1) + (13 \times 16^0) = 16 + 13$$
$$\text{or } 29$$

The largest number is 37_{eight}.

57. $(9 \times 12^2) + (10 \times 12) + 11 = 1427$ copies

59. Since A is assigned the number 65, C is assigned the number 67. Change 67 from decimal form to binary form.

$$
\begin{array}{r|l c c}
2 & 67 & & \text{Rem} \\
2 & 33 & \leftarrow & 1 \\
2 & 16 & \leftarrow & 1 \\
2 & 8 & \leftarrow & 0 \\
2 & 4 & \leftarrow & 0 \\
2 & 2 & \leftarrow & 0 \\
2 & 1 & \leftarrow & 0 \\
 & 0 & \leftarrow & 1 \\
\end{array}
$$

$$C = 1000011_{\text{two}}$$

61. Since a is assigned the number 97, k is assigned the number 107. (Since k is the eleventh letter of the alphabet, its corresponding number will be ten more than the number corresponding to a.) Change 107 to binary form.

$$
\begin{array}{r|l c c}
2 & 107 & & \text{Rem} \\
2 & 53 & \leftarrow & 1 \\
2 & 26 & \leftarrow & 1 \\
2 & 13 & \leftarrow & 0 \\
2 & 6 & \leftarrow & 1 \\
2 & 3 & \leftarrow & 0 \\
2 & 1 & \leftarrow & 1 \\
 & 0 & \leftarrow & 1 \\
\end{array}
$$

$$k = 1101011_{\text{two}}$$

63. Convert each seven-digit binary number to decimal form; then find the corresponding letters.

Base Two	Base Ten	Letter
1001000	64 + 8 = 72	H
1000101	64 + 4 + 1 = 69	E
1001100	64 + 8 + 4 = 76	L
1010000	64 + 16 = 80	P

The given number represents HELP.

65. To translate the word "New" into an ASCII string of binary digits, find the number corresponding to each letter of the word, translate each of these numbers to binary form, and finally combine these three binary numbers into a single string.

Letter	Base Ten	Base Two
N	78	1001110
e	101	1100101
w	119	1110111

The word "New" is represented by the following ASCII string: $1001110110010111110111_{two}$.

67. Writing exercise; answers will vary.

69. **(a)** The binary ones digit is 1.

(b) The binary twos digit is 1.

(c) The binary fours digit is 1.

(d) The binary eights digit is 1.

(e) The binary sixteens digit is 1.

71. In order to include all ages up to 63, we will need to add one more column to Table 14. This column will contain the numbers whose binary thirty-twos digit is 1. Thus, 6 columns would be needed.

73. In base two, every even number has 0 as its ones digit and every odd number has a 1 as its ones digit. Thus, we can distinguish odd and even numbers by looking at their ones digit. The criterion works.

75. In base four, every even number has 0 or 2 as its ones digit, while every odd number has 1 or 3 as its ones digit. Thus, we can distinguish odd and even numbers by looking at their ones digits. The criterion works.

77. In base six, every even number has 0, 2, or 4 as its ones digit, while every odd number has 1, 3, or 5 as its ones digit. The criterion works.

79. In base eight, every even number has 0, 2, 4, or 6 as its ones digit, while every odd number has 1, 3, 5, or 7 as its ones digit. The criterion works.

81. Writing exercise; answers will vary.

83. The units digit, 4, is the same for base 5 and base 10. Hence the number is not divisible by 5.

85. The units digit, 0, is the same for base 5 and base 10. Hence the number is divisible by 5.

87. Writing exercise; answers will vary.

89. Replace each base 9 digit with its 2-digit base 3 equivalent. Then combine all the base 3 equivalents into a single base 3 numeral.

$$\begin{array}{cccc} 6 & 5 & 0 & 4 \\ \downarrow & \downarrow & \downarrow & \downarrow \\ 20 & 12 & 00 & 11 \end{array}$$

$6504_{nine} = 20120011_{three}$

91. Starting at the right, break the digits into groups of two. Then convert the groups to their base 9 equivalents.

$$\begin{array}{ccccc} 2 & 12 & 20 & 12 & 21 \\ \downarrow & \downarrow & \downarrow & \downarrow & \downarrow \\ 2 & 5 & 6 & 5 & 7 \end{array}$$

$212201221_{three} = 25657_{nine}$

93. The designation for gray is (128, 128, 128).

$$16\overline{\smash{)}128}$$
$$8$$

So $128 = 80_{sixteen}$. Then gray is given by the code 808080.

95.
$$16\overline{\smash{)}255} \quad Rem$$
$$15 \leftarrow 15$$

$255 = FF_{sixteen}$

$$16\overline{\smash{)}105} \quad Rem$$
$$6 \leftarrow 9$$

$105 = 69_{sixteen}$

$$16\overline{\smash{)}180} \quad Rem$$
$$11 \leftarrow 4$$

$180 = B4_{sixteen}$

so the code for hot pink is FF69B4.

EXTENSION: **MODULAR SYSTEMS**

1. $7 + 16 = 23$ and $\dfrac{23}{12} = 1.916666667$

$1.916666667 - 1 = 0.916666667$

$12 \cdot 0.916666667 = 11$

3. $9 \cdot 7 = 63$ and $\dfrac{63}{12} = 5.25$

$12 \cdot 0.25 = 3$

5. $6 + 42 = 48$ and $\dfrac{48}{7} = 6.857142857$

$7 \cdot 0.857142857 = 6$

7. $4 \cdot 28 = 112$ and $\dfrac{112}{7} = 16$

In 7-day clock arithmetic, the product is 0.

9. $1400 + 500 = 1900$
(This equation is equivalent to stating that 5 hours after 2 P.M. is 7 P.M.

11. $0750 + 1630 = 2380$
$\qquad\qquad = 2300 + 60 + 20$
$\qquad\qquad = 2400 + 20$
$\qquad\qquad = 0020$

13. In ordinary whole number arithmetic,
$1145 + 1135 = 2280$.
In 12-hour clock arithmetic,
$1145 + 1135 = 2280 - 1200$
$\qquad\qquad\quad = 1080$
$\qquad\qquad\quad = 1000 + 60 + 20$
$\qquad\qquad\quad = 1120$.
In 24-hour clock arithmetic,
$1145 + 1135 = 2280$
$\qquad\qquad\quad = 2260 + 20$
$\qquad\qquad\quad = 2300 + 20$
$\qquad\qquad\quad = 2320$.
Therefore, all of the statements are true, each one in a different system.

15. $35 \equiv 8 \pmod 9$ is true. The difference $35 - 8 = 27$ is divisible by 9.

17. $7021 \equiv 4202 \pmod 6$ is false. The difference $7021 - 4202 = 2819$ is not divisible by 6.

19. $(62 + 95) \pmod 9$
First add 62 and 95 to get 157. Then divide 157 by 9. The remainder is 4, so $(62 + 95) \equiv 4 \pmod 9$.

21. $(82 - 45) \pmod 3$
First subtract 45 from 82 to get 37. Then divide 37 by 3. The remainder is 1, so $(82 - 45) \equiv 1 \pmod 3$.

23. $(32 \times 21) \pmod 8$
$32 \times 21 = 672$; then divide 672 by 8. The remainder is 0, so $(32 \times 21) \equiv 0 \pmod 8$.

25. $[(10 + 7) \times (5 + 3)] \pmod{10}$
$[(10 + 7) \times (5 + 3)] = 17 \cdot 8 = 136$; then divide 136 by 10. The remainder is 6, so $[(10 + 7) \times (5 + 3)] \equiv 6 \pmod{10}$.

27. $\{3, 10, 17, 24, 31, 38, ...\}$ is the solution set. To find these number, multiply 7 times the whole numbers and add 3 each time.
$(7 \times 0) + 3 = 3$
$(7 \times 1) + 3 = 10$
$(7 \times 2) + 3 = 17$

29. $6x \equiv 2 \pmod 2$
Trying a number of integers, we see that all tried work. That is, this is an identity. Thus, $\{1, 2, 3, 4, 5, 6, ...\}$ is the solution set.

31. The modulus is 100,000. After the odometer reaches 99,999, it "rolls over" to 100,000

33. A value of x must be found that will satisfy the following equations:
$x \equiv 6 \pmod 7$
$x \equiv 1 \pmod 8$
$x \equiv 3 \pmod{15}$
The solution set for the first equation is: $\{6, 13, 20, 27, 34, 41, 48, 55, 62, 69, ...\}$.
The solution set for the second equation is: $\{1, 9, 17, 25, 33, 41, 49, 57, 65, 73, ...\}$.
The solution set for the third equation is: $\{3, 18, 33, 48, 63, 78, ...\}$. Continue writing the elements of these sets until an element is found which is common to all three sets. The smallest number that satisfies all three equations, which is the only solution less than 200, is 153. Therefore, Cheryl's collection contains 153 spoons.

35. Because there are 31 days in July and 31 days in August for a total of 62, it is unnecessary to find any integers larger than 62 for the modulo systems. Megan's 21-day schedule indicates modulo 21; Michele's 30-day schedule indicates modulo 30.

Chicago
For Megan, $x \equiv 1 \pmod{21}$ has solution set July 1, July 22, and August 12.
$x \equiv 2 \pmod{21}$ has a solution set $\{2, 23, 44, ...\}$. These integers correspond to July 2, July 23, and August 13.
$x \equiv 8 \pmod{21}$ has a solution set $\{8, 29, 50, ...\}$. These integers correspond to July 8, July 29, and August 19.

For Michele, $x \equiv 23 \pmod{30}$ has solution set $\{23, 53, ...\}$. These integers correspond

to July 23 and August 22.
$x \equiv 29 \pmod{30}$ has solution set
$\{29, 59, \ldots\}$. These integers correspond to July 29 and August 28.
$x \equiv 30 \pmod{30}$ has solution set
$\{30, 60, \ldots\}$. These integers correspond to July 30 and August 29.
The only days that they will both be in Chicago are July 23 and July 29.

New Orleans
For Megan, $x \equiv 5 \pmod{21}$ has a solution set $\{5, 26, 47, \ldots\}$. These integers correspond to July 5, July 26, and August 16.
$x \equiv 12 \pmod{21}$ has solution set
$\{12, 33, 54, \ldots\}$. These integers correspond to July 12, August 2, and August 23.

For Michele, $x \equiv 5 \pmod{30}$ has solution set $\{5, 35, \ldots\}$. These integers correspond to July 5 and August 4.
$x \equiv 6 \pmod{30}$ has solution set $\{6, 36, \ldots\}$. These integers correspond to July 6 and August 5.
$x \equiv 17 \pmod{30}$ has solution set
$\{17, 47, \ldots\}$. These integers correspond to July 17 and August 16.
The only days that they will both be in New Orleans are July 5 and August 16.

San Francisco
For Megan, $x \equiv 6 \pmod{21}$ has solution set $\{6, 27, 48, \ldots\}$. These integers correspond to July 6, July 27, and August 17.
$x \equiv 18 \pmod{21}$ has solution set
$\{18, 39, 60, \ldots\}$. These integers correspond to July 18, August 8, and August 29.
$x \equiv 19 \pmod{21}$ has solution set
$\{19, 40, 61, \ldots\}$. These integers correspond to July 19, August 9, and August 30. For Michele, $x \equiv 8 \pmod{30}$ has solution set $\{8, 38, \ldots\}$. These integers correspond to July 8 and August 7.
$x \equiv 10 \pmod{30}$ has solution set
$\{10, 40, \ldots\}$. These integers correspond to July 10 and August 9.
$x \equiv 15 \pmod{30}$ has solution set
$\{15, 45, \ldots\}$. These integers correspond to July 15 and August 14.
$x \equiv 20 \pmod{30}$ has solution set
$\{20, 50, \ldots\}$. These integers correspond to July 20 and August 19. Finally, $x \equiv 25 \pmod{30}$ has the solution set $\{25, 55, \ldots\}$, which means July 25 and August 24.

The only day that they will both be in San Francisco is August 9.

37. 1865
Substitute 1865 for y in the formula given in the text.
$$a = 1865 + \left[\!\left[\frac{1864}{4}\right]\!\right] - \left[\!\left[\frac{1864}{100}\right]\!\right] + \left[\!\left[\frac{1864}{400}\right]\!\right]$$
$$= 1865 + [\![466]\!] - [\![18.64]\!] + [\![4.66]\!]$$
$$= 1865 + 466 - 18 + 4$$
$$= 2317$$
Since
$2317 = (7 \times 331) + 0$
$2317 \equiv 0 \pmod 7$
and 0 corresponds to Sunday, January 1, 1865 was a Sunday.

39. 2020
Substitute 2020 for y in the formula given in the text.
$$a = 2020 + \left[\!\left[\frac{2019}{4}\right]\!\right] - \left[\!\left[\frac{2019}{100}\right]\!\right] + \left[\!\left[\frac{2019}{400}\right]\!\right]$$
$$= 2020 + [\![504.75]\!] - [\![20.19]\!] + [\![5.0475]\!]$$
$$= 2020 + 504 - 20 + 5$$
$$= 2509$$
Since
$2509 = (7 \times 358) + 3$
$2509 \equiv 3 \pmod 7$
and 3 correspond to Wednesday, January 1, 2020 will be a Wednesday.

41. The first day of the year 2012 is a Sunday. 2012 is a leap year because 2012 is divisible by 4. The months in 2012 with a Friday the thirteenth are January, April, and July.

43. The year 2200 is not a leap year because 2200 is not divisible by 400. The first day of the year 2200 is a Wednesday, so the only month in 2200 with a Friday the thirteenth is June.

45. 0–275–98341–2
$(1 \times 0) + (2 \times 2) + (3 \times 7) + (4 \times 5)$
$\quad + (5 \times 9) + (6 \times 8) + (7 \times 3) + (8 \times 4)$
$\quad + (9 \times 1) = 200$
Because 20(mod 11) = 2, the check digit is 2. The ISBN has the correct check digit and the answer is yes.

47. 0–8050–2688–?
$(1 \times 0) + (2 \times 8) + (3 \times 0) + (4 \times 5)$
$\quad + (5 \times 0) + (6 \times 2) + (7 \times 6) + (8 \times 8)$
$\quad + (9 \times 8) = 226$
Because $226 \equiv 6 \bmod(11)$, the check digit is 6.

49. $x_{13} = 10 - (9 + 3 \cdot 7 + 8 + 3 \cdot 0 + 0 + 3 \cdot 6 + 1$
$\qquad + 3 \cdot 9 + 3 + 3 \cdot 9 + 7$
$\qquad + 3 \cdot 9)(\bmod 10)$
$\qquad = 10 - (148)(\bmod 10)$
$\qquad = 10 - 8$
$\qquad = 2$

The given check digit is incorrect.

51. $x_{13} = 10 - (9 + 3 \cdot 7 + 8 + 3 \cdot 1 + 4 + 3 \cdot 1 + 6$
$\qquad + 3 \cdot 5 + 9 + 3 \cdot 4 + 8$
$\qquad + 3 \cdot 5)(\bmod 10)$
$\qquad = 10 - (113)(\bmod 10)$
$\qquad = 10 - 3$
$\qquad = 7$

53. $\dfrac{9,512,673}{6} = 1,585,445.5$

$0.5 \cdot 6 = 3$

$\dfrac{2,583,691}{6} = 430,615.1667$

$0.1667 \cdot 6 = 1$

$3 \cdot 1 = 3$

55. $\dfrac{14,501,302,706}{19} = 763,226,458.2$

$0.2 \cdot 19 = 4$

$\dfrac{281,460,555}{19} = 14,813,713.42$

$0.42 \cdot 19 = 8$

$4 \cdot 8 = 32$ and $\dfrac{32}{19} = 1.684210526$

$0.684210526 \cdot 19 = 13$

Chapter 4 Test

1. Egyptian;
$(1 \times 1000) + (5 \times 100) + (3 \times 10) + (4 \times 1)$
$= 1534.$

2. Roman;
$1000 \times (1 \times 10) + (500 - 100) + (1 \times 50)$
$\qquad + (2 \times 10) + (5 - 1)$
$= 10,000 + 400 + 50 + 20 + 4$
$= 10,474$

3. Chinese; $(3 \times 100) + (8 \times 10) + 5 = 385$

4. Babylonian;
$(43 \times 3600) + (12 \times 60) + 20 = 155,540$

5. Mayan;
$(7 \times 7200) + (0 \times 360) + (5 \times 20) + (11 \times 1)$
$= 50,511$

6. Greek;
$(5 \times 10,000) + (3 \times 1000) + 500 + 20 + 4$
$= 53,524$

7. Using the Egyptian method:

	1	45	
	2	90	
1 + 2 + 4 + 16 = 23	4	180	45 + 90 + 180 + 720 = 1035
	8	360	
	16	720	

8.

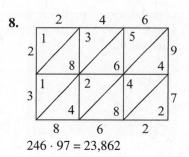

$246 \cdot 97 = 23,862$

9.

$\begin{array}{r} 21,425 \\ -\ 8,198 \end{array}$ $\begin{array}{r} 21,425 \\ -\ 08,198 \end{array}$ $\begin{array}{r} 21,425 \\ +\ 91,801 \\ \hline 113,226 \end{array}$ $\begin{array}{r} 13,226 \\ +\ \ \ \ \ \ 1 \\ \hline 13,227 \end{array}$

10. $243_{\text{five}} = (2 \times 25) + (4 \times 5) + (3 \times 1) = 73$

11. $100101_{\text{two}} = (1 \times 2^5) + (0 \times 2^4) + (0 \times 2^3)$
$\qquad + (1 \times 2^2) + (0 \times 2^1) + (1 \times 2^0)$
$\qquad = (1 \times 32) + (1 \times 4) + (1 \times 1)$
$\qquad = 37$

12. $\text{BEEF}_{\text{sixteen}} = (11 \times 16^3) + (14 \times 16^2)$
$\qquad + (14 \times 16^1) + (15 \times 16^0)$
$\qquad = (11 \times 4096) + (4 \times 256)$
$\qquad + (14 \times 16) + (15 \times 1)$
$\qquad = 48,879$

13. Using repeated division.

2	49		Rem
2	24	←	1
2	12	←	0
2	6	←	0
2	3	←	0
2	1	←	1
	0	←	1

$49 = 110001_{\text{two}}.$

14. Using repeated division,

```
5 | 2930        Rem
  5 | 586   ←    0
    5 | 117  ←   1
      5 | 23  ←  2
        5 | 4  ← 3
            0  ← 4
```

$2930 = 43210_{five}$.

15. Starting at the right, break the digits into groups of three. Then convert the groups to their octal equivalents. (See Table 6.)

```
10   101   110
 ↓    ↓     ↓
 2    5     6
```
$10101110_{two} = 256_{eight}$

16. $7215_{eight} = (7 \times 8^3) + (2 \times 8^2) + (1 \times 8^1)$
$$+ (5 \times 8^0)$$
$$= 3584 + 128 + 8 + 5$$
$$= 3725$$

```
16 | 3725        Rem
  16 | 232  ←    13
    16 | 14  ←   8
         0  ←    14
```

$7215_{eight} = E8D_{sixteen}$

17. $5041_{six} = (5 \times 6^3) + (0 \times 6^2) + (4 \times 6^1)$
$$+ (1 \times 6^0)$$
$$= 1080 + 24 + 1$$
$$= 1105$$

18.
```
B      A      D
↓      ↓      ↓
1011  1010  1101
```
$BAD_{sixteen} = 101110101101_{two}$

19. The advantage of multiplicative grouping over simple grouping is that there is less repetition of symbols.

20. The advantage of positional over multiplicative grouping is that place values are understood by position.

21. The advantage in a positional numeration system, of a smaller base over a larger base, is that there are fewer symbols to learn.

22. The advantage in a positional numeration system, of a larger base over a smaller base, is that there are fewer digits in the numerals.

23. Writing exercise; answers will vary.

24. $7_{nine} = 21_{three}$
$6_{nine} = 20_{three}$
$5_{nine} = 12_{three}$
so $765_{nine} = 212012_{three}$.

Chapter 5

1. True; remember that the natural numbers are also called the counting numbers: 1, 2, 3, 4, ...

3. True; if a number is divisible by 9, it must also be divisible by 3 since 9 is divisible by 3.

5. False; the smallest prime number is 2. The number 1 is neither prime nor composite.

7. True because 2, 4, and 8 all divide 16 evenly.

9. Remember that all natural number (factors are those that divide the given number with remainder 0: 1, 2, 3, 4, 6, 12.

11. 1, 2, 4, 7, 14, 28

13. (a) It is not divisible by 2 because it is an odd number.

    (b) It is divisible by 3 because the sum of the digits is 6, a number divisible by 3.

    (c) It is not divisible by 4 because 21, the number formed by the last two digits, is not divisible by 4.

    (d) It is not divisible by 5 because the last digit is not 0 or 5.

    (e) It is not divisible by 6 because although it is divisible by 3, it is not divisible by 2. It must be divisible by both.

    (f) It is not divisible by 8 because the three digits form a number that is not divisible by 8.

    (g) It is not divisible by 9 because the sum of the digits is not divisible by 9.

    (h) It is not divisible by 10 because the last digit is not 0.

    (i) It is not divisible by 12 because, although it is divisible by 3, it is not divisible by 4. It must be divisible by both.

15. (a) It is divisible by 2 because it is an even number.

    (b) It is divisible by 3 because the sum of the digits is 18, which is divisible by 3.

    (c) It is divisible by 4 because 60, the number formed by the last two digits, is divisible by 4.

    (d) It is divisible by 5 because the last digit is 0.

    (e) It is divisible by 6 because it is divisible by 3 and 2.

    (f) It is divisible by 8 because the last three digits form a number that is divisible by 8.

    (g) It is divisible by 9 because the sum of the digits is 18, which is divisible by 9.

    (h) It is divisible by 10 because the last digit is 0.

    (i) It is divisible by 12 because it is divisible by 3 and 4.

17. (a) Writing exercise; answers will vary.

    (b) The largest prime number whose multiples would have to be considered is 13; the square of 13 is 169, which is less than 200. The next prime number after 13 is 17, but the square of 17 is 289, a number greater than 200.

    (c) square root; square root; square root

    (d) prime

19. Two primes that are consecutive natural numbers are 2 and 3; there are no others because one of them would be an even number, which is divisible by 2.

21. The last digit must be zero, because the number must be divisible by 10.

23.

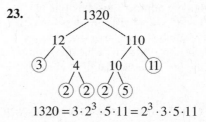

$$1320 = 3 \cdot 2^3 \cdot 5 \cdot 11 = 2^3 \cdot 3 \cdot 5 \cdot 11$$

25.

```
2 | 168
  2 | 84
    2 | 42
      3 | 21
          7
```

The prime factorization of 168 is $2^3 \cdot 3 \cdot 7$.

27.

```
2 | 468
  2 | 234
    3 | 117
      3 | 39
          13
```

The prime factorization of 468 is $2^2 \cdot 3^2 \cdot 13$.

29.
$$
\begin{array}{cccc}
21425 & 2141 & 213 & 20 \\
-\ \ 12 & -\ \ 6 & -\ 10 & -\ 6 \\
\hline
21413 & 2135 & 203 & 14
\end{array}
$$
14 is divisible by 7, so 214,256 is divisible by 7.

31.
$$
\begin{array}{cccc}
58485 & 5847 & 583 & 57 \\
-\ \ \ \ 8 & -\ \ 14 & -\ \ 6 & -14 \\
\hline
58477 & 5833 & 577 & 43
\end{array}
$$
43 is not divisible by 7, so 584,854 is not divisible by 7.

33. $6 + 2 + 8 + 6 = 22$
$5 + 4 + 4 = 13$
$22 - 13 = 9$
9 is not divisible by 11, so 6,524,846 is not divisible by 11.

35. $4 + 0 + 3 + 8 + 5 = 20$
$7 + 9 + 3 + 1 = 20$
$20 - 20 = 0$
0 is divisible by 11, so 470,933,815 is divisible by 11.

37. Based on the divisibility test for 6, which says that the number must be divisible by both 2 and 3, the divisibility test for 15 is that the number must be divisible by both 3 and 5. That is, the sum of the digits must be divisible by 3 and the last digit must be 5 or 0.

39. 0, 2, 4, 6, or 8 because they are all the even single digit numbers.

41. The last two digits must form a number that is divisible by 4. The possible values for x are 0, 4, and 8.

43. The number is divisible by 6 if it is divisible by both 2 and 3. Both 985,230 and 985,236 are divisible by 2 and 3. Then $x = 0$ or 6.

45. $42 = 2^1 \cdot 3^1 \cdot 7^1$
$2 \cdot 2 \cdot 2 = 8$
42 has 8 divisors.

47. $9 \cdot 3 = 27$
$5^8 \cdot 29^2$ has 27 divisors.

49. It is divisible by 4 and does not end in two zeros, so it is a leap year.

51. 2200 is not a leap year because it ends in two zeros and is not divisible by 400.

53. Writing exercise; answers will vary.

55. Writing exercise; answers will vary.

5.2 Exercises

1. True; the assumption made at the beginning of the proof is the negation of the statement is true, and a contradiction is reached.

3. False: Let $k = 2$, then $4k + 1 = 4(2) + 1 = 9$, which is composite.

5. False; $2^{11} - 1 = 2047$. The factors of 2047 are 23 and 89.

7. The next two primes of the form $4k + 1$ are $4(9) + 1 = 37$ and $4(10) + 1 = 41$.
$37 = 1^2 + 6^2$
$41 = 4^2 + 5^2$

9. Writing exercise; answers will vary.

11. $41^2 - 41 + 41 = 41^2$, which is not prime.

13. The answer is B. For $n > 41$, Euler's formula produces a prime sometimes. In Exercise 72, Euler's formula produces 1763 for $n = 42$. The square root of 1763 is approximately 41.99, so all of the prime numbers less than this number must be tested. If 41 is tested as a divisor of 1763, the quotient is 43. Therefore, the number is not prime. On the other hand, Euler's formula produces 1847 for $n = 43$. The square root of 1847 is approximately 42.98. If all the prime numbers less than this number are tested, none divides 1847 evenly. Therefore, the number is prime.

15. **(a)** $2^{2^4} + 1 = 2^{16} + 1 = 65,536 + 1 = 65,537$

   (b) The square root of 65,537 is approximately 256. All primes less than 256 must be checked as possible factors of this number; 251 is then the largest potential prime factor.

17. Writing exercise; answers will vary.

19. Writing exercise; answers will vary.

21. This number M is composite because it can be factored into $59 \cdot 509$.

23. $M_6 = 2^6 - 1 = 64 - 1 = 63$

25. $2^p - 1$

27. The two prime factors of 10 are 2 and 5.
$$2^2 - 1 = 4 - 1 = 3$$
$$2^5 - 1 = 32 - 1 = 31$$
Two distinct factors of M_{10} are 3 and 31.

5.3 Exercises

1. True

3. True

5. True; by definition, the only factors of a prime number are the number itself and 1. The proper divisors of a number are all the divisors except the number itself. For a prime number, then, the only proper divisor is 1. That means that any prime number must be deficient.

7. False; for every new Mersenne prime, there is a perfect number.

9. True; since $(2^7 - 1) = 128 - 1 = 127$ is a prime number, then $2^6(2^7 - 1) = 8128$ is a perfect number by Euclid. (See answer to Exercise 4.)

11. $1 + 2 + 4 + 8 + 16 + 31 + 62 + 124 + 248$
$= 496$

13. $2^{13} - 1 = 8191$, which is prime.
$$2^{13-1}(2^{13} - 1) = 4096(8192 - 1)$$
$$= 33,550,336$$
This number is even and ends in 6.

15. The divisors of 6 are 1, 2, 3, and 6. The sum of the reciprocals of these numbers is
$$\frac{1}{1} + \frac{1}{2} + \frac{1}{3} + \frac{1}{6} = \frac{6}{6} + \frac{3}{6} + \frac{2}{6} + \frac{1}{6} = \frac{12}{6} = 2.$$

17. The proper divisors of 45 are 1, 3, 5, 9, and 15. The sum of these numbers is 33. Because the sum is less than 45, the number is deficient.

19. The proper divisors of 88 are 1, 2, 4, 8, 11, 22, and 44. The sum of these numbers is 92. Because the sum is greater than 88, the number is abundant.

21. Examine the Sieve of Eratosthenes in Table 5.1 to see that the prime numbers 2, 3, 5, 7, 11, 13, 17, 19, and 23 can be deleted from the search. Examine the remaining numbers: 4 is deficient because $1 + 2 = 3$; 6 is perfect because $1 + 2 + 3 = 6$; 8 is deficient because $1 + 2 + 4 = 7$; 9 is deficient because $1 + 3 = 4$; 10 is deficient because $1 + 2 + 5 = 8$. The number 12 is abundant because the sum of its proper divisors is $1 + 2 + 3 + 4 + 6 = 16$. The numbers 14, 15, and 16 are deficient. Verify this by adding their proper divisors. The number 18 is the next abundant number; the sum of its proper divisors is 21. Then 20 is the third abundant number; the sum of its proper divisors is 22. The numbers 21 and 22 are both deficient. Verify this. Finally the fourth abundant number is 24; the sum of its proper divisors is 36.

23. The sum of the proper divisors is $975 > 945$, so 945 is abundant. Therefore, the number is abundant.

25. The sum of the proper divisors of 1,184 is 1,210. The sum of the proper divisors of 1,210 is 1,184. Thus, by definition, they are amicable.

27. $16 = 5 + 11$

29. $30 = 7 + 23$

31. **(a)** If $a = 5$ and $b = 3$, then $5 + 2 \cdot 3 = 11$.

(b) $17 = 3 + 2 \cdot 7$
$ = 7 + 2 \cdot 5$
$ = 11 + 2 \cdot 3$
$ = 13 + 2 \cdot 2$

33. 41 and 43

35. (a) $a^{p-1} - 1 = 3^{5-1} - 1 = 3^4 - 1 = 81 - 1 = 80$
Then, $80 \div 5 = 16$.

(b) $a^{p-1} - 1 = 2^{7-1} - 1 = 2^6 - 1 = 64 - 1 = 63$
Then, $63 \div 7 = 9$.

37. $5^2 + 2 = 3^3$; $25 + 2 = 27$

39. Examine the numbers.

41. No; for the first six, the sequence is 6, 8, 6, 8, 6, 6.

43. $1^2 + 1^2 + 8^2 + 4^2 = 82$, $8^2 + 2^2 = 68$, $6^2 + 8^2 = 100$, $1^2 + 0^2 + 0^2 = 1$ (1184 is happy.)
$1^2 + 2^2 + 1^2 + 0^2 = 6$ (1210 is not happy.)
Only one is happy and we do not have a happy amicable pair.

45. $3^2 + 5^2 + 3^2 + 6^2 + 1^2 + 3^2 + 2^2 + 6^2 = 129$
$1^2 + 2^2 + 9^2 = 86$, $8^2 + 6^2 = 100$, $1^2 + 0^2 + 0^2 = 1$ (35,361,326 is happy.)
$4^2 + 0^2 + 1^2 + 1^2 + 7^2 + 7^2 + 1^2 + 4^2 = 133$, $1^2 + 3^2 + 3^2 = 19$, $1^2 + 9^2 = 82$, $8^2 + 2^2 = 68$, $6^2 + 8^2 = 100$,
$1^2 + 0^2 + 0^2 = 1$ (40,117,714 is happy.)
Since both are happy, we have a happy amicable pair.

47. Writing exercise; answers will vary.

49. B; sometimes

Reference table for Exercises 51–55.

	p	$2p + 1$	*Sophie Germaine prime?*
51.	3	7	*yes*
53.	7	15	*no*
55.	13	27	*no*

51. For $p = 3$, $2p + 1 = 7$, a Sophie Germaine prime.

53. For $p = 7$, $2p + 1 = 15$, not a Sophie Germaine prime.

55. For $p = 13$, $2p + 1 = 27$, not a Sophie Germaine prime.

57.

n	$n!$	$n!-1$	$n!+1$	$n!-1$ prime?	$n!+1$ prime?
2	2	1	3	no	yes
57. 4	24	23	25	yes	no

To calculate the underlined numerical values in the table, recall the meaning of n-factorial: $n! = n(n-1)(n-2)...(3)(2)(1)$.

n	$n!$	$n!-1$	$n!+1$
3	$3 \cdot 2 \cdot 1 = 6$	5	7
4	$4 \cdot 3 \cdot 2 \cdot 1 = 24$	23	25
5	$5 \cdot 4 \cdot 3 \cdot 2 \cdot 1 = 120$	119	121

The numbers 5 and 7 are both prime. Although 23 is prime, the number 25 is not as it can be written as $5 \cdot 5$. The number 119 can be written as $7 \cdot 17$, and 121 can be written as $11 \cdot 11$.

59. Writing exercise; answers will vary.

61. B; sometimes

5.4 Exercises

1. True; they would have a common factor of at least 2. Two numbers that are relatively prime have a greatest common factor of 1.

3. True; consider the prime number 5; $5^2 = 25$. The greatest common factor of 5 and 25 is 5.

5. False; the variable $p = 2$ proves this statement false.

7. True; all natural numbers have 1 as a common factor.

9. True; consider some examples: 25 and 9, 14 and 33, 16 and 39, etc.

11. To find the greatest common factor by the prime factors method, first write the prime factorization of each number:

$60 = 2^2 \cdot 3 \cdot 5$

$140 = 2^2 \cdot 5 \cdot 7$

Now find the primes common to all factorizations, with each prime raised to the smallest exponent from either factorization:

$2^2, 5$

Finally, form the product of these numbers:

$2^2 \cdot 5 = 4 \cdot 5 = 20.$

13. To find the greatest common factor by the prime factors method, first write the prime factorization of each number:

$540 = 2^2 \cdot 3^3 \cdot 5$

$1200 = 2^4 \cdot 3 \cdot 5^2$

Now find the primes common to all factorizations, with each prime raised to the smallest exponent from either factorization:

$2^2, 3, 5$

Finally, form the product of these numbers:

$2^2 \cdot 3 \cdot 5 = 4 \cdot 3 \cdot 5 = 60.$

15. To find the greatest common factor by the prime factors method, first write the prime factorization of each number:

$52 = 2^2 \cdot 13$

$39 = 3 \cdot 13$

$78 = 2 \cdot 3 \cdot 13$

Now find the primes common to all factorizations, with each prime raised to the smallest exponent from either factorization:
13

Finally, form the product of these numbers:
13.

17. To find the greatest common factor by the method of dividing by prime factors, first write the numbers in a row and divide by their smallest common prime factor. Repeat until the numbers have no common prime factor.

2	72	90
3	36	45
3	12	15
	4	5

To find the greatest common factor, find the product of the primes on the left.

$2 \cdot 3 \cdot 3 = 18$

19. To find the greatest common factor by the method of dividing by prime factors, first write the numbers in a row and divide by their smallest common prime factor. Repeat until the numbers have no common prime factor.

2	410	360
5	205	180
	41	36

To find the greatest common factor, find the product of the primes on the left.

$2 \cdot 5 = 10$

21. To find the greatest common factor by the method of dividing by prime factors, first write the numbers in a row and divide by their smallest common prime factor. Repeat until the numbers have no common prime factor.

2	12	18	36
3	6	9	18
	2	3	6

To find the greatest common factor, find the product of the primes on the left.

$2 \cdot 3 = 6$

23.

	Remainder
$56 \div 32 = 1$	24
$32 \div 24 = 1$	8
$24 \div 8 = 3$	0

Then 8, the last positive remainder, is the greatest common factor.

25.

	Remainder
$108 \div 24 = 4$	12
$24 \div 12 = 2$	0

Then 12, the last positive remainder, is the greatest common factor.

27.

	Remainder
$480 \div 400 = 1$	80
$400 \div 80 = 5$	0

Then 80, the last positive remainder, is the greatest common factor.

29. Writing exercise; answer swill vary.

31. To find the least common multiple by the prime factors method, first write the prime factorization of each number.

$54 = 2 \cdot 3^3$

$30 = 2 \cdot 3 \cdot 5$

Choose all primes belonging to any factorization, with each prime raised to the largest exponent that it has in any

factorization: $2, 3^3, 5$. Finally, form the product of these numbers:

$2 \cdot 3^3 \cdot 5 = 270$.
The least common multiple of 54 and 30 is 270.

33. Prime factorizations:

$90 = 2 \cdot 3^2 \cdot 5$

$50 = 2 \cdot 5^2$

Least common multiple $= 2 \cdot 3^2 \cdot 5^2 = 450$

35. Prime factorizations:

$30 = 2 \cdot 3 \cdot 5$

$40 = 2^3 \cdot 5$

$70 = 2 \cdot 5 \cdot 7$.

Least common multiple $= 2^3 \cdot 3 \cdot 5 \cdot 7 = 840$.

37.

2	24	32
2	12	16
2	6	8
	3	4

Therefore, the product $2 \cdot 2 \cdot 2 \cdot 3 \cdot 4 = 96$ is the LCM.

39.

5	45	75
3	9	15
	3	5

Therefore, the product $5 \cdot 3 \cdot 3 \cdot 5 = 225$ is the LCM.

41.

2	16	120	216
2	8	60	108
2	4	30	54
3	2	15	27
	2	5	9

Therefore, the product
$2 \cdot 2 \cdot 2 \cdot 3 \cdot 2 \cdot 5 \cdot 9 = 2160$ is the LCM.

43. Use the result of Exercise 23 and the following formula.

$$LCM = \frac{m \cdot n}{GCF \text{ of } m \text{ and } n}.$$

Least common multiple of 32 and 56

$= \dfrac{32 \cdot 56}{8}$

$= 224$

45. Use the result of Exercise 25 and the following formula.

$$LCM = \frac{m \cdot n}{GCF \text{ of } m \text{ and } n}.$$

Least common multiple of 24 and 108

$= \dfrac{24 \cdot 108}{12}$

$= 216$

47. Use the result of Exercise 27 and the following formula.

$$LCM = \frac{m \cdot n}{GCF \text{ of } m \text{ and } n}.$$

Least common multiple of 480 and 400

$= \dfrac{480 \cdot 400}{80}$

$= 2400$

49. (a) In creating the GCF choose the smallest exponent on common factors. Note that $a > b > c$. Choose p^b (since $a > b$), q^c (since $a > c$), and r^c (since $b > c$). The GCF is $p^b q^c r^c$.

(b) In creating the LCM choose the largest exponent on common factors. Since $a > b > c$. Choose p^a (since $a > b$), q^a (since $a > c$), and r^b (since $b > c$). The LCM is $p^a q^a r^b$.

51. 150, 210, and 240

1. $210 \div 150 = 1$ with a remainder of 60.

2. $150 \div 60 = 2$ with a remainder of 30.

3. $60 \div 30 = 2$ with remainder of 0.

4. Then 30, the last positive remainder, is the greatest common factor of 150 and 210.

5. $240 \div 30 = 8$ with a remainder of 0.

6. Then 30 is the greatest common factor of the three given numbers.

53. 90, 105, and 315

1. $105 \div 90 = 1$ with a remainder of 15.

2. $90 \div 15 = 6$ with a remainder of 0.

3. Then 15, the last positive remainder, is the greatest common factor of 90 and 105.

4. $315 \div 15 = 21$ with a remainder of 0.

5. Then 15 is the greatest common factor of the three given numbers.

55. 144, 180, and 192

1. $180 \div 144 = 1$ with a remainder of 36.

2. $144 \div 36 = 4$ with a remainder of 0.

3. Then 36, the last positive remainder, is the greatest common factor of 144 and 180.

4. $192 \div 36 = 5$ with a remainder of 12.

5. $36 \div 12 = 3$ with a remainder of 0.

6. Then 12 is the greatest common factor of the three given numbers.

57. (a) The GCF is found in the intersection:
    $2 \cdot 3 = 6$

    (b) The LCM is found in the union:
    $2^2 \cdot 3^2 = 36$

59. (a) The GCF is found in the intersection:
    $2 \cdot 3^2 = 18$

    (b) The LCM is found in the union:
    $2^3 \cdot 3^3 = 216$

61. The numbers p and q must be relatively prime.

63. Writing exercise; answers will vary.

65. Find the least common multiple of 16 and 36. $16 = 2^4$; $36 = 2^2 \cdot 3^2$. The LCM is $2^4 \cdot 3^2 = 144$. The 144th calculator is the first that they will both inspect.

67. Find the greatest common factor of 240 and 288.
    $240 = 2^4 \cdot 3 \cdot 5$; $288 = 2^5 \cdot 3^2$
    The GCF is $2^4 \cdot 3 = 48$.

69. To answer the first question, find the least common multiple of the two dollar amounts.
    $24 = 2^3 \cdot 3$ and $50 = 2 \cdot 5^2$
    The LCM is $2^3 \cdot 3 \cdot 5^2 = 600$. The answer is $600. To answer the second question, divide $600 by $24, the price per book to obtain 25. he would have sold 25 books at this price.

***EXTENSION:* MODERN CRYPTOGRAPHY**

1. $45 \div 6 = 7.5$
    $7.5 - 7 = .5$
    $.5 \times 6 = 3$

Note that although computational results below are rounded, greater accuracy for computed answers is stored in the calculator giving accurate integer answers for the last calculation.

3. $225 \div 13 = 17.307692$
    $17.307692 \div 17 = .307692$
    $.307692 \times 13 = 4$

5. $(5^9) \div 12 = 162760.4167$
    $162760.4167 - 162760 = .4167$
    $.4167 \times 12 = 5$

7. $(8^7) \div 11 = 190650.1818$
    $190650.1818 - 190650 = .1818$
    $.1818 \times 11 = 2$

9. $M = 8^{27} \pmod{17}$
    $= 8^1 \cdot 8^2 \cdot 8^8 \cdot 8^{16} \pmod{17}$
    "Mod" each factor individually:
    $8^1 = 8 = 8$
    $8^2 = 64 = 13$
    $(8^8) = (8^2)^4 = 13^4 = 28561 = 1$
    $(8^8)^2 = 1^2 = 1$
    Thus, $M = 8 \cdot 13 \cdot 1 \cdot 1 = 104 = 2$.

11. $M = 11^{14} \pmod{18} = 11^2 \cdot 11^4 \cdot 11^8 \pmod{18}$
    "Mod" each factor individually:
    $11^2 = 121 = 13$
    $11^4 = (11^2)^2 = 13^2 = 169 = 7$
    $11^8 = (11^4)^2 = 7^2 = 49 = 13$
    Thus, $m = 13 \cdot 7 \cdot 13 = 1183 = 13$.

13. Compute α and β.

For Alice: For Bob:

$\alpha = M^a \pmod n$ $\beta = M^b \pmod n$

$= 5^7 \pmod{13}$ $= 5^6 \pmod{13}$

$= 78125 \pmod{13}$ $= 15625 \pmod{13}$

$= 8$ $= 12$

Compute the key.

Alice: $K = \beta^a \pmod n$ Bob: $K = \alpha^b \pmod n$

$= 12^7 \pmod{13}$ $= 8^6 \pmod{13}$

$= 35831808 \pmod{13}$ $= 262144 \pmod{13}$

$= 12$ $= 12$

15. Compute α and β.

For Alice: For Bob:

$\alpha = M^a \pmod n$ $\beta = M^b \pmod n$

$= 5^6 \pmod{11}$ $= 5^7 \pmod{11}$

$= 15625 \pmod{11}$ $= 78125 \pmod{11}$

$= 5$ $= 3$

Compute the key.

Alice: Bob:

$K = \beta^a \pmod n$ $K = \alpha^b \pmod n$

$= 3^6 \pmod{11}$ $= 5^7 \pmod{11}$

$= 729 \pmod{11}$ $= 78125 \pmod{11}$

$= 3$ $= 3$

	p	q	$n = p \cdot q$	$l = (p-1)(q-1)$
17.	5	11	$5 \cdot 11 = \underline{55}$	$4 \cdot 10 = \underline{40}$
19.	5	13	$5 \cdot 13 = \underline{65}$	$4 \cdot 12 = \underline{48}$

21. Using $n = 55$, $e = 7$, and $M = 15$: $C = M^e \pmod n$

$= 15^7 \pmod{55}$

$= 170859375$

$= 5.$

23. Using $n = 65$, $e = 5$, and $M = 16$: $C = M^e \pmod n$

$= 16^5 \pmod{65}$

$= 1048576$

$= 61.$

25. (a) Since d must satisfy $e \cdot d = 1 \pmod l$ we have $3 \cdot d = 1 \pmod{40}$ (*l from exercise 17 above*).
Similar to Example 3 we will look for powers of 3 until we find one congruent to 1, modulo 40:
$3^1 = 3$, $3^2 = 9$, $3^3 = 27$, $3^4 = 81 = 1$. So take $d = 3^3 = 27$.
The private key is
$n = p \cdot q = 5 \cdot 11 = 55$, $d = 27$.

(b) Using above results along with $e = 3$ and $C = 30$:

$$M = C^d \pmod{n}$$
$$= 30^{27} \pmod{55}$$
$$= 30^{1+2+8+16} \pmod{55}$$
$$= 30 \cdot 30^2 \cdot 30^8 \cdot 30^{16} \pmod{55}$$

Mod each power individually.
$30 = 30$
$30^2 = 900 = 20$
$30^8 = (30^2)^4 = 20^4 = 160000 = 5$
$30^{16} = (30^8)^2 = 5^2 = 25$
Thus, $M = 30 \cdot 20 \cdot 5 \cdot 25 = 75000 = 35$.

27. (a) Since d must satisfy $e \cdot d = 1 \pmod{l}$ we have $35 \cdot d = 1 \pmod{48}$. Similar to Example 3 we will look for powers of 35 until we find one congruent to 1, modulo 48:

$$35^1 = 35, \quad 35^2 = 1225 = 25,$$
$$35^3 = 42875 = 11, \quad 35^4 = 1500625 = 1.$$

So take $d = 35^3 = 42875 = 11$.
The private key is
$n = p \cdot q = 5 \cdot 13 = 65$, $d = 11$.

(b) Using above results along with $e = 35$ and $C = 17$:

$$M = C^d \pmod{n}$$
$$= 17^{11} \pmod{65}$$
$$= 17^{1+2+8} \pmod{65}$$
$$= 17 \cdot 17^2 \cdot 17^8 \pmod{65}$$

Mod each power individually.
$17 = 17$
$17^2 = 289 = 29$
$17^8 = (17^2)^4 = 29^4 = 707281 = 16$
Thus, $M = 17 \cdot 29 \cdot 16 = 7888 = 23$.

29. Writing exercise; answers will vary.

5.5 Exercises

1. Find the sum of the sixteenth and seventeenth terms to obtain the eighteenth: $987 + 1597 = 2584$.

3. Find the difference between F_{24} and F_{22}: $46,368 - 17,711 = 28,657$

5. $\dfrac{1+\sqrt{5}}{2}$

Reference table for Exercises 7–11

Term	All	Even	Odd
F_1	1		1
F_2	1	1	
F_3	2		2
F_4	3	3	
F_5	5		5
F_6	8	8	
F_7	13		13
F_8	21	21	
F_9	34		34
F_{10}	55	55	
F_{11}	89		89
F_{12}	144	144	
F_{13}	233		233

7. The pattern on the left side of the nth equation consists of the sum of the first n Fibonacci numbers. The pattern on the right side of the nth equation consists of the $(n + 2)$ number less one. Therefore, the 6th equation is $1 + 1 + 2 + 3 + 5 + 8 = 21 - 1$, which checks.

9. The pattern on the left side of the equations consists of the beginning consecutive odd terms of the Fibonacci sequence. The pattern on the right side of the equations consists of consecutive individual even terms of the Fibonacci sequence. On both sides add the next consecutive odd or even term of the Fibonacci sequence to produce the next equation. The next term on the left side is 89, and the next term on the right side is 144. Therefore, the next equation is $1 + 2 + 5 + 13 + 34 + 89 = 144$, which checks.

11. The pattern on the left side of the equations consists of the two terms of the Fibonacci sequence in reverse order. The pattern on the right side of the equations consists of consecutive individual even terms of the Fibonacci sequence. It is the difference of the squares of the terms on the left side that equals the term on the right side. Notice also that on the left side, the differences alternate between the squares of consecutive odd terms and consecutive even terms. On the left side, then, write the difference of squares of the next consecutive odd terms, in reverse order to produce the next equation. The next terms on the left side are 13^2 and 5^2, and the next term on the right side is 144. Therefore, the next equation is $13^2 - 5^2 = 144$, which checks.

13. The left side of each equation is built with alternating terms of the Fibonacci sequence with alternating signs. The right side of each equation consists of Fibonacci numbers squared with alternating signs. The next equation should be

$1 - 2 + 5 - 13 + 34 - 89 = -8^2$. The left side of this equation has a sum of –64. The right side of the equations also equals –64.

15. (a) $37 = 34 + 3$
Another possibility is
$37 = 21 + 8 + 5 + 3$. Can you find more?

(b) $40 = 34 + 3 + 2 + 1$

(c) $52 = 21 + 13 + 8 + 5 + 3 + 2$

17. (a) $m = 10$ and $n = 4$. The greatest common factor of 10 and 4 is 2. Also, $F_{10} = 55$ and $F_4 = 3$; $F_2 = 1$, which is the greatest common factor of 55 and 3.

(b) $m = 12$ and $n = 6$. The greatest common factor of 12 and 6 is 6. Also $F_{12} = 144$ and $F_6 = 8$; the greatest common factor of 144 and 8 is 8.

(c) $m = 14$ and $n = 6$. The greatest common factor of 14 and 6 is 2. Also, $F_{14} = 377$ and $F_6 = 8$; $F_2 = 1$, which is the greatest common factor of 377 and 8.

19. (a) Square 13. Multiply the terms of the sequence two positions away from 13 (i.e., 5 and 34). Subtract the smaller result from the larger, and record your answer. $13^2 = 169$; $5 \cdot 34 = 170$; $170 - 169 = 1$

(b) Square 13. Multiply the terms of the sequence three positions away from 13. Once again subtract the smaller result from the larger, and record your

answer. $13^2 = 169$; $3 \cdot 55 = 165$; $169 - 165 = 4$

(c) Repeat the process, moving four terms away from 13. $13^2 = 169$; $2 \cdot 89 = 178$; $178 - 169 = 9$

(d) Make a conjecture about what will happen when you repeat the process, moving five terms away. Verify your answer. The results in parts a, b, and c are 1, 4, and 9. These are the squares of the Fibonacci numbers. Because the next Fibonacci number is 5, the result should be $5^2 = 25$. Verification: $13^2 = 169$; $1 \cdot 144 = 144$, $169 - 144 = 25$

21. This term is found by adding the ninth and tenth terms: $76 + 123 = 199$.

23. 1, 3, 4, 7, 11, 18, ...
Add the first and third terms: $1 + 4 = 5$
Add the first, third, and fifth terms:
$1 + 4 + 11 = 16$
Add the first, third, fifth, and seventh terms:
$1 + 4 + 11 + 29 = 45$
Also, $1 + 4 + 11 + 29 + 76 = 121$
Each of the sums is 2 less than a Lucas number.

25. (a) $1 \cdot 1 = 1$
$1 \cdot 3 = 3$
$2 \cdot 4 = 8$
$3 \cdot 7 = 21$
$5 \cdot 11 = 55$
Notice that reading downward, the first numbers in each equation are the first five members of the Fibonacci sequence, and the second number in each equation is a Lucas number. The products, reading downward, are the 2nd, 4th, 6th, 8th, and 10th terms of the Fibonacci sequence. The next equation should be $8 \cdot 18 = 144$.

(b) $1 + 2 = 3$
$1 + 3 = 4$
$2 + 5 = 7$
$3 + 8 = 11$
$5 + 13 = 18$
Reading downward, the first terms on the left side of the equation are the first five Fibonacci numbers. The second terms on the left side are also ascending

Fibonacci numbers, beginning with 2. The results are Lucas numbers. The next equation should be 8 + 21 = 29.

(c) $1 + 1 = 2 \cdot 1$
$1 + 3 = 2 \cdot 2$
$2 + 4 = 2 \cdot 3$
$3 + 7 = 2 \cdot 5$
$5 + 11 = 2 \cdot 8$

Read downward to see the first five Fibonacci numbers on the extreme left side of the equation added to the first five Lucas numbers. Read downward on the extreme right side to see the second through sixth Fibonacci numbers. The next equation should be $8 + 18 = 2 \cdot 13$.

(d) $1 + 4 = 5 \cdot 1$
$3 + 7 = 5 \cdot 2$
$4 + 11 = 5 \cdot 3$
$7 + 18 = 5 \cdot 5$
$11 + 29 = 5 \cdot 8$

Read downward on the extreme right side of the equations to see the second through sixth Fibonacci numbers. The second number in each equation is a Lucas number. The next equation should be $18 + 47 = 5 \cdot 13$.

27. $1, 1, 2, 3$
Multiply the first and fourth: $1 \cdot 3 = 3$
Double the product of the second and third: $2(1 \cdot 2) = 4$
Add the squares of the second and third: $1^2 + 2^2 = 5$
We have obtained the triple 3, 4, 5. We can substitute these numbers into the Pythagorean theorem to verify that this is a Pythagorean triple:
$3^2 + 4^2 = 5^2$
$9 + 16 = 25$

29. $2, 3, 5, 8$
Multiply the first and fourth: $2 \cdot 8 = 16$
Double the product of the second and third: $2(3 \cdot 5) = 30$
Add the squares of the second and third: $3^2 + 5^2 = 34$
We have obtained the triple 16, 30, 34. We can substitute these numbers into the Pythagorean theorem to verify that this is a

Pythagorean triple:
$16^2 + 30^2 = 34^2$
$256 + 900 = 1156$

31. The sum of the terms on the diagonals are the Fibonacci numbers: 1, 1, 2, 3, 5, 8, 13, ...

33. $\dfrac{1 + \sqrt{5}}{2} \approx 1.618033989$;

$\dfrac{1 - \sqrt{5}}{2} \approx -0.618033989$

The digits in the decimal positions are the same.

35. $n = 14$

$$\dfrac{\left(\frac{1+\sqrt{5}}{2}\right)^{14} - \left(\frac{1-\sqrt{5}}{2}\right)^{14}}{\sqrt{5}} = 377$$

The 14th Fibonacci number is 377.

37. $n = 22$
Use a calculator to find

$$\dfrac{\left(\frac{1+\sqrt{5}}{2}\right)^{22} - \left(\frac{1-\sqrt{5}}{2}\right)^{22}}{\sqrt{5}} = 17,711.$$

The 22nd Fibonacci number is 17,711.

Chapter 5 Test

1. False; 2 and 3 differ by one and both are prime numbers.

2. True

3. True, because 9 is divisible by 3.

4. True

5. True; 1 is a factor of all numbers. The smallest multiple of a given number is found by multiplying that number by 1.

6. False; the number may be deficient.

7. 656,723,600

(a) It is divisible by 2 because it is even.

(b) It is not divisible by 3 because the sum of the digits is 35, a number not divisible by 3.

(c) It is divisible by 4 because the last two digits taken as the number 0 is divisible by 4.

(d) It is divisible by 5 because it ends in 0.

(e) It is not divisible by 6 because it is not divisible by both 2 and 3.

(f) It is divisible by 8 because the last three digits form the number 600, which is divisible by 8.

(g) It is not divisible by 9 because the sum of the digits, 35, is not divisible by 9.

(h) It is divisible by 10 because it ends in 0.

(i) It is not divisible by 12 because it is not divisible by both 3 and 4.

8. (a) Composite; $87 = 3 \cdot 29$

(b) prime; its only factors are itself and 1.

(c) The number 1 is neither prime nor composite.

9.
```
2 | 2520
2 | 1260
2 |  630
3 |  315
3 |  105
5 |   35
        7
```

$$2520 = 2^3 \cdot 3^2 \cdot 5 \cdot 7$$

10. Writing exercise; answers will vary.

11. (a) The proper divisors of 36 are 1, 2, 3, 4, 6, 9, 12, and 18. The sum of these numbers is 55. Because the sum is greater than 36, the number is abundant.

(b) The proper divisors of 8 are 1, 2, and 4. The sum of these numbers is 7. Because the sum is less than 8, the number is deficient.

(c) The proper divisors of 28 are 1, 2, 4, 7, and 14. The sum of these numbers is 28. Because the sum equals 28, the number is perfect.

12. Statement C is false. Goldbach's Conjecture for the number 8 is verified by the equation $8 = 3 + 5$. The number 1 is not prime.

13. A pair of twin primes between 50 and 70 is 59 and 61.

14.
```
3 | 135   216
3 |  45    72
3 |  15    24
       5     8
```

$$GCF = 3 \cdot 3 \cdot 3 = 27$$

15. $15 = 3 \cdot 5$

$45 = 3^2 \cdot 5$

$50 = 2 \cdot 5^2$

$LCM = 2 \cdot 3^2 \cdot 5^2 = 450$

16. This exercise is similar to Exercise 66 in section 5.4. If Katherine is off every 6 days and Josh is off every 4 days, this corresponds to modulo 6 and modulo 4. Also, the days of the week, beginning with Sunday, correspond to modulo 7.
For Katherine, start with $x = 3 \pmod 6$. This is the set of integers {3, 9, 15, 21, ...}. For Josh, $x = 3 \pmod 4$.
This is the set of integers {3, 7, 11, 15, 19, ...}.
The next common number for Katherine and Josh is 15, which corresponds to Monday.

17.
$$\begin{array}{r} 28,657 \\ + 46,368 \\ \hline 75,025 \end{array}$$

18. Notice that the numbers that are being subtracted are four Fibonacci numbers, in order, and each "set" of them deletes the first and adds another as the equations progress. The final values on the right side of the equation are also Fibonacci numbers. Finally, the first number in each equation is also a Fibonacci number, with the first several of them omitted. The next equation should be $89 - (8 + 13 + 21 + 34) = 13$.

19. Option B; sometimes; see section 5.1.

20. The sequence is obtained by adding two successive terms to obtain the next:
1, 5, 6, 11, 17, 28, 45, 73.

21. First let us choose 11.

$$11^2 = 121$$
$$6 \cdot 17 = 102$$
$$121 - 102 = 19$$

Now let us choose 45.

$$45^2 = 2025$$
$$28 \cdot 73 = 2044$$
$$2044 - 2025 = 19$$

It appears that this process yields 19 each time.

22. Option A is the exact value of the golden ratio. Options C and D are approximations.

23. Writing exercise; answers will vary.

24. Writing exercise; answers will vary.

Chapter 6

1. An integer between 4.5 and 5.5 is 5.

3. A whole number that is not positive and is less than 1 is 0.

5. An irrational number that is between $\sqrt{13}$ and $\sqrt{15}$ is $\sqrt{14}$. There are many others.

7. It is true that every natural number is positive. The natural numbers consist of $\{1, 2, 3, 4, ...\}$.

9. True; the set of integers is included in the set of rational numbers.

11. (a) $3, 7$

    (b) $0, 3, 7$

    (c) $-9, 0, 3, 7$

    (d) $-9, -1\frac{1}{4}, -\frac{3}{5}, 0, 3, 5.9, 7$

    (e) $-\sqrt{7}, \sqrt{5}$

    (f) All are real numbers.

13. Writing exercise; answers will vary.

15. 1046

17. 5436

19. $-220°$

21. $30; 15°; -5°$

23. (a) Pacific Ocean, Indian Ocean, Caribbean Sea, South China Sea, Gulf of California

    (b) Point Success, Ranier, Matlalcueyetl, Steele, McKinley

    (c) This statement is true because the absolute value of each number is its non-negative value.

    (d) This statement is false. The absolute value of the depth of the Gulf of California is 2375; the absolute value of the depth of the Caribbean Sea is 8448.

25.

27.

29. (a) $|-7| = 7$, which is choice A.

    (b) $-(-7) = 7$, which is choice A.

    (c) $-|-7| = -7$, which is choice B.

    (d) $-|-(-7)| = -|7| = -7$, which is choice B.

31. -2

    (a) Additive inverse is 2.

    (b) Absolute value is 2.

33. 6

    (a) Additive inverse is -6.

    (b) Absolute value is 6.

35. $7 - 4 = 3$

    (a) Additive inverse is -3.

    (b) Absolute value is 3.

37. $7 - 7 = 0$

    (a) Additive inverse is 0.

    (b) Absolute value is 0.

39. If $a - b > 0$, then the absolute value of $a - b$ in terms of a and b is $a - b$ because this expression produces a nonnegative number.

41. -12

43. -8

45. The smaller number is 3 because $|-4| = 4$.

47. $|-3|$ is the smaller number.

49. $-|-6| = -6$, the smaller number.

51. $|5 - 3| = |2| = 2; |6 - 2| = |4| = 4$
    The first is the smaller number.

53. $6 > -(-2)$ is a true statement because $6 > 2$.

55. $-4 \leq -(-5)$ is true because $-4 \leq 5$.

57. $|-6| < |-9|$ is true because $6 < 9$.

59. $-|8| > |-9|$ is false because $-8 < 9$.

61. $-|-5| \geq -|-9|$ is true because $-5 \geq -9$.

63. $|6 - 5| \geq |6 - 2|$ is false because $1 < 4$.

65. (a) Louisiana had the greatest change in population. It decreased 4.1%.

     (b) West Virginia had the least change in population. It increased 0.6%.

67. Fabric Mills shows the greatest change (without regard to sign).

Answers may vary in Exercises 69–73.

69. Three positive real numbers but not integers between -6 and 6 are $\dfrac{1}{2}, \dfrac{5}{8}$, and $1\dfrac{3}{4}$.

71. Three real numbers but not whole numbers between -6 and 6 are $-3\dfrac{1}{2}, -\dfrac{2}{3}$, and $\dfrac{3}{7}$.

73. Three real numbers but not rational numbers between -6 and 6 are $\sqrt{5}$, π, and $-\sqrt{3}$.

6.2 Exercises

1. The sum of two negative numbers will always be a <u>negative</u> number.

3. To simplify the expression $8 + [-2 + (-3 + 5)]$, I should begin by adding $\underline{-3}$ and $\underline{5}$, according to the rule for order of operations.

5. Writing exercise; answers will vary.

7. $-12 + (-8) = -20$

9. $12 + (-16) = -4$

11. $-12 - (-1) = -12 + 1 = -11$

13. $-5 + 11 + 3 = 6 + 3 = 9$

15. $12 - (-3) - (-5) = 12 + 3 + 5 = 20$

17. $-9 - (-11) - (4 - 6) = -9 + 11 - (-2)$
$$= -9 + 11 + 2$$
$$= 2 + 2$$
$$= 4$$

19. $(-12)(-2) = 24$

21. $9(-12)(-4)(-1)3 = -1296$

23. $\dfrac{-18}{-3} = 6$

25. $\dfrac{36}{-6} = -6$

27. $\dfrac{0}{12} = 0$

29. $-6 + [5 - (3 + 2)] = -6 + [5 - 5] = -6 + 0 = -6$

31. $-4 - 3(-2) + 5^2 = -4 + 6 + 25 = 2 + 25 = 27$

33. $(-8 - 5)(-2 - 1) = (-13)(-3) = 39$

35. $-8(-2) - [(4^2) + (7 - 3)] = 16 - [16 + (4)]$
$$= 16 - [20]$$
$$= -4$$

37. $\dfrac{(-6 + 3) \cdot (-4)}{-5 - 1} = \dfrac{(-3) \cdot (-4)}{-6} = \dfrac{12}{-6} = -2$

39. $\dfrac{2(-5) + (-3)(-2^2)}{-(3^2) + 9} = \dfrac{-10 + (-3)(-4)}{-9 + 9}$
$$= \dfrac{-10 + (-3)(-4)}{0}$$
Division by zero is undefined.

41. $-\dfrac{1}{4}[3(-5) + 7(-5) + 1(-2)]$
$$= -\dfrac{1}{4}[-15 - 35 - 2]$$
$$= -\dfrac{1}{4}[-52]$$
$$= 13$$

43. Division by zero is undefined, so A, B, and C are all undefined.

45. Commutative property of addition

47. Inverse property of addition

49. Identity property of multiplication

51. Associative property of addition

53. Identity property of addition

55. Distributive property

57. Inverse property of addition

59. Closure property of multiplication

61. **(a)** $6 - 8 = -2$ and $8 - 6 = 2$

     (b) By the results of part (a), we may conclude that subtraction is not a <u>commutative</u> operation.

     (c) When $a = b$, it is a true statement. For example, let $a = b = 5$. Then $5 - 5 = 5 - 5$ or 0.

63. **(a)** The inverse of cleaning up your room would be messing up your room.

     (b) The inverse of earning money would be spending money.

     (c) The inverse of increasing the volume on your MP3 player would be decreasing the volume.

65. Jack recognized the identity property for addition.

67. Use the given hint: Let $a = 2$, $b = 3$, $c = 4$. Now test $a + (b \cdot c) = (a + b) \cdot (a + c)$.
$a + (b \cdot c) = 2 + (3 \cdot 4) = 2 + 12 = 14.$
However, $(a + b) \cdot (a + c) = (2 + 3) \cdot (2 + 4)$
$$= 5 \cdot 6$$
$$= 30.$$
The two expressions are not equivalent. The distributive property for addition with respect to multiplication does not hold.

69. $-3^4 = -81$

The notation indicates the opposite of 3^4.

71. $(-3)^4 = (-3)(-3)(-3)(-3) = 81$

73. $-(-3)^4 = -81$

75. $-[-(-3)]^4 = -[3]^4 = -81$

77. **(a)** The change in outlay from 2006 to 2007 was $490.6 - 464.7 = 25.9$ (billion dollars).

(b) The change in outlay from 2007 to 2008 was $548.8 - 490.6 = 58.2$ (billion dollars).

(c) The change in outlay from 2008 to 2009 was $701.8 - 548.8 = 153.0$ (billion dollars).

(d) The change in outlay from 2006 to 2009 was $701.8 - 464.7 = 237.1$ (billion dollars).

79. **(a)** The difference between tax revenue and cost of benefits in the year 2000: $\$538 - 409 = \$129$ billion.
The difference between projected tax revenue and cost of benefits in the year 2010: $\$916 - 710 = \$206$ billion.
The difference between projected tax revenue and cost of benefits in the year 2020: $\$1479 - 1405 = \$74$ billion.
The difference between projected tax revenue and cost of benefits in the year 2030: $\$2041 - 2542 = -\$501$ billion.

(b) The cost of Social Security will exceed revenue in 2030 by $501 billion.

81. $-3 + (-2) + 9 + 5 + 3 + 2 + 2 = 16$

83. $\$904.89 - \$35.84 - \$26.14 - \$3.12$
$+ \$85.00 + \$120.76 = \$1045.55$

85. $14,494 - (-282) = 14,494 + 282$
$$= 14,776 \text{ feet}$$

87. $-4 + 49 = 45°\text{F}$

89. $-27 + 139 = 112°\text{F}$

91. $-36 - 24 = -60°\text{F}$

93. $15 - (-12) = 15 + 12 = 27$ feet

95. Because these years are similar to negative numbers on a number line:
$-428 + (-41) = -469$, which is 469 B.C.

97. $\$195,200 - \$180,200 = \$15,000$

99. $\$221,900 - \$219,000 = \$2900$

6.3 Exercises

1. $\dfrac{4}{8} = \dfrac{1}{2} = 0.5 = .5\overline{0}$ which are choices A, C, and D. The fractions $\dfrac{4}{8}$ can be simplified to its equivalent fraction $\dfrac{1}{2}$. It can also be changed to decimal form by dividing 4 by 8. Remember that the overline on zero indicates that this digit repeats indefinitely.

3. $\dfrac{5}{9} = 0.\overline{5}$, which is choice C. When 5 is divided by 9, the digit 5 repeats.

5. $\dfrac{16}{48} = \dfrac{16 \cdot 1}{16 \cdot 3} = \dfrac{1}{3}$

7. $-\dfrac{15}{35} = -\dfrac{5 \cdot 3}{5 \cdot 7} = -\dfrac{3}{7}$

9. $\dfrac{3}{8} = \dfrac{3 \cdot 2}{8 \cdot 2} = \dfrac{6}{16}$
   $\dfrac{3}{8} = \dfrac{3 \cdot 3}{8 \cdot 3} = \dfrac{9}{24}$
   $\dfrac{3}{8} = \dfrac{3 \cdot 4}{8 \cdot 4} = \dfrac{12}{32}$

11. $-\dfrac{5}{7} = -\dfrac{5 \cdot 2}{7 \cdot 2} = -\dfrac{10}{14}$
    $-\dfrac{5}{7} = -\dfrac{5 \cdot 3}{7 \cdot 3} = -\dfrac{15}{21}$
    $-\dfrac{5}{7} = -\dfrac{5 \cdot 4}{7 \cdot 4} = -\dfrac{20}{28}$

13. (a) $\dfrac{2}{6} = \dfrac{1}{3}$

    (b) $\dfrac{2}{8} = \dfrac{1}{4}$

    (c) $\dfrac{4}{10} = \dfrac{2}{5}$

    (d) $\dfrac{3}{9} = \dfrac{1}{3}$

15. These are the dots in the intersection of the triangle and the rectangle as a part of the dots in the entire figure.

17. (a) Christine O'Brien had 12 hits out of 36 at-bats. The fraction $\dfrac{12}{36}$ simplifies to $\dfrac{1}{3}$.

    (b) Leah Goldberg had 5 hits out of 11 at-bats. The fraction $\dfrac{5}{11}$ is a little less than $\dfrac{1}{2}$.

    (c) Leah Goldberg had 1 home run out of 11 at-bats. The fraction $\dfrac{1}{11}$ is just less than $\dfrac{1}{10}$.

    (d) Anne Kelly made 9 hits out of 40 times at bat. The fraction $\dfrac{9}{40}$ is just less than $\dfrac{10}{40}$, which equals $\dfrac{1}{4}$.

    (e) Otis Taylor made 8 hits out of 16 times at bat; Carol Britz made 10 hits out of 20 times at bat. The fractions $\dfrac{8}{16}$ and $\dfrac{10}{20}$ both reduce to $\dfrac{1}{2}$.

19. $\dfrac{3}{8} + \dfrac{1}{8} = \dfrac{3+1}{8} = \dfrac{4}{8} = \dfrac{4 \cdot 1}{4 \cdot 2} = \dfrac{1}{2}$

21. $\dfrac{5}{16} \cdot \dfrac{3}{3} + \dfrac{7}{12} \cdot \dfrac{4}{4} = \dfrac{15}{48} + \dfrac{28}{48} = \dfrac{43}{48}$

23. $\dfrac{2}{3} \cdot \dfrac{8}{8} - \dfrac{7}{8} \cdot \dfrac{3}{3} = \dfrac{16}{24} - \dfrac{21}{24} = -\dfrac{5}{24}$

25. $\dfrac{5}{8} \cdot \dfrac{7}{7} - \dfrac{3}{14} \cdot \dfrac{4}{4} = \dfrac{35}{56} - \dfrac{12}{56} = \dfrac{23}{56}$

27. $\dfrac{3}{4} \cdot \dfrac{9}{5} = \dfrac{27}{20}$

29. $-\dfrac{2}{3} \cdot -\dfrac{5}{8} = \dfrac{10}{24} = \dfrac{5 \cdot 2}{12 \cdot 2} = \dfrac{5}{12}$

31. $\dfrac{5}{12} \div \dfrac{15}{4} = \dfrac{5}{12} \cdot \dfrac{4}{15} = \dfrac{20}{180} = \dfrac{20 \cdot 1}{20 \cdot 9} = \dfrac{1}{9}$

33. $-\dfrac{9}{16} \div -\dfrac{3}{8} = -\dfrac{9}{16} \cdot -\dfrac{8}{3} = \dfrac{72}{48} = \dfrac{3 \cdot 24}{2 \cdot 24} = \dfrac{3}{2}$

35. $\left(\dfrac{1}{3} \div \dfrac{1}{2}\right) + \dfrac{5}{6} = \left(\dfrac{1}{3} \cdot \dfrac{2}{1}\right) + \dfrac{5}{6}$

$= \dfrac{2}{3} + \dfrac{5}{6}$

$= \dfrac{2}{3} \cdot \dfrac{2}{2} + \dfrac{5}{6}$

$= \dfrac{4}{6} + \dfrac{5}{6}$

$= \dfrac{9}{6}$

The fraction $\dfrac{9}{6}$ can be simplified: $\dfrac{3 \cdot 3}{3 \cdot 2} = \dfrac{3}{2}$.

37. (a) $6 \cdot \dfrac{3}{4} = \dfrac{6}{1} \cdot \dfrac{3}{4} = \dfrac{18}{4}$ or $4\dfrac{1}{2}$ cups

   (b) $\dfrac{1}{2}$ of $\left(\dfrac{3}{4} + 1\right) = \dfrac{1}{2} \cdot \left(\dfrac{3}{4} + \dfrac{4}{4}\right)$

   $= \dfrac{1}{2} \cdot \dfrac{7}{4}$

   $= \dfrac{7}{8}$ cup

39. $4 + \dfrac{1}{3} = \dfrac{4}{1} + \dfrac{1}{3} = \dfrac{12}{3} + \dfrac{1}{3} = \dfrac{13}{3}$

41. $2\dfrac{9}{10} = 2 + \dfrac{9}{10} = \dfrac{2}{1} + \dfrac{9}{10} = \dfrac{20}{10} + \dfrac{9}{10} = \dfrac{29}{10}$

43. $27 \div 4 = 6\dfrac{3}{4}$

45. $3\dfrac{1}{4} + 2\dfrac{7}{8} = \dfrac{13}{4} + \dfrac{23}{8} = \dfrac{26}{8} + \dfrac{23}{8} = \dfrac{49}{8} = 6\dfrac{1}{8}$

47. $-4\dfrac{7}{8} \cdot 3\dfrac{2}{3} = -\dfrac{39}{8} \cdot \dfrac{11}{3}$

$= -\dfrac{429}{24}$

$= -17\dfrac{21}{24}$

$= -17\dfrac{7}{8}$

49. $\dfrac{3}{4} - \dfrac{3}{16} = \dfrac{3 \cdot 4}{4 \cdot 4} - \dfrac{3}{16} = \dfrac{12 - 3}{16} = \dfrac{9}{16}$ inch

51. Using Method 1,

$\dfrac{\frac{1}{2} + \frac{1}{4}}{\frac{1}{2} - \frac{1}{4}} = \dfrac{3}{4} \div \dfrac{1}{4} = \dfrac{3}{4} \cdot \dfrac{4}{1} = \dfrac{12}{4} = 3.$

53. Using Method 1,

$\dfrac{\frac{5}{8} - \frac{1}{4}}{\frac{1}{8} + \frac{3}{4}} = \dfrac{3}{8} \div \dfrac{7}{8} = \dfrac{3}{8} \cdot \dfrac{8}{7} = \dfrac{24}{56} = \dfrac{3}{7}.$

55. Using Method 2,

$\dfrac{\frac{7}{11} + \frac{3}{10}}{\frac{1}{11} - \frac{9}{10}} \cdot \dfrac{110}{110} = \dfrac{70 + 33}{10 - 99} = \dfrac{103}{-89} = -\dfrac{103}{89}.$

57. $2 + \cfrac{1}{1 + \cfrac{1}{3 + \frac{1}{2}}} = 2 + \cfrac{1}{1 + \cfrac{1}{\frac{6}{2} + \frac{1}{2}}}$

$= 2 + \cfrac{1}{1 + \cfrac{1}{\frac{7}{2}}}$

$= 2 + \cfrac{1}{1 + \frac{2}{7}}$

$= 2 + \cfrac{1}{\frac{7}{7} + \frac{2}{7}}$

$= 2 + \cfrac{1}{\frac{9}{7}}$

$= 2 + \dfrac{7}{9}$

$= \dfrac{18}{9} + \dfrac{7}{9}$

$= \dfrac{25}{9}$

59. $\dfrac{\frac{1}{2} + \frac{3}{4}}{2} = \dfrac{\frac{2}{4} + \frac{3}{4}}{2} = \dfrac{\frac{5}{4}}{2} = \dfrac{5}{4} \div \dfrac{2}{1} = \dfrac{5}{4} \cdot \dfrac{1}{2} = \dfrac{5}{8}$

61. $\dfrac{\frac{3}{5} + \frac{2}{3}}{2} = \dfrac{\frac{9}{15} + \frac{10}{15}}{2} = \dfrac{\frac{19}{15}}{2} = \dfrac{19}{15} \div \dfrac{2}{1} = \dfrac{19}{15} \cdot \dfrac{1}{2} = \dfrac{19}{30}$

63. $\dfrac{-\frac{2}{3}+\left(-\frac{5}{6}\right)}{2} = \dfrac{-\frac{4}{6}+\left(-\frac{5}{6}\right)}{2}$

$= \dfrac{-\frac{9}{6}}{2}$

$= -\dfrac{9}{6} \div \dfrac{2}{1}$

$= -\dfrac{9}{6} \cdot \dfrac{1}{2}$

$= -\dfrac{9}{12}$

$= -\dfrac{3}{4}$

65. $\dfrac{\substack{63056+57708+56602+51170+50457 \\ +49562+48781+48430}}{8}$

$= \dfrac{425766}{8}$

$\approx \$53,221$

67. $\dfrac{5+9}{6+13} = \dfrac{14}{19}$

69. $\dfrac{4+9}{13+16} = \dfrac{13}{29}$

71. $\dfrac{7+9}{6+8} = \dfrac{16}{14}$, or $\dfrac{8}{7}$

73. $\dfrac{2+3}{1+1} = \dfrac{5}{2}$

75. Using the consecutive integers 6 and 7.
$\dfrac{6+7}{1+1} = \dfrac{13}{2}$ or $6\dfrac{1}{2}$.
The number will be halfway between the integers.

77. $\dfrac{3}{4} = 0.75$

79. $\dfrac{3}{16} = 0.1875$

81. $\dfrac{3}{11} = 0.\overline{27}$

83. $\dfrac{2}{7} = 0.\overline{285714}$

85. $0.4 = \dfrac{4}{10} = \dfrac{2 \cdot 2}{2 \cdot 5} = \dfrac{2}{5}$

87. $0.85 = \dfrac{85}{100} = \dfrac{5 \cdot 17}{5 \cdot 20} = \dfrac{17}{20}$

89. $0.934 = \dfrac{934}{1000} = \dfrac{2 \cdot 467}{2 \cdot 500} = \dfrac{467}{500}$

91. $\dfrac{8}{15}$: $15 = 3 \cdot 5$
Because 3 is one of the prime factors of the denominator, the fraction will yield a repeating decimal.

93. $\dfrac{13}{125}$: $125 = 5^3$
Because 5 is the only prime number that is a factor of the denominator, the fraction will yield a terminating decimal.

95. $\dfrac{22}{55} = \dfrac{2 \cdot 11}{5 \cdot 11} = \dfrac{2}{5}$
Because 5 is the only prime number that is a factor of the denominator, the fraction will yield a terminating decimal.

97. (a) The decimal representation for $\dfrac{1}{3}$ is
$0.333....$

(b) The decimal representation for $\dfrac{2}{3}$ is
$0.666....$

(c) $0.333... + 0.666... = 0.999....$

(d) $1 = 0.\overline{9}$

99. (a) Let $x = 0.8$
$10x = 8$
$x = \dfrac{8}{10} = \dfrac{4}{5}$

(b) Let $x = 0.799999...$
$10x = 7.99999...$ Then
$10x = 7.99999...$
$\underline{-\ \ x = 0.799999...}$
$9x = 7.200000...$
$x = \dfrac{7.2}{9} = \dfrac{72}{90} = \dfrac{4}{5}$.

101. (a) Let $x = 0.66$
$$100x = 66$$
$$x = \frac{66}{100} = \frac{33}{50}$$

(b) Let $x = 0.659999...$
$10x = 6.59999...$ Then,

$$\begin{array}{r} 10x = 6.59999... \\ -x = 0.659999... \\ \hline 9x = 5.9400000... \end{array}$$

$$x = \frac{5.94}{9} = \frac{594}{900} = \frac{33}{50}.$$

6.4 Exercises

1. This number is rational because it can be written as the ratio of one integer to another.

3. This number is irrational because it cannot be written as the ratio of one integer to another; only an approximation of the number can be written in this form.

5. $1.618 = 1\frac{618}{1000} = \frac{1618}{1000}$, a rational number.

7. $0.\overline{41}$, a rational number; use Example 9 from Section 6.3 to show that it is equivalent to the rational number $\frac{41}{99}$.

9. The number symbolized by π is irrational. Its value is a non terminating, nonrepeating decimal. The rational values given in Exercises 13 and 14 are approximations of the value of π.

11. This number is rational. It can be written as the ratio of one integer to another.
$$3\frac{14159}{100000} \text{ or } \frac{314159}{100000}$$

13. This number is irrational; it is non terminating and non repeating.

15. (a)
$$\begin{array}{r} 0.272772777277772... \\ + 0.616116111611116... \\ \hline 0.888888888888888... \end{array}$$

(b) Based on the result of part (a), we can conclude that the sum of two <u>irrational</u> numbers may be a <u>rational</u> number.

17. $\sqrt{39} \approx 6.244997998$

19. $\sqrt{15.1} \approx 3.885871846$

21. $\sqrt{884} \approx 29.73213749$

23. First find $9 \div 8 = 1.125$ on your calculator. Then take the square root of the resulting quotient.
$$\sqrt{1.125} \approx 1.060660172$$

25. $\sqrt{50} = \sqrt{25 \cdot 2} = \sqrt{25} \cdot \sqrt{2} = 5\sqrt{2}$
Using a calculator, $\sqrt{50} \approx 7.071067812$ and $5\sqrt{2} \approx 7.071067812$.

27. $\sqrt{75} = \sqrt{25 \cdot 3} = \sqrt{25} \cdot \sqrt{3} = 5\sqrt{3}$
Using a calculator, $\sqrt{75} \approx 8.660254038$ and $5\sqrt{3} \approx 8.660254038$.

29. $\sqrt{288} = \sqrt{144 \cdot 2} = \sqrt{144} \cdot \sqrt{2} = 12\sqrt{2}$
Using a calculator, $\sqrt{288} \approx 16.97056275$ and $12\sqrt{2} \approx 16.97056275$.

31. $\dfrac{5}{\sqrt{6}} = \dfrac{5}{\sqrt{6}} \cdot \dfrac{\sqrt{6}}{\sqrt{6}} = \dfrac{5\sqrt{6}}{6}$
Using a calculator, $\dfrac{5}{\sqrt{6}} \approx 2.041241452$ and $\dfrac{5\sqrt{6}}{6} \approx 2.041241452$.

33. $\sqrt{\dfrac{7}{4}} = \dfrac{\sqrt{7}}{\sqrt{4}} = \dfrac{\sqrt{7}}{2}$
Using a calculator, $\sqrt{\dfrac{7}{4}} \approx 1.322875656$ and $\dfrac{\sqrt{7}}{2} \approx 1.322875656$.

35. $\sqrt{\dfrac{7}{3}} = \dfrac{\sqrt{7}}{\sqrt{3}} \cdot \dfrac{\sqrt{3}}{\sqrt{3}} = \dfrac{\sqrt{21}}{3}$
Using a calculator, $\sqrt{\dfrac{7}{3}} \approx 1.527525232$ and $\dfrac{\sqrt{21}}{3} \approx 1.527525232$.

37. $\sqrt{17} + 2\sqrt{17} = (1+2)\sqrt{17} = 3\sqrt{17}$

39. $5\sqrt{7} - \sqrt{7} = (5-1)\sqrt{7} = 4\sqrt{7}$

41. $3\sqrt{18} + \sqrt{2} = 3\sqrt{9 \cdot 2} + \sqrt{2}$
$= 3\sqrt{9} \cdot \sqrt{2} + \sqrt{2}$
$= 3 \cdot 3\sqrt{2} + \sqrt{2}$
$= 9\sqrt{2} + \sqrt{2}$
$= (9 + 1)\sqrt{2}$
$= 10\sqrt{2}$

43. $-\sqrt{12} + \sqrt{75} = -\sqrt{4 \cdot 3} + \sqrt{25 \cdot 3}$
$= -\sqrt{4} \cdot \sqrt{3} + \sqrt{25} \cdot \sqrt{3}$
$= -2\sqrt{3} + 5\sqrt{3}$
$= (-2 + 5)\sqrt{3}$
$= 3\sqrt{3}$

45.

47. $355 \div 113 = 3.1415929$, which agrees with the first seven digits in the decimal for π.

49. $30 = \pi d = 10\pi$
$\pi = \dfrac{30}{10} = 3$

51. By Internet search, the answer is 4.

53. $\approx (1.67) \cdot 2 = 3.3$ to one decimal place.

55. $\phi = \dfrac{1 + \sqrt{5}}{2} \approx 1.618033989;$

$\dfrac{1 - \sqrt{5}}{2} \approx -0.618033989$

ϕ is positive while its conjugate is negative. The units digit of ϕ is 1 and the units digit of its conjugate is 0. The decimal digits agree.

57. It is just a coincidence that 1828 appears back-to-back early in the decimal. There is no repetition indefinitely, which would be indicative of an irrational number.

59. Using 3.14 for π and the given formula:
$P = 2\pi\sqrt{\dfrac{L}{32}}$
$P \approx 2(3.14)\sqrt{\dfrac{5.1}{32}}$
$P \approx 6.28\sqrt{0.159375}$
$P \approx 2.5$ seconds

61. Stated as a formula, the author's response indicates:
$d = \sqrt{h} \cdot 1.224 = \sqrt{156} \cdot 1.224 \approx 15.3$ miles

63. The semiperimeter, s, of the Bermuda Triangle is $\dfrac{1}{2}(850 + 925 + 1300)$ or 1537.5 miles.
$$\sqrt{\dfrac{1537.5(1537.5 - 850)(1537.5 - 925)}{(1537.5 - 1300)}}$$
$= \sqrt{1537.5(687.5)(612.5)(237.5)}$
$\approx 392{,}000$ square miles

65. The perimeter $P = 9 + 10 + 17 = 36$. The semiperimeter, $s = \dfrac{1}{2}(36) = 18$.
$A = \sqrt{18(18 - 9)(18 - 10)(18 - 17)}$
$= \sqrt{18(9)(8)(1)}$
$= \sqrt{1296}$
$= 36$
Since the perimeter and the area are both equal, the triangle is perfect.

67. $D = \sqrt{L^2 + W^2 + H^2}$
$= \sqrt{4^2 + 3^2 + 2^2}$
$= \sqrt{16 + 9 + 4}$
$= \sqrt{29} \approx 5.4$ feet

69. (a) $s = 30\sqrt{\dfrac{862}{156}} \approx 70.5$ mph

(b) $s = 30\sqrt{\dfrac{382}{96}} \approx 59.8$ mph

(c) $s = 30\sqrt{\dfrac{84}{26}} \approx 53.9$ mph

71. $\sqrt[3]{64} = 4$

73. $\sqrt[3]{343} = 7$

75. $\sqrt[3]{216} = 6$

77. $\sqrt[4]{1} = 1$

79. $\sqrt[4]{256} = 4$

81. $\sqrt[4]{4096} = 8$

83. $\sqrt[3]{43} \approx 3.50339806$

85. $\sqrt[3]{198} \approx 5.828476683$

87. $\sqrt[4]{10265.2} \approx 10.06565066$

89. $\sqrt[4]{968.1} \approx 5.578019845$

6.5 Exercises

1. True; $3.00(12) = 36$.

3. False; when 759.367 is rounded to the nearest hundredth, the result is 759.37.

5. True; $50\% = 0.5 = \dfrac{1}{2}$, and multiplying by one half yields the same result as dividing by 2.

7. True; $0.70(50) = 35$.

9. False; $.99\cent = \dfrac{99}{100}$ cent, meaning that it is less than the value of one penny.

11. $8.53 + 2.785 = 11.315$

13. $8.74 - 12.955 = -4.215$

15. $25.7 \times .032 = .8224$

17. $1019.825 \div 21.47 = 47.5$

19. $\dfrac{118.5}{1.45 + 2.3} = \dfrac{118.5}{3.75} = 31.6$

21. $21.0\% \cdot 2500$ billion $= 0.21 \cdot 2500$ billion
$= \$525$ billion

23. Because $3 \cdot (.33\cent) = .99\cent$, one could buy three stamps for $1\cent$ (and you would have $.01\cent$ left over).

25. $10(.10\cent) + 10(0.5\cent) = 1\cent + 5\cent = 6\cent$ or $\$0.06$

27. $\$1 \div .10\cent = 100\cent \div .10\cent = 1000$

29. (a) $\text{BAC} = \dfrac{[11.52]}{190} - (0.03) \approx 0.031$

   (b) $\text{BAC} = \dfrac{[10.8]}{135} - (0.045) = 0.035$

31. Substitute given values and follow order of operations.

$\text{Horsepower} = \dfrac{195 \times 302 \times 4000}{792,000}$

$\text{Horsepower} = \dfrac{235,560,000}{792,000}$

$\text{Horsepower} \approx 297$

33. (a) 78.4

   (b) 78.41

35. (a) 0.1

   (b) 0.08

37. (a) 12.7

   (b) 12.69

39. $0.42 = \dfrac{42}{100} = 42\%$

41. $0.365 = \dfrac{365}{1000} \div \dfrac{10}{10} = \dfrac{36.5}{100} = 36.5\%$

43. $0.008 = \dfrac{8}{1000} \div \dfrac{10}{10} = \dfrac{0.8}{100} = 0.8\%$

45. $2.1 = 2\dfrac{1}{10} \cdot \dfrac{10}{10} = 2\dfrac{10}{100} = \dfrac{210}{100} = 210\%$

47. $\dfrac{1}{5} = 1 \div 5 = 0.2$, which is 20%.

49. $\dfrac{1}{100} = 1 \div 100 = 0.01$, which is 1%.

51. $\dfrac{3}{8} = 3 \div 8 = 0.375$, which is 37.5%.

53. $\dfrac{3}{2} = 3 \div 2 = 1.5$, which is 150%.

55. Writing exercise; answers will vary.

57. (a) 5% means <u>5</u> in every 100.

   (b) 25% means 6 in every <u>24</u>.

   (c) 200% means <u>8</u> for every 4.

   (d) 0.5% means <u>0.5</u> in every 100.

(e) 600% means 12 for every 2.

59. No; if the item is discounted 20%, its new price is $60 − 0.2 × $60 = $48. Then, if the new price is marked up 20%, the price becomes $48 + 0.2 × $48 = $57.60.

61. (a) Boston: $\dfrac{95}{95+67} = \dfrac{95}{162} = 0.586$

(b) Tampa Bay: $\dfrac{84}{84+78} = \dfrac{84}{162} = 0.519$

(c) Toronto: $\dfrac{75}{75+87} = \dfrac{75}{162} = 0.463$

(d) Baltimore: $\dfrac{64}{64+98} = \dfrac{64}{162} = 0.395$

63. Using Method 1: $(0.26)(480) = 124.8.$

65. Using Method 1: $(0.105)(28) = 2.94.$

67. Using Method 2:
$$\frac{x}{100} = \frac{45}{30}$$
$$30x = 4500$$
$$x = \frac{4500}{30}$$
$$x = 150\%.$$

69. Using Method 2:
$$\frac{25}{100} = \frac{150}{x}$$
$$25x = 15000$$
$$x = \frac{15000}{25}$$
$$x = 600$$

71. Using Method 1:
$$(x)(28) = 0.392$$
$$x = \frac{0.392}{28}$$
$$x = 0.014$$
$$x = 1.4\%.$$

73. Find the increase and divide by the original wage.
$$11.34 - 10.50 = 0.84$$
$$\frac{0.84}{10.50} = 0.08 = 8\%$$
The hourly wage increased by 8%.

75. Find the difference and divide by the original population.
$$134,953 - 129,798 = 5155$$
$$\frac{5155}{134,953} = 0.038 = 3.8\%$$
The population decreased by 3.8%.

77. Find the difference and divide by the original price.
$$18.98 - 9.97 = 9.01$$
$$\frac{9.01}{18.98} = 0.475 = 47.5\%$$
The discount was 47.5%.

79. Choice A; the difference between $5 and $4 is $1; $1 compared to the original $4 is $\dfrac{1}{4}$ or 25%.

81. Choice C; rounding, the population of Alabama is approximately 4,000,000. About 25% of this number is
$$\frac{1}{4} \times 4,000,000 = 1,000,000.$$

83. Price increase was $3.25 - 1.35 = 1.90$. The percent increase is $\dfrac{1.90}{1.35} = 1.41 = 141\%$.
This value is approximate because we estimated the values from the graph.

85. (a) $14.7 - 40 \cdot 0.13$

(b) $14.7 - 40 \cdot 0.13 = 9.5$

(c) $0.85 \times 9.5 = 8.075$; walking (5 mph)

87. $\dfrac{6200-625}{625} \times 100 = 8.92 \times 100 = 892\%$

89. $29.57
1. Rounded to the nearest dollar, the amount of the bill is $30.
2. 10% of $30 is $3
3. $\dfrac{1}{2}$ of $3 is $1.50. $3 + 1.50 = $4.50.

91. $5.15
1. Rounded to the nearest dollar, the amount of the bill is $5.
2. 10% of $5 is $0.50.
3. $\dfrac{1}{2}$ of $0.50 is $0.25. $0.50 + 0.25 = $0.75.

93. $59.96 \approx 60$
   $10\% \text{ of } 60 = 6$
   $2 \times 6 = \$12.00$

95. $180.43 \approx 180$
   $10\% \text{ of } 180 = 18$
   $2 \times 18 = \$36.00$

97. Writing exercise; answers will vary.

***EXTENSION:* COMPLEX NUMBERS**

1. $\sqrt{-144} = i\sqrt{144} = 12i$

3. $-\sqrt{-225} = -i\sqrt{225} = -15i$

5. $\sqrt{-3} = i\sqrt{3}$

7. $\sqrt{-75} = i\sqrt{25 \cdot 3} = i\sqrt{25} \cdot \sqrt{3} = 5i\sqrt{3}$

9. $\sqrt{-5} \cdot \sqrt{-5} = i\sqrt{5} \cdot i\sqrt{5}$
   $= i^2\sqrt{5 \cdot 5}$
   $= i^2\sqrt{25}$
   $= 5i^2$
   $= 5(-1)$
   $= -5$

11. $\sqrt{-9} \cdot \sqrt{-36} = i\sqrt{9} \cdot i\sqrt{36}$
   $= 3i \cdot 6i$
   $= 18i^2$
   $= 18(-1)$
   $= -18$

13. $\sqrt{-16} \cdot \sqrt{-100} = i\sqrt{16} \cdot i\sqrt{100}$
   $= i^2 \cdot 4 \cdot 10$
   $= 40i^2$
   $= 40(-1)$
   $= -40$

15. $\dfrac{\sqrt{-200}}{\sqrt{-100}} = \dfrac{i\sqrt{200}}{i\sqrt{100}} = \sqrt{\dfrac{200}{100}} = \sqrt{2}$

17. $\dfrac{\sqrt{-54}}{\sqrt{6}} = \dfrac{i\sqrt{54}}{\sqrt{6}} = i\sqrt{\dfrac{54}{6}} = i\sqrt{9} = 3i$

19. $\dfrac{\sqrt{-288}}{\sqrt{-8}} = \dfrac{i\sqrt{288}}{i\sqrt{8}} = \sqrt{\dfrac{288}{8}} = \sqrt{36} = 6$

21. $i^8 = i^4 \cdot i^4 = 1 \cdot 1 = 1$

23. i^{42}
   $42 \div 4 = 10$, with a remainder of 2. This means that $i^{42} = i^2$, which is -1.

25. i^{47}
   $47 \div 4 = 11$, with a remainder of 3. This means that $i^{47} = i^3$, which is $-i$.

27. i^{101}
   $101 \div 4 = 25$, with a remainder of 1. This means that $i^{101} = i$.

Chapter 6 Test

1. $\left\{ -4, -\sqrt{5}, -\dfrac{3}{2}, -0.5, 0, \sqrt{3}, 4.1, 12 \right\}$

   (a) The only natural number is 12.

   (b) Whole numbers are 0 and 12.

   (c) Integers are $-4, 0,$ and 12.

   (d) Rational numbers are $-4, -\dfrac{3}{2}, -0.5, 0,$ 4.1, and 12.

   (e) Irrational numbers are $-\sqrt{5}$ and $\sqrt{3}$.

   (f) All the numbers in the set are real numbers.

2. (a) C

   (b) B

   (c) D

   (d) A

3. (a) False; the absolute value of a number is always nonnegative; the absolute value of zero is zero.

   (b) True; both sides of the equation are equal to 7.

   (c) True

   (d) False; zero, a real number, is neither positive nor negative.

4. $6^2 - 4(9-1) = 36 - 4(8) = 36 - 32 = 4$

5. $\dfrac{(-8+3)-(5+10)}{7-9}=\dfrac{-5-15}{-2}=\dfrac{-20}{-2}=10$

6. $(-3)(-2)-[5+(8-10)]=6-[5+(-2)]$
$=6-[3]$
$=3$

7. **(a)** Minnesota had the greatest change in foreclosures. The change was 52.74%.

   (b) Maine had the least change in foreclosures. The change was 10.63%.

   (c) False; $|-14.38|<|-14.76|$

   (d) True; $4(10.63)=42.52$, and $42.52<52.74$.

8. $(3852+225)-(-1299+80)=5296$ feet

9. **(a)** $221,900 - $219,000 = $2900

   (b) $217,900 - $221,900 = -$4000

   (c) $196,600 - $217,900 = -$21,300

   (d) $177,500 - $196,600 = -$19,100

10. **(a)** E, Associative property

    (b) A, Distributive property

    (c) B, Identity property

    (d) D, Commutative property

    (e) F, Inverse property

    (f) C, Closure property

11. **(a)** Whitney made 4 out of 6 and McElwain made 6 out of 7, which are more than $\dfrac{1}{2}$.

    (b) Moura made 13 out of 40; because 13 out of 39 would be 13, this is just less than $\dfrac{1}{3}$. Dawking made 2 out of 7; because 2 out of 6 would be $\dfrac{1}{3}$, this is just less than $\dfrac{1}{3}$.

    (c) Whitney made 4 out of 6, which is the same ratio as 2 out of 3.

(d) Pritchard made 4 out of 10; Miller made 8 out of 20.
$\dfrac{4}{10}=\dfrac{8}{20}=\dfrac{2}{5}$

(e) McElwain ("J-Mac"), with 6 completions out of 7 attempts, had the greatest fractional part of shots made.

12. $\dfrac{3}{16}+\dfrac{1}{2}=\dfrac{3}{16}+\dfrac{1}{2}\cdot\dfrac{8}{8}=\dfrac{3}{16}+\dfrac{8}{16}=\dfrac{11}{16}$

13. $\dfrac{9}{20}-\dfrac{3}{32}=\dfrac{9}{20}\cdot\dfrac{8}{8}-\dfrac{3}{32}\cdot\dfrac{5}{5}=\dfrac{72}{160}-\dfrac{15}{160}=\dfrac{57}{160}$

14. $\dfrac{3}{8}\cdot\left(-\dfrac{16}{15}\right)=-\dfrac{48}{120}=-\dfrac{2\cdot24}{5\cdot24}=-\dfrac{2}{5}$

15. $\dfrac{7}{9}\div\dfrac{14}{27}=\dfrac{7}{9}\cdot\dfrac{27}{14}=\dfrac{7\cdot3\cdot9}{7\cdot2\cdot9}=\dfrac{3}{2}$

16. **(a)** $\dfrac{9}{20}=0.45$

    (b) $\dfrac{5}{12}=0.41\overline{6}$

17. **(a)** $0.72=\dfrac{72}{100}=\dfrac{4\cdot18}{4\cdot25}=\dfrac{18}{25}$

    (b) $0.\overline{58}$
    Let $x=0.\overline{58}$.
    $100x=58.585858...$
    $\underline{-\ \ x=0.585858...}$
    $99x=58$
    $x=\dfrac{58}{99}$

18. **(a)** $\sqrt{10}$ is irrational because it cannot be written as the ratio of two integers.

    (b) $\sqrt{16}$ is rational. Its value is 4, which can be written as $\dfrac{4}{1}$.

    (c) $0.01=\dfrac{1}{100}$, a rational number.

    (d) $0.\overline{01}$ can be converted to $\dfrac{1}{99}$, a rational number.

(e) 0.0101101110... is an irrational number.

(f) π is irrational because it cannot be written as the ratio of two integers.

19. (a) $\sqrt{150} \approx 12.24744871$

(b) $\sqrt{150} = \sqrt{25} \cdot \sqrt{6} = 5\sqrt{6}$

20. (a) $\dfrac{13}{\sqrt{7}} \approx 4.913538149$

(b) $\dfrac{13}{\sqrt{7}} \cdot \dfrac{\sqrt{7}}{\sqrt{7}} = \dfrac{13\sqrt{7}}{\sqrt{49}} = \dfrac{13\sqrt{7}}{7}$

21. (a) $2\sqrt{32} - 5\sqrt{128} \approx -45.254834$

(b) $2\sqrt{32} - 5\sqrt{128} = 2\sqrt{16 \cdot 2} - 5\sqrt{64 \cdot 2}$
$= 2 \cdot 4\sqrt{2} - 5 \cdot 8\sqrt{2}$
$= 8\sqrt{2} - 40\sqrt{2}$
$= -32\sqrt{2}$

22. Writing exercise; answers may vary.

23. (a) $4.6 + 9.21 = 13.81$

(b) $12 - 3.725 - 8.59 = -.315$

(c) $86(0.45) = 38.7$

(d) $236.439 \div (-9.72) = -24.3$

24. (a) 9.04

(b) 9.045

25. (a) $0.185(90) = 16.65$

(b) $\dfrac{145}{100} = \dfrac{x}{70}$
$100x = 10150$
$x = \dfrac{10150}{100}$
$x = 101.5$

26. (a) 4 out of 15; $4 \div 15 = 0.26\dfrac{2}{3} = 26\dfrac{2}{3}\%$

(b) 10 out of 15; $10 \div 15 = 0.66\dfrac{2}{3} = 66\dfrac{2}{3}\%$

27. Choice D; since $300,000 is 100%, $600,000 is 200%, and $900,000 is 300% of the original value.

28. $0.69(2400) = 1656$
$0.42(2400) = 1008$
$x \cdot 2400 = 384$
$x = \dfrac{384}{2400}$
$x = 0.16 = 16\%$
$x \cdot 2400 = 144$
$x = \dfrac{144}{2400}$
$x = 0.06 = 6\%$

29. 11.7% of $148,847,000 = 0.117 \times 148847000$
$\approx 17,415,000$

30. The funding changed by the amount $-7.9 - 5.9 = -13.8$ (billion dollars).

Chapter 7

7.1 Exercises

1. Equations A and C are linear in x (or represent "first-degree" equations in x) because the highest power on the variable x is one.

3. $3(x+4) = 4x$
$3(12+4) = 4 \cdot 12$
$\qquad 3(16) = 48$
$\qquad\quad 48 = 48$
Both sides are evaluated as 48, so 12 is a solution.

5. If two equations are equivalent, they have the same <u>solution set</u>.

7. $0.06(10-x)(100) = (100)0.06(10-x)$
$\qquad\qquad\qquad\quad = 6(10-x)$
$\qquad\qquad\qquad\quad = 6 \cdot 10 - 6 \cdot x$
$\qquad\qquad\qquad\quad = 60 - 6x$
The left side of the equation is equivalent to choice B.

9. $\qquad 7x+8 = 1$
$\quad 7x+8-8 = 1-8$
$\qquad\quad 7x = -7$
$\qquad\quad \dfrac{7x}{7} = \dfrac{-7}{7}$
$\qquad\qquad x = -1; \; \{-1\}$

11. $\qquad 8-8x = -16$
$-8+8-8x = -8+(-16)$
$\qquad\quad -8x = -24$
$\qquad\quad \dfrac{-8x}{-8} = \dfrac{-24}{-8}$
$\qquad\qquad x = 3; \; \{3\}$

13. $7x-5x+15 = x+8$
$\quad 2x+15 = x+8$
$\quad 2x+15-15 = x+8-15$
$\qquad\quad 2x = x-7$
$\qquad 2x-x = x-x-7$
$\qquad\qquad x = -7; \; \{-7\}$

15. $12x+15x-9+5 = -3x+5-9$
$\qquad 27x-4 = -3x-4$
$\quad 27x+3x-4 = -3x+3x-4$
$\qquad 30x-4 = -4$
$\quad 30x-4+4 = -4+4$
$\qquad\qquad 30x = 0$
$\qquad\qquad \dfrac{30x}{30} = \dfrac{0}{30}$
$\qquad\qquad x = 0; \; \{0\}$

17. $\qquad 2(x+3) = -4(x+1)$
$\quad 2 \cdot x + 2 \cdot 3 = -4 \cdot x - 4 \cdot 1$
$\qquad 2x+6 = -4x-4$
$\quad 2x+6-6 = -4x-4-6$
$\qquad\quad 2x = -4x-10$
$\quad 2x+4x = -4x+4x-10$
$\qquad\quad 6x = -10$
$\qquad\quad \dfrac{6x}{6} = \dfrac{-10}{6}$
$\qquad\qquad x = \dfrac{-2 \cdot 5}{2 \cdot 3}$
$\qquad\qquad x = -\dfrac{5}{3}; \; \left\{ -\dfrac{5}{3} \right\}$

19. $\qquad 3(2x+1)-2(x-2) = 5$
$\quad 3 \cdot 2x + 3 \cdot 1 - 2 \cdot x + 2 \cdot 2 = 5$
$\qquad 6x+3-2x+4 = 5$
$\qquad 6x-2x+3+4 = 5$
$\qquad\qquad 4x+7 = 5$
$\qquad\quad 4x+7-7 = 5-7$
$\qquad\qquad 4x = -2$
$\qquad\qquad \dfrac{4x}{4} = \dfrac{-2}{4}$
$\qquad\qquad x = -\dfrac{2 \cdot 1}{2 \cdot 2}$
$\qquad\qquad x = -\dfrac{1}{2}; \; \left\{ -\dfrac{1}{2} \right\}$

21. $\qquad 2x+3(x-4) = 2(x-3)$
$\quad 2x+3 \cdot x - 3 \cdot 4 = 2 \cdot x - 2 \cdot 3$
$\qquad 2x+3x-12 = 2x-6$
$\qquad 5x-12 = 2x-6$
$\quad 5x-12+12 = 2x-6+12$
$\qquad\quad 5x = 2x+6$
$\qquad 5x-2x = 2x-2x+6$
$\qquad\qquad 3x = 6$
$\qquad\qquad \dfrac{3x}{3} = \dfrac{6}{3}$
$\qquad\qquad x = 2; \; \{2\}$

23.
$$6x - 4(3 - 2x) = 5(x - 4) - 10$$
$$6x - 4 \cdot 3 + 4 \cdot 2x = 5 \cdot x - 5 \cdot 4 - 10$$
$$6x - 12 + 8x = 5x - 20 - 10$$
$$6x + 8x - 12 = 5x - 30$$
$$14x - 12 = 5x - 30$$
$$14x - 12 + 12 = 5x - 30 + 12$$
$$14x = 5x - 18$$
$$14x - 5x = 5x - 5x - 18$$
$$9x = -18$$
$$\frac{9x}{9} = \frac{-18}{9}$$
$$x = -2; \{-2\}$$

25.
$$-[2x - (5x + 2)] = 2 + (2x + 7)$$
$$-[2x - 5x - 2] = 2 + 2x + 7$$
$$-[-3x - 2] = 2 + 7 + 2x$$
$$-[-3x - 2] = 9 + 2x$$
$$3x + 2 = 9 + 2x$$
$$3x + 2 - 2 = 9 - 2 + 2x$$
$$3x = 7 + 2x$$
$$3x - 2x = 7 + 2x - 2x$$
$$x = 7; \{7\}$$

27.
$$-3x + 6 - 5(x - 1) = -(2x - 4) - 5x + 5$$
$$-3x + 6 - 5x + 5 = -2x + 4 - 5x + 5$$
$$-8x + 11 = -7x + 9$$
$$-8x + 7x + 11 = -7x + 7x + 9$$
$$-1x + 11 = 9$$
$$-1x + 11 - 11 = 9 - 11$$
$$-x = -2$$
$$-1 \cdot (-x) = -1 \cdot (-2)$$
$$x = 2; \{-2\}$$

29.
$$-[3x - (2x + 5)] = -4 - [3(2x - 4) - 3x]$$
$$-[3x - 2x - 5] = -4 - [3 \cdot 2x - 3 \cdot 4 - 3x]$$
$$-[x - 5] = -4 - [6x - 12 - 3x]$$
$$-x + 5 = -4 - [6x - 3x - 12]$$
$$-x + 5 = -4 - [3x - 12]$$
$$-x + 5 = -4 - 3x + 12$$
$$-x + 5 = -4 + 12 - 3x$$
$$-x + 5 = 8 - 3x$$
$$-x + 5 - 5 = 8 - 5 - 3x$$
$$-x = 3 - 3x$$
$$-x + 3x = 3 - 3x + 3x$$
$$2x = 3$$
$$\frac{2x}{2} = \frac{3}{2}$$
$$x = \frac{3}{2}; \left\{\frac{3}{2}\right\}$$

31.
$$-(9 - 3x) - (4 + 2x) - 4 = -(2 - 5x) - x$$
$$-9 + 3x - 4 - 2x - 4 = -2 + 5x - x$$
$$3x - 2x - 9 - 4 - 4 = -2 + 5x - x$$
$$x - 17 = -2 + 4x$$
$$x - 17 + 17 = -2 + 17 + 4x$$
$$x = 15 + 4x$$
$$x - 4x = 15 + 4x - 4x$$
$$-3x = 15$$
$$\frac{-3x}{-3} = \frac{15}{-3}$$
$$x = -5; \{-5\}$$

33.
$$(2x - 6) - (3x - 4) = -(-4 + x) - 4x + 6$$
$$2x - 6 - 3x + 4 = 4 - x - 4x + 6$$
$$-x - 2 = -5x + 10$$
$$4x - 2 = 10$$
$$4x = 12$$
$$\frac{4x}{4} = \frac{12}{4}$$
$$x = 3; \{3\}$$

35. The smallest power of 10 is 10^2 or 100. This value will move the decimal point two places to the right in the coefficients 0.05 and 0.12.

37.
$$4 \cdot \left(\frac{3x}{4} + \frac{5x}{2}\right) = (13) \cdot 4$$
$$\frac{4}{1} \cdot \frac{3x}{4} + \frac{4}{1} \cdot \frac{5x}{2} = 52$$
$$3x + 2 \cdot 5x = 52$$
$$3x + 10x = 52$$
$$13x = 52$$
$$\frac{13x}{13} = \frac{52}{13}$$
$$x = 4; \{4\}$$

39.
$$15 \cdot \left(\frac{x - 8}{5} + \frac{8}{5}\right) = -\frac{x}{3} \cdot 15$$
$$\frac{15}{1} \cdot \frac{x - 8}{5} + \frac{15}{1} \cdot \frac{8}{5} = -\frac{x}{3} \cdot \frac{15}{1}$$
$$3 \cdot (x - 8) + 3 \cdot 8 = -5 \cdot x$$
$$3x - 24 + 24 = -5x$$
$$3x = -5x$$
$$3x + 5x = -5x + 5x$$
$$8x = 0$$
$$\frac{8x}{8} = \frac{0}{8}$$
$$x = 0; \{0\}$$

41.
$$6 \cdot \frac{4x+1}{3} = \left(\frac{x+5}{6} + \frac{x-3}{6}\right) \cdot 6$$
$$\frac{6}{1} \cdot \frac{4x+1}{3} = \frac{6}{1} \cdot \frac{x+5}{6} + \frac{6}{1} \cdot \frac{x-3}{6}$$
$$2 \cdot (4x+1) = (x+5) + (x-3)$$
$$8x+2 = 2x+2$$
$$8x+2-2 = 2x+2-2$$
$$8x = 2x$$
$$8x-2x = 2x-2x$$
$$6x = 0$$
$$\frac{6x}{6} = \frac{0}{6}$$
$$x = 0; \{0\}$$

43.
$$100 \cdot [0.05x + 0.12(x+5000)] = 940 \cdot 100$$
$$100 \cdot 0.05x + 100 \cdot 0.12 \cdot (x+5000) = 94{,}000$$
$$5x + 12 \cdot (x+5000) = 94{,}000$$
$$5x + 12x + 60{,}000 = 94{,}000$$
$$17x + 60{,}000 = 94{,}000$$
$$17x = 34{,}000$$
$$\frac{17x}{17} = \frac{34{,}000}{17}$$
$$x = 2000;$$
$$\{2000\}$$

45.
$$100 \cdot [0.02(50) + 0.08x] = 0.04(50+x) \cdot 100$$
$$100 \cdot 0.02(50) + 100 \cdot 0.08x = 100 \cdot 0.04(50+x)$$
$$2(50) + 8x = 4(50+x)$$
$$100 + 8x = 4 \cdot 50 + 4 \cdot x$$
$$100 + 8x = 200 + 4x$$
$$100 + 8x - 4x = 200 + 4x - 4x$$
$$100 + 4x = 200$$
$$100 - 100 + 4x = 200 - 100$$
$$4x = 100$$
$$\frac{4x}{4} = \frac{100}{4}$$
$$x = 25; \{25\}$$

47.
$$100 \cdot [0.05x + 0.10(200-x)] = 0.45x \cdot 100$$
$$100 \cdot 0.05x + 100 \cdot 0.10(200-x) = 45x$$
$$5x + 10(200-x) = 45x$$
$$5x + 2000 - 10x = 45x$$
$$2000 - 5x = 45x$$
$$2000 - 5x + 5x = 45x + 5x$$
$$2000 = 50x$$
$$\frac{2000}{50} = \frac{50x}{50}$$
$$40 = x; \{40\}$$

49. The equation $x + 2 = x + 2$ is called an <u>identity</u>, because its solution set is {all real numbers}. The equation $x + 1 = x + 2$ is called a <u>contradiction</u>, because it has no solutions.

51.
$$-2x + 5x - 9 = 3(x-4) - 5$$
$$3x - 9 = 3x - 12 - 5$$
$$3x - 9 = 3x - 17$$
$$3x - 9 + 9 = 3x - 17 + 9$$
$$3x = 3x - 8$$
$$3x - 3x = 3x - 3x - 8$$
$$0 = -8$$
This is a contradiction; the solution is the empty set, which can be symbolized by $\varnothing$.

53.
$$6x + 2(x-2) = 9x + 4$$
$$6x + 2x - 4 = 9x + 4$$
$$8x - 4 = 9x + 4$$
$$8x - 4 + 4 = 9x + 4 + 4$$
$$8x = 9x + 8$$
$$8x - 9x = 9x - 9x + 8$$
$$-x = 8$$
$$\frac{-x}{-1} = \frac{8}{-1}$$
$$x = -8; \{-8\}$$
This is a conditional equation.

55.
$$-11x + 4(x-3) + 6x = 4x - 12$$
$$-11x + 4x - 12 + 6x = 4x - 12$$
$$-11x + 4x + 6x - 12 = 4x - 12$$
$$-x - 12 = 4x - 12$$
$$-x - 12 + 12 = 4x - 12 + 12$$
$$-x = 4x$$
$$-x - 4x = 4x - 4x$$
$$-5x = 0$$
$$\frac{-5x}{-5} = \frac{0}{-5}$$
$$x = 0; \{0\}$$
This is a conditional equation.

57.
$$7[2 - (3+4x)] - 2x = -9 + 2(1-15x)$$
$$7[2 - 3 - 4x] - 2x = -9 + 2 - 30x$$
$$7[-1 - 4x] - 2x = -7 - 30x$$
$$-7 - 28x - 2x = -7 - 30x$$
$$-7 - 30x = -7 - 30x$$
This equation is an identity. The solution set is {all real numbers}.

59. The following algebraic manipulations show that letter A, B, and C are all equivalent.

A. $h = 2\left(\dfrac{A}{b}\right)$

$h = \dfrac{2}{1} \cdot \dfrac{A}{b}$

$h = \dfrac{2A}{b}$

B. $h = 2A\left(\dfrac{1}{b}\right)$

$h = \dfrac{2A}{1} \cdot \dfrac{1}{b}$

$h = \dfrac{2A}{b}$

C. $h = \dfrac{A}{\frac{1}{2}b}$

$h = \dfrac{A}{1} \div \dfrac{1}{2}b$

$h = \dfrac{A}{1} \div \dfrac{b}{2}$

$h = \dfrac{A}{1} \cdot \dfrac{2}{b}$

$h = \dfrac{2A}{b}$

D. Here is the simplification of this equation:

$h = \dfrac{\frac{1}{2}A}{b}$

$h = \dfrac{1}{2}A \div b$

$h = \dfrac{A}{2} \div \dfrac{b}{1}$

$h = \dfrac{A}{2} \cdot \dfrac{1}{b}$

$h = \dfrac{A}{2b}$.

This equation is not equivalent to the given equation.

61. $d = rt$

$\dfrac{d}{r} = \dfrac{rt}{r}$

$\dfrac{d}{r} = t$

63. $A = bh$

$\dfrac{A}{h} = \dfrac{bh}{h}$

$\dfrac{A}{h} = b$

65. $P = a + b + c$

$P - b = a + b - b + c$

$P - b - c = a + c - c$

$P - b - c = a$

67. $A = \dfrac{1}{2}bh$

$2 \cdot A = 2 \cdot \dfrac{1}{2}bh$

$2A = bh$

$\dfrac{2A}{h} = \dfrac{bh}{h}$

$\dfrac{2A}{h} = b$

69. $S = 2\pi rh + 2\pi r^2$

$S - 2\pi r^2 = 2\pi rh + 2\pi r^2 - 2\pi r^2$

$S - 2\pi r^2 = 2\pi rh$

$\dfrac{S - 2\pi r^2}{2\pi r} = \dfrac{2\pi rh}{2\pi r}$

$\dfrac{S - 2\pi r^2}{2\pi r} = h$

The last equation can be simplified further:

$h = \dfrac{S - 2\pi r^2}{2\pi r} = \dfrac{S}{2\pi r} - \dfrac{2\pi r^2}{2\pi r} = \dfrac{S}{2\pi r} - r.$

71. $C = \dfrac{5}{9}(F - 32)$

$\dfrac{9}{5} \cdot C = \dfrac{9}{5} \cdot \dfrac{5}{9}(F - 32)$

$\dfrac{9}{5}C = (F - 32)$

$\dfrac{9}{5}C + 32 = F - 32 + 32$

$\dfrac{9}{5}C + 32 = F$

73. $V = \dfrac{1}{3}\pi r^2 h$

$3 \cdot V = 3 \cdot \dfrac{1}{3}\pi r^2 h$

$3V = \pi r^2 h$

$\dfrac{3V}{\pi r^2} = \dfrac{\pi r^2 h}{\pi r^2}$

$\dfrac{3V}{\pi r^2} = h$

75. (a) $y = 226.9x - 449,700$
$$y = 226.9(2006) - 449,700$$
$$= 455,161.4 - 449,700$$
$$= 5461.4$$
$$\approx \$5460$$

(b) $y = 226.9x - 449,700$
$$7150 = 226.9x - 449,700$$
$$7150 + 449,700 = 226.9x$$
$$456,850 = 226.9x$$
$$\frac{456,850}{226.9} = \frac{226.9x}{226.9}$$
$$2013 \approx x$$

77. (a) $k = \dfrac{0.132B}{W}$
$$k = \frac{0.132(20)}{75} = 0.0352$$

(b) $R = kd$
$$R = 0.0352(0.42) \approx 0.015 = 1.5\%$$

(c) 1.5% which translates to 1.5 per hundred during an average 72-years lifetime. Thus, one would expect $1.5\% \times 5000 \div 72 \approx 1$ death each year.

7.2 Exercises

1. Expression

3. Equation: $\dfrac{2}{3} \cdot x = 36$

5. Expression

7. Let $x =$ the number of quarters; then $25 - x =$ the number of nickels.

Number of coins	Denomination	Value
x	$0.25	$0.25x$
$25 - x$	$0.05	$0.05(25 - x)$

$$0.25x + 0.05(25 - x) = 5.65$$
$$25x + 125 - 5x = 565$$
$$20x = 440$$
$$\frac{20x}{20} = \frac{440}{20}$$
$$x = 22$$

The number of quarters is 22; the number of nickels is $25 - 22 = 3$. The answer has not changed.

9. $x - 12$

11. $(x - 6)(x + 4)$

13. $\dfrac{25}{x}$, $(x \neq 0)$

15. Writing exercise; answers will vary.

17. $5x + 2 = 4x + 5$
$$5x - 4x + 2 - 2 = 4x - 4x + 5 - 2$$
$$x = 3$$

19. $3(x - 2) = x + 6$
$$3x - 6 + 6 = x + 6 + 6$$
$$3x = x + 12$$
$$3x - x = x - x + 12$$
$$\frac{2x}{2} = \frac{12}{2}$$
$$x = 6$$

21. $3x + (x + 7) = -11 - 2x$
$$4x + 7 - 7 = -11 - 2x - 7$$
$$4x = -18 - 2x$$
$$4x + 2x = -18 - 2x + 2x$$
$$\frac{6x}{6} = \frac{-18}{6}$$
$$x = -3$$

23. Step 1: Let $x =$ the revenue (in millions of dollars) of Bon Jovi.
Step 2: Then $x - 6.1 =$ the revenue (in millions of dollars) of Bruce Springsteen.
Step 3: $x + (x - 6.1) = 415.3$
Step 4: $x + x - 6.1 = 415.3$
$$2x - 6.1 = 415.3$$
$$2x = 415.3 + 6.1$$
$$2x = 421.4$$
$$\frac{2x}{2} = \frac{421.4}{2}$$
$$x = 210.7$$
Step 5: Thus Bon Jovi's tour generated $210.7 million. Then Bruce Springsteen's tour generated $210.7 - \$6.1 = \$204.6$ million.
Step 6: Since the sum of the tours was $415.3 million, we can check by adding: $210.7 + \$204.6 = \$415.3$.

25. Let x = the number of games lost.
Let $3x + 2$ = the number of games won.
Since losses plus wins must equal total number of games, we have
$$x + (3x + 2) = 82$$
$$4x + 2 = 82$$
$$4x = 80$$
$$x = 20.$$
Thus, the Celtics had 20 losses and
$3x + 2 = 3(20) + 2 = 62$ wins.

27. Let x = the number of Democrats in the Senate. Since there were a total of 98 Democrats and Republicans, we have:
$$x + (x - 18) = 98$$
$$2x - 18 = 98$$
$$2x = 116$$
$$x = 58$$
There were 58 Democrats and
$58 - 18 = 40$ Republicans.

29. Let x = the length, in inches, of the shortest piece. Then the middle piece = $x + 5$ inches long and the longest piece = $x + 9$ inches long.
$$x + (x + 5) + (x + 9) = 59$$
$$3x + 14 = 59$$
$$3x = 45$$
$$x = 15$$
$x + 5 = 20$
$x + 9 = 24$
The shortest piece is 15 inches, the middle length piece is 20 inches and the longest piece is 24 inches.

31. Let x = the number of silver medals. Then $x + 7$ = the number of bronze medals, and $(x + 7) + 23 = x + 30$ = the number of gold medals. Since the total number of medals was 100, we have
$$x + (x + 7) + (x + 30) = 100$$
$$3x + 37 = 100$$
$$3x = 63$$
$$x = 21$$
Then China won 21 silver medals,
$21 + 7 = 28$ bronze medals, and
$21 + 30 = 51$ gold medals.

33. 14% of 500
$0.14(500) = 70$ ml

35. 2.5% of 10,000
$0.025(10,000) = \$250$

37. $497(0.05) = \$24.85$

39. Let x = the number of liters of 20% solution.

Strength	L of solution	L of alcohol
12%	12	0.12(12)
20%	x	0.20(x)
14%	$12 + x$	0.14($12 + x$)

Create an equation by adding the first two algebraic expressions in the last column to total the third:
$$0.12(12) + 0.20(x) = 0.14(12 + x)$$
$$100[0.12(12) + 0.20(x)] = 100[0.14(12 + x)]$$
$$12(12) + 20x = 14(12 + x)$$
$$144 - 144 + 20x = 168 - 144 + 14x$$
$$20x - 14x = 24 + 14x - 14x$$
$$6x = 24$$
$$\frac{6x}{6} = \frac{24}{6}$$
$$x = 4 \text{ liters of } 20\%$$
$$\text{solution}$$

41. Let x = number of liters of 20% solution to be replaced with 100% solution. The amount of 20% solution that is replaced must be subtracted while the amount of 100% solution must be added.

Strength	L of solution	L of chemical
20%	20	0.20(20)
20%	x	$-0.20(x)$
100%	x	1.00(x)
40%	20	0.40(20)

Create an equation by adding the first three algebraic expressions in the last column to total the last. To clear the decimal in this equation, multiply by 10 to move the decimal point only one place to the right:
$$0.20(20) - 0.20(x) + 1.00(x) = 0.40(20)$$
$$10[0.20(20) - 0.20(x) + 1.00(x)] = 10[0.40(20)]$$
$$2(20) - 2x + 10x = 4(20)$$
$$40 - 2x + 10x = 80$$
$$40 + 8x = 80$$
$$40 - 40 + 8x = 80 - 40$$
$$8x = 40$$
$$\frac{8x}{8} = \frac{40}{8}$$
$$x = 5 \text{ liters}$$

43. Let x = the number of gallons of water to be added.

Strength	Gallons of solution	Gallons of insecticide
0%	x	$0.00(x)$
4%	3	$0.04(3)$
3%	$x + 3$	$0.03(x + 3)$

Create an equation by adding the first two algebraic expressions in the last column to total the third.

$$0.00(x) + 0.04(3) = 0.03(x + 3)$$
$$100[0.00(x) + 0.04(3)] = 100[0.03(x + 3)]$$
$$0 + 4(3) = 3(x + 3)$$
$$12 = 3x + 9$$
$$12 - 9 = 3x + 9 - 9$$
$$3 = 3x$$
$$\frac{3}{3} = \frac{3x}{3}$$
$$1 = x$$

One gallon of water must be added.

45. Let x = the amount invested at 3%

% as decimal	Amount Invested	Interest in one year
0.03	x	$0.03x$
0.04	$12{,}000 - x$	$0.04(12{,}000 - x)$
Totals	12,000	440

Create an equation by adding the first two algebraic expressions in the last column to total the third:

$$0.03x + 0.04(12{,}000 - x) = 440$$
$$100[0.03x + 0.04(12{,}000 - x)] = 100(440)$$
$$3x + 4(12{,}000 - x) = 44{,}000$$
$$3x + 48{,}000 - 4x = 44{,}000$$
$$48{,}000 - x = 44{,}000$$
$$48{,}000 - 48{,}000 - x = 44{,}000$$
$$-48{,}000$$
$$-x = -4000$$
$$\frac{-x}{-1} = \frac{-4000}{-1}$$
$$x = 4000$$

The amount invested at 3% is $4000; the amount invested at 4% is:
$12{,}000 - 4000 = \$8000$.

47.

% as decimal	Amount Invested	Interest in one year
0.045	x	$0.045x$
0.03	$2x - 1000$	$0.03(2x - 1000)$
Totals		1020

Create an equation by adding the first two algebraic expressions in the last column to total the third. To clear the decimals, both sides of the equation must be multiplied by 1000 to move the decimal point 3 places to the right.

$$0.045x + 0.03(2x - 1000) = 1020$$
$$1000[0.045x + 0.03(2x - 1000)] = 1000(1020)$$
$$45x + 30(2x - 1000) = 1{,}020{,}000$$
$$45x + 60x - 30{,}000 = 1{,}020{,}000$$
$$105x - 30{,}000 = 1{,}020{,}000$$
$$105x - 30{,}000 + 30{,}000 = 1{,}020{,}000$$
$$+ 30{,}000$$
$$105x = 1{,}050{,}000$$
$$\frac{105x}{105} = \frac{1{,}050{,}000}{105}$$
$$x = 10{,}000$$

The amount invested at 4.5% is $10,000; the amount invested at 3% is:
$2 \cdot 10{,}000 - 1000 = \$19{,}000$.

49.

% as decimal	Amount Invested	Interest in one year
0.05	29,000	$0.05(29{,}000)$
0.02	x	$0.02x$
0.03	$29{,}000 + x$	$0.03(29{,}000 + x)$

Create an equation by adding the first two algebraic expressions in the last column to total the third:

$$0.05(29{,}000) + 0.02x = 0.03(29{,}000 + x)$$
$$100[0.05(29{,}000) + 0.02x] = 100[0.03(29{,}000 + x)]$$
$$5(29{,}000) + 2x = 3(29{,}000 + x)$$
$$145{,}000 + 2x = 87{,}000 + 3x$$
$$145{,}000 + 2x - 2x = 87{,}000 + 3x - 2x$$
$$145{,}000 - 87{,}000 = 87{,}000 - 87{,}000 + x$$
$$58{,}000 = x$$

He should invest $58,000 at 2% in order to have an average return of 3%.

51. Complete the table in the text.

Denomination	Number of Coins	Value
0.01	x	$0.01x$
0.10	x	$0.10x$
0.25	$44 - 2x$	$0.25(44 - 2x)$
Totals	44	4.37

Because there are 44 coins altogether, the number of quarters is represented by subtracting the total number of pennies and dimes from 44. Now create an equation by adding the algebraic expressions in the last column; the total value of all the coins is $4.37.

$$0.01x + 0.10x + 0.25(44 - 2x) = 4.37$$
$$100[0.01x + 0.10x + 0.25(44 - 2x)] = 100(4.37)$$
$$x + 10x + 25(44 - 2x) = 437$$
$$x + 10x + 1100 - 50x = 437$$
$$x + 10x - 50x + 1100 = 437$$
$$-39x + 1100 - 1100 = 437 - 1100$$
$$-39x = -663$$
$$\frac{-39x}{-39} = \frac{-663}{-39}$$
$$x = 17$$

Mike has 17 pennies, 17 dimes, and $44 - 2 \cdot 17 = 10$ quarters.

53.

Number of tickets	Value/ticket	Total Value
x	3	$3x$
$410 - x$	7	$7(410 - x)$
410	Totals	1650

Now create an equation by adding the algebraic expressions in the last column; the total value of all the tickets is $1650.

$$3x + 7(410 - x) = 1650$$
$$3x + 2870 - 7x = 1650$$
$$2870 - 2870 - 4x = 1650 - 2870$$
$$-4x = -1220$$
$$\frac{-4x}{-4} = \frac{-1220}{-4}$$
$$x = 305$$

There were 305 students who attended and $410 - 305 = 105$ non students who attended.

55.

Number of tickets	Value/ticket	Total Value
x	35	$35x$
$105 - x$	30	$30(105 - x)$
105	Totals	3420

Now create an equation by adding the algebraic expressions in the last column; the total value of all tickets is $3420.

$$35x + 30(105 - x) = 3420$$
$$35x + 3150 - 30x = 3420$$
$$5x + 3150 = 3420$$
$$5x + 3150 - 3150 = 3420 - 3150$$
$$5x = 270$$
$$\frac{5x}{5} = \frac{270}{5}$$
$$x = 54$$

There were 54 seats sold in Row 1 and $105 - 54 = 51$ seats sold in Row 2.

57.

		Value
Number of 44-cent stamps	x	$0.44x$
Number of 17-cent stamps	$55 - x$	$0.17(55 - x)$
Totals	55	17.45

Now create an equation by adding the algebraic expressions in the last column; the total value of the stamps is $17.45.

$$0.44x + 0.17(55 - x) = 17.45$$
$$0.44x + 9.35 - 0.17x = 17.45$$
$$0.27x + 9.35 = 17.45$$
$$0.27x = 8.1$$
$$x = 30$$

Sabrina purchased 30 44-cent stamps and $55 - 30 = 25$ 17-cent stamps.

59. Use the formula $rt = d$.
$164 \cdot 2 = 328$ miles

61. No, it is not correct. The distance is
$$55\left(\frac{1}{2}\right) = 27.5 \text{ miles.}$$

63.

	Rate	Time	Distance
First train	85	t	$85t$
Second train	95	t	$95t$

Because the trains are traveling in opposite directions, the sum of their distances will equal the total distance apart of 315 kilometers. Use this information to create an equation.

$$85t + 95t = 315$$
$$180t = 315$$
$$\frac{180t}{180} = \frac{315}{180}$$
$$t = 1\frac{3}{4} \text{ hours}$$

65.

	Rate	Time	Distance
Nancy	35	t	$35t$
Mark	40	$t - \frac{1}{4}$	$40\left(t - \frac{1}{4}\right)$

Because Nancy and Mark are traveling in opposite directions, the sum of their distances will equal the total distance apart of 140 miles. Use this information to create an equation.

$$35t + 40\left(t - \frac{1}{4}\right) = 140$$
$$35t + 40t - 10 = 140$$
$$75t - 10 = 140$$
$$75t - 10 + 10 = 140 + 10$$
$$75t = 150$$
$$\frac{75t}{75} = \frac{150}{75}$$
$$t = 2 \text{ hours}$$

The question asks at what time will they be 140 miles apart. The value of t tells us that in two hours they will be 140 miles apart. Because Nancy left the house at 9:00, the time would be 11:00 A.M.

67.

	Rate	Time	Distance
First part	10	t	$10t$
Second part	15	t	$15t$

The sum of the distances is 100.

$$10t + 15t = 100$$
$$25t = 100$$
$$\frac{25t}{25} = \frac{100}{25}$$
$$t = 4 \text{ hours}$$

The time for each part of the trip is 4 hours, so the total time for the trip is 8 hours.

69.

	Rate	Time	Distance
Car	r	$\frac{1}{2}$	$\frac{1}{2}r$
Bus	$r - 12$	$\frac{3}{4}$	$\frac{3}{4}(r-12)$

The distance to and from work is the same, so set the distances equal to each other.

$$\frac{1}{2}r = \frac{3}{4}(r-12)$$
$$\frac{4}{1} \cdot \frac{1}{2}r = \frac{4}{1} \cdot \frac{3}{4}(r-12)$$
$$2r = 3(r-12)$$
$$2r = 3r - 36$$
$$2r - 3r = 3r - 3r - 36$$
$$-r = -36$$
$$\frac{-r}{-1} = \frac{-36}{-1}$$
$$r = 36 \text{ miles per hour}$$

Glen's rate of speed when driving is 36 mph. The distance to work is

$$\frac{1}{2} \cdot 36 = 18 \text{ miles.}$$

71. Solve the formula $rt = d$ for t

$$rt = d$$
$$\frac{rt}{r} = \frac{d}{r}$$
$$t = \frac{d}{r}$$

Now substitute the given values for d and r.

$$t = \frac{500}{150.318} \approx 3.326 \text{ hours}$$

73. Substitute the given values for d and r.

$$t = \frac{255}{148.725} \approx 1.715 \text{ hours}$$

75. Solve the formula $rt = d$ for r.

$$rt = d$$
$$\frac{rt}{t} = \frac{d}{t}$$
$$r = \frac{d}{t}$$

Now substitute the given values for d and t.

$$r = \frac{100}{12.54} \approx 7.97 \text{ meters per second}$$

77. Substitute the given values for d and t.

$$r = \frac{400}{47.25} \approx 8.47 \text{ meters per second}$$

7.3 Exercises

1. $\dfrac{50}{80} = \dfrac{5 \cdot 10}{8 \cdot 10} = \dfrac{5}{8}$

3. $\dfrac{17}{68} = \dfrac{1 \cdot 17}{4 \cdot 17} = \dfrac{1}{4}$

5. The units of measure must be the same in order to factor the numerator and denominator to simplify the fraction. This simplification can be done by dimensional analysis:

$$\frac{288 \text{ inches}}{12 \text{ feet}} \cdot \frac{1 \text{ foot}}{12 \text{ inches}} = \frac{2 \cdot 12 \cdot 12}{12 \cdot 12} = \frac{2}{1}.$$

7. $\dfrac{5 \text{ days}}{40 \text{ hours}} \cdot \dfrac{24 \text{ hours}}{1 \text{ day}} = \dfrac{5 \cdot 8 \cdot 3}{40} = \dfrac{3}{1}$

9. A. $0.4 = \dfrac{4}{10} = \dfrac{2}{5}$

B. 4 to 10 means $\dfrac{4}{10}$, which is equivalent to $\dfrac{2}{5}$.

C. 20 to 50 means $\dfrac{20}{50}$, which simplifies to $\dfrac{2}{5}$.

D. 5 to 2 means $\dfrac{5}{2}$, which is not the same ratio as 2 to 5.

11. Writing exercise; answers will vary.

In exercises 13 through 17, check to see if the cross products are equal to determine if the proportions are true or false.

13. $5 \cdot 56 = 280; \ 35 \cdot 8 = 280$
The proportion is true.

15. $120 \cdot 10 = 1200; \ 82 \cdot 7 = 574$
The proportion is false.

17. $\dfrac{1}{2} \cdot 10 = 5; \ 5 \cdot 1 = 5$
The proportion is true.

19. $\dfrac{x}{4} = \dfrac{175}{20}$
$20x = 4 \cdot 175$
$20x = 700$
$\dfrac{20x}{20} = \dfrac{700}{20}$
$x = 35; \ \{35\}$

21. $\dfrac{3x-2}{5} = \dfrac{6x-5}{11}$
$5(6x-5) = 11(3x-2)$
$30x - 25 = 33x - 22$
$30x - 25 + 25 = 33x - 22 + 25$
$30x - 33x = 33x - 33x + 3$
$-3x = 3$
$\dfrac{-3x}{-3} = \dfrac{3}{-3}$
$x = -1; \ \{-1\}$

23. $\dfrac{3x+1}{7} = \dfrac{2x-3}{6}$
$6(3x+1) = 7(2x-3)$
$18x + 6 = 14x - 21$
$18x + 6 - 6 = 14x - 21 - 6$
$18x = 14x - 27$
$18x - 14x = 14x - 27 - 14x$
$4x = -27$
$x = \dfrac{-27}{4}; \ \left\{-\dfrac{27}{4}\right\}$

25. Let x = cost of 24 candy bars.
$\dfrac{\$20.00}{16 \text{ bars}} = \dfrac{x}{24 \text{ bars}}$
$16 \cdot x = (20)(24)$
$\dfrac{16x}{16} = \dfrac{480}{16}$
$x = \$30.00$

27. Let x = cost of 5 quarts of oil.
$\dfrac{5 \text{ quarts}}{x} = \dfrac{8 \text{ quarts}}{\$14.00}$
$8 \cdot x = (5)(14)$
$\dfrac{8x}{8} = \dfrac{70}{8}$
$x = \$8.75$

Note 29 below represents one way (using cross products) to solve the proportion.

29. Let x = cost 5 pairs of jeans.

$$\frac{x}{5 \text{ pairs}} = \frac{\$121.50}{9 \text{ pairs}}$$
$$x \cdot 9 = (5)(121.5)$$
$$\frac{9x}{9} = \frac{607.5}{9}$$
$$x = \$67.50$$

Note 31 below represents an alternate approach (to Exercise 29) to solve the proportion.

31. Let x = cost to fill a 15 gallon tank.

$$\frac{x}{15 \text{ gallons}} = \frac{\$17.82}{6 \text{ gallons}}$$
$$15 \cdot \frac{x}{15} = 15 \cdot \frac{17.82}{6}$$
$$x = \frac{267.3}{6}$$
$$x = \$44.55$$

33. Let x = distance on the map between Memphis and Philadelphia.

$$\frac{x}{1000 \text{ mi}} = \frac{2.4 \text{ ft}}{600 \text{ mi}}$$
$$1000 \cdot \frac{x}{1000} = 1000 \cdot \frac{2.4}{600}$$
$$x = \frac{2400}{600}$$
$$x = 4 \text{ feet}$$

35. Let x = distance on the map between St. Louis and Des Moines.

$$\frac{x}{333 \text{ mi}} = \frac{8.5 \text{ in.}}{1040 \text{ mi}}$$
$$333 \cdot \frac{x}{333} = 333 \cdot \frac{8.5}{1040}$$
$$x = \frac{2830.5}{1040}$$
$$x \approx 2.7 \text{ inches}$$

37. Let x = distance apart, on the globe, between Moscow and Berlin.

$$\frac{10080 \text{ km}}{12.4 \text{ in.}} = \frac{1610 \text{ km}}{x}$$
$$10080 \cdot x = (12.4)(1610)$$
$$\frac{10080x}{10080} = \frac{19964}{10080}$$
$$x \approx 2 \text{ inches}$$

39. $\dfrac{1}{4}\dfrac{\text{cup}}{\text{gallon}} \cdot 10\dfrac{1}{2} \text{ gallons} = \dfrac{1}{4} \cdot \dfrac{21}{2}$

$$= \frac{21}{8}$$
$$= 2\frac{5}{8} \text{ cups}$$

41. $\dfrac{1 \text{ euro}}{\$1.4294} = \dfrac{300 \text{ euros}}{x}$

$$1 \cdot x = (1.4294)(300)$$
$$x = \$428.82$$

43. Let x = number of fish in lake.

$$\frac{250 \text{ tagged}}{x} = \frac{7 \text{ tagged}}{350 \text{ fish}}$$
$$7 \cdot x = (250)(350)$$
$$7x = 87,500$$
$$\frac{7x}{7} = \frac{87,500}{7}$$
$$x = 12,500 \text{ fish}$$

Create the unit pricing for each commodity (45–51).

45. $\dfrac{\$1.79}{4\text{-lb}} = \$0.4475/\text{lb}$

$\dfrac{\$4.29}{10\text{-lb}} = \$0.429/\text{lb}$

Thus, the best buy is the 10-lb size at $0.429/lb.

47. $\dfrac{\$2.44}{16\text{-oz}} \approx \$0.153/\text{oz}$

$\dfrac{\$2.98}{32\text{-oz}} \approx \$0.093/\text{oz}$

$\dfrac{\$4.95}{48\text{-oz}} \approx \$0.103/\text{oz}$

Thus, the best buy is the 32-oz size at $0.093/oz.

49. $\dfrac{\$1.66}{16\text{-oz}} \approx \$0.104/\text{oz}$

$\dfrac{\$2.59}{32\text{-oz}} \approx \$0.081/\text{oz}$

$\dfrac{\$4.29}{64\text{-oz}} \approx \$0.067/\text{oz}$

$\dfrac{\$6.49}{128\text{-oz}} \approx \$0.051/\text{oz}$

Thus, the best buy is 128-oz @ $0.051/oz.

51. $\dfrac{\$1.39}{14\text{-oz}} \approx \$0.099/\text{oz}$

$\dfrac{\$1.55}{24\text{-oz}} \approx \$0.065/\text{oz}$

$\dfrac{\$1.78}{36\text{-oz}} \approx \$0.049/\text{oz}$

$\dfrac{\$3.99}{64\text{-oz}} \approx \$0.062/\text{oz}$

Thus, the best buy is the 36-oz size at $0.049/oz.

53. $\dfrac{12}{x} = \dfrac{9}{3}$

$9 \cdot x = 12 \cdot 3$

$9x = 36$

$\dfrac{9x}{9} = \dfrac{36}{9}$

$x = 4$

55. $\dfrac{3}{x} = \dfrac{6}{2}$

$6 \cdot x = 3 \cdot 2$

$6x = 6$

$\dfrac{6x}{6} = \dfrac{6}{6}$

$x = 1$

$\dfrac{y}{\frac{4}{3}} = \dfrac{6}{2}$

$2 \cdot y = \left(\dfrac{4}{3}\right) \cdot 6$

$2y = 8$

$\dfrac{2y}{2} = \dfrac{8}{2}$

$y = 4$

57. (a)

(b) Let x = height of the chair.

$\dfrac{18}{4} = \dfrac{x}{12}$

$4 \cdot x = 18 \cdot 12$

$4x = 216$

$\dfrac{4x}{4} = \dfrac{216}{4}$

$x = 54$ feet

59. $\dfrac{225}{152.4} = \dfrac{x}{160.5}$

$152.4x = 225(160.5)$

$152.4x = 36{,}112.5$

$\dfrac{152.4x}{152.4} = \dfrac{36{,}112.5}{152.4}$

$x \approx \$237$

61. $\dfrac{225}{152.4} = \dfrac{x}{184.0}$

$152.4x = 225(184.0)$

$152.4x = 41{,}400$

$\dfrac{152.4x}{152.4} = \dfrac{41{,}400}{152.4}$

$x \approx \$272$

63. If $x = ky$

$27 = k \cdot 6$

$\dfrac{27}{6} = k,$

then $x = \dfrac{27}{6} \cdot \dfrac{2}{1} = 9.$

65. If $m = kp^2$

$20 = k \cdot 2^2$

$20 = 4k$

$\dfrac{20}{4} = \dfrac{4k}{4}$

$5 = k,$

then $m = 5 \cdot 5^2 = 5 \cdot 25 = 125.$

67. If $p = \dfrac{k}{q^2}$

$4 = \dfrac{k}{\left(\frac{1}{2}\right)^2}$

$4 = \dfrac{k}{\frac{1}{4}}$

$4 = k \div \dfrac{1}{4}$

$4 = k \cdot \dfrac{4}{1}$

$4 = 4k$

$1 = k,$

then $p = \dfrac{1}{\left(\frac{3}{2}\right)^2}$

$p = 1 \div \dfrac{9}{4}$

$p = 1 \cdot \dfrac{4}{9}$

$p = \dfrac{4}{9}.$

69. Let i = the amount of interest and r = rate of interest.

$i = kr$

$48 = k(0.05)$

$\dfrac{48}{0.05} = k$

$960 = k$

Then $i = 960(0.042) = \$40.32.$

71. Let r = rate and t = time.

$r = \dfrac{k}{t}$

$30 = \dfrac{k}{0.5}$

$30(0.5) = \dfrac{k}{0.5} \cdot \dfrac{0.5}{1}$

$15 = k$

Then $r = \dfrac{15}{0.75} = 20$ miles per hour.

73. Let m = the weight of an object on the moon and e = the weight of the object on the earth.

$m = ke$

$59 = k \cdot 352$

$\dfrac{59}{352} = \dfrac{k \cdot 352}{352}$

$\dfrac{59}{352} = k$

Then $m = \dfrac{59}{352} \cdot 1800 \approx 302$ pounds.

75. Let p = pressure and d = depth.

$p = kd$

$50 = k \cdot 10$

$\dfrac{50}{10} = \dfrac{k \cdot 10}{10}$

$5 = k$

Then $p = 5 \cdot 20 = 100$ pounds per square inch.

77. Let p = pressure and v = volume.

$p = \dfrac{k}{v}$

$10 = \dfrac{k}{3}$

$10 \cdot 3 = \dfrac{k}{3} \cdot 3$

$30 = k$

Then $p = \dfrac{30}{1.5} = 20$ pounds per square foot.

79. Let d = distance and t = time.

$d = kt^2$

$400 = k \cdot 5^2$

$400 = k \cdot 25$

$\dfrac{400}{25} = \dfrac{k \cdot 25}{25}$

$16 = k$

Then $d = 16 \cdot 3^2 = 16 \cdot 9 = 144$ feet.

81. Let V = volume, T = temperature, and P = the pressure of the gas.

$V = \dfrac{kT}{P}$

$1.3 = \dfrac{k \cdot 300}{18}$

$18(1.3) = \dfrac{k \cdot 300}{18} \cdot 18$

$23.4 = 300k$

$k = 0.078$

Then $V = \dfrac{(0.078)340}{24} = 1.105$ liters.

$f = \dfrac{kws^2}{r}$

$242 = \dfrac{k \cdot 2000 \times 30^2}{500}$

$242 = \dfrac{k \cdot 1800000}{500}$

$242 = 3600k$

$k \approx 0.06722$

Thus $f \approx \dfrac{0.06722ws^2}{r}$

$f = \dfrac{0.06722 \times 2000 \times 50^2}{750}$

$f \approx 448.1$ pounds

83. Let L = maximum load, D = diameter of the cross section and H = the height.

$$L = \frac{kD^4}{H^2}$$

$$8 = \frac{k \cdot 1^4}{9^2}$$

$$81 \cdot 8 = \frac{k \cdot 1}{81} \cdot 81$$

$$648 = k$$

Thus $L = \dfrac{648D^4}{H^2}$

$$L = \frac{648 \cdot \left(\frac{2}{3}\right)^4}{12^2}$$

$$= \frac{648 \cdot \left(\frac{16}{81}\right)}{144}$$

$$= 648 \cdot \frac{16}{81} \cdot \frac{1}{144}$$

$$= \frac{8}{9} \text{ metric ton.}$$

85. Let w = weight, and g = girth, and l = length of trout.

$$w = klg^2$$

$$10.5 = k \cdot 26 \cdot 18^2$$

$$10.5 = k \cdot 8424$$

$$\frac{10.5}{8424} = \frac{k \cdot 8424}{8424}$$

$$k \approx 0.001246$$

Thus $w = 0.001246 \cdot 22 \cdot 15^2 \approx 6.2$ pounds.

7.4 Exercises

1. $\{x \mid x \le 3\}$ matches with letter D, $(-\infty, 3]$. The bracket is used in interval notation because 3 is included in the set.

3. $\{x \mid x < 3\}$ matches with letter B, the number line showing all values less than 3.

5. $\{x \mid -3 \le x \le 3\}$ matches with letter F, $[-3, 3]$. The brackets indicate that the endpoints, -3 and 3, are included.

7. Use parentheses when the symbol is < or >. Use brackets when the symbol is ≤ or ≥.

9.
$$4x + 1 \ge 21$$
$$4x + 1 - 1 \ge 21 - 1$$
$$4x \ge 20$$
$$\frac{4x}{4} \ge \frac{20}{4}$$
$$x \ge 5$$
Interval notation: $[5, \infty)$
Graph:

11.
$$\frac{3x-1}{4} > 5$$
$$\frac{4}{1} \cdot \frac{3x-1}{4} > 5 \cdot 4$$
$$3x - 1 > 20$$
$$3x - 1 + 1 > 20 + 1$$
$$3x > 21$$
$$\frac{3x}{3} > \frac{21}{3}$$
$$x > 7$$
Interval notation: $(7, \infty)$
Graph:

13. In this exercise, remember to reverse the inequality sign when multiplying or dividing both sides by a negative number.
$$-4x < 16$$
$$\frac{-4x}{-4} > \frac{16}{-4}$$
$$x > -4$$
Interval notation: $(-4, \infty)$
Graph:

15. In this exercise remember to reverse the inequality sign when multiplying or dividing both sides by a negative number.
$$-\frac{3}{4}x \ge 30$$
$$-\frac{4}{3} \cdot -\frac{3}{4}x \le \frac{30}{1} \cdot -\frac{4}{3}$$
$$x \le -\frac{30 \cdot 4}{3}$$
$$x \le -40$$
Interval notation: $(-\infty, -40]$
Graph:

17.
$$-1.3x \ge -5.2$$
$$\frac{-1.3x}{-1.3} \le \frac{-5.2}{-1.3}$$
$$x \le 4$$
Interval notation: $(-\infty, 4]$
Graph:

19.
$$\frac{2x-5}{-4} > 5$$
$$\frac{-4}{1} \cdot \frac{2x-5}{-4} < 5 \cdot -4$$
$$2x-5 < -20$$
$$2x-5+5 < -20+5$$
$$2x < -15$$
$$\frac{2x}{2} < \frac{-15}{2}$$
$$x < \frac{-15}{2}$$

Interval notation: $\left(-\infty, \dfrac{-15}{2}\right)$

Graph:

21.
$$x+4(2x-1) \ge x$$
$$x+8x-4 \ge x$$
$$9x-4 \ge x$$
$$9x-4+4 \ge x+4$$
$$9x \ge x+4$$
$$9x-x \ge x-x+4$$
$$8x \ge 4$$
$$\frac{8x}{8} \ge \frac{4}{8}$$
$$x \ge \frac{4 \cdot 1}{4 \cdot 2}$$
$$x \ge \frac{1}{2}$$

Interval notation: $\left[\dfrac{1}{2}, \infty\right)$

Graph:

23.
$$-(4+x)+2-3x < -14$$
$$-4-x+2-3x < -14$$
$$-2-4x < -14$$
$$-2+2-4x < -14+2$$
$$-4x < -12$$
$$\frac{-4x}{-4} > \frac{-12}{-4}$$
$$x > 3$$

Interval notation: $(3, \infty)$

Graph:

25.
$$-3(x-6) > 2x-2$$
$$-3x+18 > 2x-2$$
$$-3x-2x+18 > 2x-2x-2$$
$$-5x+18 > -2$$
$$-5x+18-18 > -2-18$$
$$-5x > -20$$
$$\frac{-5x}{-5} < \frac{-20}{-5}$$
$$x < 4$$

Interval notation: $(-\infty, 4)$

Graph:

27. Clear the fractions by multiplying both sides of the equation by the LCM, 6.
$$\frac{2}{3}(3x-1) \ge \frac{3}{2}(2x-3)$$
$$\frac{6}{1} \cdot \frac{2}{3}(3x-1) \ge \frac{6}{1} \cdot \frac{3}{2}(2x-3)$$
$$2 \cdot 2(3x-1) \ge 3 \cdot 3(2x-3)$$
$$4(3x-1) \ge 9(2x-3)$$
$$12x-4 \ge 18x-27$$
$$12x-4+4 \ge 18x-27+4$$
$$12x \ge 18x-23$$
$$12x-18x \ge 18x-18x-23$$
$$-6x \ge -23$$
$$\frac{-6x}{-6} \le \frac{-23}{-6}$$
$$x \le \frac{23}{6}$$

Interval notation: $\left(-\infty, \dfrac{23}{6}\right]$

Graph:

29. Clear the fractions by multiplying both sides of the equation by the LCM, 8.
$$-\frac{1}{4}(x+6)+\frac{3}{2}(2x-5) < 10$$
$$\frac{8}{1} \cdot -\frac{1}{4}(x+6)+\frac{8}{1} \cdot \frac{3}{2}(2x-5) < 10 \cdot 8$$
$$2 \cdot -1(x+6)+4 \cdot 3(2x-5) < 80$$
$$-2(x+6)+12(2x-5) < 80$$
$$-2x-12+24x-60 < 80$$
$$22x-72 < 80$$
$$22x-72+72 < 80+72$$
$$22x < 152$$
$$\frac{22x}{22} < \frac{152}{22}$$
$$x < \frac{76 \cdot 2}{11 \cdot 2}$$
$$x < \frac{76}{11}$$

Interval notation: $\left(-\infty, \dfrac{76}{11}\right)$

Graph:

31. $3(2x-4)-4x < 2x+3$
$6x-12-4x < 2x+3$
$2x-12 < 2x+3$
$2x-2x-12 < 2x-2x+3$
$-12 < 3$

Because it is always true that negative 12 is less than 3, x can be replaced by any real number and the inequality will be true. This is called an identity.
Interval notation: $(-\infty, +\infty)$
Graph:

33. $8\left(\dfrac{1}{2}x+3\right) < 8\left(\dfrac{1}{2}x-1\right)$
$4x+24 < 4x-8$
$4x-4x+24 < 4x-4x-8$
$24 < -8$

Because it is not true that 24 is less than negative 8, there is no real number that can replace x to obtain a true statement. This is called a contradiction and there is no solution. This can be symbolized by the empty set: $\varnothing$.

35. Writing exercise; answers will vary.

37. In a three-part inequality, remember to work toward isolating the variable by applying the same operation to all three parts of the inequality.
$-4 < x-5 < 6$
$-4+5 < x-5+5 < 6+5$
$1 < x < 11$
Interval notation: $(1, 11)$
Graph:

39. $-9 \le x+5 \le 15$
$-9-5 \le x+5-5 \le 15-5$
$-14 \le x \le 10$
Interval notation: $[-14, 10]$
Graph:

41. $-6 \le 2x+4 \le 16$
$-6-4 \le 2x+4-4 \le 16-4$
$-10 \le 2x \le 12$
$\dfrac{-10}{2} \le \dfrac{2x}{2} \le \dfrac{12}{2}$
$-5 \le x \le 6$
Interval notation: $[-5, 6]$
Graph:

43. $-19 \le 3x-5 \le 1$
$-19+5 \le 3x-5+5 \le 1+5$
$-14 \le 3x \le 6$
$\dfrac{-14}{3} \le \dfrac{3x}{3} \le \dfrac{6}{3}$
$\dfrac{-14}{3} \le x \le 2$
Interval notation: $\left[\dfrac{-14}{3}, 2\right]$
Graph:

45. $-1 \le \dfrac{2x-5}{6} \le 5$
$6\cdot-1 \le \dfrac{6}{1}\cdot\dfrac{2x-5}{6} \le 6\cdot5$
$-6 \le 2x-5 \le 30$
$-6+5 \le 2x-5+5 \le 30+5$
$-1 \le 2x \le 35$
$\dfrac{-1}{2} \le \dfrac{2x}{2} \le \dfrac{35}{2}$
$\dfrac{-1}{2} \le x \le \dfrac{35}{2}$
Interval notation: $\left[\dfrac{-1}{2}, \dfrac{35}{2}\right]$
Graph:

47. Remember to reverse the inequality symbols in this exercise when dividing by -9.
$4 \le 5-9x < 8$
$4-5 \le 5-5-9x < 8-5$
$-1 \le -9x < 3$
$\dfrac{-1}{-9} \ge \dfrac{-9x}{-9} > \dfrac{3}{-9}$
$\dfrac{1}{9} \ge x > -\dfrac{1}{3}$
To make this more meaningful, restate the inequality as $-\dfrac{1}{3} < x \le \dfrac{1}{9}$.
Interval notation: $\left(-\dfrac{1}{3}, \dfrac{1}{9}\right]$
Graph:

49. The following months show percentages greater than 7.7%: April, 12.9%; May, 22.1%; June, 20.7%, and July, 11.1%.

51. First find what percent 1500 is of 17,252. Let r = the percent in decimal form.
$$r \cdot 17,252 = 1500$$
$$\frac{r \cdot 27,252}{17,252} = \frac{1500}{17,252}$$
$$r = 0.087$$
This decimal value, 0.087, is equivalent to 8.7%. Look for the months for which the percent is less than 8.7%. These months are: January, February, March, August, September, October, November, and December.

53. Read the values at the ends of the bars. The bars that have values greater than or equal to 6687 are PA in 2006, OH in 2005 and 2006, and IA in 2005 and 2006.

55. The egg production for TX in 2006 was $x = 5039$ million. The egg production for CA in 2006 was $y = 4962$ million.
$$x > y$$

57. Let d = number of additional $\frac{1}{5}$ miles.

Frank must pay $3.00 plus $0.50 for each additional $\frac{1}{5}$ mile, and the amount cannot exceed $7.50. Solve the inequality.
$$3.00 + 0.50d \le 7.50$$
$$3.00 - 3.00 + 0.50d \le 7.50 - 3.00$$
$$0.50d \le 4.50$$
$$\frac{0.50d}{0.50} \le \frac{4.50}{0.50}$$
$$d \le 9$$

Then Frank can travel $9 \cdot \frac{1}{5} = \frac{9}{5}$ or

$1\frac{4}{5}$ miles in addition to the first $\frac{1}{5}$ mile for a total for 2 miles.

59. Let x = score needed on the third test. Solve the inequality:
$$\frac{90 + 82 + x}{3} \ge 84$$
$$\frac{3}{1} \cdot \frac{90 + 82 + x}{3} \ge 84 \cdot 3$$
$$90 + 82 + x \ge 252$$
$$172 + x \ge 252$$
$$172 - 172 + x \ge 252 - 172$$
$$x \ge 80$$
Hollis must score at least a grade of 80 on the third test.

61. Let x = number of miles. Then the cost to rent each vehicle is:
Agency A: $29.95 + 0.28x$
Agency B: $34.95 + 0.25x$
To determine the number of miles at which the price to rent from Agency B would exceed the price to rent from Agency A, solve the inequality:
$$\text{Agency B} > \text{Agency A}$$
$$34.95 + 0.25x > 29.95 + 0.28x$$
$$34.95 - 29.95 + 0.25x > 29.95 - 29.95 + 0.28x$$
$$5 + 0.25x > 0.28x$$
$$5 + 0.25x - 0.25x > 0.28x - 0.25x$$
$$5 > 0.03x$$
$$\frac{5}{0.03} > \frac{0.03x}{0.03}$$
$$166.67 > x$$
The price to rent from Agency B would exceed the price to rent from Agency A for mileage less than 167 miles.

63. Solve the following inequalities for weight (w) given each value of height (h).
$$19 < \text{BMI} < 25$$
$$19 < \frac{703 \times w}{h^2} < 25$$

(a) $h = 72$ inches
$$19 < \frac{703 \times w}{72^2} < 25$$
$$72^2 \cdot 19 < 703 \times w < 72^2 \cdot 25$$
$$98496 < 704 \times w < 129600$$
$$\frac{98496}{704} < \frac{704w}{704} < \frac{129600}{704}$$
$$140 < w < 184 \text{ pounds}$$

(b) Answers will vary.

For 65 we want R > C for each company to show a profit.

65.
$$R > C$$
$$24x > 20x + 100$$
$$4x > 100$$
$$\frac{4x}{4} > \frac{100}{4}$$
$$x > 25$$

Thus, they must sell at least 26 DVDs to make a profit.

7.5 Exercises

1. $\left(\frac{5}{3}\right)^2 = \frac{25}{9}$, which is choice A.

3. $\left(-\frac{3}{5}\right)^{-2} = \frac{1}{\left(-\frac{3}{5}\right)^2}$

$$= \frac{1}{\frac{9}{25}}$$

$$= 1 \div \frac{9}{25}$$

$$= 1 \cdot \frac{25}{9}$$

$$= \frac{25}{9}, \text{ which is choice A.}$$

5. $-\left(-\frac{3}{5}\right)^2 = -\left(-\frac{3}{5} \cdot -\frac{3}{5}\right) = -\left(\frac{9}{25}\right) = -\frac{9}{25}$,

which is choice D.

7. $5^4 = 5 \cdot 5 \cdot 5 \cdot 5 = 625$

9. $(-2)^5 = -2 \cdot -2 \cdot -2 \cdot -2 \cdot -2 = -32$

11. $-2^3 = -(2 \cdot 2 \cdot 2) = -8$

13. $-(-3)^4 = -(-3 \cdot -3 \cdot -3 \cdot -3) = -81$

15. $7^{-2} = \frac{1}{7^2} = \frac{1}{49}$

17. $-7^{-2} = -\frac{1}{7^2} = -\frac{1}{49}$

19. $\frac{2}{(-4)^{-3}} = 2 \div (-4)^{-3}$

$$= 2 \div \frac{1}{(-4)^3}$$

$$= 2 \cdot (-4)^3$$

$$= 2 \cdot -64$$

$$= -128$$

21. $\frac{5^{-1}}{4^{-2}} = \frac{\frac{1}{5}}{\frac{1}{4^2}} = \frac{1}{5} \div \frac{1}{4^2} = \frac{1}{5} \cdot \frac{4^2}{1} = \frac{16}{5}$

23. $\left(\frac{1}{5}\right)^{-3} = \frac{1^{-3}}{5^{-3}} = \frac{5^3}{1^3} = 125$

25. $\left(\frac{4}{5}\right)^{-2} = \frac{4^{-2}}{5^{-2}} = \frac{5^2}{4^2} = \frac{25}{16}$

27. $4^{-1} + 5^{-1} = \frac{1}{4} + \frac{1}{5}$

$$= \frac{5}{5} \cdot \frac{1}{4} + \frac{1}{5} \cdot \frac{4}{4}$$

$$= \frac{5}{20} + \frac{4}{20}$$

$$= \frac{9}{20}$$

29. $12^0 = 1$

31. $(-4)^0 = 1$

33. $3^0 - 4^0 = 1 - 1 = 0$

35. In order to raise a fraction to a negative power, we may change the fraction to its <u>reciprocal</u> and change the exponent to the <u>additive inverse</u> of the original exponent.

37. A. Simplify each side of the equation to see if the statement is true:

$$-\frac{3}{4} = \left(\frac{3}{4}\right)^{-1} = \frac{3^{-1}}{4^{-1}} = \frac{4}{3}$$

The equation is not correct.

B. $\frac{3^{-1}}{4^{-1}} = \left(\frac{4}{3}\right)^{-1}$

$$\frac{4}{3} = \frac{4^{-1}}{3^{-1}} = \frac{3}{4}$$

This equation is not correct.

C. $\dfrac{3^{-1}}{4} = \dfrac{3}{4^{-1}}$

$\dfrac{1}{3\cdot 4} = 3\cdot 4$

$\dfrac{1}{12} = 12$

This equation is not correct.

D. $\dfrac{3^{-1}}{4^{-1}} = \left(\dfrac{3}{4}\right)^{-1} = \dfrac{3^{-1}}{4^{-1}}$

This equation is correct.

39. $x^{12}\cdot x^4 = x^{12+4} = x^{16}$

41. $\dfrac{5^{17}}{5^{16}} = 5^{17-16} = 5$

43. $\dfrac{3^{-5}}{3^{-2}} = 3^{-5-(-2)} = 3^{-5+2} = 3^{-3} = \dfrac{1}{3^3} = \dfrac{1}{27}$

45. $\dfrac{9^{-1}}{9} = 9^{-1-1} = 9^{-2} = \dfrac{1}{9^2} = \dfrac{1}{81}$

47. $t^5 t^{-12} = t^{5+(-12)} = t^{-7} = \dfrac{1}{t^7}$

49. $(3x)^2 = 3^2\cdot x^2 = 9x^2$

51. $a^{-3}a^2 a^{-4} = a^{-3+2+(-4)}$

$= a^{-3+2-4}$

$= a^{-5}$

$= \dfrac{1}{a^5}$

53. $\dfrac{x^7}{x^{-4}} = x^{7-(-4)} = x^{7+4} = x^{11}$

55. $\dfrac{r^3 r^{-4}}{r^{-2} r^{-5}} = \dfrac{r^{3+(-4)}}{r^{-2+(-5)}}$

$= \dfrac{r^{-1}}{r^{-7}}$

$= r^{-1-(-7)}$

$= r^{-1+7}$

$= r^6$

57. $7k^2(-2k)(4k^{-5}) = (7\cdot -2\cdot 4)k^{2+1+(-5)}$

$= -56k^{-2}$

$= -\dfrac{56}{k^2}$

59. $(z^3)^{-2} z^2 = z^{3\cdot -2} z^2$

$= z^{-6} z^2$

$= z^{-6+2}$

$= z^{-4}$

$= \dfrac{1}{z^4}$

61. $-3r^{-1}(r^{-3})^2 = -3r^{-1} r^{-3\cdot 2}$

$= -3r^{-1} r^{-6}$

$= -3r^{-1+(-6)}$

$= -3r^{-7}$

$= -\dfrac{3}{r^7}$

63. $(3a^{-2})^3(a^3)^{-4} = 3^3 a^{-6}\cdot a^{-12}$

$= 27a^{-6+(-12)}$

$= 27a^{-18}$

$= \dfrac{27}{a^{18}}$

65. $(x^{-5} y^2)^{-1} = x^{-5\cdot -1} y^{2\cdot -1} = x^5 y^{-2} = \dfrac{x^5}{y^2}$

67. The reciprocal of x is $\dfrac{1}{x}$.

(a) $x^{-1} = \dfrac{1}{x}$, by definition.

(b) $\dfrac{1}{x}$ is the reciprocal.

(c) $\left(\dfrac{1}{x^{-1}}\right)^{-1} = \dfrac{1^{-1}}{x^1} = \dfrac{1}{x}$

(d) $-x$ is the opposite of x. It is not the reciprocal.

69. $230 = 2.3\times 10^2$

71. $0.02 = 2\times 10^{-2}$

73. $6.5 \times 10^3 = 6500$

75. $1.52 \times 10^{-2} = 0.0152$

77.
$$\frac{0.002 \times 3900}{0.000013} = \frac{(2 \times 10^{-3}) \times (3.9 \times 10^3)}{1.3 \times 10^{-5}}$$
$$= \frac{(2 \times 3.9) \times (10^{-3} \times 10^3)}{1.3 \times 10^{-5}}$$
$$= \frac{(7.8) \times (10^0)}{1.3 \times 10^{-5}}$$
$$= \frac{7.8}{1.3} \times \frac{10^0}{10^{-5}}$$
$$= 6 \times 10^{0-(-5)}$$
$$= 6 \times 10^{0+5}$$
$$= 6 \times 10^5$$

79.
$$\frac{0.0004 \times 56,000}{0.000112} = \frac{(4 \times 10^{-4}) \times (5.6 \times 10^4)}{1.12 \times 10^{-4}}$$
$$= \frac{(4 \times 5.6) \times (10^{-4} \times 10^4)}{1.12 \times 10^{-4}}$$
$$= \frac{22.4 \times 10^0}{1.12 \times 10^{-4}}$$
$$= \frac{22.4}{1.12} \times \frac{10^0}{10^{-4}}$$
$$= 20 \times 10^{0-(-4)}$$
$$= 20 \times 10^4$$
$$= 2.0 \times 10^5$$

(Note that 20×10^4 is not in scientific notation. As 20 is reduced by a power of ten, the exponent on 10 must increase by a power of ten.)

81.
$$\frac{840,000 \times 0.03}{0.00021 \times 600} = \frac{(8.4 \times 10^5) \times (3 \times 10^{-2})}{(2.1 \times 10^{-4}) \times (6 \times 10^2)}$$
$$= \frac{(8.4 \times 3) \times (10^5 \times 10^{-2})}{(2.1 \times 6) \times (10^{-4} \times 10^2)}$$
$$= \frac{25.2 \times 10^3}{12.6 \times 10^{-2}}$$
$$= 2 \times 10^{3-(-2)}$$
$$= 2 \times 10^{3+2}$$
$$= 2 \times 10^5$$

83. $\$1,000,000,000 = \$1 \times 10^9$

$\$1,000,000,000,000 = \$1 \times 10^{12}$

$\$3,100,000,000,000 = \$3.1 \times 10^{12}$

$210,385 = 2.10385 \times 10^5$

85. $\$18.69$ billion $= \$18,690,000,000$
$$= \$1.869 \times 10^{10}$$

87. 10 billion $= 10,000,000,000 = 1 \times 10^{10}$

89. $2 \times 10^9 = 2,000,000,000$

91. **(a)** 304.1 million $= 304,100,000$
$$= 3.041 \times 10^8$$

(b) $\$1,000,000,000,000 = \$1 \times 10^{12}$

(c) $\dfrac{\$1 \times 10^{12}}{3.041 \times 10^8} \approx 0.3288 \times 10^{12-8}$
$$= 0.3288 \times 10^4$$
$$= \$3288$$

93.
$$\frac{1.8 \times 10^7}{1.9 \times 10^{13}} = \frac{1.8}{1.9} \times \frac{10^7}{10^{13}}$$
$$= 0.947368 \times 10^{7-13}$$
$$= 0.947368 \times 10^{-6} \text{ or}$$
$$\approx 9.474 \times 10^{-7} \text{ parsec}$$

95. $\dfrac{9 \times 10^{12}}{3 \times 10^{10}} = \dfrac{9}{3} \times 10^{12-10} = 3 \times 10^2$, which is 300 sec.

97.
$$\frac{1.86 \times 10^5 \text{ mi}}{1 \text{ s}} \cdot \frac{60 \text{ s}}{1 \text{ min}} \cdot \frac{60 \text{ min}}{1 \text{ hr}} \cdot \frac{24 \text{ hr}}{1 \text{ day}}$$
$$\cdot \frac{365 \text{ days}}{1 \text{ year}}$$

This simplifies to $58,656,960 \times 10^5$ miles; in scientific notation this is approximately 5.87×10^{12} miles.

99. Use the formula $d = rt$. The total distance that the spacecraft must travel is the difference between 6.7×10^7 and 3.6×10^7. The rate of the spacecraft is 1.55×10^3. Insert this information into the formula and solve for t.

$$6.7 \times 10^7 - 3.6 \times 10^7 = (1.55 \times 10^3)t$$
$$3.1 \times 10^7 = (1.55 \times 10^3)t$$
$$\frac{3.1 \times 10^7}{1.55 \times 10^3} = \frac{(1.55 \times 10^3)t}{1.55 \times 10^3}$$
$$\frac{3.1}{1.55} \times \frac{10^7}{10^3} = t$$
$$2 \times 10^4 = t$$

The time would be 20,000 hours.

7.6 Exercises

1. $(3x^2 - 4x + 5) + (-2x^2 + 3x - 2)$
    $= 3x^2 - 2x^2 - 4x + 3x + 5 - 2$
    $= x^2 - x + 3$

3. Remember that the negative in front of the parenthesis affects all the terms within the grouping symbol; i.e., all the signs change.
    $(12y^2 - 8y + 6) - (3y^2 - 4y + 2)$
    $= 12y^2 - 3y^2 - 8y + 4y + 6 - 2$
    $= 9y^2 - 4y + 4$

5. $(6m^4 - 3m^2 + m) - (2m^3 + 5m^2 + 4m)$
    $\quad + (m^2 - m)$
    $= 6m^4 - 3m^2 + m - 2m^3 - 5m^2 - 4m + m^2$
    $\quad - m$
    $= 6m^4 - 2m^3 - 3m^2 - 5m^2 + m^2 + m - 4m$
    $\quad - m$
    $= 6m^4 - 2m^3 - 7m^2 - 4m$

7. $5(2x^2 - 3x + 7) - 2(6x^2 - x + 12)$
    $= 10x^2 - 15x + 35 - 12x^2 + 2x - 24$
    $= 10x^2 - 12x^2 - 15x + 2x + 35 - 24$
    $= -2x^2 - 13x + 11$

9. Use the FOIL method:
    $(x+3)(x-8) = x \cdot x + x \cdot (-8) + 3 \cdot x + 3 \cdot (-8)$
    $\quad\quad\quad\quad = x^2 - 8x + 3x - 24$
    $\quad\quad\quad\quad = x^2 - 5x - 24$

11. $(4r-1)(7r+2)$
    $= 4r \cdot 7r + 4r \cdot 2 + (-1) \cdot 7r + (-1) \cdot 2$
    $= 28r^2 + 8r - 7r - 2$
    $= 28r^2 + r - 2$

13. Use the distributive property. Also remember to add exponents when multiplying variables that have the same base.
    $4x^2(3x^3 + 2x^2 - 5x + 1)$
    $= 4x^2 \cdot 3x^3 + 4x^2 \cdot 2x^2 + 4x^2 \cdot (-5x) + 4x^2 \cdot 1$
    $= 12x^5 + 8x^4 - 20x^3 + 4x^2$

15. The FOIL method can always be used to multiply two binomials.
    $(2m+3)(2m-3)$
    $= 2m \cdot 2m + (-3) \cdot 2m + 3 \cdot 2m + 3 \cdot (-3)$
    $= 4m^2 - 6m + 6m - 9$
    $= 4m^2 - 9$
    However, it is helpful to recognize that it is the product of the sum and difference of two terms.

17. It is important to remember that the binomial $4m + 2n$ is the base that is being squared.
    $(4m + 2n)^2$
    $= 4m \cdot 4m + 4m \cdot 2n + 2n \cdot 4m + 2n \cdot 2n$
    $= 16m^2 + 16mn + 4n^2$
    it is also helpful to recognize the pattern of a binomial squared.

19. It is important to remember that the binomial $5r + 3t^2$ is the base that is being squared:
    $(5r + 3t^2)^2$
    $= (5r + 3t^2)(5r + 3t^2)$
    $= 5r \cdot 5r + 5r \cdot 3t^2 + 3t^2 \cdot 5r + 3t^2 \cdot 3t^2$
    $= 25r^2 + 15rt^2 + 15rt^2 + 9t^4$
    $= 25r^2 + 30rt^2 + 9t^4$
    It is also helpful to recognize the pattern of a binomial squared.

21. Vertical multiplication is often less confusing when multiplying two polynomials of more than two terms. Multiply from the right as in number multiplication; i.e., start with -1 times -4. Line up like terms in columns.

$$
\begin{array}{r}
-z^2 + 3z - 4 \\
2z - 1 \\
\hline
z^2 - 3z + 4 \\
-2z^3 + 6z^2 - 8z \\
\hline
-2z^3 + 7z^2 - 11z + 4
\end{array}
$$

23. Using vertical multiplication, these polynomials have been multiplied beginning at the right: First multiply $-3k$ from the second polynomial times each term of the first polynomial; then multiply $+2n$ times each term of the first polynomial; finally multiply $+m$ times each term of the first polynomial. Place like terms in columns as you multiply to simplify being added later.

$$
\begin{array}{r}
m \quad -n \quad +k \\
m+2n \quad -3k \\
\hline
-3km+3kn \qquad\qquad -3k^2 \\
+2mn \qquad +2kn \qquad -2n^2 \\
-mn \ +km \qquad +m^2 \\
\hline
+mn-2km+5kn+m^2-2n^2-3k^2
\end{array}
$$

25. Vertical multiplication might be the preferred method as in the previous exercise.

$$
\begin{array}{r}
a-b+2c \\
a-b+2c \\
\hline
+2ac-2bc+4c^2 \\
-ab+b^2 \qquad -2bc \\
+a^2 \ -ab \qquad +2ac \\
\hline
a^2-2ab+b^2+4ac-4bc+4c^2
\end{array}
$$

27. The answer is (A). Choices (B) and (C) are trinomials of degree 6; but they are not in descending powers; choice (D) is not a trinomial.

29. Writing exercise; answers will vary.

31. $8m^4+6m^3-12m^2$
$= 2m^2 \cdot 4m^2 + 2m^2 \cdot 3m - 2m^2 \cdot 6$
$= 2m^2(4m^2+3m-6)$

33. $4k^2m^3+8k^4m^3-12k^2m^4$
$= 4k^2m^3 \cdot 1 + 4k^2m^3 \cdot 2k^2 - 4k^2m^3 \cdot 3m$
$= 4k^2m^3(1+2k^2-3m)$

35. In this exercise, the greatest common factor is $2(a+b)$.
$2(a+b)+4m(a+b)$
$= 2 \cdot (a+b) + 2m \cdot 2 \cdot (a+b)$
$= 2(a+b)(1+2m)$

37. In this exercise, the greatest common factor is $m-1$.
$2(m-1)-3(m-1)^2+2(m-1)^3$
$= (m-1) \cdot 2 - (m-1) \cdot 3(m-1)^1$
$\qquad\qquad\qquad + (m-1) \cdot 2(m-1)^2$
$= (m-1)[2-3(m-1)+2(m-1)^2]$
$= (m-1)[2-3m+3+2(m-1)(m-1)]$
$= (m-1)[2-3m+3+2(m^2-2m+1)]$
$= (m-1)[2-3m+3+2m^2-4m+2]$
$= (m-1)[2m^2-7m+7]$

39. $6st+9t-10s-15 = 3t \cdot (2s+3) - 5 \cdot (2s+3)$
$\qquad\qquad\qquad\qquad = (2s+3)(3t-5)$

Remember that the binomial factors can be written either as shown above or as $(3t-5)(2s+3)$.

41. $rt^3+rs^2-pt^3-ps^2$
$= r \cdot (t^3+s^2) - p \cdot (t^3+s^2)$
$= (t^3+s^2)(r-p)$

43. $16a^2+10ab-24ab-15b^2$
$= 2a \cdot (8a+5b) - 3b \cdot (8a+5b)$
$= (8a+5b)(2a-3b)$

45. $20z^2-8zx-45zx+18x^2$
$= 4z \cdot (5z-2x) - 9x \cdot (5z-2x)$
$= (5z-2x)(4z-9x)$

47. $1-a+ab-b = (1-a)+b(a-1)$
$\qquad\qquad\qquad = (1-a)-b(1-a)$
$\qquad\qquad\qquad = (1-a)(1-b)$

49. Recall that the mental process is to think of two numbers whose product is -15 and whose sum is -2. Use these numbers, -5 and $+3$ to rename the middle term of the trinomial.
After creating the 4-term polynomial in the second line, factor by grouping.

$x^2-2x-15 = x^2-5x+3x-15$
$\qquad\qquad\quad = x(x-5)+3(x-5)$
$\qquad\qquad\quad = (x-5)(x+3)$

51. $y^2+2y-35 = y^2+7y-5y-35$
$\qquad\qquad\quad = y(y+7)-5(y+7)$
$\qquad\qquad\quad = (y+7)(y-5)$

53. First, factor out the greatest common factor 6; then proceed to factor the trinomial.

$$6a^2 - 48a - 120 = 6(a^2 - 8a - 20)$$
$$= 6[a^2 - 10a + 2a - 20]$$
$$= 6[a(a - 10) + 2(a - 10)]$$
$$= 6(a - 10)(a + 2)$$

55. $3m^3 + 12m^2 + 9m = 3m[m^2 + 4m + 3]$
$$= 3m[m^2 + 3m + m + 3]$$
$$= 3m[m(m + 3) + 1(m + 3)]$$
$$= 3m(m + 3)(m + 1)$$

57. When the leading coefficient is not 1, remember that the product to consider is found by multiplying the leading coefficient by the last term. In this exercise multiply 6 times −6, so that a product of −36 is needed along with a sum of +5.

$$6k^2 + 5kp - 6p^2$$
$$= 6k^2 + 9kp - 4kp - 6p^2$$
$$= 3k(2k + 3p) - 2p(2k + 3p)$$
$$= (2k + 3p)(3k - 2p)$$

59. $5a^2 - 7ab - 6b^2 = 5a^2 - 10ab + 3ab - 6b^2$
$$= 5a(a - 2b) + 3b(a - 2b)$$
$$= (a - 2b)(5a + 3b)$$

61. $21x^2 - xy - 2y^2 = 21x^2 - 7xy + 6xy - 2y^2$
$$= 7x(3x - y) + 2y(3x - y)$$
$$= (3x - y)(7x + 2y)$$

63. $24a^4 + 10a^3b - 4a^2b^2$
$$= 2a^2[12a^2 + 5ab - 2b^2]$$
$$= 2a^2[12a^2 - 3ab + 8ab - 2b^2]$$
$$= 2a^2[3a(4a - b) + 2b(4a - b)]$$
$$= 2a^2(4a - b)(3a + 2b)$$

65. $15x^2y^5 - 20x^3y^3 + 15xy^2$
$$= 5xy^2(3xy^3 - 4x^2y + 3)$$

67. $9m^2 - 12m + 4 = (3m)^2 - 2(3m)(2) + (2)^2$
$$= (3m - 2)^2$$

69. $32a^2 - 48ab + 18b^2$
$$= 2[16a^2 - 24ab + 9b^2]$$
$$= 2[(4a)^2 - 2(4a)(3b) + (3b)^2]$$
$$= 2(4a - 3b)^2$$

71. $4x^2y^2 + 28xy + 49$
$$= (2xy)^2 + 2(2xy)(7) + (7)^2$$
$$= (2xy + 7)^2$$

73. $x^2 - 36 = (x)^2 - (6)^2 = (x + 6)(x - 6)$

75. $y^2 - w^2 = (y)^2 - (w)^2 = (y + w)(y - w)$

77. $9a^2 - 16 = (3a)^2 - (4)^2 = (3a + 4)(3a - 4)$

79. $25s^4 - 9t^2 = (5s^2)^2 - (3t)^2$
$$= (5s^2 + 3t)(5s^2 - 3t)$$

81. This exercise requires factoring twice.
$$p^4 - 625 = (p^2)^2 - (25)^2$$
$$= (p^2 + 25)(p^2 - 25)$$
$$= (p^2 + 25)[(p)^2 - (5)^2]$$
$$= (p^2 + 25)(p + 5)(p - 5)$$

83. $8 - a^3 = (2)^3 - (a)^3$
$$= (2 - a)(2^2 + 2 \cdot a + a^2)$$
$$= (2 - a)(4 + 2a + a^2)$$

85. $125x^3 - 27 = (5x)^3 - (3)^3$
$$= (5x - 3)[(5x)^2 + 5x \cdot 3 + 3^2]$$
$$= (5x - 3)(25x^2 + 15x + 9)$$

87. $27y^9 + 125z^6$
$$= (3y^3)^3 + (5z^2)^3$$
$$= (3y^3 + 5z^2)((3y^3)^2 - 3y^3 \cdot 5z^2 + (5z^2)^2)$$
$$= (3y^3 + 5z^2)(9y^6 - 15y^3z^2 + 25z^4)$$

89. Factor by grouping.
$$x^2 + xy - 5x - 5y = x \cdot (x + y) - 5 \cdot (x + y)$$
$$= (x + y)(x - 5)$$

91. Factor this trinomial by multiplying the leading coefficient, 12, by −35 to obtain a product of −420. Then, find two numbers whose product is −420 and whose sum is +16.

$$12m^2 + 16mn - 35n^2$$
$$= 12m^2 - 14mn + 30mn - 35n^2$$
$$= 2m \cdot (6m - 7n) + 5n \cdot (6m - 7n)$$
$$= (6m - 7n)(2m + 5n)$$

93. This is a perfect square trinomial that can be factored by following the pattern.

$$4z^2 + 28z + 49 = (2z)^2 + 2(2z)(7) + (7)^2$$
$$= (2z + 7)^2$$

95. This is a sum of cubes.

$$1000x^3 + 343y^3$$
$$= (10x)^3 + (7y)^3$$
$$= (10x + 7y)((10x)^2 - 10x \cdot 7y + (7y)^2)$$
$$= (10x + 7y)(100x^2 - 70xy + 49y^2)$$

97. This is the difference of cubes.

$$125m^6 - 216$$
$$= (5m^2)^3 - (6)^3$$
$$= (5m^2 - 6)((5m^2)^2 + 5m^2 \cdot 6 + 6^2)$$
$$= (5m^2 - 6)(25m^4 + 30m^2 + 36)$$

99. Factor out the greatest common factor, $(m - 2n)$.

$$p^4(m - 2n) + q(m - 2n) = (m - 2n)(p^4 + q)$$

7.7 Exercises

1. For the quadratic equation $5x^2 + 4x - 8 = 0$, the values of a, b, and c are respectively $\underline{5}$, $\underline{4}$, and $\underline{-8}$.

3. Yes, the quadratic formula can be used to solve the equation. Write the equation in standard form: $2x^2 - 5 = 0$. Now $a = 2$, $b = 0$, $c = -5$. Substitute these values into the quadratic formula.

$$x = \frac{-b \pm \sqrt{b^2 - 4ac}}{2a}$$

$$x = \frac{-(0) \pm \sqrt{(0)^2 - 4 \cdot 2 \cdot (-5)}}{2 \cdot 2}$$

$$= \frac{\pm\sqrt{40}}{4}$$

$$= \frac{\pm\sqrt{4 \cdot 10}}{4}$$

$$= \frac{\pm 2\sqrt{10}}{4}$$

$$= \pm\frac{\sqrt{10}}{2}$$

The solution set is $\left\{\pm\dfrac{\sqrt{10}}{2}\right\}$.

5. Set each factor equal to zero and solve each equation.

$$(x + 3)(x - 9) = 0$$
$$x + 3 = 0 \quad \text{or} \quad x - 9 = 0$$
$$x = -3 \qquad\qquad x = 9$$

The solution set is $\{-3, 9\}$.

7. Set each factor equal to zero and solve each equation.

$$(2x - 7)(5x + 1) = 0$$
$$2x - 7 = 0 \quad \text{or} \quad 5x + 1 = 0$$
$$2x = 7 \qquad\qquad 5x = -1$$
$$x = \frac{7}{2} \qquad\qquad x = -\frac{1}{5}$$

The solution set is $\left\{\dfrac{7}{2}, -\dfrac{1}{5}\right\}$.

9. Factor the trinomial to obtain the two factors; then set each one equal to zero.

$$x^2 - x - 12 = 0$$
$$(x - 4)(x + 3) = 0$$
$$x - 4 = 0 \quad \text{or} \quad x + 3 = 0$$
$$x = 4 \qquad\qquad x = -3$$

The solution set is $\{-3, 4\}$.

11. Factor the trinomial to obtain the two factors; then set each one equal to zero.

$$x^2 + 9x + 14 = 0$$
$$(x + 2)(x + 7) = 0$$
$$x + 2 = 0 \quad \text{or} \quad x + 7 = 0$$
$$x = -2 \qquad\qquad x = -7$$

The solution set is $\{-7, -2\}$.

13. Add -1 to both sides of the equation and then factor the left side.

$$12x^2 + 4x = 1$$
$$12x^2 + 4x - 1 = 0$$
$$(2x + 1)(6x - 1) = 0$$
$$2x + 1 = 0 \quad \text{or} \quad 6x - 1 = 0$$
$$2x = -1 \qquad\qquad 6x = 1$$
$$x = -\frac{1}{2} \qquad\qquad x = \frac{1}{6}$$

The solution set is $\left\{-\dfrac{1}{2}, \dfrac{1}{6}\right\}$.

15. FOIL the left side of the equation and combine like terms. Then add 16 to both sides to set the equation equal to zero.

$$(x+4)(x-6) = -16$$
$$x^2 - 6x + 4x - 24 = -16$$
$$x^2 - 2x - 24 = -16$$
$$x^2 - 2x - 8 = 0$$
$$(x-4)(x+2) = 0$$
$$x - 4 = 0 \quad \text{or} \quad x + 2 = 0$$
$$x = 4 \qquad\qquad x = -2$$

The solution set is $\{-2, 4\}$.

17.
$$x^2 = 64$$
$$\sqrt{x^2} = \pm\sqrt{64}$$
$$x = \pm 8$$

The solution set is $\{\pm 8\}$.

19.
$$x^2 = 24$$
$$\sqrt{x^2} = \pm\sqrt{24}$$
$$x = \pm\sqrt{4 \cdot 6}$$
$$x = \pm 2\sqrt{6}$$

The solution set is $\left\{\pm 2\sqrt{6}\right\}$.

21. $x^2 = -5$; the solution set is $\varnothing$. There is no real number that will produce a negative number when it is squared.

23.
$$(x-4)^2 = 9$$
$$\sqrt{(x-4)^2} = \pm\sqrt{9}$$
$$x - 4 = \pm 3$$
$$x - 4 + 4 = 4 \pm 3$$
$$x = 7, 1$$

The solution set is $\{1, 7\}$.

25.
$$(4-x)^2 = 3$$
$$\sqrt{(4-x)^2} = \pm\sqrt{3}$$
$$4 - x = \pm\sqrt{3}$$
$$4 - 4 - x = -4 \pm\sqrt{3}$$
$$-x = -4 \pm\sqrt{3}$$
$$x = 4 \pm\sqrt{3}$$

The solution set is $\left\{4 \pm\sqrt{3}\right\}$.

27.
$$(2x-5)^2 = 13$$
$$\sqrt{(2x-5)^2} = \pm\sqrt{13}$$
$$2x - 5 = \pm\sqrt{13}$$
$$2x - 5 + 5 = 5 \pm\sqrt{13}$$
$$2x = 5 \pm\sqrt{13}$$
$$x = \frac{5 \pm\sqrt{13}}{2}$$

The solution set is $\left\{\dfrac{5 \pm\sqrt{13}}{2}\right\}$.

29. For the equation $4x^2 - 8x + 1 = 0$, $a = 4$, $b = -8$, and $c = 1$. Substitute these values into the quadratic formula.

$$x = \frac{-b \pm\sqrt{b^2 - 4ac}}{2a}$$
$$x = \frac{-(-8) \pm\sqrt{(-8)^2 - 4 \cdot 4 \cdot 1}}{2 \cdot 4}$$
$$= \frac{8 \pm\sqrt{64 - 16}}{8}$$
$$= \frac{8 \pm\sqrt{48}}{8}$$
$$= \frac{8 \pm\sqrt{16 \cdot 3}}{8}$$
$$= \frac{8 \pm 4\sqrt{3}}{8}$$
$$= \frac{4\left(2 \pm 1\sqrt{3}\right)}{8}$$
$$= \frac{2 \pm\sqrt{3}}{2}$$

The solution set is $\left\{\dfrac{2 \pm\sqrt{3}}{2}\right\}$.

31. First write the equation in standard form, which is $2x^2 - 2x - 1 = 0$. Then $a = 2$, $b = -2$, $c = -1$. Substitute these values into the quadratic formula.

$$x = \frac{-b \pm\sqrt{b^2 - 4ac}}{2a}$$

$$x = \frac{-(-2) \pm \sqrt{(-2)^2 - 4 \cdot 2 \cdot -1}}{2 \cdot 2}$$

$$= \frac{2 \pm \sqrt{4+8}}{4}$$

$$= \frac{2 \pm \sqrt{12}}{4}$$

$$= \frac{2 \pm \sqrt{4 \cdot 3}}{4}$$

$$= \frac{2 \pm 2\sqrt{3}}{4}$$

$$= \frac{2(1 \pm \sqrt{3})}{4}$$

$$= \frac{1 \pm \sqrt{3}}{2}$$

The solution set is $\left\{ \frac{1 \pm \sqrt{3}}{2} \right\}$.

33. First write the equation in standard form, which is $x^2 - x - 1 = 0$. Then, $a = 1$, $b = -1$, $c = -1$. Substitute these values into the quadratic formula.

$$x = \frac{-b \pm \sqrt{b^2 - 4ac}}{2a}$$

$$x = \frac{-(-1) \pm \sqrt{(-1)^2 - 4 \cdot 1 \cdot -1}}{2 \cdot 1}$$

$$= \frac{1 \pm \sqrt{1+4}}{2}$$

$$= \frac{1 \pm \sqrt{5}}{2}$$

The solution set is $\left\{ \frac{1 \pm \sqrt{5}}{2} \right\}$.

35. Write the equation in standard form by expanding the left side of the equation to obtain $4x^2 + 4x = 1$. Then subtract 1 from both sides to obtain $4x^2 + 4x - 1 = 0$. Now $a = 4$, $b = 4$, $c = -1$. Substitute these values into the quadratic formula.

$$x = \frac{-b \pm \sqrt{b^2 - 4ac}}{2a}$$

$$x = \frac{-(4) \pm \sqrt{(4)^2 - 4 \cdot 4 \cdot -1}}{2 \cdot 4}$$

$$= \frac{-4 \pm \sqrt{16+16}}{8}$$

$$= \frac{-4 \pm \sqrt{32}}{8}$$

$$= \frac{-4 \pm \sqrt{16 \cdot 2}}{8}$$

$$= \frac{-4 \pm 4\sqrt{2}}{8}$$

$$= \frac{4(-1 \pm 1\sqrt{2})}{8}$$

$$= \frac{-1 \pm \sqrt{2}}{2}$$

The solution set is $\left\{ \frac{-1 \pm \sqrt{2}}{2} \right\}$.

37. FOIL the left side of the equation to obtain $x^2 - x - 6 = 1$. Then subtract 1 from both sides to obtain $x^2 - x - 7 = 0$. Now $a = 1$, $b = -1$, $c = -7$. Substitute these values into the quadratic formula.

$$x = \frac{-b \pm \sqrt{b^2 - 4ac}}{2a}$$

$$x = \frac{-(-1) \pm \sqrt{(-1)^2 - 4 \cdot 1 \cdot (-7)}}{2 \cdot 1}$$

$$= \frac{1 \pm \sqrt{1+28}}{2}$$

$$= \frac{1 \pm \sqrt{29}}{2}$$

The solution set is $\left\{ \frac{1 \pm \sqrt{29}}{2} \right\}$.

39. Write the equation in standard form by adding 14 to both sides of the equation: $x^2 - 6x + 14 = 0$. Now $a = 1$, $b = -6$, $c = 14$. Substitute these values into the quadratic formula.

$$x = \frac{-b \pm \sqrt{b^2 - 4ac}}{2a}$$

$$x = \frac{-(-6) \pm \sqrt{(-6)^2 - 4 \cdot 1 \cdot 14}}{2 \cdot 1}$$

$$= \frac{6 \pm \sqrt{36-56}}{2}$$

$$= \frac{6 \pm \sqrt{-20}}{2}$$

A negative under the radical creates an imaginary number. The final solutions to the equation are not real numbers. Thus the solution set is $\varnothing$.

41. The presence of $2x^3$ makes it a <u>cubic</u> equation (degree 3).

43. $a = 1, b = 6, c = 9$

$b^2 - 4ac = 6^2 - 4 \cdot 1 \cdot 9 = 36 - 36 = 0$

(c); the equation has one rational solution because the quantity under the radical is zero. The solution will be $x = \dfrac{-b}{2a}$.

45. $a = 6, b = 7, c = -3$

$b^2 - 4ac = 7^2 - 4 \cdot 6 \cdot (-3) = 49 + 72 = 121$

(a); the equation has two different rational solutions. $\sqrt{121} = 11$

47. $a = 9, b = -30, c = 15$

$b^2 - 4ac = (-30)^2 - 4 \cdot 9 \cdot 15$

$= 900 - 540$

$= 360$

(b); the equation has two different irrational solutions because 360 is not a perfect square.

49. When using the quadratic formula, when $b^2 - 4ac$ is positive, the equation has <u>two</u> real solutions. Consider the quadratic formula $x = \dfrac{-b \pm \sqrt{b^2 - 4ac}}{2a}$.

If the quantity under the radical is positive, there will be two real solutions:

$x_1 = \dfrac{-b + \sqrt{\text{positive number}}}{2a}$

$x_2 = \dfrac{-b - \sqrt{\text{positive number}}}{2a}$

51. Replace s with 200:

$200 = -16t^2 + 45t + 400$

$200 - 200 = -16t^2 + 45t + 400 - 200$

$0 = -16t^2 + 45t + 200$

Then $a = -16, b = 45, c = 200$.

$t = \dfrac{-45 \pm \sqrt{(45)^2 - 4 \cdot (-16) \cdot 200}}{2 \cdot -16}$

$t = \dfrac{-45 \pm \sqrt{2025 + 12,800}}{-32}$

$t = \dfrac{-45 \pm \sqrt{14,825}}{-32}$

$t \approx \dfrac{-45 \pm 121.76}{-32}$

Then $t \approx \dfrac{-45 + 121.76}{-32}$ or $t \approx \dfrac{-45 - 121.76}{-32}$.

The first equation produces a negative value of t, which is not meaningful. The second equation produces $t \approx 5.2$ seconds.

53. At time zero, the object is at its initial height above the ground. Therefore, by letting $t = 0$, the value of s can be found. This is the height of the building.

55. (a) Replace s with 128. Obtain standard form for the equation by adding $16t^2$ and subtracting $144t$ from both sides of the equation.

$s = 144t - 16t^2$

$128 = 144t - 16t^2$

$16t^2 - 144t + 128 = 0$

Now solve for t by factoring.

$16t^2 - 144t + 128 = 0$

$16(t^2 - 9t + 8) = 0$

$16(t - 1)(t - 8) = 0$

Set each factor containing t equal to zero.

$t - 1 = 0$ or $t - 8 = 0$

$t = 1$ $t = 8$

The object will be 128 feet above the ground at 1 second and at 8 seconds. As it travels upward it will reach this height at 1 second; as it falls back to earth, it will be at the same height at 8 seconds.

(b) The object will strike the ground when $s = 0$.

$0 = 144t - 16t^2$

$0 = 16t(9 - t)$

Set each factor equal to zero.

$16t = 0$ or $9 - t = 0$

$t = 0$ $9 = t$

The object is on the ground at time zero when it is first projected, and it falls back to the ground 9 seconds after it is projected.

57. Use the Pythagorean theorem with the legs represented by the algebraic expressions $2x$ and $4x - 1$. The hypotenuse has the value $4x + 1$.

$$(2x)^2 + (4x-1)^2 = (4x+1)^2$$
$$4x^2 + 16x^2 - 8x + 1 = 16x^2 + 8x + 1$$
$$20x^2 - 8x + 1 = 16x^2 + 8x + 1$$
$$4x^2 - 16x = 0$$
$$4x(x-4) = 0$$
$$4x = 0 \text{ or } x - 4 = 0$$
$$x = 0 \text{ or } \quad x = 4$$

The value $x = 0$ does not make sense because then the leg of length $2x$ has length 0. So using $x = 4$, the lengths of the sides of the triangle are $2x = 2 \cdot 4 = 8$, $4x - 1 = 4 \cdot 4 - 1 = 15$, and $4x + 1 = 4 \cdot 4 + 1 = 17$.

59. Let 100 = length of the shorter leg, 400 = length of the longer leg (height of the building), c = length of the hypotenuse.

Use the Pythagorean theorem: $a^2 + b^2 = c^2$.

$$(100)^2 + (400)^2 = c^2$$
$$10,000 + 160,000 = c^2$$
$$170,000 = c^2$$
$$\sqrt{170,000} = \sqrt{c^2}$$
$$412.3 \approx c$$

The length of the wire is about 412.3 feet.

61. Examine the figure in the text to see that the two legs of the right triangle can be represented by x and $x + 70$. If the two ships are 170 miles apart, this value is the length of the hypotenuse. Again use the Pythagorean theorem with $a = x$, $b = x + 70$, and $c = 170$.

$$x^2 + (x+70)^2 = (170)^2$$
$$x^2 + (x^2 + 140x + 4900) = 28900$$
$$2x^2 + 140x + 4900 = 28900$$
$$2x^2 + 140x - 24000 = 0$$

Now use the quadratic formula with $a = 2$, $b = 140$, $c = -24,000$.

$$x = \frac{-(140) \pm \sqrt{(140)^2 - 4 \cdot 2 \cdot (-24,000)}}{2 \cdot 2}$$
$$= \frac{-140 \pm \sqrt{19,600 + 192,000}}{4}$$
$$= \frac{-140 \pm \sqrt{211,600}}{4}$$
$$= \frac{-140 \pm 460}{4}$$

Use $\dfrac{-140 + 460}{4} = 80$ miles. Then the ship traveling due east had traveled 80 miles, and the ship traveling south had traveled 80 + 70 or 150 miles.

63. Let a = length of the shorter leg, $2a + 2$ = length of the longer leg, $2a + 2 + 1$ = length of the hypotenuse. Using the Pythagorean theorem: $a^2 + b^2 = c^2$.

$$a^2 + (2a+2)^2 = (2a+2+1)^2$$
$$a^2 + (4a^2 + 8a + 4) = (2a+3)^2$$
$$5a^2 + 8a + 4 = 4a^2 + 12a + 9$$
$$a^2 - 4a - 5 = 0$$

Now use the quadratic formula with $a = 1$, $b = -4$, and $c = -5$.

$$a = \frac{-(-4) \pm \sqrt{(-4)^2 - 4 \cdot 1 \cdot (-5)}}{2 \cdot 1}$$
$$= \frac{4 \pm \sqrt{16 + 20}}{2}$$
$$= \frac{4 \pm \sqrt{36}}{2}$$
$$= \frac{4 \pm 6}{2}$$

Use $\dfrac{4 + 6}{2} = 5$ to obtain a positive value for a. Then the shorter leg is 5 cm, the longer leg is $2 \cdot 5 + 2 = 12$ cm, and the hypotenuse is $12 + 1 = 13$ cm.

65. Let w = width of the rectangle. Then $2w - 1$ = length of rectangle. Use the Pythagorean theorem with the width and length as the legs and 2.5 as the hypotenuse of the right triangle.

$$w^2 + (2w-1)^2 = (2.5)^2$$
$$w^2 + (4w^2 - 4w + 1) = 6.25$$
$$5w^2 - 4w + 1 = 6.25$$
$$5w^2 - 4w - 5.25 = 0$$

Now use the quadratic formula with $a = 5$,

$b = -4$, and $c = -5.25$.

$$w = \frac{-(-4) \pm \sqrt{(-4)^2 - 4 \cdot 5 \cdot (-5.25)}}{2 \cdot 5}$$

$$= \frac{4 \pm \sqrt{16 + 105}}{10}$$

$$= \frac{4 \pm \sqrt{121}}{10}$$

$$= \frac{4 \pm 11}{10}$$

Use $\dfrac{4 + 11}{10} = 1.5$ to obtain a positive value

for w. The second value $\dfrac{4 - 11}{10}$ is negative, which is not meaningful. The width is 1.5 cm and the length is $2(1.5) - 1 = 2$ cm.

67. The area of the floor is $15 \cdot 20$ or 300 square feet, and the area of the rug is given as 234 square feet. The remaining area of the strip around the rug will be $300 - 234 = 66$ square feet. Draw a sketch of the rug surrounded by the flooring. Let w equal the width of the border; that is, the distance from the rug to the wall.

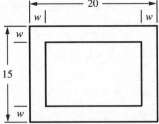

Four rectangles can be formed: The two rectangles on the left and right sides of the figure each have an area of $15 \cdot w$. The rectangles on the top and bottom of the figure each have an area of $(20 - 2w) \cdot w$. Because the total area of these rectangles is 66, an equation can be written:

$$2(15w) + 2[(20 - 2w) \cdot w] = 66$$

$$30w + 2[20w - 2w^2] = 66$$

$$30w + 40w - 4w^2 = 66$$

$$70w - 4w^2 = 66$$

$$-4w^2 + 70w - 66 = 66 - 66$$

$$-4w^2 + 70w - 66 = 0$$

Before using the quadratic formula, both sides of the equation can be divided by 2 or -2 in order to make the coefficients smaller.

$$\frac{-4w^2}{-2} + \frac{70w}{-2} - \frac{66}{-2} = \frac{0}{-2}$$

Now use the quadratic formula with $a = 2$,

$b = -35$, $c = 33$.

$$w = \frac{-(-35) \pm \sqrt{(-35)^2 - 4 \cdot 2 \cdot 33}}{2 \cdot 2}$$

$$= \frac{35 \pm \sqrt{1225 - 264}}{4}$$

$$= \frac{35 \pm \sqrt{961}}{4}$$

$$= \frac{35 \pm 31}{4}$$

The first value of w, $\dfrac{35 + 31}{4}$ yields $\dfrac{66}{4}$ or $16\dfrac{1}{2}$ feet. The second value of w, $\dfrac{35 - 31}{4}$, yields $\dfrac{4}{4}$ or 1 foot. Only the second value is meaningful in the context of the problem. That is, the width of the border is 1 foot rather than $16\dfrac{1}{2}$ feet.

69. It is helpful to make a sketch of the proposed garden and grass strip surrounding it. In the figure, the inner rectangle that represents the garden has dimensions $20 - 2x$ meters by $30 - 2x$ meters, where x represents the width of the strip of grass surrounding the garden. The outer rectangle, which represents the backyard has dimensions 20 by 30. Also, the area of the garden added to the area of the border will equal the total area of the backyard, or the area of the surrounding rectangle. This can be written as an equation.

$$(30 - 2x)(20 - 2x) + 184 = 30 \cdot 20$$

$$600 - 100x + 4x^2 + 184 = 600$$

$$784 - 100x + 4x^2 = 600$$

$$784 - 600 - 100x + 4x^2 = 600 - 600$$

$$184 - 100x + 4x^2 = 0$$

To make computations with the quadratic formula a little easier, divide both sides of the equation by 4 to give $46 - 25x + x^2 = 0$. Now use the quadratic formula with $a = 1$; $b = -25$; $c = 46$.

$$x = \frac{-(-25) \pm \sqrt{(-25)^2 - 4 \cdot 1 \cdot 46}}{2 \cdot 1}$$
$$= \frac{25 \pm \sqrt{625 - 184}}{2}$$
$$= \frac{25 \pm \sqrt{441}}{2}$$
$$= \frac{25 \pm 21}{2}$$

The first fraction yields 23 and the second fraction yields 2. However, one dimension of the garden, $20 - 2x$, would be a negative number if 23 is used to replace x. When 2 is used to replace x, the dimensions are $20 - 2 \cdot 2 = 16$ meters by $30 - 2 \cdot 2 = 26$ meters, which are reasonable answers.

71. Let $w =$ the width of the sheet of metal. Then the length of the sheet of metal can be represented by $l = 2w - 4$. In creating the box, 2 inches is taken away on each side of the length and on each side of the width. (See the diagram in the text.) The dimensions of the box are:
Length:
$L = l - 2 \cdot 2 = 2w - 4 - 2 \cdot 2 = 2w - 8.$
Width: $W = w - 2 \cdot 2 = w - 4.$
Height: $H = 2$ (the depth or height of the box when the corners have been turned up). The volume of box is $256 = LWH$. Substitute these algebraic expressions into the formula and solve for w.
$$256 = (2w - 8)(w - 4) \cdot 2$$
$$256 = (2w^2 - 8w - 8w + 32) \cdot 2$$
$$256 = (2w^2 - 16w + 32) \cdot 2$$
$$256 = 4w^2 - 32w + 64$$
$$0 = 4w^2 - 32w + 64 - 256$$
$$0 = 4w^2 - 32w - 192$$
To make computations with the quadratic formula easier, divide both sides of the equation by 4 to give $w^2 - 8w - 48 = 0$. Now use the quadratic formula with $a = 1$, $b = -8$, $c = -48$.

$$w = \frac{-(-8) \pm \sqrt{(-8)^2 - 4 \cdot 1 \cdot (-48)}}{2 \cdot 1}$$
$$= \frac{8 \pm \sqrt{64 + 192}}{2}$$
$$= \frac{8 \pm \sqrt{256}}{2}$$
$$= \frac{8 \pm 16}{2}$$
$$= 4 \pm 8$$

Since w must be positive, use $w = 4 + 8 = 12$ inches and $l = 2 \cdot 12 - 4 = 20$ inches.

73. Write a proportion and solve for x.
$$\frac{3x - 19}{x - 3} = \frac{x - 4}{4}$$
$$(x - 3)(x - 4) = 4(3x - 19)$$
$$x^2 - 7x + 12 = 12x - 76$$
$$x^2 - 19x + 88 = 0$$
$$(x - 8)(x - 11) = 0$$
$$x - 8 = 0 \text{ or } x - 11 = 0$$
$$x = 8 \text{ or } \qquad x = 11$$

Check these values in all sides of the triangles to make sure they do not produce negative values.
$x = 8$:
$3x - 19 = 3(8) - 19 = 24 - 19 = 5$
$x - 4 = 8 - 4 = 4$
$x - 3 = 8 - 3 = 5$
$x = 11$:
$3x - 19 = 3(11) - 19 = 33 - 19 = 14$
$x - 4 = 11 - 4 = 7$
$x - 3 = 11 - 3 = 8$
Since both values of x are acceptable, the length of AC could be $3x - 19 = 5$ or $3x - 19 = 14$.

75. $x^3 + 9x = 26$

$m = 9$

$n = 26$

$$x = \sqrt[3]{\frac{n}{2} + \sqrt{\left(\frac{n}{2}\right)^2 + \left(\frac{m}{3}\right)^3}} - \sqrt[3]{-\left(\frac{n}{2}\right) + \sqrt{\left(\frac{n}{2}\right)^2 + \left(\frac{m}{3}\right)^3}}$$

$$= \sqrt[3]{\frac{26}{2} + \sqrt{\left(\frac{26}{2}\right)^2 + \left(\frac{9}{3}\right)^3}} - \sqrt[3]{-\left(\frac{26}{2}\right) + \sqrt{\left(\frac{26}{2}\right)^2 + \left(\frac{9}{3}\right)^3}}$$

$$= \sqrt[3]{13 + \sqrt{13^2 + 3^3}} - \sqrt[3]{-13 + \sqrt{13^2 + 3^3}}$$

$$= \sqrt[3]{13 + \sqrt{196}} - \sqrt[3]{-13 + \sqrt{196}}$$

$$= \sqrt[3]{13 + 14} - \sqrt[3]{-13 + 14}$$

$$= \sqrt[3]{27} - \sqrt[3]{1}$$

$$= 3 - 1$$

$$= 2$$

The solution set is $\{2\}$.

EXTENSION: COMPLEX SOLUTIONS OF QUADRATIC EQUATIONS

1. $x^2 + 12 = 0$

$a = 1,\ b = 0,\ c = 12$

$$x = \frac{-b \pm \sqrt{b^2 - 4ac}}{2a}$$

$$x = \frac{-0 \pm \sqrt{0^2 - 4(1)(12)}}{2(1)}$$

$$x = \frac{\pm\sqrt{-48}}{2}$$

$$x = \frac{\pm 4i\sqrt{3}}{2}$$

$$x = \pm 2i\sqrt{3}$$

The solution set is $\left\{-2i\sqrt{3},\ 2i\sqrt{3}\right\}$.

3. $9x(x - 2) = -13$

$9x^2 - 18x + 13 = 0$

$a = 9,\ b = -18,\ c = 13$

$$x = \frac{-b \pm \sqrt{b^2 - 4ac}}{2a}$$

$$x = \frac{18 \pm \sqrt{(-18)^2 - 4(9)(13)}}{2(9)}$$

$$x = \frac{18 \pm \sqrt{-144}}{18}$$

$$x = \frac{18 \pm 12i}{18}$$

$$x = 1 \pm \frac{2}{3}i$$

The solution set is $\left\{1 + \frac{2}{3}i,\ 1 - \frac{2}{3}i\right\}$.

5. $x^2 - 6x + 14 = 0$

$a = 1, b = -6, c = 14$

$x = \dfrac{-b \pm \sqrt{b^2 - 4ac}}{2a}$

$x = \dfrac{6 \pm \sqrt{(-6)^2 - 4(1)(14)}}{2(1)}$

$x = \dfrac{6 \pm \sqrt{-20}}{2}$

$x = \dfrac{6 \pm 2i\sqrt{5}}{2}$

$x = 3 \pm i\sqrt{5}$

The solution set is $\left\{ 3 + i\sqrt{5},\ 3 - i\sqrt{5} \right\}$.

7. $4x^2 - 4x = -7$

$4x^2 - 4x + 7 = 0$

$a = 4, b = -4, c = 7$

$x = \dfrac{-b \pm \sqrt{b^2 - 4ac}}{2a}$

$x = \dfrac{4 \pm \sqrt{(-4)^2 - 4(4)(7)}}{2(4)}$

$x = \dfrac{4 \pm \sqrt{-96}}{8}$

$x = \dfrac{4 \pm 4i\sqrt{6}}{8}$

$x = \dfrac{1}{2} \pm \dfrac{\sqrt{6}}{2}i$

The solution set is $\left\{ \dfrac{1}{2} + \dfrac{\sqrt{6}}{2}i,\ \dfrac{1}{2} - \dfrac{\sqrt{6}}{2}i \right\}$.

9. $x(3x + 4) = -2$

$3x^2 + 4x + 2 = 0$

$a = 3, b = 4, c = 2$

$x = \dfrac{-b \pm \sqrt{b^2 - 4ac}}{2a}$

$x = \dfrac{-4 \pm \sqrt{4^2 - 4(3)(2)}}{2(3)}$

$x = \dfrac{-4 \pm \sqrt{-8}}{6}$

$x = \dfrac{-4 \pm 2i\sqrt{2}}{6}$

$x = -\dfrac{2}{3} \pm \dfrac{\sqrt{2}}{3}i$

The solution set is $\left\{ -\dfrac{2}{3} + \dfrac{\sqrt{2}}{3}i,\ -\dfrac{2}{3} - \dfrac{\sqrt{2}}{3}i \right\}$.

Chapter 7 Test

1. $5x - 3 + 2x = 3(x - 2) + 11$

$7x - 3 = 3x - 6 + 11$

$7x - 3 = 3x + 5$

$7x - 3 + 3 = 3x + 5 + 3$

$7x = 3x + 8$

$7x - 3x = 3x - 3x + 8$

$4x = 8$

$\dfrac{4x}{4} = \dfrac{8}{4}$

$x = 2;\ \{2\}$

2. $\dfrac{2x - 1}{3} + \dfrac{x + 1}{4} = \dfrac{43}{12}$

$\dfrac{12}{1} \cdot \dfrac{2x - 1}{3} + \dfrac{12}{1} \cdot \dfrac{x + 1}{4} = \dfrac{12}{1} \cdot \dfrac{43}{12}$

$4(2x - 1) + 3(x + 1) = 43$

$8x - 4 + 3x + 3 = 43$

$11x - 1 = 43$

$11x - 1 + 1 = 43 + 1$

$11x = 44$

$\dfrac{11x}{11} = \dfrac{44}{11}$

$x = 4;\ \{4\}$

3. $3x - (2 - x) + 4x = 7x - 2 - (-x)$

$3x - 2 + x + 4x = 7x - 2 + x$

$8x - 2 = 8x - 2$

This is an identity; the solution set is {all real numbers}.

4. $S = vt - 16t^2$

$S + 16t^2 = vt - 16t^2 + 16t^2$

$S + 16t^2 = vt$

$\dfrac{S + 16t^2}{t} = \dfrac{vt}{t}$

$\dfrac{S + 16t^2}{t} = v$ or $v = \dfrac{S}{t} + 16t$

5. Let k = the area of Kauai; $k + 177$ = the area of Maui; and $(k + 177) + 3293$ = the area of Hawaii.

$k + k + 177 + (k + 177) + 3293 = 5300$

$3k + 3647 = 5300$

$3k + 3647 - 3647 = 5300 - 3647$

$3k = 1653$

$\dfrac{3k}{3} = \dfrac{1653}{3}$

$k = 551$

Then the area of Kauai is 551 square miles; the area of Maui is 551 + 177 = 728 square

miles; and the area of Hawaii is
728 + 3293 = 4021 square miles.

6.

Strength	L of solution	L of alcohol
20%	x	$0.20(x)$
50%	10	$0.50(10)$
40%	$x + 10$	$0.40(x + 10)$

Create an equation by adding the first two
algebraic expressions in the last column to
total the third:
$$0.2(x)+0.5(10)=0.4(x+10)$$
$$10[0.2(x)+0.5(10)]=10[0.4(x+10)]$$
$$2x+50=4(x+10)$$
$$2x+50=4x+40$$
$$2x+50-50=4x+40-50$$
$$2x=4x-10$$
$$2x-4x=4x-4x-10$$
$$-2x=-10$$
$$\frac{-2x}{-2}=\frac{-10}{-2}$$
$$x=5 \text{ liters of 20\%}$$
$$\text{solution.}$$

7.

	Rate	Time	Distance
Passenger Train	60	t	$60t$
Freight train	75	t	$75t$

Because the trains are traveling in opposite
directions, the sum of their distances will
equal the total distance apart of 297 miles.
Use this information to create an equation:
$$60t+75t=297$$
$$135t=297$$
$$\frac{135t}{135}=\frac{297}{135}$$
$$t=2.2 \text{ hours}$$

8. $\frac{\$4.38}{16 \text{ slices}} \approx \$0.274/\text{slice}$

$\frac{\$3.30}{12 \text{ slices}} = \$0.275/\text{slice}$

16 slices for $4.38 is the better buy.

9. Let z = the actual distance between Seattle
and Cincinnati.
$$\frac{z}{46}=\frac{1050}{21}$$
$$21z=46\cdot1050$$
$$21z=48,300$$
$$\frac{21z}{21}=\frac{48,300}{21}$$
$$z=2300 \text{ miles}$$

10. $I=\frac{k}{r}$
$$80=\frac{k}{30}$$
$$30\cdot80=k$$
$$2400=k$$
Then $I=\frac{2400}{12}$
$$I=200 \text{ amps.}$$

11. $-4x+2(x-3)\ge 4x-(3+5x)-7$
$$-4x+2x-6\ge4x-3-5x-7$$
$$-2x-6\ge-x-10$$
$$-2x-6+6\ge-x-10+6$$
$$-2x\ge-x-4$$
$$-2x+x\ge-x+x-4$$
$$-x\ge-4$$
$$\frac{-x}{-1}\le\frac{-4}{-1}$$
$$x\le4$$
Interval notation: $(-\infty, 4]$
Graph:

12. $-10<3x-4\le14$
$$-10+4<3x-4+4\le14+4$$
$$-6<3x\le18$$
$$\frac{-6}{3}<\frac{3x}{3}\le\frac{18}{3}$$
$$-2<x\le6$$
Interval notation: $(-2, 6]$
Graph:

13. A. $-3x<9$
$$\frac{-3x}{-3}>\frac{9}{-3}$$
$$x>-3$$
This is not equivalent.

B. $-3x > -9$

$$\frac{-3x}{-3} < \frac{-9}{-3}$$

$$x < 3$$

This is not equivalent.

C. $-3x > 9$

$$\frac{-3x}{-3} < \frac{9}{-3}$$

$$x < -3$$

This is equivalent.

D. $-3x < -9$

$$\frac{-3x}{-3} > \frac{-9}{-3}$$

$$x > 3$$

This is not equivalent.

14. Let x = the possible scores on the fourth test.

$$\frac{83 + 76 + 79 + x}{4} \geq 80$$

$$\frac{4}{1} \cdot \frac{238 + x}{4} \geq 80 \cdot 4$$

$$238 + x \geq 320$$

$$x \geq 82$$

He must score on 82 or better.

15. $\left(\dfrac{4}{3}\right)^2 = \dfrac{4}{3} \cdot \dfrac{4}{3} = \dfrac{16}{9}$

16. $-(-2)^6 = -(-2 \cdot -2 \cdot -2 \cdot -2 \cdot -2 \cdot -2)$

$$= -(64)$$

$$= -64$$

17. $\left(\dfrac{3}{4}\right)^{-3} = \dfrac{3^{-3}}{4^{-3}}$

$$= \dfrac{\frac{1}{3^3}}{\frac{1}{4^3}}$$

$$= \dfrac{\frac{1}{27}}{\frac{1}{64}}$$

$$= \dfrac{1}{27} \div \dfrac{1}{64}$$

$$= \dfrac{1}{27} \cdot \dfrac{64}{1}$$

$$= \dfrac{64}{27}$$

18. $-5^0 + (-5)^0 = -1 + 1 = 0$

19. $9(4p^3)(6p^{-7}) = 9 \cdot 4 \cdot 6 \cdot p^{3+(-7)}$

$$= 216p^{-4}$$

$$= \dfrac{216}{1} \cdot \dfrac{1}{p^4}$$

$$= \dfrac{216}{p^4}$$

20. $\dfrac{m^{-2}(m^3)^{-3}}{m^{-4}m^7} = \dfrac{m^{-2} \cdot m^{-9}}{m^{-4+7}}$

$$= \dfrac{m^{-11}}{m^3}$$

$$= m^{-11-3}$$

$$= m^{-14}$$

$$= \dfrac{1}{m^{14}}$$

21. (a) Moving the decimal 8 places to the right gives: $6.93 \times 10^8 = 693,000,000$.

(b) Moving the decimal 7 places to the left gives: $1.25 \times 10^{-7} = 0.000000125$.

22. $\dfrac{(2,500,000)(0.00003)}{(0.05)(5,000,000)}$

$$= \dfrac{(2.5 \times 10^6)(3 \times 10^{-5})}{(5 \times 10^{-2})(5 \times 10^6)}$$

$$= \dfrac{(2.5 \times 3) \times (10^{6-5})}{(5 \times 5) \times (10^{-2+6})}$$

$$= \dfrac{7.5 \times 10^1}{25 \times 10^4}$$

$$= \dfrac{7.5}{25} \times \dfrac{10^1}{10^4}$$

$$= 0.3 \times 10^{1-4}$$

$$= 0.3 \times 10^{-3}$$

$$= 3 \times 10^{-4}$$

Remember that scientific notation has one non zero digit to the left of the decimal point. As 0.3 is made larger by a power of ten to become 3, 10^{-3} must become smaller by a power of ten.

23. Solve the formula $D = rt$ for t by dividing both sides of the equation by r.

$$\dfrac{D}{r} = \dfrac{rt}{r} = t$$

Then replace the given distance and rate and simplify.

$$\frac{4.58 \times 10^9}{3.00 \times 10^5} = \frac{4.58}{3.00} \times \frac{10^9}{10^5}$$
$$\approx 1.53 \times 10^{9-5}$$
$$= 1.53 \times 10^4$$
$$= 15,300 \text{ seconds}$$

24. $(3k^3 - 5k^2 + 8k - 2) - (3k^3 - 9k^2 + 2k - 12)$
$$= 3k^3 - 5k^2 + 8k - 2 - 3k^3 + 9k^2 - 2k + 12$$
$$= (3k^3 - 3k^3) + (-5k^2 + 9k^2) + (8k - 2k)$$
$$\quad + (-2 + 12)$$
$$= 4k^2 + 6k + 10$$

25. $(5x + 2)(3x - 4)$
$$= 5x \cdot 3x + 5x \cdot (-4) + 2 \cdot 3x + 2 \cdot (-4)$$
$$= 15x^2 - 20x + 6x - 8$$
$$= 15x^2 - 14x - 8$$

26. $(4x^2 - 3)(4x^2 + 3) = (4x^2)^2 - (3)^2$
$$= 16x^4 - 9$$

If the pattern is recognized, this multiplication of binomials can be done quickly. Otherwise, use the FOIL method as in Exercise 25.

27. Using vertical multiplication:

$$3x^2 + 8x - 9$$
$$\underline{x + 4}$$
$$12x^2 + 32x - 36$$
$$\underline{3x^3 + 8x^2 - 9x}$$
$$3x^3 + 20x^2 + 23x - 36$$

28. One of many possibilities is
$$2t^5 + 8t^4 + t^3 - 7t^2 + 2t + 1.$$

29. Find two quantities whose product is $2 \cdot 3q^2 = 6q^2$ and whose sum is $-5q$, the coefficient of the middle term. The two quantities are $-2q$ and $-3q$. Use these two algebraic expressions as coefficients for p in place of the middle term; then factor the four-term polynomial by grouping.
$$2p^2 - 5pq + 3q^2 = 2p^2 - 2pq - 3pq + 3q^2$$
$$= 2p(p - q) - 3q(p - q)$$
$$= (p - q)(2p - 3q)$$
Remember that the two factors can also be expressed as $(2p - 3q)(p - q)$.

30. This is the difference of squares.
$$(10x)^2 - (7y)^2 = (10x + 7y)(10x - 7y)$$

31. This is the difference of cubes.
$$(3y)^3 - (5x)^3$$
$$= (3y - 5x)[(3y)^2 + 3y \cdot 5x + (5x)^2]$$
$$= (3y - 5x)(9y^2 + 15xy + 25x^2)$$

32. Factor by grouping.
$$4x + 4y - mx - my = 4(x + y) - m(x + y)$$
$$= (4 - m)(x + y)$$

33. In this equation $a = 6, b = 7, c = -3$. Substitute these values into the quadratic formula.
$$x = \frac{-b \pm \sqrt{b^2 - 4ac}}{2a}$$
$$x = \frac{-(7) \pm \sqrt{(7)^2 - 4 \cdot 6 \cdot (-3)}}{2 \cdot 6}$$
$$= \frac{-7 \pm \sqrt{49 + 72}}{12}$$
$$= \frac{-7 \pm \sqrt{121}}{12}$$
$$= \frac{-7 \pm 11}{12}$$

The two solutions are $\dfrac{-7 + 11}{12} = \dfrac{4}{12} = \dfrac{1}{3}$ or

$\dfrac{-7 - 11}{12} = \dfrac{-18}{12} = -\dfrac{3}{2}$. The solution set is

$\left\{-\dfrac{3}{2}, \dfrac{1}{3}\right\}$.

This equation could also be solved by factoring.

34. $x^2 - 13 = 0$
$$x^2 = 13$$
$$\sqrt{x^2} = \pm\sqrt{13}$$
$$x = \pm\sqrt{13}$$
The solution set is $\left\{\pm\sqrt{13}\right\}$.

35. First write the equation in standard form.
$$x^2 - x - 7 = 0$$
In this equation $a = 1, b = -1, c = -7$. Substitute these values into the quadratic formula.
$$x = \frac{-b \pm \sqrt{b^2 - 4ac}}{2a}$$

$$x = \frac{-(-1) \pm \sqrt{(-1)^2 - 4 \cdot 1 \cdot -7}}{2 \cdot 1}$$

$$= \frac{1 \pm \sqrt{1 + 28}}{2}$$

$$= \frac{1 \pm \sqrt{29}}{2}$$

The solution set is $\left\{ \dfrac{1 \pm \sqrt{29}}{2} \right\}$.

36. Replace s with 25 and solve for t.

$$25 = 16t^2 + 15t$$

$$25 - 25 = 16t^2 + 15t - 25$$

$$0 = 16t^2 + 15t - 25$$

In this equation $a = 16$, $b = 15$, $c = -25$.
Substitute these values into the quadratic
formula.

$$x = \frac{-b \pm \sqrt{b^2 - 4ac}}{2a}$$

$$x = \frac{-(15) \pm \sqrt{(15)^2 - 4 \cdot 16 \cdot (-25)}}{2 \cdot 16}$$

$$= \frac{-15 \pm \sqrt{225 + 1600}}{32}$$

$$= \frac{-15 \pm \sqrt{1825}}{32}$$

$$= \frac{-15 \pm 42.7}{32}$$

The first fraction simplifies to about
0.87 seconds. The second fraction produces
a negative number, which is not meaningful.

Chapter 8

8.1 Exercises

1. For any value of x, the point $(x, 0)$ lies on the <u>x</u>-axis.

3. The circle $x^2 + y^2 = 9$ has the point <u>$(0, 0)$</u> as its center.

5. **(a)** The point $(1, 6)$ is located in quadrant I since the x-value and the y-value are both positive.

   (b) The point $(-4, -2)$ is located in quadrant III since the x-value and the y-value are both negative.

   (c) The point $(-3, 6)$ is located in quadrant II since the x-value is negative and the y-value is positive.

   (d) The point $(7, -5)$ is located in quadrant IV since the x-value is positive and the y-value is negative.

   (e) The point $(-3, 0)$ is located between quadrants II and III (i.e., on the negative x-axis) since the x-value is negative and the y-value is 0.

7. **(a)** If $xy > 0$, then x and y must have same signs. Thus, the point must lie in Quadrant I or III.

   (b) If $xy < 0$, then x and y must have opposite signs. Thus, the point must lie in Quadrant II or IV.

   (c) If $\dfrac{x}{y} < 0$, then x and y must have opposite signs. Thus, the point must lie in Quadrant II or IV.

   (d) If $\dfrac{x}{y} > 0$, then x and y must have same signs. Thus, the point must lie in Quadrant I or III.

9–18. *Note: graph shows even and odd answers.*

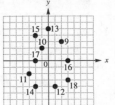

19. **(a)** The variable x represents the year, and the variable y represents federal tax revenue (in billions of dollars).

    (b) The revenue that corresponds to $x = 2006$ is about $2400 billion.

    (c) $(2006, 2400)$

21. **(a)**
$$d = \sqrt{(x_2 - x_1)^2 + (y_2 - y_1)^2}$$
$$= \sqrt{(-2-3)^2 + (1-4)^2}$$
$$= \sqrt{(-5)^2 + (-3)^2}$$
$$= \sqrt{25+9}$$
$$= \sqrt{34}$$

    (b) Using $\left(\dfrac{x_1 + x_2}{2}, \dfrac{y_1 + y_2}{2} \right)$, we have
$$\left(\frac{3+(-2)}{2}, \frac{4+1}{2} \right) = \left(\frac{1}{2}, \frac{5}{2} \right).$$

23. **(a)**
$$d = \sqrt{(x_2 - x_1)^2 + (y_2 - y_1)^2}$$
$$= \sqrt{[3-(-2)]^2 + (-2-4)^2}$$
$$= \sqrt{5^2 + (-6)^2}$$
$$= \sqrt{25+36}$$
$$= \sqrt{61}$$

    (b) Using $\left(\dfrac{x_1 + x_2}{2}, \dfrac{y_1 + y_2}{2} \right)$, we have
$$\left(\frac{(-2)+3}{2}, \frac{4+(-2)}{2} \right) = \left(\frac{1}{2}, 1 \right).$$

25. **(a)**
$$d = \sqrt{(x_2 - x_1)^2 + (y_2 - y_1)^2}$$
$$= \sqrt{[2-(-3)]^2 + (-4-7)^2}$$
$$= \sqrt{(5)^2 + (-11)^2}$$
$$= \sqrt{25+121}$$
$$= \sqrt{146}$$

(b) Using $\left(\dfrac{x_1+x_2}{2}, \dfrac{y_1+y_2}{2}\right)$,

$$\left(\dfrac{-3+2}{2}, \dfrac{7+(-4)}{2}\right) = \left(-\dfrac{1}{2}, \dfrac{3}{2}\right).$$

27. $(x-3)^2 + (y-2)^2 = 25$

The equation indicates a graph of a circle with a radius of 5 and centered at (3, 2). Therefore, choose graph B.

29. $(x+3)^2 + (y-2)^2 = 25$
$[x-(-3)]^2 + (y-2)^2 = 25$

The equation indicates a graph of a circle with a radius of 5 and centered at (−3, 2). Thus, choose graph D.

31. Use the equation of a circle, where $h = 0$, $k = 0$, and $r = 6$.

$$(x-h)^2 + (y-k)^2 = r^2$$
$$(x-0)^2 + (y-0)^2 = 6^2$$
$$x^2 + y^2 = 36$$

33. Use the equation of a circle, where $h = -1$, $k = 3$, and $r = 4$.

$$(x-h)^2 + (y-k)^2 = r^2$$
$$[x-(-1)]^2 + (y-3)^2 = 4^2$$
$$(x+1)^2 + (y-3)^2 = 16$$

35. Use the equation of a circle, where $h = 0$, $k = 4$, and $r = \sqrt{3}$.

$$(x-h)^2 + (y-k)^2 = r^2$$
$$(x-0)^2 + (y-4)^2 = \left(\sqrt{3}\right)^2$$
$$x^2 + (y-4)^2 = 3$$

37. The equation, $x^2 + y^2 = r^2$, $r > 0$ or

equivalently, $(x-0)^2 + (y-0)^2 = r^2$,

$r > 0$, implies that the center is located at (0, 0) and the radius is r.

39. To find the center and radius, complete the square on x and y.

$$x^2 + y^2 + 4x + 6y + 9 = 0$$
$$(x^2 + 4x +) + (y^2 + 6y +) = -9$$
$$(x^2 + 4x + 4) + (y^2 + 6y + 9) = -9 + 4 + 9$$
$$(x+2)^2 + (y+3)^2 = 4 \text{ or}$$
$$[x-(-2)]^2 + [y-(-3)]^2 = 4$$

Thus, by inspection, the center is located at (−2, −3) and the radius is given by

$r = \sqrt{4} = 2$. Remember that the added constants, 4 and 9, come from squaring $\dfrac{1}{2}$ of the coefficients of each first-degree term, i.e. $\left[\dfrac{1}{2}(4)\right]^2 = 4$ and $\left[\dfrac{1}{2}(6)\right]^2 = 9$.

41. To find the center and radius, complete the square on x and y.

$$x^2 + y^2 + 10x - 14y - 7 = 0$$
$$(x^2 + 10x +) + (y^2 - 14y +) = 7$$
$$(x^2 + 10x + 25) + (y^2 - 14y + 49) = 7 + 25$$
$$ + 49$$
$$(x+5)^2 + (y-7)^2 = 81 \text{ or}$$
$$[x-(-5)]^2 + (y-7)^2 = 9^2$$

Thus, by inspection, the center is located at (−5, 7) and the radius is given by $r = 9$.

43. To find the center and radius, complete the square on x and y.

$$3x^2 + 3y^2 - 12x - 24y + 12 = 0$$
$$x^2 + y^2 - 4x - 8y + 4 = 0 \text{ Divide by } 3.$$
$$(x^2 - 4x +) + (y^2 - 8y +) = -4$$
$$(x^2 - 4x + 4) + (y^2 - 8y + 16) = -4 + 4 + 16$$
$$(x-2)^2 + (y-4)^2 = 16$$

Thus, by inspection, the center is located at (2, 4) and the radius is given by $r = \sqrt{16} = 4$.

45. The equation, $x^2 + y^2 = 36$, is equivalent to $(x-0)^2 + (y-0)^2 = 6^2$. Thus, the center of the graph is located at (0, 0) and has a radius $r = 6$. The graph is as follows.

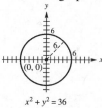

$x^2 + y^2 = 36$

47. The equation, $(x-2)^2 + y^2 = 36$, is equivalent to $(x-2)^2 + (y-0)^2 = 6^2$. Thus, the center of the graph is located at (2, 0) and has a radius $r = 6$. The graph is as follows.

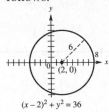

$(x-2)^2 + y^2 = 36$

49. The equation, $(x+2)^2 + (y-5)^2 = 16$, is equivalent to $[x-(-2)]^2 + (y-5)^2 = 4^2$. Thus, the center of the graph is located at $(-2, 5)$ and has a radius $r = 4$. The graph is as follows.

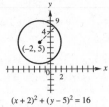

$(x+2)^2 + (y-5)^2 = 16$

51. The equation, $(x+3)^2 + (y+2)^2 = 36$, is equivalent to $[x-(-3)]^2 + [y-(-2)]^2 = 6^2$. Thus, the center of the graph is located at $(-3, -2)$ and has a radius $r = 6$. The graph is as follows.

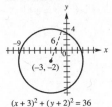

$(x+3)^2 + (y+2)^2 = 36$

53. (a)
$$d = \sqrt{(x_2 - x_1)^2 + (y_2 - y_1)^2}$$
$$= \sqrt{(2-(-4))^2 + (5-3)^2}$$
$$= \sqrt{(6)^2 + (2)^2}$$
$$= \sqrt{36+4}$$
$$= \sqrt{40}$$
$$= 2\sqrt{10}$$

(b) Using $\left(\dfrac{x_1 + x_2}{2}, \dfrac{y_1 + y_2}{2} \right)$, we have
$$\left(\frac{-4+2}{2}, \frac{3+5}{2} \right) = (-1, 4).$$

55. Using $\dfrac{y_1 + y_2}{2}$, we have
$$\frac{28.0 + 20.3}{2} = 24.15 \text{ or } 24.15\%.$$

57. Using $\dfrac{y_1 + y_2}{2}$, we have
$$\frac{17,603 + 13,359}{2} = 15,481 \text{ or } \$15,481.$$

59. Writing exercise; answers will vary.

61. Writing exercise; answers will vary.

63.

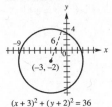

Thus, the epicenter is at $(-2, -2)$.

65. Writing exercise; answers will vary.

67. Let $(x_1, y_1) = (3, -8)$, $(x, y) = (6, 5)$, and (x_2, y_2) be the coordinates of the other endpoint. Then
$$x = \frac{x_1 + x_2}{2}$$
$$6 = \frac{3 + x_2}{2}$$
$$12 = 3 + x_2$$
$$x_2 = 9$$
and
$$y = \frac{y_1 + y_2}{2}$$
$$5 = \frac{-8 + y_2}{2}$$
$$10 = -8 + y_2$$
$$y_2 = 18.$$
Thus, the coordinates of the other endpoint are $(9, 18)$.

69. Only option (B) can be written in the form $x^2 + y^2 = r^2$ where $r^2 > 0$, and hence, is the only equation that will represent a circle.

8.2 Exercises

1. The given equation is $2x + y = 5$.

For $(0,\ )$: $2 \cdot 0 + y = 5$

$$y = 5$$

or the ordered pair $(0, 5)$.

For $(\ , 0)$: $2x + 0 = 5$

$$x = \frac{5}{2}$$

or the ordered pair $\left(\frac{5}{2}, 0\right)$.

For $(1,\ )$: $2 \cdot 1 + y = 5$

$$y = 3$$

or the ordered pair $(1, 3)$.

For $(\ , 1)$: $2x + 1 = 5$

$$2x = 4$$
$$x = 2$$

or the ordered pair $(2, 1)$.

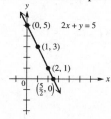

3. The given equation is $x - y = 4$.

For $(0,\ )$: $0 - y = 4$

$$-y = 4$$
$$y = -4$$

or the ordered pair $(0, -4)$.

For $(\ , 0)$: $x - 0 = 4$

$$x = 4$$

or the ordered pair $(4, 0)$.

For $(2,\ )$: $2 - y = 4$

$$-y = 2$$
$$y = -2$$

or the ordered pair $(2, -2)$.

For $(\ , -1)$: $x - (-1) = 4$

$$x + 1 = 4$$
$$x = 3$$

or the ordered pair $(3, -1)$.

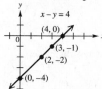

5. The given equation is $4x + 5y = 20$.

For $(0,\ )$: $4 \cdot 0 + 5y = 20$

$$5y = 20$$
$$y = 4$$

or the ordered pair $(0, 4)$.

For $(\ , 0)$: $4x + 5 \cdot 0 = 20$

$$4x = 20$$
$$x = 5$$

or the ordered pair $(5, 0)$.

For $(3,\ )$: $4 \cdot 3 + 5y = 20$

$$12 + 5y = 20$$
$$5y = 8$$
$$y = \frac{8}{5}$$

or the ordered pair $\left(3, \frac{8}{5}\right)$.

For $(\ , 2)$: $4x + 5 \cdot 2 = 20$

$$4x + 10 = 20$$
$$4x = 10$$
$$x = \frac{5}{2}$$

or the ordered pair $\left(\frac{5}{2}, 2\right)$.

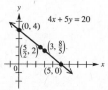

7. The given equation is $3x + 2y = 8$.
From the partially completed table in the text we want to complete the evaluation of the ordered pairs $(0,\ )$, $(\ , 0)$, $(2,\ )$, $(\ , -2)$.

For $(0,\ )$: $3 \cdot 0 + 2y = 8$

$$2y = 8$$
$$y = 4$$

or the ordered pair $(0, 4)$.

For $(\ , 0)$: $3x + 0 \cdot y = 8$

$$3x = 8$$
$$x = \frac{8}{3}$$

or the ordered pair $\left(\frac{8}{3}, 0\right)$.

For $(2,\ )$: $3 \cdot 2 + 2y = 8$

$$6 + 2y = 8$$
$$2y = 2$$
$$y = 1$$

or the ordered pair $(2, 1)$.

For $(\ , -2)$: $3x + 2 \cdot (-2) = 8$

$$3x - 4 = 8$$
$$3x = 12$$
$$x = 4$$

or the ordered pair $(4, -2)$.

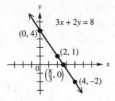

9. Writing exercise; answers will vary.

11. Option A is correct since y is constant at $y = 3$.

13. The given equation is $3x + 2y = 12$.
To find the x-intercept, let $y = 0$.
$$3x + 2 \cdot 0 = 12$$
$$3x = 12$$
$$x = 4$$
The x-intercept is $(4, 0)$.
To find the y-intercept, let $x = 0$.
$$3 \cdot 0 + 2y = 12$$
$$2y = 12$$
$$y = 6$$
The y-intercept is $(0, 6)$.

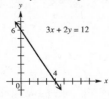

15. The given equation is $5x + 6y = 10$.
To find the x-intercept, let $y = 0$.
$$5x + 6 \cdot 0 = 10$$
$$5x = 10$$
$$x = 2$$
The x-intercept is $(2, 0)$.
To find the y-intercept, let $x = 0$.
$$5 \cdot 0 + 6y = 10$$
$$6y = 10$$
$$y = \frac{5}{3}$$
The y-intercept is $\left(0, \frac{5}{3}\right)$.

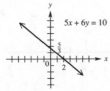

17. The given equation is $2x - y = 5$.
To find the x-intercept, let $y = 0$.
$$2x - 0 = 5$$
$$x = \frac{5}{2}$$

The x-intercept is $\left(\frac{5}{2}, 0\right)$.
To find the y-intercept, let $x = 0$.
$$2 \cdot 0 - y = 5$$
$$-y = 5$$
$$y = -5$$
The y-intercept is $(0, -5)$.

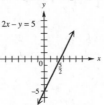

19. The given equation is $x - 3y = 2$.
To find the x-intercept, let $y = 0$.
$$x - 3 \cdot 0 = 2$$
$$x = 2$$
The x-intercept is $(2, 0)$.
To find the y-intercept, let $x = 0$.
$$0 - 3y = 2$$
$$y = -\frac{2}{3}$$
The x-intercept is $\left(0, -\frac{2}{3}\right)$.

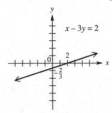

21. The given equation is $y + x = 0$.
To find the y-intercept, let $x = 0$.
$$y + 0 = 0$$
$$y = 0$$
The y-intercept is $(0, 0)$. Observe that this is also the x-intercept. Thus, the graph runs through the origin.
To find the second point let x (or y) take on a value and solve for the other variable, i.e. let $x = 2$.
$$y + 2 = 0$$
$$y = -2$$
Thus, a second point would have coordinates $(2, -2)$.

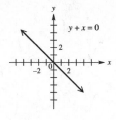

23. The given equation is $3x = y$.
To find the y-intercept, let $x = 0$.
$$3 \cdot 0 = y$$
$$y = 0$$
The y-intercept is $(0, 0)$. Observe that this is also the x-intercept. Thus, the graph runs through the origin.
To find a second point let x (or y) take on a value and solve for the other variable, i.e. let $x = 1$.
$$3 \cdot 1 = y$$
$$y = 3$$
Thus, a second point would have coordinates $(1, 3)$.

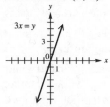

25. The given equation is $x = 2$.
The equation is represented by a vertical line where x is 2 for any value of y. Thus, when $y = 0$, x remains the value 2. The x-intercept is $(2, 0)$. There is no y-intercept.

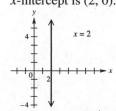

27. The given equation is $y = 4$.
The equation is represented by a horizontal line where y is 4, for any value of x. Thus, when $x = 0$, y remains the value 4. The y-intercept is $(0, 4)$. There is no x-intercept.

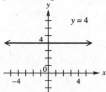

29. The graph of $y + 2 = 0$ is a horizontal line crossing the y-axis at $y = -2$. This fits option C.

31. The graph of $x + 3 = 0$ is a vertical line which crosses the x-axis at $x = -3$. This fits option A.

33. The graph of $y - 2 = 0$ is a horizontal line crossing the y-axis at $y = 2$. This fits option D.

35. The graph of $x - 3 = 0$ is a vertical line which crosses the x-axis at $x = 3$. This fits option B.

37. The diagram of the roof indicates a rise (change in y) of 6 feet and a run (change in x) of 20 feet. Thus, the slope (pitch) is given by $m = \dfrac{\text{rise}}{\text{run}} = \dfrac{6}{20} = \dfrac{3}{10}$.

39. (a) The coordinates of the given points are $(-1, -4)$ and $(3, 2)$. Let $(x_1, y_1) = (-1, -4)$ and $(x_2, y_2) = (3, 2)$. Then,
$$m = \frac{y_2 - y_1}{x_2 - x_1} = \frac{2 - (-4)}{3 - (-1)} = \frac{3}{2}.$$
Observe that either point may be chosen as (x_1, y_1) or (x_2, y_2).

(b) The coordinates of the given points are $(-3, 5)$ and $(1, -2)$. Let $(x_1, y_1) = (-3, 5)$ and $(x_2, y_2) = (1, -2)$. Then,
$$m = \frac{y_2 - y_1}{x_2 - x_1} = \frac{-2 - 5}{1 - (-3)} = -\frac{7}{4}.$$

41. Let $(x_1, y_1) = (-2, -3)$ and $(x_2, y_2) = (-1, 5)$. Then,
$$m = \frac{y_2 - y_1}{x_2 - x_1} = \frac{5 - (-3)}{-1 - (-2)} = 8.$$

43. Let $(x_1, y_1) = (8, 1)$ and $(x_2, y_2) = (2, 6)$. Then, $m = \dfrac{y_2 - y_1}{x_2 - x_1} = \dfrac{6 - 1}{2 - 8} = -\dfrac{5}{6}$.

45. Let $(x_1, y_1) = (2, 4)$ and $(x_2, y_2) = (-4, 4)$. Then, $m = \dfrac{y_2 - y_1}{x_2 - x_1} = \dfrac{4 - 4}{(-4) - 2} = \dfrac{0}{-6} = 0$.

47. Refer to graph (Figure A, in the text).

(a) Let $(x_1, y_1) = (1990, 11{,}338)$ and $(x_2, y_2) = (2005, 14{,}818)$. Then,

$$m = \frac{y_2 - y_1}{x_2 - x_1}$$
$$= \frac{14{,}818 - 11{,}338}{2005 - 1990}$$
$$= \frac{3480}{15}$$
$$= 232 \ (\textit{thousand})$$
$$= 232{,}000.$$

(b) The slope of the line in Figure A is <u>positive</u>. This means that during the period represented, enrollment <u>increased</u>.

(c) Since the slope, or *rate of change*, is $\dfrac{232{,}000 \text{ students}}{1 \text{ yr}}$. The increase in students per year is 232,000.

(d) Refer to graph (Figure B, in the text). Let $(x_1, y_1) = (1990, 20)$ and $(x_2, y_2) = (2007, 3.8)$. Then,

$$m = \frac{y_2 - y_1}{x_2 - x_1}$$
$$= \frac{3.8 - 20}{2007 - 1990}$$
$$= \frac{-16.2}{17} \approx -0.95.$$

(e) The slope of the line in Figure B is <u>negative</u>. This shows us that during the period represented, the number of students per computer <u>decreased</u>.

(f) Since the slope, or *rate of change*, is $\dfrac{-0.95 \text{ student per computer}}{1 \text{ yr}}$.

The decrease in students per computer *per year* is 0.95 student per computer.

49. Locate the point $(-3, 2)$. The slope is
$$m = \frac{1}{2} = \frac{\text{change in } y}{\text{change in } x}.$$

From $(-3, 2)$ move 1 unit up and 2 units to the right. This brings you to another point, $(-1, 3)$. Draw the line through $(-3, 2)$ and $(-1, 3)$.

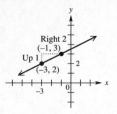

51. Locate the point $(-2, -1)$. The slope is
$$m = \frac{-5}{4} = \frac{\text{change in } y}{\text{change in } x}.$$

From $(-2, -1)$ move 5 units down and 4 units to the right. This brings you to another point, $(2, -6)$. Draw the line through $(-2, -1)$ and $(2, -6)$.

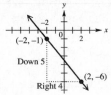

53. Locate the point $(-1, -4)$. The slope is
$$m = -2 = \frac{-2}{1} = \frac{\text{change in } y}{\text{change in } x}.$$

From $(-1, -4)$ move 2 units down and 1 unit to the right. This brings you to another point, $(0, -6)$. Draw the line through $(-1, -4)$ and $(0, -6)$.

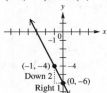

55. First locate the point $(2, -5)$. Use the definition of slope $\left(m = 0 = \dfrac{0}{n}, \text{ for any} \right.$ non-zero integer, n) to move up (or down) 0 units (change in y-values) and to the right (or left) n units (change in x-values) to locate another point on the graph. Draw a line through these two points. Locate the point $(2, -5)$. The slope is
$$m = 0 = \frac{0}{n} = \frac{\text{change in } y}{\text{change in } x}, \text{ for any non-zero}$$

integer, n. From $(2, -5)$ move 0 units up and n units to the left or right to locate another point on the line. Draw the line through these two points.

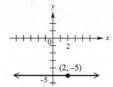

57. Locate $(4, 2)$. An undefined slope means that the line is vertical. The x-value of every point is 4.

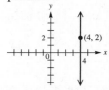

59. L_1 is through $(4, 6)$ and $(-8, 7)$, and L_2 is through $(7, 4)$ and $(-5, 5)$. For L_1,

$$m_1 = \frac{y_2 - y_1}{x_2 - x_1} = \frac{7 - 6}{-8 - 4} = \frac{1}{-12} = -\left(\frac{1}{12}\right).$$

For L_2,

$$m_2 = \frac{y_2 - y_1}{x_2 - x_1} = \frac{5 - 4}{-5 - 7} = \frac{1}{-12} = -\left(\frac{1}{12}\right).$$

Since the slopes are equal, the lines are parallel.

61. L_1 is through $(2, 0)$ and $(5, 4)$, and L_2 is through $(6, 1)$ and $(2, 4)$.

For L_1, $m_1 = \frac{y_2 - y_1}{x_2 - x_1} = \frac{4 - 0}{5 - 2} = \frac{4}{3}$. For L_2,

$$m_2 = \frac{y_2 - y_1}{x_2 - x_1} = \frac{4 - 1}{2 - 6} = \frac{3}{-4} = -\left(\frac{3}{4}\right).$$

Since the slopes are negative reciprocals of each other, the lines are perpendicular.

63. L_1 is through $(0, 1)$ and $(2, -3)$, and L_2 is through $(10, 8)$ and $(5, 3)$.

For L_1, $m_1 = \frac{y_2 - y_1}{x_2 - x_1} = \frac{-3 - 1}{2 - 0} = \frac{-4}{2} = -2$.

For L_2, $m_2 = \frac{y_2 - y_1}{x_2 - x_1} = \frac{3 - 8}{5 - 10} = \frac{-5}{-5} = 1$.

The slopes are not related and, hence, the lines are neither parallel nor perpendicular.

65. Use the two points $(160, 0)$ and $(250, 63)$ to represent the front of the upper deck and the back of the upper deck. The change in x is found by 250 feet − 160 feet = 90 feet and the change in y is the height or 63 feet.

Thus, $m = \dfrac{\text{change in } y}{\text{change in } x} = \dfrac{63 \text{ ft}}{90 \text{ ft}} = \dfrac{7}{10}$.

67. The average rate of change is the some value as the slope. Thus,

$$\begin{aligned} \text{average rate of change} &= \frac{y_2 - y_1}{x_2 - x_1} \\ &= \frac{4 - 20}{4 - 0} \\ &= \frac{-16}{4} \\ &= -4 \text{ (thousands of} \\ &\qquad \text{dollars per year)} \end{aligned}$$

The average rate of change is −$4000 per year; the value of the machine is decreasing $4000 per year during those years.

69. (a) The average rate of change is the same value as the slope. Thus,

$$\begin{aligned} \text{average rate of change} &= \frac{y_2 - y_1}{x_2 - x_1} \\ &= \frac{111.3 - 280.9}{2006 - 2000} \\ &= \frac{-169.6}{6} \\ &\approx -28.27. \end{aligned}$$

The average rate of change is −28.27 thousand per year.

(b) The negative slope means that the number of mobile homes *decreased* by an average of 28.27 thousand per year from 2000 to 2006.

71. The average rate of change is the same value as the slope. Thus,

$$\begin{aligned} \text{average rate of change} &= \frac{y_2 - y_1}{x_2 - x_1} \\ &= \frac{5705 - 1590}{2006 - 2003} \\ &= \frac{4115}{3} \\ &\approx \$1371.67 \text{ per year.} \end{aligned}$$

Sales of plasma TVs *increased* by an average of $1371.67 million per year from 2003 to 2006.

8.3 Exercises

1. By the slope-intercept form, the equation $y = 4x$ has a slope = 4 which matches D (where the slope is determined by the two point formula). Alternatively, the points $(0, 0)$ and $(1, 4)$ both satisfy the equation.

3. By the slope-intercept form, the equation $y = -2x + 1$ has slope $= -2$ and y-intercept $(0, 1)$ which matches B.

5. The equation $y = 2x + 3$ matches the graph A. Observe that the positive slope, 2, discounts options B, C, D, E, and G. Of the remaining options only A suggests the y-intercept as potentially the value 3.

7. The equation $y = -2x - 3$ matches the graph C. Observe that the negative slope, -2, discounts options A, B, E, F, and H. Of the remaining options only C suggests the y-intercept as potentially the value -3.

9. The equation $y = 2x$ matches the graph H. Observe that the y-intercept is the value 0. This means that the graph runs through the origin, $(0, 0)$. Only G and H satisfy this condition and G indicates a negative slope.

11. The equation $y = 3$ represents a horizontal line with a y-intercept at 3. These conditions match option B.

13. Using the two intercept points, the slope is given by $m = \dfrac{y_2 - y_1}{x_2 - x_1} = \dfrac{0 - (-3)}{1 - 0} = 3,$ and $b = -3$ (the ordinate of the y-intercept). Thus, the slope-intercept form of the line is $y = mx + b = 3x + (-3)$, or $y = 3x - 3$.

15. Using the two intercept points, the slope is given by $m = \dfrac{y_2 - y_1}{x_2 - x_1} = \dfrac{0 - 3}{3 - 0} = \dfrac{-3}{3} = -1,$ and $b = 3$ (the ordinate of the y-intercept). Thus, the slope-intercept form of the line is $y = mx + b = (-1)x + 3$, or $y = -x + 3$.

17. Use the point-slope form of the line to write the equation: $y - y_1 = m(x - x_1)$, or
$$y - 4 = -\frac{3}{4}[x - (-2)].$$
To write the equation in slope-intercept form:
$$y - 4 = -\frac{3}{4}x + \left(-\frac{3}{4}\right) \cdot 2$$
$$y = -\frac{3}{4}x - \frac{3}{2} + 4$$
$$y = -\frac{3}{4}x - \frac{3}{2} + \frac{8}{2}$$
$$y = -\frac{3}{4}x + \frac{5}{2}$$

19. Use the point-slope form of the line to write the equation: $y - y_1 = m(x - x_1)$, or
$$y - 8 = -2(x - 5).$$
To write the equation in slope-intercept form:
$$y - 8 = -2x + 10$$
$$y = -2x + 10 + 8$$
$$y = -2x + 18$$

21. Use the point-slope form of the line to write the equation: $y - y_1 = m(x - x_1)$, or
$$y - 4 = \frac{1}{2}[x - (-5)].$$
To write the equation in slope-intercept form:
$$y - 4 = \frac{1}{2}x + \left(\frac{1}{2}\right) \cdot 5$$
$$y = \frac{1}{2}x + \frac{5}{2} + 4$$
$$y = \frac{1}{2}x + \frac{5}{2} + \frac{8}{2}$$
$$y = \frac{1}{2}x + \frac{13}{2}$$

23. Use the point-slope form of the line to write the equation: $y - y_1 = m(x - x_1)$, or
$y - 0 = 4 \cdot (x - 3)$. This simplifies to slope-intercept form: $y = 4x - 12$.

25. Using the point-slope form of the line to write the equation, $y - y_1 = m(x - x_1)$, or
$y - 5 = 0 \cdot (x - 9)$. This simplifies to slope-intercept form: $y - 5 = 0$, or
$$y = 5.$$
Alternatively, with $m = 0$ (horizontal line), the equation takes the form of $y = k$ and by inspection we can recognize $y = 5$ as the equation of the horizontal line.

27. An undefined slope indicates that the line is vertical and therefore, the equation is of the form $x = k$. Thus, by inspection, $x = 9$ is the equation of the vertical line.

29. A vertical line is of the form $x = k$. Thus, by inspection, $x = 0.5$ is the equation of the line.

31. A horizontal line is of the form $y = k$. Thus, by inspection, $y = 8$ is the equation of the line.

33. Using the two points, the slope is given by

$$m = \frac{y_2 - y_1}{x_2 - x_1} = \frac{8-4}{5-3} = \frac{4}{2} = 2.$$

Use the point-slope form of the line (and either point) to write the equation:

$y - y_1 = m(x - x_1)$, or $y - 4 = 2(x - 3)$.

To write the equation in slope-intercept form:

$$y - 4 = 2x - 6$$
$$y = 2x - 6 + 4$$
$$y = 2x - 2$$

35. Using the two points, the slope is given by

$$m = \frac{y_2 - y_1}{x_2 - x_1} = \frac{5-1}{(-2)-6} = \frac{4}{-8} = -\frac{1}{2}.$$

Use the point-slope form of the line (and either point) to write the equation:

$y - y_1 = m(x - x_1)$, or $y - 1 = -\frac{1}{2}(x - 6)$.

To write the equation in slope-intercept form:

$$y - 1 = -\frac{1}{2}x - \left(-\frac{1}{2}\right)6$$
$$y - 1 = -\frac{1}{2}x + 3$$
$$y = -\frac{1}{2}x + 3 + 1$$
$$y = -\frac{1}{2}x + 4$$

37. Using the two points, the slope is given by

$$m = \frac{y_2 - y_1}{x_2 - x_1}$$
$$= \frac{\frac{2}{3} - \frac{2}{5}}{\frac{4}{3} - \left(-\frac{2}{5}\right)}$$
$$= \frac{\frac{10}{15} - \frac{6}{15}}{\frac{20}{15} + \frac{6}{15}}$$
$$= \frac{\frac{4}{15}}{\frac{26}{15}}$$
$$= \frac{4}{15} \cdot \frac{15}{26}$$
$$= \frac{2}{13}.$$

Use point-slope form of the line (and either point) to write the equation:

$y - y_1 = m(x - x_1)$, or

$$y - \frac{2}{5} = \frac{2}{13}\left[x - \left(-\frac{2}{5}\right)\right].$$

To write the equation in slope-intercept form:

$$y - \frac{2}{5} = \frac{2}{13}\left(x + \frac{2}{5}\right)$$
$$y - \frac{2}{5} = \frac{2}{13}x + \left(\frac{2}{13}\right)\cdot\left(\frac{2}{5}\right)$$
$$y = \frac{2}{13}x + \frac{4}{65} + \frac{2}{5}$$
$$y = \frac{2}{13}x + \frac{4}{65} + \frac{26}{65}$$
$$y = \frac{2}{13}x + \frac{30}{65}$$
$$y = \frac{2}{13}x + \frac{6}{13}$$

39. Using the two points, the slope is given by

$$m = \frac{y_2 - y_1}{x_2 - x_1} = \frac{5-5}{1-2} = \frac{0}{-1} = 0.$$

Using the point-slope form of the line to write the equation, $y - y_1 = m(x - x_1)$, or

$y - 5 = 0 \cdot (x - 2)$. This simplifies to slope-intercept form: $y - 5 = 0$, or

$$y = 5.$$

41. These points lie on a vertical line and the slope is undefined (since the denominator in the slope ratio, $x_2 - x_1 = 7 - 7$, is 0).

Therefore, one can't use the point-slope form of the line. Rather, the equation of a vertical line is in the form $x = k$. Thus, $x = 7$.

43. Using the two points, the slope is given by

$$m = \frac{y_2 - y_1}{x_2 - x_1} = \frac{-3-(-3)}{-1-1} = \frac{0}{-2} = 0.$$

Instead of using the point-slope form of the line to write the equation, we offer an alternative solution. We know that with $m = 0$, the line is horizontal and is of the form $y = k$. Thus, $y = -3$ is the equation.

45. Using the slope-intercept form of the line: $y = mx + b$, or $y = 5x + 15$.

47. Using the slope-intercept form of the line:

$y = mx + b$, or $y = -\frac{2}{3}x + \frac{4}{5}.$

49. Using the slope-intercept form of the line

with $m = \frac{2}{5}$ and $b = 5$: $y = mx + b$, or

$$y = \frac{2}{5}x + 5.$$

51. Writing exercise; answers will vary.

53. (a) To write $x + y = 12$ in slope-intercept form, solve for y: $x + y = 12$
$$y = -x + 12.$$

(b) By inspection, the slope, m, is given by $m = -1$ (the understood coefficient of the x-term).

(c) By inspection, $b = 12$ (the constant), so the y-intercept is $(0, 12)$.

55. (a) To write $5x + 2y = 20$ in slope-intercept form, solve for y: $5x + 2y = 20$
$$2y = -5x + 20$$
$$y = -\frac{5}{2}x + 10.$$

(b) By inspection, the slope, m, is given by
$$m = -\frac{5}{2}.$$

(c) By inspection, $b = 10$, so the y-intercept is $(0, 10)$.

57. (a) To write $2x - 3y = 10$ in slope-intercept form, solve for y: $2x - 3y = 10$
$$-3y = -2x + 10$$
$$y = \frac{2}{3}x - \frac{10}{3}.$$

(b) By inspection, the slope, m, is given by
$$m = \frac{2}{3}.$$

(c) By inspection, $b = -\frac{10}{3}$, so the y-intercept is $\left(0, -\frac{10}{3}\right)$.

59. Write $3x - y = 8$ in slope-intercept form in order to identify the slope of the given line:
$$3x - y = 8$$
$$-y = -3x + 8$$
$$y = 3x - 8$$
By inspection, $m = 3$. This is also the slope of the new line, since they are parallel. Using the point-slope form of the line, $y - y_1 = m(x - x_1)$, or $y - 2 = 3(x - 7)$. This simplifies to slope-intercept form:
$$y - 2 = 3x - 21, \text{ or}$$
$$y = 3x - 19.$$

61. Write $-x + 2y = 10$ in slope-intercept form in order to identify the slope of the given line: $-x + 2y = 10$
$$2y = x + 10$$
$$y = \frac{1}{2}x + 5$$
By inspection, $m = \frac{1}{2}$. This is also the slope of the new line, since they are parallel. Using the point, $(-2, -2)$ and point-slope form of the line, $y - y_1 = m(x - x_1)$, or
$$y - (-2) = \frac{1}{2}[x - (-2)]$$
$$y + 2 = \frac{1}{2}x + \frac{2}{2}$$
$$y = \frac{1}{2}x + 1 - 2$$
$$y = \frac{1}{2}x - 1$$

63. Write $2x - y = 7$ in slope-intercept form in order to identify the slope of the given line:
$$2x - y = 7$$
$$-y = -2x + 7$$
$$y = 2x - 7$$
By inspection, $m = 2$. Thus, the slope of any line perpendicular to the given line is $-\frac{1}{2}$. Using the given point $(8, 5)$ and the point-slope form of the line, $y - y_1 = m(x - x_1)$, or
$$y - 5 = -\frac{1}{2}(x - 8)$$
$$2y - 10 = -x + 8$$
$$2y = -x + 8 + 10$$
$$y = -\frac{1}{2}x + \frac{18}{2}$$
$$y = -\frac{1}{2}x + 9$$

65. Since $x = 9$ is a vertical line, any line perpendicular to it will be horizontal and have slope $= 0$. To write the equation of a horizontal line through the point $(-2, 7)$ we can use the form $y = k$ with $k = 7$. Thus, the line is $y = 7$.

67. (a) Using the points (0, 3921) and (3, 7805), the slope is given by

$$m = \frac{y_2 - y_1}{x_2 - x_1}$$
$$= \frac{7805 - 3921}{3 - 0}$$
$$= \frac{3884}{3}$$
$$\approx 1294.7.$$

Using the point-slope form for the equation, the first point, and the calculated slope we have

$$y - y_1 = m(x - x_1)$$
$$y - 3921 = 1294.7(x - 0)$$
$$y = 1294.7x + 3921.$$

Sales of digital cameras in the United States *increased* by $1294.7 million per year from 2003 to 2006.

(b) Substituting $x = 4$ (for the year 2007), we have

$y = 1294.7x + 3921$
$y = 1294.7(4) + 3921$
$y \approx 9099.8.$

In 2007, sales of digital cameras in the United States were approximately $9099.8 million.

69. (a) Using the points (0, 34.3) and (50, 84.1), the slope is given by

$$m = \frac{y_2 - y_1}{x_2 - x_1}$$
$$= \frac{84.1 - 34.3}{50 - 0}$$
$$= \frac{49.8}{50}$$
$$= 0.996.$$

Using the point-slope form for the equation, the point (0, 34.3), and the calculated slope we have

$$y - y_1 = m(x - x_1)$$
$$y - 34.3 = 0.996(x - 0)$$
$$y = 0.996x + 34.3.$$

(b) Substituting $x = 45$ (for the year 1995), we have

$y = 0.996x + 34.3$
$y = 0.996(45) + 34.3$
$y \approx 79.1$

In 1995, 79.1% of the U.S. population 25 years and older were at least high school graduates.

71. (a) Use the points (3, 99,059) and (7, 95,898) to create the slope. The slope is given by

$$m = \frac{y_2 - y_1}{x_2 - x_1}$$
$$= \frac{95,898 - 99,059}{7 - 3}$$
$$= \frac{-3161}{4}$$
$$= -790.25.$$

Using the slope and the point (3, 99,059) in conjunction with the point-slope form of the line we arrive at

$$y - y_1 = m(x - x_1)$$
$$y - 99,059 = -790.25(x - 3)$$
$$y = 99,059 + (-790.25x) + 2370.75$$
$$y = -790.25x + 101,430$$

rounding to the nearest million.

(b) In 2005, $x = 5$ so that

$y = -790.25(5) + 101,430$
$= -3951.25 + 101,430$
$\approx 97,479$ million.

The result using the model is a bit low.

73. (a) Use the data points for Virgo and Bootes, (50, 990) and (1700, 25000), to create the slope. The slope is given by

$$m = \frac{y_2 - y_1}{x_2 - x_1}$$
$$= \frac{25000 - 990}{1700 - 50}$$
$$= \frac{24010}{1650}$$
$$\approx 14.55.$$

Using the slope and any data point, we choose (50, 990), in conjunction with the point-slope form of the line we arrive at: $y - y_1 = m(x - x_1)$, or

$$y - 990 = 14.55(x - 50)$$
$$y = 990 + 14.55(x - 50)$$
$$y = 990 + 14.55x - 727.5$$
$$y = 14.55x + 262.5.$$

Note that answers may vary for Exercise 71(a) depending on round off error. The text answers come from retaining most accurate calculated value of the slope in calculator when computing the constant.

(b) Substituting $y = 37,000$ for the velocity of Hydra), we have

$$y = 14.55x + 262.5$$
$$37000 = 14.55x + 262.5$$
$$0 = 14.55x + 262.5 - 37000$$
$$0 \approx 14.55x - 36737.5$$
$$36737.5 = 14.55x$$
$$x \approx 2525 \text{ light-years.}$$

75. (a) When $C = 0°$, $F = \underline{32°}$; when $C = 100°$, $F = \underline{212°}$.

(b) The two ordered pairs are $(0, 32)$ and $(100, 212)$.

(c) Using the points $(0, 32)$ and $(100, 212)$, the slope is given by

$$m = \frac{y_2 - y_1}{x_2 - x_1} = \frac{212 - 32}{100 - 0} = \frac{180}{100} = \frac{9}{5}.$$

(d) Use the vertical axis intercept $(0, 32)$ as the point $(0, b)$ and the slope-intercept form for the equation to arrive at

$$F = mC + b, \text{ or } F = \frac{9}{5}C + 32.$$

(e) To solve for C in terms of F:

$$F = \frac{9}{5}C + 32$$
$$F - 32 = \frac{9}{5}C$$
$$\frac{5}{9}(F - 32) = C, \text{ or}$$
$$C = \frac{5}{9}(F - 32).$$

(f) When the Celsius temperature is $50°$, the Fahrenheit temperature is $122°$.

8.4 Exercises

1. Writing exercise; answers will vary.

3. The relation $\{(0, 0), (1, 1), (2, 4), (4, 16)\}$, is a function, since corresponding to each first component, there is a unique second component.
The domain is $\{0, 1, 2, 4\}$.
The range is $\{0, 1, 4, 16\}$.

5. The relation, $\{(1, 1), (1, -1), (2, 4), (2, -4), (3, 9), (3, -9)\}$, is not a function, since corresponding to the first component, there is more than one second component.
The domain is $\{1, 2, 3\}$.
The range is $\{-9, -4, -1, 1, 4, 9\}$.

7. The input-output machine would not represent a function. The domain is $(0, \infty)$. The range is $(-\infty, 0) \cup (0, \infty)$.

9. The table does represent a function since each race is paired with only one outcome. The domain is {Hispanic, Native American, Asian American, African American, White}. The range (in millions) is {21.3, 1.6, 8.2, 24.6, 152.0}.

11. The graph represents a function, since it passes the "vertical line test" (only one intersection). The domain is $(-\infty, \infty)$. The range is $(-\infty, 4]$.

13. The graph does not represent a function, since it does not pass the "vertical line test." The domain is $[-4, 4]$. The range is $[-3, 3]$.

15. The equation, $y = x^2$, represents a function, since any value for x will yield exactly one value for y (for a number squared, there is only one answer). The domain is the set of real numbers or $(-\infty, \infty)$.

17. To determine if the equation, $x = y^2$, is or is not a function, solve for y: $y = \pm\sqrt{x}$. Since any replacement for x will yield two values for y, the equation doesn't represent a function. In order to get "real" number values for y, x can be replaced only with values such that $x \geq 0$. These values represent the domain $[0, \infty)$.

19. Since ordered pairs such as $(2, 0)$ and $(2, 1)$ satisfy the inequality, $x + y < 4$, it does not represent a function. Note that the graph of any linear inequality, such as this one, will be a half-plane and, hence, will not satisfy the "vertical line test." Any real number can be used for x. Therefore, the domain is $(-\infty, \infty)$.

21. Since any value for x in the domain of the relation, $y = \sqrt{x}$, will yield exactly one value for y, the principal square root of x, the equation represents a function. To keep y "real values" (i.e., a real number), x must satisfy the inequality $x \geq 0$. Thus, the domain is given by $[0, \infty)$.

23. Since any value for x, in the domain, will yield exactly one value for y, the equation represents a function. Observe that in order to keep the fraction defined, $x \neq 0$, but all other (real) numbers will work. This implies that the domain is all real numbers except 0 or $(-\infty, 0) \cup (0, \infty)$.

25. The relation, $y = \sqrt{4x + 2}$, is a function, because for any choice of x in the domain, there is exactly one corresponding value of y. The domain is the set of values that satisfies $4x + 2 \geq 0$. Solving the inequality we arrive at $x \geq \dfrac{-2}{4}$ or $x \geq -\dfrac{1}{2}$. Thus, the domain is $\left[-\dfrac{1}{2}, \infty \right)$.

27. The relation, $y = \dfrac{2}{x - 2}$, is a function, because for any choice of x in the domain, there is exactly one corresponding value of y. We may use any real value for x except those which make the denominator 0, i.e., $x \neq 2$. Thus, the domain is given by $(-\infty, 2) \cup (2, \infty)$.

29. (a) The values along the vertical axis, representing the dependent variable, are $[0, 3000]$.

(b) The water is increasing between 0 and 25 hours, so the water is increasing for a total of $25 - 0 = 25$ hours. The water is decreasing between 50 and 75 hours, so the water is decreasing for a total of $75 - 50 = 25$ hours.

(c) The graph shows that 2000 gallons of water are left in the pool after 90 hours.

(d) The value of $g(0) = 0$, indicates the pool is empty at time zero.

31. One example: The height of a child depends on her age, so height is a function of age.

33. If $f(x) = 3 + 2x$, then $f(1) = 3 + 2(1) = 5$.

35. If $g(x) = x^2 - 2$, then $g(2) = 2^2 - 2 = 2$.

37. If $g(x) = x^2 - 2$, then
$$g(-1) = (-1)^2 - 2 = -1.$$

39. If $f(x) = 3 + 2x$, then
$$f(-8) = 3 + 2(-8) = -13.$$

41. If $f(x) = 3 + 2x$, then $f(0) = 3 + 2(0) = 3$.

43. If $f(x) = 3 + 2x$, then
$$f\left(-\frac{3}{2} \right) = 3 + 2\left(-\frac{3}{2} \right) = 3 - 3 = 0.$$

45. $f(x) = -2x + 5$

The domain and range is $(-\infty, \infty)$.

47. $h(x) = \dfrac{1}{2}x + 2$

The domain and range is $(-\infty, \infty)$.

49. $G(x) = 2x$

The domain and range is $(-\infty, \infty)$.

51. $f(x) = 5$

The domain is $(-\infty, \infty)$. The range is $\{5\}$.

53. (a) $y + 2x^2 = 3$
$$y = 3 - 2x^2$$
Thus, $f(x) = 3 - 2x^2$.

(b) $f(3) = 3 - 2(3)^2 = -15$

55. (a) $4x - 3y = 8$

$$-3y = 8 - 4x$$

$$y = \frac{8 - 4x}{-3}$$

Thus, $f(x) = \frac{8 - 4x}{-3}$.

(b) $f(3) = \frac{8 - 4(3)}{-3} = \frac{4}{3}$

57. The equation $2x + y = 4$ has a straight <u>line</u> as its graph. One point that lies on the line is $(3, \underline{-2})$. If we solve the equation for y and use function notation, we have a linear function $f(x) = \underline{-2x + 4}$. For this function, $f(3) = -2(3) + 4 = \underline{-2}$, meaning that the point $\underline{(3, -2)}$ lies on the graph of the function.

59. (a)

x	$f(x)$
0	\$0
1	\$2.50
2	\$5.00
3	\$7.50

(b) The linear function that gives a rule for the amount charged is $f(x) = \underline{2.50x}$.

(c)

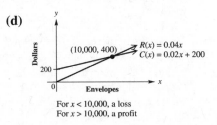

61. (a) Using $h(r) = 69.09 + 2.24r$, then
$h(56) = 69.09 + 2.24(56) = 194.53$ cm.

(b) Using $h(t) = 81.69 + 2.39t$, then
$h(40) = 81.69 + 2.39(40) = 177.29$ cm.

(c) Using $h(r) = 61.41 + 2.32r$, then
$h(50) = 61.41 + 2.32(50) = 177.41$ cm.

(d) Using $h(t) = 72.57 + 2.53t$, then
$h(36) = 72.57 + 2.53(36) = 163.65$ cm.

63. Using $f(x) = 10(x - 65) + 50$, then

(a) $f(76) = 10(76 - 65) + 50 = \$160$.

(b) $100 = 10(x - 65) + 50$

$$100 = 10x - 650 + 50$$

$$100 = 10x - 600$$

$$10x = 700$$

$$x = 70 \text{ mph.}$$

(c) Since the function is only defined for real numbers larger than 65, the smallest whole number which can be used is 66 mph.

(d) $200 < 10(x - 65) + 50$

$$200 < 10x - 650 + 50$$

$$200 < 10x - 600$$

$$10x > 800$$

$$x > 80 \text{ mph.}$$

65. Let x represent the number of envelopes stuffed. Then,

(a) $C(x) = 0.02x + 200$.

(b) $R(x) = 0.04x$.

(c) Set $C(x) = R(x)$ and solve for x.
$$0.02x + 200x = 0.04x$$

$$200 = 0.04x - 0.02x$$

$$200 = 0.02x$$

$$x = 10,000$$

(d)

For $x < 10,000$, a loss
For $x > 10,000$, a profit

67. Let x represent the number of deliveries he makes. Then,

(a) $C(x) = 3x + 2300$.

(b) $R(x) = 5.50x$.

(c) Set $C(x) = R(x)$ and solve for x.
$$3.00x + 2300 = 5.50x$$

$$2300 = 5.50x - 3.00x$$

$$2300 = 2.50x$$

$$x = 920$$

(d)

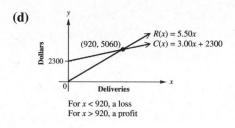

For $x < 920$, a loss
For $x > 920$, a profit

8.5 Exercises

1. The equation $g(x) = x^2 - 5$ matches F with a vertex $(0, -5)$ and opening up since the leading coefficient (1) is positive.

3. The equation $F(x) = (x-1)^2$ matches C with a vertex $(1, 0)$ and opening up since the leading coefficient (1) is positive.

5. the equation $H(x) = (x-1)^2 + 1$ matches E with a vertex $(1, 1)$ and opening up since the leading coefficient (1) is positive.

7. Writing exercise; answers will vary.

9. Write the function, $f(x) = -3x^2$, in the form $f(x) = a(x-h)^2 + k$:
$f(x) = -3(x-0)^2 + 0$.
Thus, the vertex, (h, k), is given by $(0, 0)$.

11. Write the function, $f(x) = x^2 + 4$, in the form $f(x) = a(x-h)^2 + k$:
$f(x) = (x-0)^2 + 4$.
Thus, the vertex, (h, k), is given by $(0, 4)$.

13. Write the function, $f(x) = (x-1)^2$, in the form $f(x) = a(x-h)^2 + k$:
$f(x) = (x-1)^2 + 0$.
Thus, the vertex, (h, k), is given by $(1, 0)$.

15. Write the function, $f(x) = (x+3)^2 - 4$, in the form $f(x) = a(x-h)^2 + k$:
$f(x) = [x - (-3)]^2 + (-4)$.
Thus, the vertex, (h, k), is given by $(-3, -4)$.

17. Writing exercise; answers will vary.

19. The graph $f(x) = -3x^2 + 1$ opens downward since the leading coefficient (-3) is negative. It is narrower since $|-3| > 1$.

21. the graph $f(x) = \frac{2}{3}x^2 - 4$ opens upward since the leading coefficient $\left(\frac{2}{3}\right)$ is positive. It is wider since $\left|\frac{2}{3}\right| < 1$.

23. **(a)** With $h > 0$, $k > 0$, both coordinates of the vertex are positive, $(+, +)$, which puts the vertex in quadrant I.

(b) With $h > 0$, $k < 0$, we have $(+, -)$ and the vertex is in quadrant IV.

(c) With $h < 0$, $k > 0$, we have $(-, +)$ and the vertex is in quadrant II.

(d) With $h < 0$, $k < 0$, we have $(-, -)$ and the vertex is in quadrant III.

25. Write the function, $f(x) = 3x^2$, in the form $f(x) = a(x-h)^2 + k$: $f(x) = 3(x-0)^2 + 0$.
Thus, the vertex, (h, k), is given by $(0, 0)$.
Since $|a| = 3 > 1$, the graph has narrower branches than $f(x) = x^2$ and opens upward.
To find two other points:
$f(\pm 1) = 3(\pm 1)^2 = 3$.
Thus $(-1, 3)$ and $(1, 3)$ also lie on the graph.

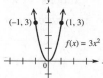

27. Write the function, $f(x) = -\frac{1}{4}x^2$, in the form $f(x) = a(x-h)^2 + k$:
$f(x) = -\frac{1}{4}(x-0)^2 + 0$.
Thus, the vertex, (h, k), is given by $(0, 0)$.
Since $|a| = \frac{1}{4} < 1$, the graph has wider branches than $f(x) = x^2$. It opens downward, since $a = -\frac{1}{4} < 0$. To find two

other points: $f(\pm 4) = -\dfrac{1}{4}(\pm 4)^2 = -4$.

Thus, $(-4, -4)$ and $(4, -4)$ also lie on the graph.

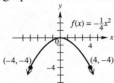

29. Write the function, $f(x) = x^2 - 1$, in the

form $f(x) = a(x - h)^2 + k$:

$f(x) = 1 \cdot (x - 0)^2 + (-1)$.

Thus, the vertex, (h, k), is given by $(0, -1)$. Since $|a| = 1$, the graph has the same branches as $f(x) = x^2$. It opens upward, since $a = 1 > 0$. To find two other points:

$f(\pm 2) = 1 \cdot (\pm 2)^2 - 1 = 3$. Thus, $(-2, 3)$ and $(2, 3)$ also lie on the graph.

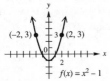

31. Write the function, $f(x) = -x^2 + 2$, in the

form $f(x) = a(x - h)^2 + k$:

$f(x) = -1 \cdot (x - 0)^2 + 2$.

Thus, the vertex, (h, k), is given by $(0, 2)$. Since $|a| = 1$, the graph has the same branches as $f(x) = x^2$. It opens downward since, $a = -1 < 0$. To find two other points:

$f(\pm) = -1 \cdot (\pm 2)^2 + 2 = -2$. Thus, $(-2, -2)$ and $(2, -2)$ also lie on the graph.

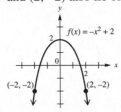

33. Write the function, $f(x) = 2x^2 - 2$, in the

form $f(x) = a(x - h)^2 + k$:

$f(x) = 2(x - 0)^2 + (-2)$.

Thus, the vertex, (h, k), is given by $(0, -2)$. Since $|a| = 2 > 1$, the graph has narrower branches than $f(x) = x^2$. It opens upward,

since $a > 0$. To find two other points:

$f(\pm 1) = 2(\pm 1)^2 - 2 = 0$. Thus, $(-1, 0)$ and $(1, 0)$ also lie on the graph.

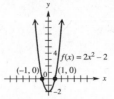

35. Write the function, $f(x) = (x - 4)^2$, in the

form $f(x) = a(x - h)^2 + k$:

$f(x) = 1 \cdot (x - 4)^2 + 0$.

Thus, the vertex, (h, k), is given by $(4, 0)$. it opens upward, since $a > 0$. Find two other points, e.g. let $x = 3$ and $x = 5$:

$f(3) = (3 - 4)^2 = 1, \; f(5) = (5 - 4)^2 = 1$.

Thus, $(3, 1)$ and $(5, 1)$ also lie on the graph.

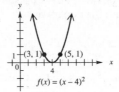

37. Write the function, $f(x) = 3(x + 1)^2$, in the

form $f(x) = a(x - h)^2 + k$:

$f(x) = 3[x - (-1)]^2 + 0$.

Thus, the vertex, (h, k), is given by $(-1, 0)$. It opens upward, since $a > 0$. Find two other points, e.g. let $x = -2$ and $x = 0$:

$f(-2) = 3(-2 + 1)^2 = 3, \; f(0) = 3(0 + 1)^2 = 3$.

Thus, $(-2, 3)$ and $(0, 3)$ also lie on the graph.

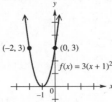

39. Write the function, $f(x) = (x + 1)^2 - 2$, in

the form $f(x) = a(x - h)^2 + k$:

$f(x) = 1 \cdot [x - (-1)]^2 - 2$.

Thus, the vertex, (h, k), is given by $(-1, -2)$. It opens upward, since $a > 0$. Find two other points, e.g. let $x = -2$ and $x = 0$:

$f(-2) = (-2 + 1)^2 - 2 = -1$,

$f(0) = (0+1)^2 - 2 = -1.$

Thus, $(-2, -1)$ and $(0, -1)$ also lie on the graph.

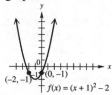

In exercises 41–45, we are finding the vertex by writing each function in the standard form of a quadratic function in order to identify directly the vertex, (h, k). Completing the square is the technique used. However, an alternate technique, using the formula $\left(-\dfrac{b}{2a}, f\left(-\dfrac{b}{2a}\right)\right)$ will yield the coordinates of each vertex.

41. Write the function, $f(x) = x^2 + 8x + 14$, in the form $f(x) = a(x-h)^2 + k$ by completing the square on x.

$f(x) = x^2 + 8x + 14$

$\quad = \left[x^2 + 8x + \left(\dfrac{8}{2}\right)^2\right] + 14 - \left(\dfrac{8}{2}\right)^2$

$\quad = (x^2 + 8x + 16) + 14 - 16$

$\quad = (x+4)^2 - 2$

$f(x) = [x - (-4)]^2 - 2$

Thus, the vertex, (h, k), is given by $(-4, -2)$. It opens upward, since $a > 0$.

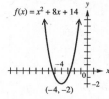

43. Write the function, $f(x) = x^2 + 2x - 4$, in the form $f(x) = a(x-h)^2 + k$ by completing the square on x.

$f(x) = x^2 + 2x - 4$

$\quad = \left[x^2 + 2x + \left(\dfrac{2}{2}\right)^2\right] - 4 - \left(\dfrac{2}{2}\right)^2$

$\quad = [x^2 + 2x + 1] - 4 - 1$

$\quad = (x+1)^2 - 5$

$f(x) = [x - (-1)]^2 - 5$

Thus, the vertex, (h, k), is given by $(-1, -5)$. It opens upward, since $a > 0$.

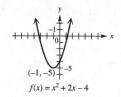

$f(x) = x^2 + 2x - 4$

45. Write the function, $f(x) = -2x^2 + 4x + 5$, in the form $f(x) = a(x-h)^2 + k$ by completing the square on x.

$f(x) = -2x^2 + 4x + 5$

$\quad = -2(x^2 - 2x) + 5$

$\quad = -2\left[x^2 - 2x + \left(\dfrac{2}{2}\right)^2\right] + 5 - (-2)\left(\dfrac{2}{2}\right)^2$

$\quad = -2(x-1)^2 + 5 + 2$

$f(x) = -2(x-1)^2 + 7$

Thus, the vertex, (h, k), is given by $(1, 7)$. It opens downward, since $a < 0$.

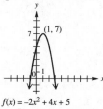

$f(x) = -2x^2 + 4x + 5$

47. If we let x represent the width dimension, then $50 - x$ represents the length dimension (length plus width dimensions will equal half of the fencing needed for perimeter). Create a function representing the area.

$A(x) = \text{length} \cdot \text{width}$

$\quad = (50 - x) \cdot x$

$\quad = -x^2 + 50x$

Use the formula $\dfrac{-b}{2a}$ to create the x-value of the turning point. Thus,

$x = \dfrac{-50}{2(-1)} = 25$ meters.

49. Observe that the vertex $(t, h(t))$ of the parabola represents the time, t, when the object reaches its maximum and the maximum height, $h(t)$. Use the formula $\dfrac{-b}{2a}$ to find at what time the object reaches the maximum height. Thus,

$t = \dfrac{-b}{2a} = \dfrac{-32}{2(-16)} = 1$ sec.

Then, $h(1) = -16(1)^2 + 32(1) = 16$ feet is the

maximum height. To find when the object hits the ground let $h = 0$. Then, solving for t, we have

$$32t - 16t^2 = 0$$
$$16t(2 - t) = 0$$
$$t = 0 \text{ and } t = 2.$$

Therefore, $t = 0$ seconds or $t = 2$ seconds. It is at ground level at $t = 0$ seconds (when it is thrown) and $t = 2$ seconds (when it hits the ground). Notice that it takes 1 second to reach maximum height and 1 more second to hit the ground.

51. The answer to both questions is given by the vertex (highest point) of the parabola suggested by the

equation $s(t) = -4.9t^2 + 40t$. To find the vertex, use the formula, $\left(-\dfrac{b}{2a}, \ s\left(-\dfrac{b}{2a} \right) \right)$, or complete the

square to reach standard form. Using the formula $\dfrac{-b}{2a}$ to find at what time the object reaches the

maximum height, we have $\dfrac{-b}{2a} = \dfrac{-40}{2(-4.9)} \approx 4.1$ seconds.

Then, $s(4.1) = -4.9(4.1)^2 + 40(4.1)$
$$\approx 81.6 \text{ meters}$$
is the maximum height.

53. Using the function $T(x) = 0.00787x^2 - 1.528x + 75.89$:

(a) $T(50) = 0.00787(50)^2 - 1.528(50)$
$$+ 75.89$$
$$= 19.675 - 76.4 + 75.89$$
$$\approx 19.2 \text{ hours}$$

(b) Solve for x in the equation $T(x) = 0.00787x^2 - 1.528x + 75.89$, where $T(x) = 3$.

$3 = 0.00787x^2 - 1.528x + 75.89$, or $0 = 0.00787x^2 - 1.528x + 72.89$.

Using the quadratic formula $x = \dfrac{-b \pm \sqrt{b^2 - 4ac}}{2a}$,

$$x = \dfrac{-(-1.528) \pm \sqrt{(-1.528)^2 - 4 \cdot (0.00787) \cdot (72.89)}}{2 \cdot (0.00787)}$$

$$\approx \dfrac{1.528 \pm \sqrt{2.3348 - 2.2946}}{0.01574}$$

$$\approx \dfrac{1.528 \pm \sqrt{0.0402}}{0.01574}$$

$$\approx \dfrac{1.528 - 0.2005}{0.01574} \text{ (calculating only the smallest solution)}$$

$$= \dfrac{1.3275}{0.01574} \approx 84.3 \text{ ppm.}$$

55. Using the model $f(x) = 0.056057x^2 + 1.066057x$,

$f(45) = 0.056057(45)^2 + 1.06657(45)$
$$= 113.515425 + 47.99565$$
$$\approx 161.5.$$
This means that when the speed is 45 mph, the stopping distance is 161.5 feet.

8.6 Exercises

1. For an exponential function $f(x) = a^x$, if $a > 1$, the graph <u>rises</u> from left to right. If $0 < a < 1$, the graph <u>falls</u> from left to right.

3. The graph of the exponential function $y = a^x$ <u>does not</u> have an x-intercept, since $a^x \neq 0$ for any value of x.

5. For a logarithmic function $g(x) = \log_a x$, if $a > 1$, the graph <u>rises</u> from left to right. If $0 < a < 1$, the graph <u>falls</u> from left to right.

7. The graph of the exponential function $g(x) = \log_a x$ <u>does not</u> have a y-intercept, since $\log_a 0$ does not exist.

9. $9^{3/7} \approx 2.56425419972$

11. $(0.83)^{-1.2} \approx 1.25056505582$

13. $\left(\sqrt{6}\right)^{\sqrt{5}} \approx 7.41309466897$

15. $\left(\dfrac{1}{3}\right)^{9.8} \approx 2.10965628481 \times 10^{-5}$

    $= 0.000021095628481$

17. Generate several ordered pairs that satisfy the function $f(x) = 3^x$ (e.g., $\left(-1, \dfrac{1}{3}\right)$, $(0, 1)$, $(1, 3)$, etc.) and plot these values. Sketch a smooth curve through these points. Remember that the graph will rise from left to right since $b > 1$ and that the x-axis acts as an asymptote.

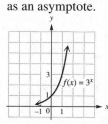

19. Generate several ordered pairs that satisfy the function $f(x) = \left(\dfrac{1}{4}\right)^x$; e.g. $(-1, 4)$, $(0, 1)$, $\left(1, \dfrac{1}{4}\right)$, etc., and plot these values. Sketch a smooth curve through these points. Remember that the graph will fall from left

to right since $b < 1$ and that the x-axis acts as an asymptote.

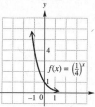

21. $e^3 \approx 20.0855369232$

23. e^{-4} or $\dfrac{1}{e^4} \approx 0.018315638889$

25. $4^2 = 16$ is equivalent to $2 = \log_4 16$.

27. $\left(\dfrac{2}{3}\right)^{-3} = \dfrac{27}{8}$ is equivalent to $-3 = \log_{2/3}\left(\dfrac{27}{8}\right)$.

29. $5 = \log_2 32$ is equivalent to $2^5 = 32$.

31. $1 = \log_3 3$ is equivalent to $3^1 = 3$.

33. $\ln 4 \approx 1.38629436112$

35. $\ln 0.35 \approx -1.0498221245$

37. By inspecting the graph, the year 2000 corresponds

    (a) to an approximate <u>0.5°C</u> increase on the exponential curve, and

    (b) to an approximate <u>0.35°C</u> increase on the linear graph.

39. By inspecting the graph, the year 2020 corresponds

    (a) to an approximate <u>1.6°C</u> increase on the exponential curve, and

    (b) to an approximate <u>0.5°C</u> increase on the linear graph.

41. Since $g(x) = \log_3 x$ is the inverse of $f(x) = 3^x$ (Exercise 17), we can reflect the graph of $f(x)$ across the line $y = x$ to get the graph of $g(x)$. This may be accomplished by interchanging the role of the x and y values

that were generated to graph $f(x)$. Thus, some generated ordered pairs for $g(x)$ would include $\left(\dfrac{1}{3}, -1\right)$, $(1, 0)$, $(3, 1)$, etc.

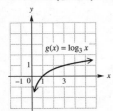

43. Since $g(x) = \log_{1/4} x$ is the inverse of

$f(x) = \left(\dfrac{1}{4}\right)^x$ (Exercise 19), we can reflect

the graph of $f(x)$ across the line $y = x$ to get the graph of $g(x)$. This may be accomplished by interchanging the role of the x and y values that were generated to graph $f(x)$. Thus, some generated ordered pairs for $g(x)$ would include $\left(\dfrac{1}{4}, 1\right)$, $(1, 0)$, $(4, -1)$, etc.

Observe that this will now give an asymptote along the y-axis.

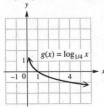

45. Using the compound interest formula

$A = P\left(1 + \dfrac{r}{n}\right)^{nt}$ where $P = \$20{,}000$,

$r = 3\% = 0.03$, and $t = 4$ years.

(a) annually, $n = 1$:

$$A = 20000\left(1 + \dfrac{0.03}{1}\right)^{1\cdot 4}$$
$$= 20000(1.03)^4$$
$$\approx 20000 \cdot (1.12550881)$$
$$\approx \$22{,}510.18.$$

(b) semiannually, $n = 2$:

$$A = 20000\left(1 + \dfrac{0.03}{2}\right)^{2\cdot 4}$$
$$= 20000(1.015)^8$$
$$\approx 20000 \cdot (1.126492587)$$
$$\approx \$22{,}529.85.$$

47. (a) Using the compound interest formula

$A = P\left(1 + \dfrac{r}{n}\right)^{nt}$ where $P = \$27{,}500$,

$r = 2.95\% = 0.0295$, $n = 365$, and $t = 5$ years we get

$$A = 27500\left(1 + \dfrac{0.0295}{365}\right)^{365\cdot 5}$$
$$= 27500(1.000080822)^{1825}$$
$$\approx 27500 \cdot (1.158926377)$$
$$\approx \$31{,}870.48.$$

(b) Using the continuous compound interest formula $A = Pe^{rt}$, where $P = \$27{,}500$, $r = 2.95\% = 0.0295$, and $t = 5$ years we get

$$A = 27500e^{0.0295\cdot 5}$$
$$= 27500e^{0.1475}$$
$$= 27500 \cdot (1.158933285)$$
$$\approx \$31{,}870.67.$$

49. Plan A: Using the compound interest

formula $A = P\left(1 + \dfrac{r}{n}\right)^{nt}$ where

$P = \$40{,}000$, $r = 1.5\% = 0.015$, $n = 4$, and $t = 3$ years we get

$$A = 40000\left(1 + \dfrac{0.015}{4}\right)^{4\cdot 3}$$
$$= 40000(1.00375)^{12}$$
$$\approx 40000 \cdot (1.0459398251)$$
$$\approx \$41{,}837.59.$$

Plan B: Using the continuous compound interest formula $A = Pe^{rt}$, where $P = \$40{,}000$, $r = 1.4\% = 0.014$, and $t = 3$ years we get

$$A = 40000e^{0.014\cdot 3}$$
$$= 40000e^{0.042}$$
$$\approx 40000 \cdot (1.042894479)$$
$$\approx \$41{,}715.78.$$

Therefore, Plan A is better by $121.81.

51. (a) Use the compound interest formula

$A = P\left(1 + \dfrac{r}{n}\right)^{nt}$, where $P = \$60{,}000$,

$r = 7\%$, $n = 4$, and $t = 5$ years.

$$A = P\left(1+\frac{r}{n}\right)^{nt}$$

$$= 60,000\left(1+\frac{0.07}{4}\right)^{4\cdot5}$$

$$= 60,000(1.0175)^{20}$$

$$\approx 60,000(1.4147782)$$

$$\approx \$84,886.69$$

Now use the continuous compound interest formula $A = Pe^{rt}$, where $P = \$60,000$, $r = 6.75\%$, and $t = 5$ years.

$A = Pe^{rt}$, where $P = \$60,000$,

$r = 6.75\%$, and $t = 5$ years.

$$A = Pe^{rt}$$

$$= 60,000e^{0.0675\cdot5}$$

$$= 60,000e^{0.3375}$$

$$\approx \$84,086.38$$

The better investment is 7% compounded quarterly.

(b) $\$84,886.69 - \$84,086.38 = \$800.31$

53. To find the doubling time of an investment P, let $A = 2P$. Using $r = 2.5\%$, solve for t.

$$A = Pe^{rt}$$

$$2P = Pe^{0.025t}$$

$$2 = e^{0.025t}$$

$$\ln(2) = 0.025t$$

$$t = \frac{\ln(2)}{0.025}$$

$$t \approx 27.73 \text{ years}$$

55. To find the time for an investment P to triple, let $A = 3P$. Using $r = 5\%$, solve for t.

$$A = Pe^{rt}$$

$$3P = Pe^{0.05t}$$

$$3 = e^{0.05t}$$

$$\ln(3) = 0.05t$$

$$t = \frac{\ln(3)}{0.05}$$

$$t \approx 21.97 \text{ years}$$

57. Using the model $f(x) = 1013e^{-0.0001341x}$,

(a) to predict the pressure at 1500 meters yields

$$f(1500) = 1013e^{-0.0001341(1500)}$$

$$= 1013e^{-0.20115}$$

$$= 1013(0.8177897539)$$

$$\approx 828 \text{ millibars.}$$

(b) to predict the pressure at 11,000 meters yields

$$f(11000) = 1013e^{-0.0001341(11000)}$$

$$= 1013e^{-1.4751}$$

$$= 1013(0.2287558503)$$

$$\approx 232 \text{ millibars.}$$

59. Using the model $f(x) = 146250(2)^{0.0176x}$, the approximate total population in

(a) 2000, where $x = 0$, is given by

$$f(0) = 146250(2)^{0.0176(0)}$$

$$= 146250(1)$$

$$= 146,250 \text{ thousand people.}$$

(b) 2025, where $x = 25$, is given by

$$f(25) = 146250(2)^{0.0176(25)}$$

$$\approx 146250(1.3566)$$

$$\approx 198,403 \text{ thousand people.}$$

(c) The population increased by $198403 - 146250 = 52,153$. Thus, the percentage increase is

$$\frac{52153}{146250} \approx 0.3566 \approx 36\%.$$

61. Use the model, $A(t) = 500e^{-0.032t}$, to find the amount of sample remaining after

(a) 4 years.

$$A(4) = 500e^{-0.032(4)}$$

$$= 500e^{-0.128}$$

$$\approx 500(0.87985333791)$$

$$\approx 440 \text{ grams}$$

(b) 8 years.

$$A(8) = 500e^{-0.032(8)}$$

$$= 500e^{-0.256}$$

$$\approx 500(0.7741419688)$$

$$\approx 387 \text{ grams}$$

(c) 20 years.

$$A(20) = 500e^{-0.032(20)}$$
$$= 500e^{-0.64}$$
$$\approx 500(0.527292424)$$
$$\approx 264 \text{ grams}$$

(d) To find the half life let the amount of sample, $A(x) = \frac{1}{2} \cdot 500 = 250$ where 500 grams is the amount of the original sample. Then solve for x:

$$250 = 500e^{-0.032x}$$
$$0.5 = e^{-0.32x}$$
$$\ln(0.5) = \ln(e^{-0.032x})$$
$$= (-0.032x)\ln e$$
$$\ln(0.5) = (-0.032x)(1)$$
$$x = \frac{\ln(0.5)}{-0.032}$$
$$\approx \frac{-0.6931471806}{-0.032}$$
$$\approx 22 \text{ years}$$

63. To find the half life let the amount of sample, $A(t) = \frac{1}{2} \cdot A_0$ where A_0 grams is the amount of the original sample. Then solve for t:

$$\frac{1}{2} \cdot A_0 = A_0 e^{-0.00043t}$$
$$0.5 = e^{-0.00043t}$$
$$\ln(0.5) = \ln(e^{-0.00043t})$$
$$= (-0.00043t)\ln e$$
$$\ln(0.5) = (-0.00043t)(1)$$
$$t = \frac{\ln(0.5)}{-0.00043}$$
$$\approx \frac{-0.6931471806}{-0.00043}$$
$$\approx 1611.97 \text{ years}$$

65. Using the mathematical model for carbon 14 dating, let $y = \frac{1}{3} \cdot y_0$ where y_0 is the amount of carbon 14 in the original sample.

Then solve for t:

$$\frac{1}{3} \cdot y_0 = y_p e^{-0.0001216t}$$
$$\frac{1}{3} = e^{-0.0001216t}$$
$$\ln\left(\frac{1}{3}\right) = \ln(e^{-0.0001216t})$$
$$= (-0.0001216t)\ln e$$
$$\ln\left(\frac{1}{3}\right) = (-0.0001216t)(1)$$
$$t = \frac{\ln\left(\frac{1}{3}\right)}{-0.0001216}$$
$$\approx \frac{-1.098612289}{-0.0001216}$$
$$\approx 9034 \text{ or about } 9000 \text{ years}$$

67. Using the mathematical model for carbon 14 dating, let $y = 20\% \cdot y_0$ where y_0 is the amount of carbon 14 in the comparable living specimen. Note the assumption that the original specimen would initially have the same amount of carbon as the comparable living specimen. Then solve for t:

$$0.20 \cdot y_0 = y_0 e^{-0.0001216t}$$
$$0.20 = e^{-0.0001216t}$$
$$\ln(0.20) = \ln(e^{-0.0001216t})$$
$$= (-0.0001216t)\ln e$$
$$\ln(0.20) = (-0.0001216t)(1)$$
$$t = \frac{\ln(0.20)}{-0.0001216}$$
$$\approx \frac{-1.609437912}{-0.0001216}$$
$$\approx 13236 \text{ or about } 13,000 \text{ years}$$

8.7 Exercises

1. If $(3, -6)$ is a solution of a linear system in two variables, then substituting <u>3</u> for x and <u>−6</u> for y leads to true statements in both equations.

3. The solution of the system is the ordered pair of the point where the graphs intersect. This point is in quadrant IV, so the correct choice is D, $(3, -3)$.

5. $x + y = 6$
$x - y = 4$

To decide if $(5, 1)$ is a solution of the system, substitute 5 for x and 1 for y in each equation to see if the results are true statements.

$$5 + 1 = 6$$
$$\quad 6 = 6 \quad \text{True}$$
$$5 - 1 = 4$$
$$\quad 4 = 4 \quad \text{True}$$

Therefore, (5, 1) is a solution to the above system.

7. $2x - y = 8$
$\quad 3x + 2y = 20$

To decide if (5, 2) is a so0lution of the system, substitute 5 for x and 2 for y in each equation to see if the results are true statements.

$$2(5) - 2 = 8$$
$$\quad\quad 8 = 8 \quad \text{True}$$
$$3(5) + 2(2) = 20$$
$$\quad\quad 19 = 20 \quad \text{False}$$

Therefore, (5, 2) is not a solution to the system.

9. The intercepts of the line $x + y = 6$ are (0, 6) and (6, 0). The line $x - y = 0$ passes through the origin. The graph matching these properties is graph B. To check, verify that the point (3, 3) satisfies both equations.

$x + y = 6$	$x - y = 0$
$3 + 3 = 6$	$3 - 3 = 0$
$6 = 6$	$0 = 0$

11. The line $x + y = 0$ passes through the origin. The intercepts of the line $x - y = -6$ are (0, 6) and (−6, 0). The graph matching these properties is graph A. To check, verify that the point (−3, 3) satisfies both equations.

$x + y = 0$	$x - y = -6$
$-3 + 3 = 0$	$-3 - 3 = -6$
$0 = 0$	$-6 = -6$

13. $x + y = 4$
$\quad 2x - y = 2$

Graph the line $x + y = 4$ through its intercepts (0, 4) and (4, 0) and the line $2x - y = 2$ through its intercepts (0, −2) and (1, 0).

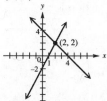

The lines appear to intersect at (2, 2). Check this ordered pair in the system.

$$2 + 2 = 4$$
$$\quad 4 = 4 \quad \text{True}$$
$$2(2) - 2 = 2$$
$$\quad\quad 2 = 2 \quad \text{True}$$

Thus, {(2, 2)} is the solution.

15. $2x - 5y = 11$
$\quad 3x + y = 8$

Multiply the second equation by 5 then add to the first equation.

$$2x - 5y = 11$$
$$\underline{15x + 5y = 40}$$
$$17x = 51$$
$$x = 3$$

Substitute this value for x into the first equation and solve for y.

$$2(3) - 5y = 11$$
$$-5y = 11 - 6$$
$$-5y = 5$$
$$y = -1$$

Since the ordered pair (3, −1) satisfies both equations, it checks. The solution set is {(3, −1)}.

17. $3x + 4y = -6$
$\quad 5x + 3y = 1$

Multiply the first equation by 3 and the second by −4 then add.

$$9x + 12y = -18$$
$$\underline{-20x - 12y = -4}$$
$$-11x = -22$$
$$x = 2$$

Substitute this value for x into the first equation and solve for y.

$$3(2) + 4y = -6$$
$$4y = -6 - 6$$
$$4y = -12$$
$$y = -3$$

Since the ordered pair (2, −3) satisfies both equations, it checks. The solution set is {(2, −3)}.

19. $3x + 3y = 0$
$\quad 4x + 2y = 3$

Simplify the first equation by dividing both sides by 3.

$$x + y = 0$$
$$4x + 2y = 3$$

Multiply the first equation by −2 then add. Solve the resulting equation for x.

$$-2x - 2y = 0$$
$$\underline{4x + 2y = 3}$$
$$2x = 3$$
$$x = \frac{3}{2}$$

Substitute this value for x into the equation $x + y = 0$ and solve for y.

$$\frac{3}{2} + y = 0$$
$$y = -\frac{3}{2}$$

Since the ordered pair $\left(\frac{3}{2}, -\frac{3}{2}\right)$ satisfies both equations, it checks. The solution set is

$$\left\{\left(\frac{3}{2}, -\frac{3}{2}\right)\right\}.$$

21. $7x + 2y = 6$
$-14x - 4y = -12$

To simplify the second equation, divide by 2. Then add both equations.

$$7x + 2y = 6$$
$$\underline{-7x - 2y = -6}$$
$$0 = 0 \qquad \text{True}$$

The equations are dependent and hence have an infinite number of solutions. Solve the first equation for x in terms of y.

$$7x + 2y = 6$$
$$7x = 6 - 2y$$
$$x = \frac{6 - 2y}{7}$$

We will leave the answer as the ordered pair $\left(\frac{6 - 2y}{7}, y\right)$. The solution set is

$\left\{\left(\frac{6 - 2y}{7}, y\right)\right\}$. (Note: This allows one to

create a specific solution as an ordered pair, for any real number replacement of y.)

23. $\dfrac{x}{2} + \dfrac{y}{3} = -\dfrac{1}{3}$
$\dfrac{x}{2} + 2y = -7$

Multiply the first equation by 6 and the second by −6 in order to eliminate fractions and to get opposite coefficients for the x-terms. Add the two resulting equations and solve for y.

$$3x + 2y = -2$$
$$\underline{-3x - 12y = 42}$$
$$-10y = 40$$
$$y = -4$$

Substitute this value for y in the equation $3x + 2y = -2$ and solve for x.

$$3x + 2(-4) = -2$$
$$3x - 8 = -2$$
$$3x = 6$$
$$x = 2$$

Since the ordered pair $(2, -4)$ satisfies both equations, it checks. The solution set is $\{(2, -4)\}$.

25. $5x - 5y = 3$
$x - y = 12$

Multiply the second equation by −5 then add both equations.

$$5x - 5y = 3$$
$$\underline{-5x + 5y = -60}$$
$$0 = -57 \quad \text{False}$$

The equations are inconsistent and, thus, have no solutions. They are parallel lines with no points of intersection. The solution set is the empty set $\varnothing$.

27. $4x + y = 6$
$y = 2x$

Substitute the value for y from the second equation into the first equation and solve for x.

$$4x + (2x) = 6$$
$$6x = 6$$
$$x = 1$$

Substitute this value of x into the equation $y = 2x$ and solve for y.

$$y = 2(1)$$
$$y = 2$$

Since the ordered pair $(1, 2)$ satisfies both equations, it checks. The solution set is $\{(1, 2)\}$.

29. $3x - 4y = -22$
$-3x + y = 0$

Solve the second equation for y.

$$-3x + y = 0$$
$$y = 3x$$

Substitute this value for y into the first equation and solve for x.

$$3x - 4(3x) = -22$$
$$3x - 12x = -22$$
$$-9x = -22$$
$$x = \frac{22}{9}$$

Replace $x = \frac{22}{9}$ in the equation $y = 3x$ and evaluate for y.

$$y = 3\left(\frac{22}{9}\right)$$
$$y = \frac{22}{3}$$

Since the ordered pair $\left(\frac{22}{9}, \frac{22}{3}\right)$ satisfies both equations, it checks. The solution set is $\left\{\left(\frac{22}{9}, \frac{22}{3}\right)\right\}$.

31. $-x - 4y = -14$
$$2x = y + 1$$

Solve the second equation for y.

$$2x = y + 1$$
$$2x - 1 = y$$

Substitute this value for y into the first equation and solve for x.

$$-x - 4(2x - 1) = -14$$
$$-x - 8x + 4 = -14$$
$$-9x = -18$$
$$x = 2$$

Substitute $x = 2$ into the equation $y = 2x - 1$ and evaluate y.

$$y = 2(2) - 1$$
$$y = 3$$

Since the ordered pair (2, 3) satisfies both equations, it checks. The solution set is {(2, 3)}.

33. $5x - 4y = 9$
$$3 - 2y = -x$$

Solve the second equation for x by multiplication of both sides by -1.

$$3 - 2y = -x$$
$$-3 + 2y = x$$

Substitute this value of x into the first equation and solve for y.

$$5(-3 + 2y) - 4y = 9$$
$$-15 + 10y - 4y = 9$$
$$6y = 24$$
$$y = 4$$

Substitute this value of y into the equation

$x = -3 + 2y$ and evaluate for x.
$$x = -3 + 2(4)$$
$$x = 5$$

Since the ordered pair (5, 4) satisfies both equations, it checks. The solution set is {(5, 4)}.

35. $x = 3y + 5$
$$x = \frac{3}{2}y$$

Replace x in the first equation by $\frac{3}{2}y$, the value of x in the second equation. Multiply both sides by 2 in order to eliminate the fraction and solve for y.

$$\frac{3}{2}y = 3y + 5$$
$$3y = 6y + 10$$
$$-3y = 10$$
$$y = -\frac{10}{3}$$

Substitute this value for y into the second equation (or the first equation) and evaluate for x.

$$x = \frac{3}{2}y$$
$$x = \frac{3}{2}\left(-\frac{10}{3}\right)$$
$$x = -5$$

Since the ordered pair $\left(-5, -\frac{10}{3}\right)$ satisfies both equations, it checks. The solution set is $\left\{\left(-5, -\frac{10}{3}\right)\right\}$.

37. $\frac{1}{2}x + \frac{1}{3}y = 3$
$$y = 3x$$

Substitute the value for y from the second equation into the first equation and solve for x.

$$\frac{1}{2}x + \frac{1}{3}(3x) = 3$$
$$\frac{1}{2}x + x = 3$$
$$\frac{3}{2}x = 3$$
$$x = \left(\frac{2}{3}\right)3$$
$$x = 2$$

Substitute this value of x into the equation $y = 3x$ and solve for y.

$y = 3(2)$

$y = 6$

Since the ordered pair (2, 6) satisfies both equations, it checks. the solution set is $\{(2, 6)\}$.

39. Writing exercise; answers will vary.

41. $3x + 2y + z = 8$

$2x - 3y + 2z = -16$

$x + 4y - z = 20$

Eliminate z from the first and third equations by adding.

$3x + 2y + z = 8$

$\underline{x + 4y - z = 20}$

$4x + 6y = 28$

Eliminate z from the second and third equations by adding 2 times the third to the second.

$2x - 3y + 2z = -16$

$\underline{2x + 8y - 2z = 40}$

$4x + 5y = 24$

We are left with two equations in x and y. Multiply the second by -1 and add to the first to eliminate the x-term.

$4x + 6y = 28$

$\underline{-4x - 5y = -24}$

$y = 4$

Substitute this value of y into the equation $4x + 5y = 24$ (either equation in x and y may be used). Solve for x.

$4x + 5(4) = 24$

$4x = 4$

$x = 1$

Replace x and y by these values in the equation $x + 4y - z = 20$ (any one of the original 3 equations may be used). Solve for z.

$(1) + 4(4) - z = 20$

$17 - z = 20$

$-z = 3$

$z = -3$

Since the ordered triple $(1, 4, -3)$ satisfies all three equations, it checks. The solution set is $\{(1, 4, -3)\}$.

43. $2x + 5y + 2z = 0$

$4x - 7y - 3z = 1$

$3x - 8y - 2z = -6$

Eliminating z from the first and third equations by adding.

$2x + 5y + 2z = 0$

$\underline{3x - 8y - 2z = -6}$

$5x - 3y = -6$

Eliminate z from the first and second equations by adding 3 times the first to 2 times the second.

$6x + 15y + 6z = 0$

$\underline{8x - 14y - 6z = 2}$

$14x + y = 2$

We are left with two equations in x and y. Multiply the second equation by 3 and add to the first to eliminate the y-term.

$5x - 3y = -6$

$\underline{42x + 3y = 6}$

$47x = 0$

$x = 0$

Substitute this value of x into the equation $14x + y = 2$ (either equation in x and y may be used). Solve for y.

$14(0) + y = 2$

$y = 2$

Replace x and y by these values in the equation $2x + 5y + 2z = 0$ (any one of the original 3 equations may be used). Solve for x.

$2(0) + 5(2) + 2z = 0$

$10 + 2z = 0$

$2z = -10$

$z = -5$

The solution set is $\{(0, 2, -5)\}$.

45. $x + y - z = -2$

$2x - y + z = -5$

$-x + 2y - 3z = -4$

Add the first and second equations and solve for x.

$x + y - z = -2$

$\underline{2x - y + z = -5}$

$3x = -7$

$x = -\dfrac{7}{3}$

Add -3 times the first equation to the third.

$-3x - 3y + 3z = 6$

$\underline{-x + 2y - 3z = -4}$

$-4x - y = 2$

Substitute $x = -\dfrac{7}{3}$ into the equation

$-4x - y = 2$.

$$-4\left(-\frac{7}{3}\right) - y = 2$$

$$\frac{28}{3} - y = 2$$

$$-y = \frac{6}{3} - \frac{28}{3}$$

$$-y = -\frac{22}{3}$$

$$y = \frac{22}{3}$$

Replace x and y by these values in the equation $x + y - z = -2$. Solve for z.

$$\left(-\frac{7}{3}\right) + \left(\frac{22}{3}\right) - z = -2$$

$$-z = -2 - \frac{15}{3}$$

$$-z = -\frac{6}{3} - \frac{15}{3}$$

$$-z = -\frac{21}{3}$$

$$z = 7$$

The solution set is $\left\{\left(-\frac{7}{3}, \frac{22}{3}, 7\right)\right\}$.

47. $2x - 3y + 2z = -1$
$x + 2y + z = 17$
$2y - z = 7$

Add -2 times the second equation to the first equation.

$$2x - 3y + 2z = -1$$
$$-2x - 4y - 2z = -34$$
$$\overline{\quad -7y \quad\quad = -35}$$
$$y = 5$$

Substitute $y = 5$ into the equation $2y - z = 7$.
$2(5) - z = 7$
$10 - z = 7$
$-z = -3$
$z = 3$

Substitute these values of y and z into the equation $x + 2y + z = 17$ and solve for x.
$x + 2(5) + 3 = 17$
$x + 13 = 17$
$x = 4$
The solution set is $\{(4, 5, 3)\}$.

49. $4x + 2y - 3z = 6$
$x - 4y + z = -4$
$-x + 2z = 2$
Add the second equation and third equation.

$$x - 4y + z = -4$$
$$-x \quad\quad + 2z = 2$$
$$\overline{\quad -4y + 3z = -2}$$

Add -4 times the second equation to the first equation.

$$4x + 2y - 3z = 6$$
$$-4x + 16y - 4z = 16$$
$$\overline{\quad 18y - 7z = 22}$$

We are left with two equations in y and z. Multiply the first equation by 7 and the second equation by 3.

$$-28y + 21z = -14$$
$$54y - 21z = 66$$
$$\overline{26y \quad\quad = 52}$$
$$y = 2$$

Replace this value of y in the equation $-4y + 3z = -2$.
$-4(2) + 3z = -2$
$3z = -2 + 8$
$3z = 6$
$z = 2$

Substitute this value for z into $-x + 2z = 2$ and solve for x.
$-x + 2(2) = 2$
$-x + 4 = 2$
$-x = -2$
$x = 2$
The solution set is $\{(2, 2, 2)\}$.

51. $2x + y = 6$
$3y - 2z = -4$
$3x - 5z = -7$

Multiply the first equation by -3 and add to the second equation in order to eliminate the y-term.

$$-6x - 3y \quad\quad = -18$$
$$3y - 2z = -4$$
$$\overline{-6x \quad\quad - 2z = -22} \text{ or}$$
$$-3x - z = -11$$

Eliminate the x-term from this equation and the third original equation. Solve for z.

$$3x - 5z = -7$$
$$-3x - z = -11$$
$$\overline{\quad -6z = -18}$$
$$z = 3$$

Substitute this value of z into the original equation $3x - 5z = -7$ to find x.
$3x - 5(3) = -7$
$3x - 15 = -7$
$3x = 8$
$$x = \frac{8}{3}$$

Substitute the value $z = 3$ into the equation
$3y - 2z = -4$ and solve for y.
$$3y - 2z = -4$$
$$3y - 2(3) = -4$$
$$3y - 6 = -4$$
$$3y = 2$$
$$y = \frac{2}{3}$$
The solution set is $\left\{ \left(\frac{8}{3}, \frac{2}{3}, 3 \right) \right\}$.

53. The graph representing sales of digital cameras is below the graph representing sales of conventional cameras for the years 2000, 2001, 2002, and the first part of 2003.

55. $\quad 2.5x + y = 19.4$
$\quad -1.7x + y = 4.4$
Multiply the second equation by -1 and add.
$$2.5x + y = 19.4$$
$$\underline{1.7x - y = -4.4}$$
$$4.2x = 15$$
$$x \approx 3.6$$
Substitute this value of x into the equation
$2.5x + y = 19.4$ and solve for y.
$$2.5x + y = 19.4$$
$$2.5(3.6) + y = 19.4$$
$$9 + y = 19.4$$
$$y = 10.4$$
The solution is approximately $(3.6, 10.5)$.

8.8 Exercises

1. Let $x =$ the number of games won and $y =$ the number of games lost.
Write an equation to represent the total number of games played.
$$x + y = 162$$
Write a second equation to represent the relationship between games won and games lost.
$$x = y + 28$$
Substituting this value of x into the first equation, we arrive at a new equation in y only.
$$(y + 28) + y = 162$$
Solve for y.
$$y + 28 + y = 162$$
$$2y + 28 = 162$$
$$2y = 134$$
$$y = 67$$
Substitute this value for y into the equation
$x = y + 28$ and solve for x.

$$x = y + 28$$
$$x = 67 + 28$$
$$x = 95$$
There were 95 wins and 67 losses.

3. Let $l =$ the length of the basketball court and $w =$ the width of the basketball court.
Use the relationship of the perimeter to the side lengths, $P = 2l + 2w$, to write the following equation.
$$2l + 2w = 288 \text{ or }$$
$$l + w = 144$$
Write a second equation to represent the relationship between the length and the width dimensions.
$$w = l - 44 \text{ or }$$
$$l - w = 44$$
Solve the resulting system of equations by adding to eliminate the w-term.
$$l - w = 44$$
$$\underline{l + w = 144}$$
$$2l = 188$$
$$l = 94$$
Substituting this value of l into the equation
$w = l - 44$, we find the width dimension.
$$w = 94 - 44 = 50$$
The dimensions are: length = 94 feet and width = 50 feet.

5. Let $x =$ the number of days at daily rate and $y =$ the number of days at weekend rate.
Write an equation to represent the total number of days that the car was rented.
$$x + y = 6$$
Write a second equation to represent the total cost of renting both weekdays and weekends.
$$53x + 35y = 264$$
Solve the resulting system of equations by multiplying the first equation by -35 and adding to the second equation.
$$-35x - 35y = -210$$
$$\underline{53x + 35y = 264}$$
$$18x = 54$$
$$x = 3$$
Substitute this value of x into the first equation and solve for y.
$$x + y = 6$$
$$3 + y = 6$$
$$y = 3$$
There were 3 weekend days and 3 weekdays for the rental.

7. Let x = the side length of the equilateral triangle and y = the side length of the square.
Write an equation to represent the relationship between the side lengths.
$$y = x + 4 \text{ or}$$
$$-x + y = 4$$

The perimeter of the square may be expressed as $4y$. The perimeter of the triangle may be expressed as $3x$.
Write a second equation to represent the relationship between the perimeters.
$$4y = 3x + 24 \text{ or}$$
$$-3x + 4y = 24$$

Solve the resulting system of equations by multiplying the first equation $-x + y = 4$ by -3 and adding to the second equation $-3x + 4y = 24$.
$$3x - 3y = -12$$
$$\underline{-3x + 4y = 24}$$
$$y = 12$$

Substituting this value of y into the equation $-x + y = 4$ gives the value for x.
$$-x + 12 = 4 \text{ or}$$
$$-x = -8$$
$$x = 8$$
The side dimension for the square is 12 centimeters, and the side dimension for the triangle is 8 centimeters.

9. Let x = the cost of a cappuccino and y = the cost of a latte.
Write two equations to represent the two sets of purchase costs.
$$2x + 3y = 10.95$$
$$x + 2y = 6.65$$

Multiply the two equations by 100 to eliminate the decimal point.
$$200x + 300y = 1095$$
$$100x + 200y = 665$$

Multiply the second equation by -2 and add to the first equation in order to eliminate the y-term.
$$200x + 300y = 1095$$
$$\underline{-200x - 400y = -1330}$$
$$-100y = -235$$
$$y = 2.35$$

Substitute this value of y into the equation $x + 2y = 6.65$ and solve for y.

$$x + 2y = 6.65$$
$$x + 2(2.35) = 6.65$$
$$x + 4.70 = 6.65$$
$$x = 1.95$$
The cost of a cappuccino is \$1.95 and that of a latte is \$2.35.

11. Let x = average total cost for one day in New York City and y = average total cost for one day in Washington, D.C.
Write two equations representing the known costs for days in these cities.
$$2x + 3y = 2772$$
$$4x + 2y = 3488$$

Multiply the first equation by -2 and add to the second equation in order to eliminate the x-term.
$$-4x - 6y = -5544$$
$$\underline{4x + 2y = 3488}$$
$$-4y = -2056$$
$$y = 514$$

Substitute this value of y into the equation $2x + 3y = 2772$ and solve for x.
$$2x + 3(514) = 2772$$
$$2x + 1542 = 2772$$
$$2x = 1230$$
$$x = 615$$
The average cost for a day in New York City was \$615, and the average cost for a day in Washington D.C. was \$514.

13. Let x = the average cost of a Yankees ticket and y = the average cost of a Red Sox ticket.
Write two equations to represent the costs of different combinations of tickets.
$$2x + 3y = 296.66$$
$$3x + 2y = 319.39$$

Multiply the first equation by -3 and the second equation by 2 and add the results to eliminate the x-term.
$$-6x - 9y = -889.98$$
$$\underline{6x + 4y = 638.78}$$
$$-5y = -251.20$$
$$y = 50.24$$

Substitute this value of y into the equation $2x + 3y = 296.66$ and solve for x.
$$2x + 3(50.24) = 296.66$$
$$2x + 150.72 = 296.66$$
$$2x = 145.94$$
$$x = 72.97$$
The average cost of a Yankees ticket was \$72.97, and the average cost of a Red Sox ticket was \$50.24.

15. Let x = the FCI for hockey and
y = the FCI for basketball.
Write an equation to represent the total FCI prices for these sports.
$x + y = 580.16$
Write an equation to represent the relationship between the FCI prices for these sports.
$y = x + 3.70$
Substituting this value of y into the first equation, we arrive at a new equation in x only.
$x + (x + 3.70) = 580.16$
Solve for x.
$$x + x + 3.70 = 580.16$$
$$2x + 3.70 = 580.16$$
$$2x = 576.46$$
$$x = 288.23$$
Substitute this value for x into the equation $y = x + 3.70$ and solve for y.
$y = 288.23 + 3.70$
$y = 291.93$
The FCI for the NHL was \$288.23, and the FCI for the NBA was \$291.93.

17. (a) $(0.10)(120) = 12$ oz

(b) $(0.25)(120) = 30$ oz

(c) $(0.40)(120) = 48$ oz

(d) $(0.50)(120) = 60$ oz

19. The cost of x pounds of ham is \$2.29x.

21. Let x = the number of liters of 15% acid and y = the number of liters of 33% acid needed.
Write an equation to represent the total liters of solution.
$x + y = 40$
Write another equation to represent the total amount of acid.
$0.15x + 0.33y = 0.21(40)$, or equivalently
$$15x + 33y = 21(40)$$
$$5x + 11y = 7(40)$$
$$5x + 11y = 280$$
Solve the resulting system of equations.
$$x + y = 40$$
$$5x + 11y = 280$$
Multiply the first equation by -5 and add to the second equation in order to eliminate the x-term.

$$-5x - 5y = -200$$
$$\underline{5x + 11y = 280}$$
$$6y = 80$$
$$y = \frac{40}{3} = 13\frac{1}{3}$$
Substitute this value of y into the equation $x + y = 40$ and solve for x.
$$x + \left(\frac{40}{3}\right) = 40$$
$$x = 40 - \frac{40}{3}$$
$$x = \frac{120}{3} - \frac{40}{3}$$
$$x = \frac{80}{33} = 26\frac{2}{3}$$
It requires $26\frac{2}{3}$ liters of 15% acid and

$13\frac{1}{3}$ liters of 33% acid to make 40 liters of 21% acid.

23. Let x = the number of liters of 100% antifreeze and y = the number of liters of 4% antifreeze needed.
Write an equation to represent the total liters of solution.
$x + y = 18$
Write (and simplify) another equation to represent the total amount of antifreeze.
$$1.00x + 0.04y = 0.20(18)$$
$$100x + 4y = 20(18)$$
$$100x + 4y = 360$$
Solve the resulting system of equations.
$$x + y = 18$$
$$100x + 4y = 360$$
Multiply the first equation by -4 and add to the second equation in order to eliminate the y-term.
$$-4x - 4y = -72$$
$$\underline{100x + 4y = 360}$$
$$96x = 288$$
$$x = 3$$
There are 3 liters of pure antifreeze needed.

25. Let x = the number of liters of 50% juice and y = the number of liters of 30% juice.
Write an equation to represent the total amount of juice mixture.
$x + y = 200$
Write (and simplify) another equation to represent the total amount of pure juice.

$$0.50x + 0.30y = 0.45(200)$$
$$50x + 30y = 45(200)$$
$$5x + 3y = 900$$

Solve the resulting system of equations.

$$x + y = 200$$
$$5x + 3y = 900$$

Multiply the first equation by -3 and add to the second equation in order to eliminate the y-term.

$$-3x - 3y = -600$$
$$\underline{5x + 3y = 900}$$
$$2x = 300$$
$$x = 150$$

Substitute this value of x into the equation $x + y = 200$ and solve for y.

$$(150) + y = 200$$
$$y = 50$$

Thus, 150 liters of 50% juice and 50 liters of 30% juice are to be used.

27. Let x = the amount of $1.20 per pound candy and y = the amount of $2.40 per pound candy.
Write an equation to represent the total pounds of candy mix.
$$x + y = 160$$
Write (and simplify) another equation to represent the total value of the candy.
$$1.20x + 2.40y = 1.65(160)$$
$$12x + 24y = 2640$$
$$x + 2y = 220$$

Solve the resulting system of equations.

$$x + y = 160$$
$$x + 2y = 220$$

Multiply the first equation by -1 and add to the second equation in order to eliminate the x-term.

$$-x - y = -160$$
$$\underline{x + 2y = 220}$$
$$y = 60$$

Substitute this value of y into the equation $x + y = 160$ and solve for x.

$$x + (60) = 160$$
$$x = 100$$

He wants to mix 100 pounds of the $1.20 per pound candy with 60 pounds of the $2.40 per pound candy.

29. Let x = the amount of money invested at 4% and y = the amount of money invested at 3%.
Write an equation to represent the total number of dollars invested.
$$x + y = 15000$$

Write another equation to represent the total return (interest earned) on the investments.
$$0.04x + 0.03y = 550$$
$$4x + 3y = 55000$$

Solve the resulting system of equations.

$$x + y = 15000$$
$$4x + 3y = 55000$$

Multiply the first equation by -3 and add to the second equation in order to eliminate the y-term.

$$-3x - 3y = -45000$$
$$\underline{4x + 3y = 55000}$$
$$x = 10000$$

Substitute this value of x into the equation $x + y = 15000$ and solve for y.

$$(10000) + y = 15000$$
$$y = 5000$$

He should invest $10,000 at 4% and $5000 at 3%.

31. (a) Going upstream, the current will slow the boat down, so the speed of the boat going upstream will be less than the speed of the boat in still water. Subtract the speed of the current: $(10 - x)$ mph

(b) Going downstream, the current will speed the boat up, so the speed of the boat going downstream will be greater than the speed of the boat in still water. Add the speed of the current: $(10 + x)$ mph.

33. Let x = the speed of the train and y = the speed of the plane.
Write an equation to show the relationship of the speed of the plane compared to that of the train.
$$y = 3x - 20$$
Because the time values are the same for the train and the plane, solve $d = rt$ for t.
The time for the train can be represented as
$$t = \frac{d}{r} = \frac{150}{x}.$$
The time for the plane can be represented as
$$t = \frac{d}{r} = \frac{400}{y}.$$
Thus, $\dfrac{150}{x} = \dfrac{400}{y}$
$$150y = 400x.$$

Substituting the value $y = 3x - 20$ into the resulting equation gives

$$150(3x - 20) = 400x$$
$$450x - 3000 = 400x$$
$$50x = 3000$$
$$x = 60.$$

Substituting this value of x into $y = 3x - 20$
gives $y = 3(60) - 20$
$$y = 160.$$

Thus, the speed of the train is 60 miles per hour, and that of the plane is 160 miles per hour.

35. Let x = the speed of the boat in still water and y = the speed of the current.
Using the relationship $d = rt$, the distance traveled by boat upstream into the current is given by $36 = (x - y)(2)$ and the same distance returning with the current is $(x + y)(1.5)$.
The resulting two equations are
$$36 = (x - y)(2)$$
$$36 = (x + y)(1.5)$$
Write the equations in standard form.
$$2x - 2y = 36$$
$$1.5x + 1.5y = 36$$
Observe that the second equation may be simplified as follows before solving the system of equations.
$$1.5x + 1.5y = 36$$
$$15x + 15y = 360$$
$$x + y = 24$$
The resulting system is $2x - 2y = 36$
$$x + y = 24.$$
Solve by multiplying the second equation by 2 and adding to the first.
$$2x - 2y = 36$$
$$2x + 2y = 48$$
$$\overline{ 4x = 84}$$
$$x = 21$$
Substitute this value of x into the equation $x + y = 24$ and solve for y.
$$(21) + y = 24$$
$$y = 3$$
Thus, the speed of the boat is 21 miles per hour, and the speed of the current is 3 miles per hour.

37. Let x = the number of gold medals,
y = the number of silver medals, and
z = the number of bronze medals won.
Write an equation to represent the total number of medals won by Russia.
$$x + y + z = 72$$
Write a second equation to represent the relationship between the number of gold and

bronze medals.
$$z = x + 5$$
$$x - z = -5$$
Write a third equation to represent the relationship between the number of silver and bronze medals.
$$y = 2z - 35$$
$$y - 2z = -35$$
The resulting system of equations is as follows, in standard form.
$$x + y + z = 72$$
$$x - z = -5$$
$$y - 2z = -35$$
Substitute $z = x + 5$ in $y = 2z - 35$.
$$y = 2(x + 5) - 35$$
$$y = 2x + 10 - 35$$
$$y = 2x - 25$$
Substitute $y = 2x - 25$ and $z = x + 5$ in
$x + y + z = 72$.
$$x + (2x - 25) + (x + 5) = 72$$
$$4x - 20 = 72$$
$$4x = 92$$
$$x = 23$$
Substitute this value of x into the equation $y = 2x - 25$ to find y.
$$y = 2(23) - 25$$
$$y = 46 - 25$$
$$y = 21$$
Substitute the value of x into the equation $z = x + 5$ to find z.
$$z = 23 + 5$$
$$z = 28$$
Russia won 23 gold medals, 21 silver medals, and 28 bronze medals.

39. Let x = the length of the shortest side,
y = the length of the middle side, and
z = the length of the longest side.
Write an equation to represent the perimeter.
$$x + y + z = 56$$
Write a second equation to represent the relationship between the longest side and the others.
$$z = (x + y) - 4$$
$$-x - y + z = -4 \text{ in standard form}$$
Write a third equation to represent the relationship between the longest and shortest sides.
$$z = 3x - 4$$
$$-3x + z = -4 \text{ in standard form}$$
The resulting system equations is as follows.
$$x + y + z = 56$$
$$-x - y + z = -4$$
$$-3x + z = -4$$

Add the first and second equation to eliminate the x and y-terms.

$$x + y + z = 56$$
$$\underline{-x - y + z = -4}$$
$$2z = 52$$
$$z = 26$$

Substitute this value of z into the equation $-3x + z = -4$ to find x.

$$-3x + (26) = -4$$
$$-3x = -30$$
$$x = 10$$

Substitute the values for x and z into the equation $x + y + z = 56$ to find the value for y.

$$(10) + y + (26) = 56$$
$$y + 36 = 56$$
$$y = 20$$

The shortest side is 10 inches, the middle side is 20 inches, and the longest side is 26 inches.

41. Let x = the number of units of type A clamps, y = the number of units of type B clamps, and z = the number of units of type C clamps.
Write an equation to represent the total number of units produced.

$$x + y + z = 490$$

Write a second equation to represent the relationship between the number of units of type C clamps and units of type A and type B clamps.

$$z = (x + y) + 10, \text{ or}$$
$$-x - y + z = 10$$

Write a third equation to represent the relationship between the number type A and type B clamps.

$$y = 2x, \text{ or}$$
$$-2x + y = 0$$

The resulting system of equations is as follows:

$$x + y + z = 490$$
$$-x - y + z = 10$$
$$-2x + y = 0.$$

Add the first and second equations to eliminate the x- and y-terms.

$$x + y + z = 490$$
$$\underline{-x - y + z = 10}$$
$$2z = 500$$
$$z = 250$$

Substitute this value of z into the equation $x + y + z = 490$ and use with the other equation $-2x + y = 0$, in x and y.

$$x + y + (250) = 490$$
$$x + y = 240$$
$$-2x + y = 0$$

Multiply the equation $-2x + y = 0$ by -1 and add to the equation $x + y = 240$.

$$2x - y = 0$$
$$\underline{x + y = 240}$$
$$3x = 240$$
$$x = 80$$

Substitute the values for x and z into the equation $x + y + z = 490$ and solve for y.

$$(80) + y + (250) = 490$$
$$y + 330 = 490$$
$$y = 160$$

The shop must make 80 type A, 160 type B, and 250 type C clamps per day.

43. Let x = the number of \$16 tickets, y = the number of \$23 tickets, and z = the number of \$40 VIP tickets.
Write an equation to represent the total sale of all tickets.

$$16x + 23y + 40z = 46,575$$

Write a second equation to represent the relationship between \$16 tickets and VIP tickets.

$$x = 9z$$
$$x - 9z = 0$$

Write a third equation to represent the relationship between \$16 tickets and the sum of \$23 tickets and VIP tickets.

$$x = (y + z) + 55$$
$$x - y - z = 55$$

The resulting system of equations is as follows.

$$16x + 23y + 40z = 46,575$$
$$x - 9z = 0$$
$$x - y - z = 55$$

Multiply the third equation by 23 and add to the first equation.

$$16x + 23y + 40z = 46,575$$
$$\underline{23x - 23y - 23z = 1265}$$
$$39x + 17z = 47,840$$

Substitute $x = 9z$ in the equation $39x + 17z = 47,840$.

$$39(9z) + 17z = 47,840$$
$$351z + 17z = 47,840$$
$$368z = 47,840$$
$$z = 130$$

Substitute this value of z in the equation $x = 9z$.

$$x = 9(130)$$
$$x = 1170$$

Substitute $x = 1170$ and $z = 130$ in the

equation $x - y - z = 55$.
$$1170 - y - 130 = 55$$
$$1040 - y = 55$$
$$-y = -985$$
$$y = 985$$

There were 1170 $16-tickets sold, 985 $23-tickets sold, and 130 $40-tickets sold.

45. Let x = number of wins,
y = number of losses, and
z = number of overtime losses.
Write an equation to represent the total number of games played.
$$x + y + z = 82$$
Write an equation to represent the total number of points awarded.
$$2x + 0y + 1z = 116$$
$$2x + z = 116$$
Write an equation to represent the relationship between the number of losses and the number of overtime losses.
$$y = z + 9$$
$$y - z = 9$$
The resulting system of equations is as follows.
$$x + y + z = 82$$
$$2x + z = 116$$
$$y - z = 9$$
Multiply the first equation by -2 and add to the second equation.
$$-2x - 2y - 2z = -164$$
$$\underline{2x \quad\quad + z = 116}$$
$$-2y \ - z = -48$$
Multiply this result by -1 and add to the third equation of the system.
$$2y + z = 48$$
$$\underline{y - z = 9}$$
$$3y \quad = 57$$
$$y = 19$$
Substitute this value of y in the equation $y = z + 9$.
$$19 = z + 9$$
$$10 = z$$
Substitute this value of z in the equation $2x + z = 116$.
$$2x + 10 = 116$$
$$2x = 106$$
$$x = 53$$
The Bruins had 53 wins, 19 losses, and 10 overtime losses.

EXTENSION: USING MATRIX ROW OPERATIONS TO SOLVE SYSTEMS

1. $x + y = 5$
$x - y = -1$
Write the augmented matrix.
$$\begin{bmatrix} 1 & 1 & | & 5 \\ 1 & -1 & | & -1 \end{bmatrix}$$
Multiply row 1 by -1 and add to row 2.
$$\begin{bmatrix} 1 & 1 & | & 5 \\ 0 & -2 & | & -6 \end{bmatrix}$$
Multiply row 2 by $-\dfrac{1}{2}$.
$$\begin{bmatrix} 1 & 1 & | & 5 \\ 0 & 1 & | & 3 \end{bmatrix}$$
Multiply row 2 by -1 and add to row 1.
$$\begin{bmatrix} 1 & 0 & | & 2 \\ 0 & 1 & | & 3 \end{bmatrix}$$
The resulting matrix represents the following system of equations.
$$1x + 0y = 2$$
$$0x + 1y = 3$$
That is, $x = 2$, $y = 3$.
Thus, $\{(2, 3)\}$ is the solution set.

3. $x + y = -3$
$2x - 5y = -6$
Form the augmented matrix.
$$\begin{bmatrix} 1 & 1 & | & -3 \\ 2 & -5 & | & -6 \end{bmatrix}$$
Multiply row 1 by -2 and add to row 2.
$$\begin{bmatrix} 1 & 1 & | & -3 \\ 0 & -7 & | & 0 \end{bmatrix}$$
Multiply row 2 by $-\dfrac{1}{7}$.
$$\begin{bmatrix} 1 & 1 & | & -3 \\ 0 & 1 & | & 0 \end{bmatrix}$$
Multiply row 2 by -1 and add to row 1.
$$\begin{bmatrix} 1 & 0 & | & -3 \\ 0 & 1 & | & 0 \end{bmatrix}$$
That is, $x = -3$, $y = 0$.
Thus, $\{(-3, 0)\}$ is the solution set.

5. $2x - 3y = 10$
   $2x + 2y = 5$
   Form the augmented matrix.
   $$\begin{bmatrix} 2 & -3 & | & 10 \\ 2 & 2 & | & 5 \end{bmatrix}$$

Multiply row 1 by $\dfrac{1}{2}$.

$$\begin{bmatrix} 1 & -\frac{3}{2} & | & 5 \\ 2 & 2 & | & 5 \end{bmatrix}$$

Multiply row 1 by -2 and add to row 2.

$$\begin{bmatrix} 1 & -\frac{3}{2} & | & 5 \\ 0 & 5 & | & -5 \end{bmatrix}$$

Multiply row 2 by $\dfrac{1}{5}$.

$$\begin{bmatrix} 1 & -\frac{3}{2} & | & 5 \\ 0 & 1 & | & -1 \end{bmatrix}$$

Multiply row 2 by $\dfrac{3}{2}$ and add to row 1.

$$\begin{bmatrix} 1 & 0 & | & \frac{7}{2} \\ 0 & 1 & | & -1 \end{bmatrix}$$

That is, $x = \dfrac{7}{2}$, $y = -1$.

Thus, $\left\{ \left(\dfrac{7}{2}, -1 \right) \right\}$ is the solution set.

7. $3x - 7y = 31$
   $2x - 4y = 18$
   Form the augmented matrix.
   $$\begin{bmatrix} 3 & -7 & | & 31 \\ 2 & -4 & | & 18 \end{bmatrix}$$

Multiply row 1 by $\dfrac{1}{3}$.

$$\begin{bmatrix} 1 & -\frac{7}{3} & | & \frac{31}{3} \\ 2 & -4 & | & 18 \end{bmatrix}$$

Multiply row 1 by -2 and add to row 2.

$$\begin{bmatrix} 1 & -\frac{7}{3} & | & \frac{31}{3} \\ 0 & \frac{2}{3} & | & -\frac{8}{3} \end{bmatrix}$$

Multiply row 2 by $\dfrac{3}{2}$.

$$\begin{bmatrix} 1 & -\frac{7}{3} & | & \frac{31}{3} \\ 0 & 1 & | & -4 \end{bmatrix}$$

Multiply row 2 by $\dfrac{7}{3}$ and add to row 1.

$$\begin{bmatrix} 1 & 0 & | & 1 \\ 0 & 1 & | & -4 \end{bmatrix}$$

That is, $x = 1$, $y = -4$.
Thus, $\{(1, -4)\}$ is the solution set.

9. $x + y - z = 6$
   $2x - y + z = -9$
   $x - 2y + 3z = 1$
   Form the augmented matrix.
   $$\begin{bmatrix} 1 & 1 & -1 & | & 6 \\ 2 & -1 & 1 & | & -9 \\ 1 & -2 & 3 & | & 1 \end{bmatrix}$$

Multiply row 1 by -1 and add to row 3.
Multiply row 1 by -2 and add to row 2.

$$\begin{bmatrix} 1 & 1 & -1 & | & 6 \\ 0 & -3 & 3 & | & -21 \\ 0 & -3 & 4 & | & -5 \end{bmatrix}$$

Multiply row 2 by -1 and add to row 3.

$$\begin{bmatrix} 1 & 1 & -1 & | & 6 \\ 0 & -3 & 3 & | & -21 \\ 0 & 0 & 1 & | & 16 \end{bmatrix}$$

Multiply row 2 by $-\dfrac{1}{3}$.

$$\begin{bmatrix} 1 & 1 & -1 & | & 6 \\ 0 & 1 & -1 & | & 7 \\ 0 & 0 & 1 & | & 16 \end{bmatrix}$$

Multiply row 2 by -1 and add to row 1.

$$\begin{bmatrix} 1 & 0 & 0 & | & -1 \\ 0 & 1 & -1 & | & 7 \\ 0 & 0 & 1 & | & 16 \end{bmatrix}$$

Add row 3 to row 2.

$$\begin{bmatrix} 1 & 0 & 0 & | & -1 \\ 0 & 1 & 0 & | & 23 \\ 0 & 0 & 1 & | & 16 \end{bmatrix}$$

That is, $x = -1$, $y = 23$, $z = 16$.
Thus, $\{(-1, 23, 16)\}$ is the solution set.

11. $2x - y + 3z = 0$
   $x + 2y - z = 5$
   $2y + z = 1$
   Form the augmented matrix.
   $$\begin{bmatrix} 2 & -1 & 3 & | & 0 \\ 1 & 2 & -1 & | & 5 \\ 0 & 2 & 1 & | & 1 \end{bmatrix}$$

Interchange row 1 and row 2.

$$\begin{bmatrix} 1 & 2 & -1 & | & 5 \\ 2 & -1 & 3 & | & 0 \\ 0 & 2 & 1 & | & 1 \end{bmatrix}$$

Multiply row 1 by -2 and add to row 2.

$$\begin{bmatrix} 1 & 2 & -1 & | & 5 \\ 0 & -5 & 5 & | & -10 \\ 0 & 2 & 1 & | & 1 \end{bmatrix}$$

Multiply row 2 by $-\dfrac{1}{5}$.

$$\begin{bmatrix} 1 & 2 & -1 & | & 5 \\ 0 & 1 & -1 & | & 2 \\ 0 & 2 & 1 & | & 1 \end{bmatrix}$$

Multiply row 2 by -2 and add to row 3.

$$\begin{bmatrix} 1 & 2 & -1 & | & 5 \\ 0 & 1 & -1 & | & 2 \\ 0 & 0 & 3 & | & -3 \end{bmatrix}$$

Multiply row 3 by $\dfrac{1}{3}$.

$$\begin{bmatrix} 1 & 2 & -1 & | & 5 \\ 0 & 1 & -1 & | & 2 \\ 0 & 0 & 1 & | & -1 \end{bmatrix}$$

Add row 3 to row 1 and to row 2.

$$\begin{bmatrix} 1 & 2 & 0 & | & 4 \\ 0 & 1 & 0 & | & 1 \\ 0 & 0 & 1 & | & -1 \end{bmatrix}$$

Add -2 times row 2 to row 1.

$$\begin{bmatrix} 1 & 0 & 0 & | & 2 \\ 0 & 1 & 0 & | & 1 \\ 0 & 0 & 1 & | & -1 \end{bmatrix}$$

That is, $x = 2$, $y = 1$, $z = -1$.
Thus, $\{(2, 1, -1)\}$ is the solution set.

13. $-x + y = -1$
$\, y - z = 6$
$\, x + z = -1$

Form the augmented matrix.

$$\begin{bmatrix} -1 & 1 & 0 & | & -1 \\ 0 & 1 & -1 & | & 6 \\ 1 & 0 & 1 & | & -1 \end{bmatrix}$$

Multiply row 1 by -1.

$$\begin{bmatrix} 1 & -1 & 0 & | & 1 \\ 0 & 1 & -1 & | & 6 \\ 1 & 0 & 1 & | & -1 \end{bmatrix}$$

Multiply row 1 by -1 and add to row 3.

$$\begin{bmatrix} 1 & -1 & 0 & | & 1 \\ 0 & 1 & -1 & | & 6 \\ 0 & 1 & 1 & | & -2 \end{bmatrix}$$

Add row 2 to row 1.

$$\begin{bmatrix} 1 & 0 & -1 & | & 7 \\ 0 & 1 & -1 & | & 6 \\ 0 & 1 & 1 & | & -2 \end{bmatrix}$$

Multiply row 2 by -1 and add to row 3.

$$\begin{bmatrix} 1 & 0 & -1 & | & 7 \\ 0 & 1 & -1 & | & 6 \\ 0 & 0 & 2 & | & -8 \end{bmatrix}$$

Multiply row 3 by $\dfrac{1}{2}$.

$$\begin{bmatrix} 1 & 0 & -1 & | & 7 \\ 0 & 1 & -1 & | & 6 \\ 0 & 0 & 1 & | & -4 \end{bmatrix}$$

Add row 3 to row 1.

$$\begin{bmatrix} 1 & 0 & 0 & | & 3 \\ 0 & 1 & -1 & | & 6 \\ 0 & 0 & 1 & | & -4 \end{bmatrix}$$

Add row 3 to row 2.

$$\begin{bmatrix} 1 & 0 & 0 & | & 3 \\ 0 & 1 & 0 & | & 2 \\ 0 & 0 & 1 & | & -4 \end{bmatrix}$$

That is, $x = 3$, $y = 2$, $z = -4$.
Thus, $\{(3, 2, -4)\}$ is the solution set.

15. $2x - y + 4z = -1$
$-3x + 5y - z = 5$
$2x + 3y + 2z = 3$

Form the augmented matrix.

$$\begin{bmatrix} 2 & -1 & 4 & | & -1 \\ -3 & 5 & -1 & | & 5 \\ 2 & 3 & 2 & | & 3 \end{bmatrix}$$

Add row 2 to row 1.

$$\begin{bmatrix} -1 & 4 & 3 & | & 4 \\ -3 & 5 & -1 & | & 5 \\ 2 & 3 & 2 & | & 3 \end{bmatrix}$$

Multiply row 1 by 2 and add to row 3.

$$\begin{bmatrix} -1 & 4 & 3 & | & 4 \\ -3 & 5 & -1 & | & 5 \\ 0 & 11 & 8 & | & 11 \end{bmatrix}$$

Multiply row 1 by -1.

$$\begin{bmatrix} 1 & -4 & -3 & | & -4 \\ -3 & 5 & -1 & | & 5 \\ 0 & 11 & 8 & | & 11 \end{bmatrix}$$

Multiply row 1 by 3 and add to row 2.

$$\begin{bmatrix} 1 & -4 & -3 & | & -4 \\ 0 & -7 & -10 & | & -7 \\ 0 & 11 & 8 & | & 11 \end{bmatrix}$$

Multiply row 2 by $-\dfrac{1}{7}$.

$$\begin{bmatrix} 1 & -4 & -3 & | & -4 \\ 0 & 1 & \frac{10}{7} & | & 1 \\ 0 & 11 & 8 & | & 11 \end{bmatrix}$$

Multiply row 2 by 4 and add to row 1.

$$\begin{bmatrix} 1 & 0 & \frac{19}{7} & 0 \\ 0 & 1 & \frac{10}{7} & 1 \\ 0 & 11 & 8 & 11 \end{bmatrix}$$

Multiply row 2 by −11 and add to row 3.

$$\begin{bmatrix} 1 & 0 & \frac{19}{7} & 0 \\ 0 & 1 & \frac{10}{7} & 1 \\ 0 & 0 & -\frac{54}{7} & 0 \end{bmatrix}$$

Multiply row 3 by $-\dfrac{7}{54}$.

$$\begin{bmatrix} 1 & 0 & \frac{19}{7} & 0 \\ 0 & 1 & \frac{10}{7} & 1 \\ 0 & 0 & 1 & 0 \end{bmatrix}$$

Multiply row 3 by $-\dfrac{19}{7}$ and add to row 1.

Multiply row 3 by $-\dfrac{10}{7}$ and add to row 2.

$$\begin{bmatrix} 1 & 0 & 0 & 0 \\ 0 & 1 & 0 & 1 \\ 0 & 0 & 1 & 0 \end{bmatrix}$$

That is, $x = 0$, $y = 1$, $z = 0$.
Thus, $\{(0, 1, 0)\}$ is the solution set.

17. $x + y - 2z = 1$
$2x - y - 4z = -4$
$3x - 2y + z = -7$

Form the augmented matrix.

$$\begin{bmatrix} 1 & 1 & -2 & 1 \\ 2 & -1 & -4 & -4 \\ 3 & -2 & 1 & -7 \end{bmatrix}$$

Multiply row 1 by −3 and add to row 3.

$$\begin{bmatrix} 1 & 1 & -2 & 1 \\ 2 & -1 & -4 & -4 \\ 0 & -5 & 7 & -10 \end{bmatrix}$$

Multiply row 1 by −2 and add to row 2.

$$\begin{bmatrix} 1 & 1 & -2 & 1 \\ 0 & -3 & 0 & -6 \\ 0 & -5 & 7 & -10 \end{bmatrix}$$

Interchange row 2 and row 3.

$$\begin{bmatrix} 1 & 1 & -2 & 1 \\ 0 & -5 & 7 & -10 \\ 0 & -3 & 0 & -6 \end{bmatrix}$$

Thus $x + y - 2z = 1$
$-5y + 7z = -10$
$-3y = -6$ or
$y = 2$.

Substitute $y = 2$ into the 2nd equation and

solve for z.
$-5(2) + 7z = -10$
$7z = 0$
$z = 0$

Substitute $y = 2$ and $z = 0$ into the equation
$x + y - 2z = 1$ and solve for x.
$x + y - 2z = 1$
$x + 2 - 2(0) = 1$
$x = -1$
The solution set is $\{(-1, 2, 0)\}$.

19. Let $x =$ revenue for AT&T, and
$y =$ revenue for Verizon.
Write an equation to represent the total
revenue of the companies.
$x + y = 221.4$
Write an equation to represent the
relationship between the revenues of the
companies.
$x + y = 221.4$
$x - y = 26.6$
Form the augmented matrix.

$$\begin{bmatrix} 1 & 1 & 221.4 \\ 1 & -1 & 26.6 \end{bmatrix}$$

Multiply row 1 by −1 and add to row 2.

$$\begin{bmatrix} 1 & 1 & 221.4 \\ 0 & -2 & -194.8 \end{bmatrix}$$

Multiply row 2 by $-\dfrac{1}{2}$.

$$\begin{bmatrix} 1 & 1 & 221.4 \\ 0 & 1 & 97.4 \end{bmatrix}$$

Thus $x + y = 221.4$
$y = 97.4$

Substitute $y = 97.4$ into the first equation
and solve for x.
$x + 97.4 = 221.4$
$x = 124.0$
The revenue for AT&T was $124.0 billion,
and the revenue for Verizon was
$97.4 billion.

8.9 Exercises

1. The answer is C since this represents the
region where $x \geq 5$ and (at the same time)
$y \leq -3$.

3. The answer is B since this represents the
region where $x > 5$ and (at the same time)
$y < -3$.

5. $x + y \leq 2$

Graph the boundary line $x + y = 2$.
Since the inequality is of the form "$\leq$," i.e.,
includes the equality, the graph is a solid
line. Next, try a test point not on the line,
such as (0, 0), in the inequality.

$x + y \leq 2$
$0 + 0 \leq 2$ True

Thus, shade the region containing the point
(0, 0) and all the other points in the region
below and including the line itself.

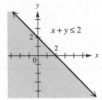

7. $4x - y \leq 5$

Graph the boundary line $4x - y = 5$.
Try a test point, such as (0, 0), in the
inequality.

$4x - y \leq 5$
$4(0) - 0 \leq 5$ True

Thus, the half-plane including (0, 0) is to be
shaded. Observe that the line itself is a part
of the solution set.

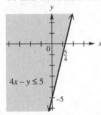

9. $x + 3y \geq -2$

Graph the boundary line $x + 3y = -2$.
Try a test point, such as (0, 0), in the
inequality.

$x + 3y \geq -2$
$0 + 3(0) \geq -2$ True

Thus, the half-plane including (0, 0) is to be
shaded. The line itself is also a part of the
solution set.

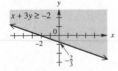

11. $x + 2y \leq -5$

Graph the boundary line $x + 2y = -5$.
Try a test point, such as (0, 0), in the
inequality.

$x + 2y \leq -5$
$0 + 2(0) \leq -5$
$0 \leq -5$ False

Since the point does not work, the other
half-plane represents the solution and is to
be shaded. The line itself is a part of the
solution set.

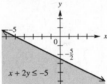

13. $4x - 3y < 12$

Graph the boundary line $4x - 3y = 12$.
Try a test point, such as (0, 0), in the
inequality.

$4x - 3y < 12$
$4(0) - 3(0) < 12$
$0 < 12$ True

Thus, the half-plane including (0, 0) is to be
shaded. The line itself is not a part of the
solution set and, therefore, is indicated with
a dashed line.

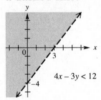

15. $y > -x$

Use the alternate equivalent form, $x + y > 0$.
Graph the boundary line $x + y = 0$.
Try a test point, such as (1, 1), which does
not lie on the line.

$1 + 1 > 0$
$2 > 0$ True

Thus, the half-plane to be shaded is that
above the line and includes the point (1, 1).
The line must be dashed since "=" is not
included in the strict inequality ">."

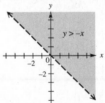

17. $x + y \leq 1$
    $x \geq 0$

Using the boundary equations $x + y = 1$ and
$x = 0$, (y-axis) sketch the graph of each
individual inequality, shading the

appropriate half-planes. The intersection of these regions (area in common) represents the solution set as below.

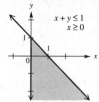

19. $2x - y \geq 1$
$3x + 2y \geq 6$

Using the boundary equations $2x - y = 1$ and $3x + 2y = 6$, sketch the graph of each individual inequality, shading the appropriate half-planes. The intersection of these regions represents the solution set as shown below.

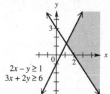

21. $-x - y < 5$
$x - y \leq 3$

Use the boundary equations $-x - y = 5$ and $x - y = 3$ to sketch the graph of each individual inequality, shading the appropriate half-planes. Note that the line itself is not include with the first inequality. The intersection of these regions represents the solution set as shown below.

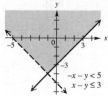

23.   Evaluate $3x + 5y$ at each vertex.

Point	Value = $3x + 5y$	
(1, 1)	$3(1) + 5(1) = 8$	←minimum
(6, 3)	$3(6) + 5(3) = 33$	
(5, 10)	$3(5) + 5(10) = 65$	←maximum
(2, 7)	$3(2) + 5(7) = 41$	

Thus, there is a maximum of 65 at (5, 10) and a minimum of 8 at (1, 1).

25.   Find $x \geq 0$ and $y \geq 0$ such that
$2x + 3y \leq 6$
$4x + y \leq 6$
and $5x + 2y$ is maximized.
To find the vertex points, solve the system of boundary equations.
$2x + 3y = 6$
$4x + y = 6$

Multiply the second equation by -3 and add to the first equation.

$$2x + 3y = 6$$
$$\underline{-12x - 3y = -18}$$
$$-10x = -12$$
$$x = \frac{-12}{-10} = \frac{6}{5}$$

Substitute this value for x in the second equation $4x + y = 6$.

$$4\left(\frac{6}{5}\right) + y = 6$$
$$y = 6 - \frac{24}{5}$$
$$y = \frac{30}{5} - \frac{24}{5}$$
$$y = \frac{6}{5}$$

Thus, a corner point is $\left(\frac{6}{5}, \frac{6}{5}\right)$.

Sketch the graph of the feasible region representing the intersection of all of the constraints (i.e., system of inequalities).

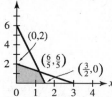

Evaluate $5x + 2y$ at each vertex.

Point	Value = $5x + 2y$	
(0, 2)	$5(0) + 2(2) = 4$	
$\left(\frac{6}{5}, \frac{6}{5}\right)$	$5\left(\frac{6}{5}\right) + 2\left(\frac{6}{5}\right) = \frac{42}{5}$	←maximum
$\left(\frac{3}{2}, 0\right)$	$5\left(\frac{3}{2}\right) + 2(0) = \frac{15}{2}$	
(0, 0)	$5(0) + 2(0) = 0$	

Thus, the maximum value occurs at the vertex point $\left(\frac{6}{5}, \frac{6}{5}\right)$ and has the value $\frac{42}{5}$.

27. Find $x \geq 2$ and $y \geq 5$ such that
$$3x - y \geq 12$$
$$x + y \leq 15$$
and $2x + y$ is minimized.
Sketch a graph of the solution to the system (feasible region).

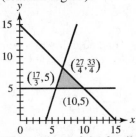

The vertices of the feasible region are the intersection points of each pair of boundary lines. The top vertex point is the intersection of the lines $3x - y = 12$ and $x + y = 15$. Solve this system.
$$3x - y = 12$$
$$x + y = 15$$
Add the equations to eliminate y.
$$\begin{array}{r} 3x - y = 12 \\ x + y = 15 \\ \hline 4x = 27 \end{array}$$
$$x = \frac{27}{4}$$

Substitute this value of x into the second equation $x + y = 15$ and solve for y.
$$\left(\frac{27}{4}\right) + y = 15$$
$$y = 15 - \frac{27}{4}$$
$$y = \frac{60}{4} - \frac{27}{4} = \frac{33}{4}$$

The resulting vertex point is $\left(\frac{27}{4}, \frac{33}{4}\right)$.

The bottom left vertex can be found by solving the system.
$$3x - y = 12$$
$$y = 5$$
Substitute this value of y into the first equation $3x - y = 12$ and solve for x.
$$3x - (5) = 12$$
$$3x = 17$$
$$x = \frac{17}{3}$$

This gives the vertex $\left(\frac{17}{3}, 5\right)$.

To find the remaining vertex solve the following system.

$$x + y = 15$$
$$y = 5$$
Substitute this value of y into the first equation $x + y = 15$ and solve for x.
$$x + (5) = 15$$
$$x = 10$$
This gives the vertex (10, 5).
Evaluate $2x + y$ at each vertex.

Point	Value $= 2x + y$	
$\left(\frac{27}{4}, \frac{33}{4}\right)$	$2\left(\frac{27}{4}\right) + \left(\frac{33}{4}\right) = \frac{87}{4}$ $= 21.75$	
$\left(\frac{17}{3}, 5\right)$	$2\left(\frac{17}{3}\right) + (5) = \frac{49}{3}$ ≈ 16.33	←minimum
(10, 5)	$2(10) + (5) = 25$	

Thus, vertex point $\left(\dfrac{17}{3}, 5\right)$ gives the minimum value, $\dfrac{49}{3}$.

Note: In the following exercises, $x \geq 0$ and $y \geq 0$ are understood constraints.

29. Let $x =$ the number of refrigerators shipped to warehouse A and $y =$ the number of refrigerators shipped to warehouse B.
Since at least 100 refrigerators must be shipped, $x + y \geq 100$.
Since warehouse A holds a maximum of 100 refrigerators and has 25 already, it has room for at most 75 more, so $0 \leq x \leq 75$. Similarly, warehouse B has room for at most 80 more refrigerators, so $0 \leq y \leq 80$.
The cost function is given by $12x + 10y$.
The linear programming problem may be stated as follows: Find $x \geq 0$ and $y \geq 0$ such that $x + y \geq 100$
$$x \leq 75$$
$$y \leq 80$$
and $12x + 10y$ is minimized. Graph the feasible region.

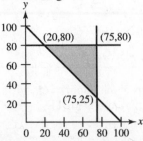

Evaluate the objective function at each vertex.

Point	Cost = 12x + 10y	
(20, 80)	12(20) + 10(80) = 1040	←minimum
(75, 80)	12(75) + 10(80) = 1700	
(75, 25)	12(75) + 10(25) = 1150	

The minimum value of the objective function is 1040 occurring at (20, 80). Therefore, 20 refrigerators should be shipped to warehouse A and 80 to warehouse B, for a minimum cost of $1040.

31. Let x = the number of red pills and y = the number of blue pills.
Use a table to summarize the given information.

	Vitamin A	Vitamin B_1	Vitamin C
Red pills	8	1	2
Blue pills	2	1	7
Daily requirement	16	5	20

The linear programming problem may be stated as follows: Find $x \geq 0$ and $y \geq 0$ such that $8x + 2y \geq 16$ (Vitamin A)
$$x + y \geq 5 \quad \text{(Vitamin } B_1\text{)}$$
$$2x + 7y \geq 20 \quad \text{(Vitamin C)}$$
and the Cost = $0.1x + 0.2y$ is minimized. Solve the corresponding equations to find the vertex points, and graph the feasible region.

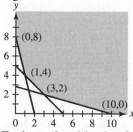

Evaluate the objective function at each vertex.

Point	Cost = 0.1x + 0.2y	
(0, 8)	0.1(0) + 0.2(8) = 1.6	
(1, 4)	0.1(1) 1+ 0.2(4) = 0.9	
(3, 2)	0.1(3) + 0.2(2) = 0.7	←minimum
(10, 0)	0.1(10) + 0.2(0) = 1	

The minimum value of the objective function is 0.7 (i.e., 70¢) occurring at (3, 2). Therefore, she should take 3 red pills and 2 blue pills, for a minimum cost of 70¢ per day.

33. Let x = number of barrels of gasoline and y = number of barrels of oil.
The linear programming problem may be stated as follows: Find $x \geq 0$ and $y \geq 0$ with the following constraints:
$x \geq 2y$
$y \geq 3$ million
$x \leq 6.4$ million,
where revenue = $2.9x + 3.5y$ and is to be maximized.
Solve the corresponding system of boundary equations for their points of intersection to find the vertices, and sketch the graph representing the intersection of the constraints. The vertices of the feasible region are (6.4, 3), (6, 3), and (6.4, 3.2).

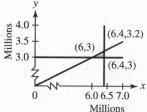

Evaluate the objective function at each vertex.

Point	Revenue = 2.9x + 3.5y (in millions)	
(6.4, 3)	2.9(6.4) + 3.5(3) = 29.06	
(6, 3)	2.9(6) + 3.5(3) = 27.9	
(6.4, 3.2)	2.9(6.4) + 3.5(3.2) = 29.76	←maximum

Thus, producing 6.4 million barrels of gasoline and 3.2 million barrels of fuel oil should yield the maximum revenue of $29.76 million.

35. Let x = number of medical kits and y = number of containers of water. Use a table to summarize the given information.

	Volume	Weight
Medical Kits	1	10
Containers of water	1	20
Maximum allowed	6000	80000

Weight constraint: $10x + 20y \leq 80000$.
Volume constraint: $(1)x + (1)y \leq 6000$.
Objective function: $6x + 10y$
The problem may be stated as follows.
For $x \geq 0$ and $y \geq 0$,
$10x + 20y \leq 80000$ (Pounds)
$x + y \leq 6000$. (Cubic feet)
These constraints may be written in the simpler equivalent form: $x + 2y \leq 8000$
$x + y \leq 6000$.
The number of people aided $= 6x + 10y$ and is to be maximized.
Solve the corresponding equations for their points of intersection to find the vertices, and sketch the graph representing the intersection of the constraints. The vertex points are (0, 4000), (4000, 2000) and (6000, 0).

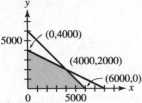

Evaluate the objective function at each vertex.

Point	People aided $= 6x + 10y$	
(0, 4000)	$6(0) + 10(4000)$ $= 40000$	
(4000, 2000)	$6(4000) + 10(2000)$ $= 44000$	←maximum
(6000, 0)	$6(6000) + 10(0)$ $= 36000$	

Thus, ship 4000 medical kits and 2000 containers of water in order to maximize the number of people.

Chapter 8 Test

1. $d = \sqrt{(x_2 - x_1)^2 + (y_2 - y_1)^2}$
$= \sqrt{(2 - (-3))^2 + (1 - 5)^2}$
$= \sqrt{(5)^2 + (-4)^2}$
$= \sqrt{25 + 16}$
$= \sqrt{41}$

2. Use the equation of a circle, where $h = -1$, $k = 2$, and $r = 3$.
$(x - h)^2 + (y - k)^2 = r^2$
$(x - (-1))^2 + (y - 2)^2 = 3^2$
$(x + 1)^2 + (y - 2)^2 = 9$

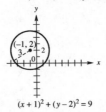

$(x + 1)^2 + (y - 2)^2 = 9$

3. $3x - 2y = 8$
To find the x-intercept, let $y = 0$.
$3x - 2(0) = 8$
$3x = 8$
$x = \dfrac{8}{3}$

The x-intercept is $\left(\dfrac{8}{3}, 0\right)$. To find the y-intercept, let $x = 0$.
$3(0) - 2y = 8$
$-2y = 8$
$y = -4$
The y-intercept is $(0, -4)$.

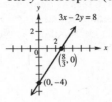

4. Let $(x_1, y_1) = (6, 4)$ and $(x_2, y_2) = (-1, 2)$.
Then, $m = \dfrac{y_2 - y_1}{x_2 - x_1} = \dfrac{2 - 4}{-1 - 6} = \dfrac{-2}{-7} = \dfrac{2}{7}$.

5. The slope-intercept form of the equation of the line

(a) passing through the point $(-1, 3)$, with slope $-\frac{2}{5}$. Use the point-slope form of the line to write the equation:
$$y - y_1 = m(x - x_1)$$
$$y - 3 = -\frac{2}{5}[x - (-1)]$$
$$y = -\frac{2}{5}x - \frac{2}{5} + 3$$
$$y = -\frac{2}{5}x + \frac{13}{5}$$

(b) passing through $(-7, 2)$ and perpendicular to $y = 2x$. Use the slope of the given line, 2, to find the slope of any perpendicular line, $-\frac{1}{2}$. Use the point-slope form of the line to write the equation:
$$y - y_1 = m(x - x_1), \text{ or }$$
$$y - 2 = -\frac{1}{2}[x - (-7)]$$
$$y = -\frac{1}{2}x - \frac{7}{2} + 2$$
$$y = -\frac{1}{2}x - \frac{3}{2}.$$

(c) The line displays show two points on the line, $(-2, 3)$ and $(6, -1)$. Use these points to create the slope of the line:
$$m = \frac{y_2 - y_1}{x_2 - x_1} = \frac{-1 - 3}{6 - (-2)} = \frac{-4}{8} = -\frac{1}{2}.$$
Use this slope and either known point to write the equation:
$$y - y_1 = m(x - x_1)$$
$$y - 3 = -\frac{1}{2}[x - (-2)]$$
$$y = -\frac{1}{2}x - 1 + 3$$
$$y = -\frac{1}{2}x + 2.$$

6. Option B shows the graph of a line with a positive slope and a negative y-intercept.

7. (a) Use the data points $(0, 63.3)$ and $(60, 77.6)$ to find the slope.
$$m = \frac{y_2 - y_1}{x_2 - x_1}$$
$$= \frac{77.6 - 63.3}{60 - 0}$$
$$= \frac{14.3}{60}$$
$$\approx 0.238$$
Use the slope and the slope-intercept form of the line.
$$y = mx + b$$
$$y = 0.238x + 63.3$$

(b) Let $x = 70$ and solve for y.
$$y = 0.238x + 63.3$$
$$y = 0.238(70) + 63.3$$
$$y = 16.66 + 63.3$$
$$y = 79.96$$
Life expectancy in 2013 will be about 80 years.

(c) The midpoint will occur at 1978 and is
$$\frac{71.4 + 74.6}{2} = \frac{146}{2} = 73 \text{ years.}$$

8. The average rate of change is the same value as the slope. Thus,
$$\text{average rate of change} = \frac{y_2 - y_1}{x_2 - x_1}$$
$$= \frac{614.9 - 942.5}{2006 - 2000}$$
$$= \frac{-327.6}{6}$$
$$= -54.6 \text{ million}$$
CDs per year.

9. Let x = the number of days the book is overdue and y = the total fine. Then $y = 0.05x + 0.50$ will model the situation. To find three ordered pairs, evaluate the equation at $x = 1, 5,$ and 10.
$y = 0.05(1) + 0.50 = 0.55$ or $(1, 0.55)$
$y = 0.05(5) + 0.50 = 0.75$ or $(5, 0.75)$
$y = 0.05(10) + 0.50 = 1.00$ or $(10, 1.00)$

10. Estimate the values for two points on the line such as $(0, 1)$ and $(3, 3)$. Use these values to create the slope.
$$m = \frac{y_2 - y_1}{x_2 - x_1} = \frac{3 - 1}{3 - 0} = \frac{2}{3}$$
Because one of our points is the y-intercept, $(0, 1)$, we may choose to use the slope-intercept form of the line to write the

equation.
$$y = mx + b$$
$$y = \frac{2}{3}x + 1$$

11. Given the function $f(x) = x^2 - 3x + 12$, then

  (a) the domain is $(-\infty, \infty)$, and

  (b) $f(-2) = (-2)^2 - 3(-2) + 12 = 22$.

12. Given the function $f(x) = \dfrac{2}{x-3}$, then

  (a) the domain is $(-\infty, 3) \cup (3, \infty)$.

  (b) $f(3)$ is undefined.

13. To find the break-even point with
$C(x) = 50x + 5000$ and $R(x) = 60x$, set
$C(x) = R(x)$ and solve for x.
$$50x + 5000 = 60x$$
$$5000 = 10x$$
$$x = 500$$
Evaluate $R(x)$ at $x = 500$ to find the
corresponding revenue.
$R(500) = 60(500) = 30000$
Thus, 500 calculators, which produces
$30,000 in revenue, is the break-even point.

14. $f(x) = -(x+3)^2 + 4$
By inspection of the given standard form of
the equation, the axis is at $x = -3$ and the
vertex is at $(-3, 4)$. The domain is $(-\infty, \infty)$
and the range is $(-\infty, 4]$.

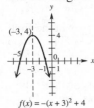

$f(x) = -(x+3)^2 + 4$

15. Let x = the width dimension of the lot and
$(320 - 2x)$ = length dimension of the lot.
If y = area function, then
$$y = l \cdot w = (320 - 2x) \cdot x = -2x^2 + 320x.$$
Use the formula for the x-value of the
vertex. This value of x will give the largest
value for y, the area.
$$x = \frac{-b}{2a} = \frac{-320}{2(-2)} = 80$$
Evaluate $(320 - 2x)$ at this value of x to

determine the length dimension.
length = $320 - 2(80) = 320 - 160 = 160$
Thus, the dimensions are 80 feet by 160
feet.

16. (a) $5.1^{4.7} \approx 2116.31264888$

  (b) $e^{-1.85} \approx 0.157237166314$

  (c) $\ln 23.56 \approx 3.15955035878$

17. $P = \$12,000$, $n = 4$, $r = 2\% = 0.02$, and
$t = 3$ years.

  (a) Use the compounding formula
  $$A = P\left(1 + \frac{r}{n}\right)^{nt} :$$
  $$A = 12000\left(1 + \frac{0.02}{4}\right)^{4 \cdot 3}$$
  $$= 12000(1.005)^{12}$$
  $$\approx 12,740.13$$

  (b) Use the compounding formula
  $A = Pe^{rt}$ for continuous compounding:
  $$A = 12000e^{(0.02)3}$$
  $$= 12000e^{0.06}$$
  $$\approx 12,742.04$$

18. $A(t) = 2.00e^{-0.053t}$

  (a) The amount present in 4 years is given
  by $A(4) = 2.00e^{-0.053(4)}$
  $$= 2.00e^{-0.212}$$
  $$\approx 1.62 \text{ grams.}$$

  (b) The amount present in 10 years is given
  by $A(10) = 2.00e^{-0.053(10)}$
  $$= 2.00e^{-0.53}$$
  $$\approx 1.18 \text{ grams.}$$

  (c) The amount present in 20 years is given
  by $A(20) = 2.00e^{-0.053(20)}$
  $$= 2.00e^{-1.06}$$
  $$\approx 0.69 \text{ gram.}$$

  (d) The initial amount present given by
  $A(0) = 2.00e^{-0.053(0)}$
  $$= 2.00e^0$$
  $$\approx 2.00 \text{ grams.}$$

19. $2x + 3y = 2$
$3x - 4y = 20$

Multiply the first equation by 4 and the second by 3 then add.

$$8x + 12y = 8$$
$$\underline{9x - 12y = 60}$$
$$17x = 68$$
$$x = 4$$

Substitute this value for x into the first equation and solve for y.

$$2(4) + 3y = 2$$
$$3y = 2 - 8$$
$$3y = -6$$
$$y = -2$$

Since the ordered pair $(4, -2)$ satisfies both equations, it checks. The solution set is $\{(4, -2)\}$.

20. $2x + y + z = 3$
$x + 2y - z = 3$
$3x - y + z = 5$

Eliminate z from the first and second equations by adding.

$$2x + y + z = 3$$
$$\underline{x + 2y - z = 3}$$
$$3x + 3y = 6$$
$$x + y = 2$$

Eliminate z from first and third equations by multiplying the first equation by -1 and adding to the third.

$$-2x - y - z = -3$$
$$\underline{3x - y + z = 5}$$
$$x - 2y = 2$$

We are left with two equations in x and y. Multiply the first by 2 and add to the second in order to eliminate the y-term.

$$2x + 2y = 4$$
$$\underline{x - 2y = 2}$$
$$3x = 6$$
$$x = 2$$

Substitute this value of x into the equation $x + y = 2$ (either equation in x and y may be used). Solve for y.

$$(2) + y = 2$$
$$y = 0$$

Replace x and y by these values in the first equation $2x + y + z = 3$ (any one of the original 3 equations may be used). Solve for z.

$$2(2) + (0) + z = 3$$
$$z = 3 - 4$$
$$z = -1$$

The solution set is $\{(2, 0, -1)\}$.

21. $2x + 3y - 6z = 11$
$x - y + 2z = -2$
$4x + y - 2z = 7$

Eliminate y and z from the second and third equations by adding.

$$x - y + 2z = -2$$
$$\underline{4x + y - 2z = 7}$$
$$5x = 5$$
$$x = 1$$

It is not possible to find a unique value for y and z corresponding to the value $x = 1$. This means that the system is dependent and has an infinite number of answers. Therefore, replace $x = 1$ into the third equation (this may be done in any of the original three equations) and solve for y in terms of z.

$$4(1) + y - 2z = 7$$
$$y - 2z = 7 - 4$$
$$y = 2z + 3$$

The solution set may now be expressed as $\{(1, 2z + 3, z)\}$.

22. Let x = gross receipts for *Star Wars Episode IV: A New Hope* and
y = gross receipts for *Indiana Jones and the Kingdom of the Crystal Skull*.
Write an equation representing the gross receipts for both movies.
$x + y = 778$
Write a second equation that relates gross receipts for both movies.
$y = x - 144$
Substitute this value for y in the first equation.

$$x + (x - 144) = 778$$
$$2x - 144 = 778$$
$$2x = 922$$
$$x = 461$$

Substitute this value for x into the equation $y = x - 144$ and solve for y.
$y = 461 - 144$
$y = 317$
The gross receipts for *Star Wars Episode IV: A New Hope* were \$461 million. the gross receipts for *Indiana Jones and the Kingdom of the Crystal Skull* were \$317 million.

23. Let x = the sale price with a 10% commission, y = the sale price with a 6% commission and z = the sale price with a 5% commission.

The total property sold is then given by $x + y + z = 280,000$.

The total commission is given by $0.10x + 0.06y + 0.05z = 17,000$.

In addition, there is the following relationship between the sales: $z = x + y$. Simplifying the second equation and writing all in standard form results in the system

$$x + y + z = 280,000$$
$$10x + 6y + 5z = 1,700,000$$
$$-x - y + z = 0.$$

Adding the first and last equation results in the elimination of x and y terms.

$$x + y + z = 280,000$$
$$\underline{-x - y + z = 0}$$
$$2z = 280,000$$
$$z = 140,000$$

Multiply the first equation $x + y + z = 280000$ by -6 and add to the second equation to eliminate the y terms.

$$-6x - 6y - 6z = -1,680,000$$
$$\underline{10x + 6y + 5z = 1,700,000}$$
$$4x - z = 20,000$$

Replace the value for z in this last equation and solve for x.

$$4x - 140,000 = 20,000$$
$$4x = 160,000$$
$$x = 40,000$$

Replace x and z in the equation $z = x + y$ by their values and solve for y.

$$140,000 = 40,000 + y$$
$$140,000 - 40,000 = y$$
$$y = 100,000$$

Thus, Keshon Grant sold a property for $40,000 with a 10% commission, another property for $100,000 with a 6% commission, and a third for $140,000 with a 5% commission.

24. $x + y \leq 6$
$2x - y \geq 3$

Use the boundary equations $x + y = 6$ and $2x - y = 3$ to sketch the graph of each individual inequality, shading the appropriate half-planes. The intersections of these regions represents the solution set as

shown below.

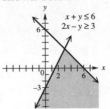

25. Let x = the number of pigs raised, and y = the number of geese raised.

The constraints are
$$x + y \leq 16$$
$$y \leq 12$$

$50x + 20y \leq 500$, or $5x + 2y \leq 50$.

The profit function is given by $40x + 80y$. The linear programming problem may be stated as follows: Find $x \geq 0$ and $y \geq 0$ such that $x + y \leq 16$
$$y \leq 12$$
$$5x + 2y \leq 50$$

and Profit = $40x + 80y$ is maximized. Solve the system of boundary equations for their points of intersection to find the vertices, and sketch the graph representing the intersection of the constraints.

The vertices of the feasible region are $(0, 12)$, $(4, 12)$, $(6, 10)$, and $(10, 0)$.

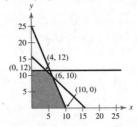

Evaluate the objective function at each vertex.

Point	Profit = $40x + 80y$	
$(0, 12)$	$40(0) + 80(12) = 960$	
$(4, 12)$	$40(4) + 80(12) = 1120$	←maximum
$(6, 10)$	$40(6) + 80(10) = 1040$	
$(10, 0)$	$40(10) + 80(0) = 400$	

The maximum profit will be $1120 if Callie raises 4 pigs and 12 geese.

Chapter 9

1. The sum of the measures of two complementary angles is <u>90</u> degrees.

3. The measures of two vertical angles are <u>equal</u>.

5. It is true that a line segment has two endpoints.

7. It is false that if A and B are distinct points on a line, then ray AB and ray BA represent the same set of points. The initial point of ray AB is point A; the initial point of ray BA is point B.

9. It is true that if two lines are parallel, they lie in the same plane.

11. It is true that segment AB and segment BA represent the same set of points. A and B are the endpoints of the line segment and can be named in either order.

13. It is false that there is no angle that is its own supplement. A 90° or right angle is supplementary to a 90° angle.

15. (a) $\overset{\bullet\quad\bullet}{AB}$

    (b) ●————————————●
        A   B

17. (a) $\overset{\longleftarrow\bullet}{CB}$

    (b) ◄————●———●————●——
           A   B   C

19. (a) $\overset{\circ\quad\longrightarrow}{BC}$

    (b) ○————●——●————————►
        B   C D

21. (a) $\overset{\bullet\quad\longrightarrow}{BA}$

    (b) ◄————●————●—
          A   B

23. (a) $\overset{\bullet\quad\bullet}{CA}$

    (b) ●————●————●
        A   B   C

25. Letter F; line segment PQ is the same as line segment QP.

27. Letter D; ray QR names the same set of points as ray QS. The initial point is Q, and the set of points passes through both R and S.

29. Letter B; ray RP is the same as ray RQ. The initial point for both is R.

31. Letter E; line segment PS is the same as line segment SP.

33. $\overleftrightarrow{MN} \cup \overrightarrow{NO}$ names the same set of points as $\overrightarrow{MO}$. The union symbol joins the two line segments.

35. $\overrightarrow{MO} \cap \overleftarrow{OM}$ indicates the intersection or overlap of two rays. The intersection is the line segment $\overline{MO}$.

37. $\overrightarrow{OP} \cap O$ have no points in common because point O is not part of the half-line $\overrightarrow{OP}$. Therefore, the intersection is the empty set, symbolized by $\varnothing$.

39. $\overleftrightarrow{NP} \cap \overrightarrow{OP}$ indicates the same set of points as $\overrightarrow{OP}$.

41. $90 - 28 = 62°$

43. $90 - 89 = 1°$

45. $(90 - x)°$

47. $180 - 132 = 48°$

49. $180 - 26 = 154°$

51. $(180 - y)°$

53. $\angle ABE$ and $\angle CBD$; $\angle ABD$ and $\angle CBE$

55. (a) 52°; they are vertical angles.

    (b) $180 - 52 = 128°$; they are supplementary angles.

57. The designated angles are supplementary; their sum is 180°.
$$(10x + 7) + (7x + 3) = 180$$
$$17x + 10 = 180$$
$$17x = 170$$
$$x = 10$$
Then one angle measure is $10 \cdot 10 + 7 = 107°$, and the other angle measure is $7 \cdot 10 + 3 = 73°$.

59. The angles are vertical, so they have the same measurement. Set the algebraic expressions each to each other.
$$3x + 45 = 7x + 5$$
$$-4x = -40$$
$$x = 10$$
Then one angle measure is
$3 \cdot 10 + 45 = 75°$, and the other angle measure is $7 \cdot 10 + 5 = 75°$.

61. The angles are vertical, so they have the same measurement. Set the algebraic expressions each to each other.
$$11x - 37 = 7x + 27$$
$$4x = 64$$
$$x = 16$$
Then one angle measure is
$11 \cdot 16 - 37 = 139°$, and the other angle measure is $7 \cdot 16 + 27 = 139°$.

63. The designated angles are supplementary; their sum is 180°.
$$(3x + 5) + (5x + 15) = 180$$
$$8x + 20 = 180$$
$$8x = 160$$
$$x = 20$$
Then one angle measure is $3 \cdot 20 + 5 = 65°$, and the other angle measure is
$5 \cdot 20 + 15 = 115°$.

65. The designated angles are complementary; their sum is 90°.
$$(5k + 5) + (3k + 5) = 90$$
$$8k + 10 = 90$$
$$8k = 80$$
$$k = 10$$
Then one angle measure is $5 \cdot 10 + 5 = 55°$, and the other angle measure is
$3 \cdot 10 + 5 = 35°$.

67. Alternate exterior angles have equal measures.
$$2x + 61 = 6x - 51$$
$$-4x = -112$$
$$x = 28$$
Then the measure of each angle is
$2 \cdot 28 + 61 = 117°$, and $6 \cdot 28 - 51 = 117°$.

69. Alternate exterior angles have equal measures.
$$10x + 11 = 15x - 54$$
$$-5x = -65$$
$$x = 13$$
Then the measure of each angle is
$10 \cdot 13 + 11 = 141°$, and
$15 \cdot 13 - 54 = 141°$.

71. Let x = measure of the angle,
$180 - x$ = measure of its supplement,
$90 - x$ = measure of its complement.
$$90 - x = \frac{1}{5}(180 - x) - 10$$
$$5 \cdot (90 - x) = \frac{5}{1} \cdot \frac{1}{5}(180 - x) - 5 \cdot 10$$
$$450 - 5x = (180 - x) - 50$$
$$450 - 5x = 180 - x$$
$$450 - 5x + x = 130 - x + x$$
$$450 - 4x = 130$$
$$450 - 450 - 4x = 130 - 450$$
$$-4x = -320$$
$$x = 80°$$

73. Let x = measure of the angle,
$180 - x$ = measure of its supplement,
$90 - x$ = measure of its complement.
$$\frac{1}{2}(180 - x) = 2(90 - x) - 12$$
$$\frac{2}{1} \cdot \frac{1}{2}(180 - x) = 2 \cdot 2(90 - x) - 2 \cdot 12$$
$$180 - x = 4(90 - x) - 24$$
$$180 - x = 360 - 4x - 24$$
$$180 - x = 336 - 4x$$
$$180 - x + 4x = 336 - 4x + 4x$$
$$180 + 3x = 336$$
$$180 - 180 + 3x = 336 - 180$$
$$3x = 156$$
$$\frac{3x}{3} = \frac{156}{3}$$
$$x = 52°$$

75. Some of the unknown angles must be solved before other unknown angles. Here is one order that they can be solved.

$\angle 1 = 55°$; vertical angle to 55°.

$\angle 8 = 180 - 120 = 60°$; supplementary angle to 120°.

$\angle 6 = 180 - 60 = 120°$; supplementary angle to $\angle 8$.

$\angle 7 = 60°$; vertical angle to $\angle 8$.

$\angle 5 = 60°$; alternate interior angle to $\angle 7$.

$\angle 3 = 60°$; vertical angle to $\angle 5$.

$\angle 2 = 180 - (55 + 60) = 180 - 55 - 60 = 65°$.
Angles 2, 3, and 55° all add to 180°.

$\angle 4 = 180 - (55 + 60) = 180 - 55 - 60 = 65°$;
straight angle composed of $\angle 4$, $\angle 5$, and 55°.

0 = 55°; alternate interior angle to the 55° angle.

$\angle 9 = 55°$; vertical angle to $\angle 10$.

9.2 Exercises

1. A segment joining two points on a circle is called a <u>chord</u>.

3. A regular triangle is called an <u>equilateral</u> or <u>equiangular</u> triangle.

5. False; a rhombus does not have equal angle measures.

7. False; the sum of the angle measures of a triangle always equals 180°. If a triangle had two obtuse angles, then the sum of the measures of the angles would exceed 180° which is impossible. Therefore, a triangle has at most one obtuse angle.

9. True; a square is a rhombus with four 90° angles.

11. Writing exercise; answers will vary.

13. Both; it is closed, and there are no intersecting curves.

15. Closed

17. Closed

19. Neither

21. Convex

23. Convex

25. Not convex

27. Right; scalene

29. Acute; equilateral

31. Right; scalene

33. Right; isosceles

35. Obtuse; scalene

37. Acute; isosceles

39. An isosceles right triangle is a triangle having a 90° angle and two perpendicular sides of equal length.

41. The sum of the measures of the three angles of any triangle is always 180°.
$$(x-10)+(2x-50)+x = 180$$
$$4x-60 = 180$$
$$4x = 240$$
$$x = 60$$
The measure of angle A is $60 - 10 = 50°$; angle B is $2 \cdot 60 - 50 = 70°$; angle $C = 60°$.

43. $(x-30)+(2x-120)+\left(\dfrac{1}{2}x+15\right) = 180$
$$3\frac{1}{2}x - 135 = 180$$
$$\frac{7}{2}x = 315$$
$$\frac{2}{7} \cdot \frac{7}{2}x = 315 \cdot \frac{2}{7}$$
$$x = 90$$
The measure of angle A is $90 - 30 = 60°$; angle B is $2 \cdot 90 - 120 = 60°$; angle C is $\dfrac{1}{2} \cdot 90 + 15 = 60°$.

45. Let x = the angle measure of A (or B), $x + 24$ = the angle measure of C.
$$x + x + (x+24) = 180$$
$$3x + 24 = 180$$
$$3x = 156$$
$$x = 52$$
The measure of angle A (or B) is $52°$; the measure of angle C is $52 + 24 = 76°$.

47. The measure of the exterior angle of a triangle is equal to the sum of the measures of the two opposite interior angles.
$$10x + (15x-10) = 20x + 25$$
$$25x - 10 = 20x + 25$$
$$25x - 20x - 10 = 20x - 20x + 25$$
$$5x - 10 = 25$$
$$5x - 10 + 10 = 25 + 10$$
$$5x = 35$$
$$x = 7$$
Then $(20 \cdot 7) + 25 = 165°$.

49. $(2-7x)+(100-10x) = 90-20x$
$$-17x + 102 = 90 - 20x$$
$$-17x + 102 - 102 = 90 - 102 - 20x$$
$$-17x = -12 - 20x$$
$$3x = -12$$
$$x = -4$$
Then $90 - 20 \cdot (-4) = 90 + 80 = 170°$.

51. **(a)** The center is at point O.

(b) There are four line segments that are radii: $\overset{\bullet\ \bullet}{OA}$, $\overset{\bullet\ \bullet}{OC}$, $\overset{\bullet\ \bullet}{OB}$, and $\overset{\bullet\ \bullet}{OD}$.

(c) There are two diameters: $\overset{\bullet\ \ \bullet}{AC}$ and $\overset{\bullet\ \ \bullet}{BD}$.

(d) There are four chords: $\overset{\bullet\ \ \bullet}{AC}$, $\overset{\bullet\ \ \bullet}{BD}$, $\overset{\bullet\ \ \bullet}{BC}$, and $\overset{\bullet\ \bullet}{AB}$.

(e) There are two secants: $\overset{\leftrightarrow}{AB}$ and $\overset{\leftrightarrow}{BC}$

(f) There is one tangent: $\overset{\leftrightarrow}{AE}$

53. (e) The sum of the measures of the angles of a triangle equals 180° (because the pencil has gone through one-half of a complete rotation).

EXTENSION: GEOMETRIC CONSTRUCTIONS

1. With the radius of the compasses greater than one-half the length PQ, place the point of the compasses at P and swing arcs above and below line r. Then with the same radius and the point of the compasses at Q, swing two more arcs above and below line r. Locate the two points of intersections of the arcs above and below, and call them A and B. With a straightedge, join A and B. AB is the perpendicular bisector of PQ.

3. With the radius of the compasses greater than the distance from P to r, place the point of the compasses at P and swing an arc intersecting line r in two points. Call these points A and B. Swing arcs of equal radius to the left of line r, with the point of the compasses at A and at B, intersecting at point Q. With a straightedge, join P and Q. PQ is the perpendicular from P to r.

5. With any radius, place the point of the compasses at P and swing arcs to the left and right, intersecting line r in two points. Call these points A and B. With an arc of sufficient length, place the point of the compasses first at A and then at B, and swing arcs either both above or both below line r, intersecting at point Q. With a straightedge, join P and Q. PQ is perpendicular to line r at P.

7. Using a radius of arbitrary length, place the point of the compasses at A and swing an arc intersecting the sides of angle A at two points. Call the point of intersection on the horizontal side B and call the other point of

intersection C. Draw a horizontal working line and locate any point A' on this line. With the same radius used earlier, place the point of the compasses at A' and swing an arc intersecting the working line at B'. Return to angle A, and set the radius of the compass equal to BC. On the working line, place the point of the compasses at B' and swing an arc intersecting the first arc at C'. Now draw line $A'C'$. Angle A' is equal to angle A.

9. Writing exercise; answers will vary.

9.3 Exercises

1.

STATEMENTS	REASONS
1. $AC = BD$	1. Given
2. $AD = BC$	2. Given
3. $AB = AB$	3. Reflexive property
4. $\triangle ABD \cong \triangle BAC$	4. SSS Congruence Property

3.

STATEMENTS	REASONS
1. $\overset{\bullet\ \bullet}{DB}$ is perpendicular to $\overset{\bullet\ \bullet}{AC}$	1. Given
2. $AC = BC$	2. Given
3. $\angle ABD = \angle CBD$	3. Both are right angles by definition of perpendicularity.
4. $DB = DB$	4. Reflexive property
5. $\triangle ABD \cong \triangle CBD$	5. SAS Congruence Property

5.

STATEMENTS	REASONS
1. $\angle BAC = \angle DAC$	1. Given
2. $\angle BCA = \angle DCA$	2. Given
3. $AC = AC$	3. Reflexive property
4. $\triangle ABC \cong \triangle ADC$	4. ASA Congruence Property

7. If $\angle B$ measures $46°$, then $\angle A$ measures $\underline{67°}$ and $\angle C$ measures $\underline{67°}$. In an isosceles triangle, the angles opposite the equal sides are also equal in measure. thus, $\angle A = \angle C$. The sum of the angles of the triangle is 180. Then $180 - 46 = 134$ and $134 \div 2 = 67°$.

9. If $\angle B$ measures $40°$, then the sum of $\angle A$ and $\angle C$ is $180 - 40 = 140°$. Because the triangle is isosceles, $\angle A = \angle C$, so each of these angles measures $70°$. Then $\angle BCD$ measures $180 - 70 = 110°$, because $\angle C$ and $\angle BCD$ are supplementary.

11. Writing exercise; answers will vary.

13. Corresponding angles are equal in measure.
$\angle A$ and $\angle P$
$\angle B$ and $\angle Q$
$\angle C$ and $\angle R$
Corresponding sides are proportional in length.
$\overleftrightarrow{AB}$ and $\overleftrightarrow{PQ}$
$\overleftrightarrow{AC}$ and $\overleftrightarrow{PR}$
$\overleftrightarrow{CB}$ and $\overleftrightarrow{RQ}$

15. Sometimes it is helpful to sketch the triangles, drawing them side by side. It is easier to determine the corresponding sides. Corresponding angles are equal in measure.
$\angle HGK$ and $\angle EGF$ because they are vertical.
$\angle H$ and $\angle F$
$\angle K$ and $\angle E$
Corresponding sides are proportional in length.
$\overleftrightarrow{HK}$ and $\overleftrightarrow{EF}$
$\overleftrightarrow{GH}$ and $\overleftrightarrow{GF}$
$\overleftrightarrow{GK}$ and $\overleftrightarrow{GE}$

17. $\angle P = \angle C = 78°$
$\angle M = \angle B = 46°$
$$\angle N = \angle A = 180 - (78 + 46)$$
$$= 180 - 124$$
$$= 56°$$
because the sum of the angle measures must equal $180°$.

19. $\angle T = \angle X = 74°$
$\angle V = \angle Y = 28°$
$$\angle W = \angle Z = 180 - (74 + 28)$$
$$= 180 - 102$$
$$= 78°$$
because the sum of the angle measures must equal $180°$.

21. $\angle T = \angle P = 20°$
$\angle V = \angle Q = 64°$
$$\angle U = \angle R = 180 - (20 + 64)$$
$$= 180 - 84$$
$$= 96°$$
because the sum of the angle measures must equal $180°$.

23. Corresponding sides must be proportional.
$$\frac{a}{8} = \frac{25}{10} \qquad \text{and} \qquad \frac{b}{6} = \frac{25}{10}$$
$$10 \cdot a = 8 \cdot 25 \qquad\qquad 10 \cdot b = 6 \cdot 25$$
$$\frac{10a}{10} = \frac{200}{10} \qquad\qquad \frac{10b}{10} = \frac{150}{10}$$
$$a = 20 \qquad\qquad\qquad b = 15$$

25. Corresponding sides must be proportional.
$$\frac{a}{12} = \frac{6}{12}$$
$a = 6$ because denominators are equal
and
$$\frac{b}{15} = \frac{6}{12}$$
$$12 \cdot b = 15 \cdot 6$$
$$\frac{12b}{12} = \frac{90}{12}$$
$$b = \frac{6 \cdot 15}{6 \cdot 2}$$
$$b = \frac{15}{2}.$$

27. Corresponding sides must be proportional.
$$\frac{x}{4} = \frac{9}{6}$$
$$6 \cdot x = 4 \cdot 9$$
$$\frac{6x}{6} = \frac{36}{6}$$
$$x = 6$$

29. Corresponding sides must be proportional.

$$\frac{x}{50} = \frac{220}{100}$$
$$100 \cdot x = 50 \cdot 220$$
$$\frac{100x}{100} = \frac{11000}{100}$$
$$x = 110$$

31. Corresponding sides must be proportional. In the third step, reduce the fraction on the right to lowest terms to make computations easier.

$$\frac{c}{100} = \frac{10+90}{90}$$
$$\frac{c}{100} = \frac{100}{90}$$
$$\frac{c}{100} = \frac{10}{9}$$
$$9 \cdot c = 100 \cdot 10$$
$$\frac{9c}{9} = \frac{1000}{9}$$
$$c = 111\frac{1}{9}$$

33. Let h = height of the tree.

$$\frac{h}{45} = \frac{2}{3}$$
$$3 \cdot h = 45 \cdot 2$$
$$\frac{3h}{3} = \frac{90}{3}$$
$$h = 30 \text{ m}$$

35. Let x = length of the mid-length side.

$$\frac{5}{x} = \frac{4}{400}$$
$$4 \cdot x = 5 \cdot 400$$
$$\frac{4x}{4} = \frac{2000}{4}$$
$$x = 500 \text{ m}$$

Now let y = the longest side.

$$\frac{7}{y} = \frac{4}{400}$$
$$4 \cdot y = 7 \cdot 400$$
$$\frac{4y}{4} = \frac{2800}{4}$$
$$y = 700 \text{ m}$$

The lengths of the other two sides are 500 meters and 700 meters.

37. Let h = height of the building. In step 2 the fraction on the right is reduced to make further computations easier.

$$\frac{h}{15} = \frac{300}{40}$$
$$\frac{h}{15} = \frac{15}{2}$$
$$2 \cdot h = 15 \cdot 15$$
$$\frac{2h}{2} = \frac{225}{2}$$
$$h = 112.5 \text{ ft}$$

39. The two right triangles in the figure are similar, so corresponding sides are proportional. Write and solve a proportion to find r.

$$\frac{7}{6} = \frac{11+7}{r}$$
$$7r = 6(11+7)$$
$$7r = 6(18)$$
$$7r = 108$$
$$r = \frac{108}{7}$$

41. Use the Pythagorean theorem $a^2 + b^2 = c^2$ with $a = 8$ and $b = 15$.

$$8^2 + 15^2 = c^2$$
$$64 + 225 = c^2$$
$$289 = c^2$$
$$\sqrt{289} = \sqrt{c^2}$$
$$17 = c$$

43. Use the Pythagorean theorem $a^2 + b^2 = c^2$ with $b = 84$ and $c = 85$.

$$a^2 + 84^2 = 85^2$$
$$a^2 + 7056 = 7225$$
$$a^2 = 169$$
$$\sqrt{a^2} = \sqrt{169}$$
$$a = 13$$

45. Use the Pythagorean theorem $a^2 + b^2 = c^2$ with $a = 14$ and $b = 48$.

$$14^2 + 48^2 = c^2$$
$$196 + 2304 = c^2$$
$$2500 = c^2$$
$$\sqrt{2500} = \sqrt{c^2}$$
$$50 \text{ m} = c$$

47. Use the Pythagorean theorem $a^2 + b^2 = c^2$ with $b = 21$ and $c = 29$.

$$a^2 + 21^2 = 29^2$$
$$a^2 + 441 = 841$$
$$a^2 = 400$$
$$\sqrt{a^2} = \sqrt{400}$$
$$a = 20 \text{ in.}$$

49. The sum of the squares of the two shorter sides of a right triangle is equal to the square of the longest side.

51. Given $r = 2$ and $s = 1$,
$$a = r^2 - s^2 = 2^2 - 1^2 = 4 - 1 = 3$$
$$b = 2rs = 2 \cdot 2 \cdot 1 = 4$$
$$c = r^2 + s^2 = 2^2 + 1^2 = 4 + 1 = 5$$
The Pythagorean triple is (3, 4, 5).

53. Given $r = 4$ and $s = 3$,
$$a = r^2 - s^2 = 4^2 - 3^2 = 16 - 9 = 7$$
$$b = 2rs = 2 \cdot 4 \cdot 3 = 24$$
$$c = r^2 + s^2 = 4^2 + 3^2 = 16 + 9 = 25$$
The Pythagorean triple is (7, 24, 25).

55. Given $r = 4$ and $s = 2$,
$$a = r^2 - s^2 = 4^2 - 2^2 = 16 - 4 = 12$$
$$b = 2rs = 2 \cdot 4 \cdot 2 = 16$$
$$c = r^2 + s^2 = 4^2 + 2^2 = 16 + 4 = 20$$
The Pythagorean triple is (12, 16, 20).

57. Substitute the expressions in r and s for a and b.
$$a^2 + b^2 = (r^2 - s^2)^2 + (2rs)^2$$
$$= r^4 - 2r^2s^2 + s^4 + 4r^2s^2$$
$$= r^4 + 2r^2s^2 + s^4$$
$$= (r^2 + s^2)^2$$
$$= c^2$$

59. When $m = 3$, $\dfrac{m^2 - 1}{2} = \dfrac{3^2 - 1}{2} = \dfrac{8}{2} = 4$,
$$\dfrac{m^2 + 1}{2} = \dfrac{3^2 + 1}{2} = \dfrac{10}{2} = 5.$$
The Pythagorean triple is (3, 4, 5).

61. When $m = 7$, $\dfrac{m^2 - 1}{2} = \dfrac{7^2 - 1}{2} = \dfrac{48}{2} = 24$,
$$\dfrac{m^2 + 1}{2} = \dfrac{7^2 + 1}{2} = \dfrac{50}{2} = 25.$$
The Pythagorean triple is (7, 24, 25).

63. Replace a^2 with m^2, replace b^2 with $\left(\dfrac{m^2 - 1}{2}\right)^2$, and show that their sum simplifies to c^2.

$$a^2 + b^2 = m^2 + \left(\frac{m^2 - 1}{2}\right)^2$$
$$= m^2 + \frac{m^4 - 2m^2 + 1}{4}$$
$$= \frac{4m^2}{4} + \frac{m^4 - 2m^2 + 1}{4}$$
$$= \frac{m^4 + 2m^2 + 1}{4}$$
$$= \frac{(m^2 + 1)^2}{4}$$
$$= \left(\frac{m^2 + 1}{2}\right)^2$$
$$= c^2$$

65. For $n = 2$, $2n = 2 \cdot 2 = 4$, $n^2 - 1 = 2^2 - 1 = 3$, $n^2 + 1 = 2^2 + 1 = 5$.
The Pythagorean triple is (4, 3, 5).

67. For $n = 4$, $2n = 2 \cdot 4 = 8$, $n^2 - 1 = 4^2 - 1 = 15$, $n^2 + 1 = 4^2 + 1 = 17$.
The Pythagorean triple is (8, 15, 17).

69. Replace a with $2n$ and b with $n^2 - 1$ in the Pythagorean theorem, and show that the expression simplifies to c^2.
$$a^2 + b^2 = (2n)^2 + (n^2 - 1)^2$$
$$= 4n^2 + n^4 - 2n^2 + 1$$
$$= n^4 + 2n^2 + 1$$
$$= (n^2 + 1)^2$$
$$= c^2$$

71. Let b = length of longer leg, $b + 1$ = length of hypotenuse c, 7 = length of shorter leg, a.
Substitute these expressions into the Pythagorean theorem, $a^2 + b^2 = c^2$, and

solve for *b*.

$$7^2 + b^2 = (b+1)^2$$
$$49 + b^2 = b^2 + 2b + 1$$
$$49 + b^2 - b^2 = b^2 - b^2 + 2b + 1$$
$$49 = 2b + 1$$
$$48 = 2b$$
$$24 = b$$

The longer leg is 24 m.

73. Let h = the height of the tree, one of the legs of the triangle, $2h + 2$ = another leg of the triangle. Substitute these expressions into the Pythagorean theorem, $a^2 + b^2 = c^2$, and solve for h.

$$30^2 + h^2 = (2h+2)^2$$
$$900 + h^2 = 4h^2 + 8h + 4$$
$$900 + h^2 - h^2 = 4h^2 - h^2 + 8h + 4$$
$$900 = 3h^2 + 8h + 4$$
$$900 - 900 = 3h^2 + 8h + 4 - 900$$
$$0 = 3h^2 + 8h - 896$$
$$0 = (3h + 56)(h - 16)$$

$3h + 56 = 0$ or $h - 16 = 0$
$h = -\dfrac{56}{3}$ $h = 16$

A negative height is not meaningful. The height is 16 feet.

75. Let h = the height of the break, $a = 3$ ft, one leg of the triangle, $c = 10 - h$, the hypotenuse of the triangle. Substitute these expressions into the Pythagorean theorem, $a^2 + b^2 = c^2$, and solve for h.

$$3^2 + h^2 = (10 - h)^2$$
$$9 + h^2 = 100 - 20h + h^2$$
$$9 + h^2 - h^2 = 100 - 20h + h^2 - h^2$$
$$9 = 100 - 20h$$
$$9 - 100 = 100 - 100 - 20h$$
$$-91 = -20h$$
$$4.55 = h$$

The height of the break is 4.55 feet.

77. Let c = the length of the diagonal.
$$12^2 + 15^2 = c^2$$
$$144 + 225 = c^2$$
$$369 = c^2$$
$$\sqrt{369} = \sqrt{c^2}$$
$$19.21 \approx c$$

Then $0.21(12) = 2.52$, which is 3 inches to the nearest inch. The diagonal should be 19 feet, 3 inches.

79. Let c = the length of the diagonal.
$$16^2 + 24^2 = c^2$$
$$256 + 576 = c^2$$
$$832 = c^2$$
$$\sqrt{832} = \sqrt{c^2}$$
$$28.84 \approx c$$

Then $0.84(12) = 10.08$, which is 10 inches to the nearest inch. The diagonal should be 28 feet, 10 inches.

81. (a) By proportion, we have $\dfrac{c}{b} = \dfrac{b}{j}$.

   (b) By proportion, we also have $\dfrac{c}{a} = \dfrac{a}{k}$.

   (c) From part (a), $b^2 = cj$. Cross multiply to obtain this equation.

   (d) From part (b), $a^2 = ck$. Again, cross multiply to obtain this equation.

   (e) From the results of parts (c) and (d) and factoring, $a^2 + b^2 = c(j + k)$. But since $j + k = c$, it follows that $a^2 + b^2 = c^2$. Obtain the first underlined statement from $a^2 + b^2 = cj + ck = c(j + k)$.

83. Right triangle ABC had $AD = DB + 8$, $AC = 12$ and $AB = 20$. By the Pythagorean theorem
$$(AC)^2 + (BC)^2 = (AB)^2$$
$$12^2 + (BC)^2 = 20^2$$
$$144 + (BC)^2 = 400$$
$$(BC)^2 = 256$$
$$BC = 16.$$

Therefore, $CD = CB - DB = 16 - DB$. We may also apply the Pythagorean theorem to right triangle ACD.
$$(AC)^2 + (CD)^2 = (AD)^2$$

Substitute 12 for AC, $16 - DB$ for CD, and $DB + 8$ for AD in the preceding equation.

$$12^2 + (16 - DB)^2 = (DB + 8)^2$$
$$144 + 256 - 32(DB) + (DB)^2 = (DB)^2$$
$$+ 16(DB)$$
$$+ 64$$
$$336 = 48(DB)$$
$$DB = 7$$

$CD + DB = CB$, so $CD = CB - DB$, or
$DC = 16 - 7 = 9$.

85. Because $\overleftrightarrow{XY}$ is parallel to $\overleftrightarrow{VW}$, triangle VWZ is similar to triangle XYZ, so corresponding side lengths are proportional.

$$\frac{VZ}{XZ} = \frac{WZ}{YZ}$$
$$\frac{10}{8} = \frac{WZ}{WZ - 4}$$
$$10(WZ - 4) = 8(WZ)$$
$$10(WZ) - 40 = 8(WZ)$$
$$2(WZ) = 40$$
$$WZ = 20$$

Then $ZY = WZ - 4 = 20 \cdot 4 = 16$.

87. $AD = AC$, so triangle DAC is an isosceles right triangle. Then

$$\angle DCA = \frac{1}{2}(180 - 90) = \frac{1}{2}(90) = 45°.$$

$\angle ACB = 180 - 45 = 135°$ since $\angle ACB$ and $\angle DCA$ are supplementary angles. $AC = CB$, so triangle ACB is an isosceles triangle, and $\angle CAB = \angle CBA$.

$$\angle CAB = \frac{1}{2}(180 - 135) = \frac{1}{2}(45) = 22.5°.$$

89. $STRY$ is a square, so $\angle STR = 90°$. TOR is an equilateral triangle, so $\angle RTO = 60°$.

$\angle STO = 90 + 60 = 150°$. $ST = TR$ and $TR = TO$, so $ST = TO$ and STO is an equilateral triangle.

$$\angle TOS = \frac{1}{2}(180 - 150) = \frac{1}{2}(30) = 15°.$$

91. Answers will vary.

93. Answers will vary.

9.4 Exercises

1. The perimeter of an equilateral triangle with side length equal to <u>24</u> inches is the same as the perimeter of a rectangle with length 20 inches and width 16 inches. The perimeter of the rectangle is

$2 \cdot 20 + 2 \cdot 16 = 72$. The perimeter of the triangle is also 72. If all three sides must have the same length, then one side has length $72 \div 3 = 24$ inches.

3. If the area of a certain triangle is 48 square inches, and the base measures 8 inches, then the height must measure <u>12</u> inches. The formula for the area of a triangle is $A = \frac{1}{2}bh$. Substitute the given values into the formula and solve for h.

$$48 = \frac{1}{2} \cdot 8h$$
$$48 = 4h$$
$$h = 12$$

5. Circumference

7. $A = lw$
$A = 4 \cdot 3$
$A = 12 \text{ cm}^2$

9. $A = lw$
$A = 2\frac{1}{2} \cdot 2$
$A = 5 \text{ cm}^2$

11. $A = bh$
$A = 4 \cdot 2$
$A = 8 \text{ in}^2$

13. $A = bh$
$A = 3 \cdot 1.5$
$A = 4.5 \text{ cm}^2$

15. $A = \frac{1}{2}bh$
$A = \frac{1}{2} \cdot 22 \cdot 38$
$A = \frac{1}{2} \cdot \frac{22}{1} \cdot \frac{38}{1}$
$A = 418 \text{ mm}^2$

17. $A = \frac{1}{2}h(b + B)$
$A = \frac{1}{2} \cdot 2(3 + 5)$
$A = 1(8)$
$A = 8 \text{ cm}^2$

19. $A = \pi r^2$

$A = (3.14)(1)^2$

$A = 3.14 \text{ cm}^2$

21. The diameter is 36, so the radius is 18 m.

$A = \pi r^2$

$A = (3.14)(18)^2$

$A = (3.14)(324)$

$A = 1017.36 \text{ m}^2$

23. Let s = length of a side of the window. Use the formula $P = 4s$. Replace P with $7s - 12$ and solve for s.

$$P = 4s$$
$$7s - 12 = 4s$$
$$7s - 7s - 12 = 4s - 7s$$
$$-12 = -3s$$
$$\frac{-12}{-3} = \frac{-3s}{-3}$$
$$4 = s$$

The length of a side of the window is 4 m.

25. The formula for perimeter of a triangle is $P = a + b + c$. Translating the problem, let a be the shortest side. Then $b = 100 + a$ and $c = 200 + a$. Replace a, b, and c in the formula with these expressions; replace P with 1200 and solve for a.

$$P = a + b + c$$
$$1200 = a + (100 + a) + (200 + a)$$
$$1200 = 3a + 300$$
$$900 = 3a$$
$$300 = a$$

Side a is 300 ft; $b = 100 + 300 = 400$ ft; side $c = 200 + 300 = 500$ ft.

27. One formula for circumference is $C = 2\pi r$. Translating the second sentence of the problem, $C = 6r + 12.88$. Equate these two expressions for C and solve the equation for r.

$$6r + 12.88 = 2\pi r$$
$$6r + 12.88 = 2(3.14)r$$
$$6r + 12.88 = 6.28r$$
$$6r - 6r + 12.88 = 6.28r - 6r$$
$$12.88 = 0.28r$$
$$\frac{12.88}{0.28} = \frac{0.28r}{0.28}$$
$$46 = r$$

The radius is 46 ft.

29. The formula for the area of a trapezoid is $A = \frac{1}{2}h(b + B)$. Substitute the numerical values given in the problem and compute to find the area.

$$A = \frac{1}{2}h(b + B)$$
$$A = \frac{1}{2}(165.97)(115.80 + 171.00)$$
$$A = \frac{1}{2}(165.97)(286.8)$$
$$A = \frac{47600.196}{2}$$
$$A = 23,800.098$$

Rounded to the nearest hundredth, the area is 23,800.10 sq ft.

31. Use perimeter since it measures distance around.

33. $d = 2r = 2 \cdot 6 = 12$ in.

$C = 2\pi r = 2\pi \cdot 6 = 12\pi$ in.

$A = \pi r^2 = \pi \cdot 6^2 = 36\pi$ in^2

35. $r = \frac{1}{2} \cdot 10 = 5$ ft

$C = \pi d = \pi \cdot 10 = 10\pi$ ft

$A = \pi r^2 = \pi \cdot 5^2 = 25\pi$ ft^2

37. $d = \dfrac{C}{\pi} = \dfrac{12\pi}{\pi} = 12$ cm

$r = \frac{1}{2} \cdot 12 = 6$ cm

$A = \pi \cdot 6^2 = 36\pi$ cm^2

39. $r^2 = \dfrac{A}{\pi} = \dfrac{100\pi}{\pi} = 100$

Then $r = \sqrt{100} = 10$ in.

$d = 2r = 2 \cdot 10 = 20$ in.

$C = 2\pi r = 2 \cdot \pi \cdot 10 = 20\pi$ in.

41. Use the formula $P = 4s$, replacing s with x and P with 58.

$$P = 4x$$
$$58 = 4x$$
$$\frac{58}{4} = \frac{4x}{4}$$
$$x = 14.5$$

43. Use the formula $P = 2l + 2w$, replacing l and w with the expressions in x and replacing P with 38.

$$38 = 2(2x - 3) + 2(x + 1)$$
$$38 = 4x - 6 + 2x + 2$$
$$38 = 6x - 4$$
$$42 = 6x$$
$$\frac{42}{6} = \frac{6x}{6}$$
$$x = 7$$

45. Use the formula $A = s^2$, replacing s with x and A with 26.01.

$$26.01 = x^2$$
$$\sqrt{26.01} = \sqrt{x^2}$$
$$5.1 = x$$

47. Use the formula $A = \frac{1}{2}bh$, replacing b and h with the expressions in x and replacing A with 15.

$$15 = \frac{1}{2}x(x + 1)$$
$$2 \cdot 15 = \frac{2}{1} \cdot \frac{1}{2}x(x + 1)$$
$$30 = x(x + 1)$$
$$30 = x^2 + x$$
$$0 = x^2 + x - 30$$

Now factor the trinomial and set each factor equal to zero.

$$0 = (x + 6)(x - 5)$$
$$x + 6 = 0 \quad \text{or} \quad x - 5 = 0$$
$$x = -6 \qquad\qquad x = 5$$

The solution -6 is not meaningful because the base of a triangle must be positive a number. The answer then is 5.

49. Use the formula $C = 2\pi r$, replacing C with 37.68 and r with the expression in x.

$$37.68 = 2(3.14)(x + 1)$$
$$37.68 = 6.28(x + 1)$$
$$\frac{37.68}{6.28} = \frac{6.28(x + 1)}{6.28}$$
$$6 = x + 1$$
$$x = 5$$

51. Use the formula $A = \pi r^2$, replacing A with 28.26. The diameter is $4x$, so this expression must be divided in half for r.

$$28.26 = 3.14(2x)^2$$
$$\frac{28.26}{3.14} = \frac{3.14(2x)^2}{3.14}$$
$$9 = (2x)^2$$
$$9 = 4x^2$$
$$\frac{9}{4} = \frac{4x^2}{4}$$
$$\frac{9}{4} = x^2$$
$$\sqrt{\frac{9}{4}} = \sqrt{x^2}$$
$$x = \frac{3}{2} = 1.5$$

53. (a) $A = 4 \cdot 5 = 20 \text{ cm}^2$

(b) $A = 8 \cdot 10 = 80 \text{ cm}^2$

(c) $A = 12 \cdot 15 = 180 \text{ cm}^2$

(d) $A = 16 \cdot 20 = 320 \text{ cm}^2$

(e) The rectangle in part (b) had sides twice as long as the sides of the rectangle in part (a). Divide the larger area by the smaller ($80 \div 20 = 4$). By doubling the sides, the area increased $\underline{4}$ times.

(f) To get the rectangle in part (c) each side of the rectangle of part (a) was multiplied by $\underline{3}$. This made the larger area $\underline{9}$ times the smaller area ($180 \div 20 = 9$).

(g) To get the rectangle of part (d) each side of the rectangle of part (a) was multiplied by $\underline{4}$. This made the area increase to $\underline{16}$ times what it was originally ($320 \div 20 = 16$).

(h) In general, if the length of each side of a rectangle is multiplied by n, the area is multiplied by $\underline{n}^2$.

55. Because each measurement is multiplied by 2, the area will increase by $2^2 = 4$. Then $4 \cdot 200 = \$800$.

57. If the radius of a circle is multiplied by n, then the area of the circle is multiplied by $\underline{n^2}$.

59. Find the area of the parallelogram and the area of the triangle. Then add the two area values.

Parallelogram $A = 6 \cdot 10 = 60$

Triangle $A = \frac{1}{2}(10)(4) = 20$

Total area $60 + 20 = 80$

61. There are 2 semicircles or equivalently 1 full circle with radius of 3. Find the area of this circle and of the rectangle.

Rectangle $A = 8 \cdot 6 = 48$

Circle $A = (3.14) \cdot 3^2 = 28.26$

Total area $48 + 28.26 = 76.26$

63. Find the area of the trapezoid that surrounds the triangle. Subtract the area of the triangle.

Trapezoid $A = \frac{1}{2}(12)(18+11) = 174$

Triangle $A = \frac{1}{2}(12)(7) = 42$

Shaded area $174 - 42 = 132 \text{ ft}^2$

65. Find the area of the rectangle that surrounds the triangles. Subtract the areas of the triangles. The length of the rectangle is $48 + 48 = 96$.

Rectangle $A = 74 \cdot 96 = 7104$

One triangle $A = \frac{1}{2}(48)(36) = 864$

Shaded area $7104 - 2(864) = 5376 \text{ cm}^2$

67. Find the area of the square that surrounds the circle. Subtract the area of the circle. The diameter is 26; therefore, the radius is $26 \div 2 = 13$.

Square $A = 26^2 = 676$

Circle $A = (3.14)(13)^2 = 530.66$

Shaded area $676 - 530.66 = 145.34 \text{ m}^2$

69. The best buy is the pizza with the lowest cost per square inch or unit price if you have enough money and you can eat all of it!

10" pizza $A = (3.14)(5^2) = 78.5 \text{ in}^2$

Unit price $\frac{5.99}{78.5} \approx \$0.076$

12" pizza $A = (3.14)(6^2) = 113.04 \text{ in}^2$

Unit price $\frac{7.99}{113.04} \approx \$0.071$

14" pizza $A = (3.14)(7^2)$ or 153.86 in^2

Unit price $\frac{8.99}{153.86} \approx \$0.058$

The best buy is the 14" pizza.

71. The best buy is the pizza with the lowest cost per square inch or unit price.

10" pizza $A = (3.14)(5^2) = 78.5 \text{ in}^2$

Unit price $\frac{9.99}{78.5} \approx \$0.127$

12" pizza $A = (3.14)6^2 = 113.04 \text{ in}^2$

Unit price $\frac{11.99}{113.04} \approx \$0.106$

14" pizza $A = (3.14)(7^2)$ or 153.86 in^2

Unit price $\frac{12.99}{153.86} \approx \$0.084$

The best buy is the 14" pizza.

73. $A = \frac{1}{2}(a+b)(a+b)$

75. $\frac{1}{2}(a+b)(a+b) = \frac{1}{2}ab + \frac{1}{2}ab + \frac{1}{2}c^2$

77. The key is to construct OB and to realize that the diagonals of a rectangle are equal in length. So by inspection, $OB = AC = 13$ in. OB is a radius. Therefore, the diameter $= 2 \cdot 13 = 26$ inches.

79. The key is to construct TV and UW to create more triangles. By inspection, all the small triangles are equal. $PQRS$ has 8 triangles. $TUVW$ has 4 triangles. Therefore $TUVW$ has half the area of $PQRS$, which is 625 ft^2. Otherwise, find the area by first solving for the length of one side using the Pythagorean theorem.

81. The key is to construct a perpendicular line from E to side DC. Let the point of intersection of side DC to be labeled point F. By inspection there are two sets of equal triangles: $\triangle DAE$ and $\triangle EFD$; $\triangle CBE$ and $\triangle EFC$. Then the area of the shaded region is half the area of the square. Therefore,

Area $= \frac{36^2}{2} = \frac{1296}{2} = 648 \text{ in}^2$.

83. The key is to construct a square using two radii from O and bounding the shaded region. The area of the small square is r^2, and the area of the quarter circle is $\dfrac{\pi r^2}{4}$. Therefore, the area of the shaded region is
$$r^2 - \frac{\pi r^2}{4} = r^2\left(1 - \frac{\pi}{4}\right) = \frac{(4-\pi)r^2}{4}.$$

85. Draw the segment $\overline{BD}$ to divide the figure into two right triangles.

Area of triangle $DAB = \dfrac{1}{2}(8)(6) = 24$

$(DB)^2 = (AD)^2 + (AB)^2$
$(DB)^2 = 6^2 + 8^2$
$(DB)^2 = 36 + 64$
$(DB)^2 = 100$

$(DB)^2 = (DC)^2 + (BC)^2$
$100 = (DC)^2 + 2^2$
$96 = (DC)^2$
$4\sqrt{6} = DC$

Area of triangle $DCB = \dfrac{1}{2}\left(4\sqrt{6}\right)(2) = 4\sqrt{6}$

The area of the quadrilateral is $24 + 4\sqrt{6}$.

87. First find the area of the given triangle. To do so, we need to find the height of the triangle. Draw the altitude h from the vertex between the two sides of length 13 to the base of length 24. This altitude bisects the base, so we have a right triangle.
$$\left[\frac{1}{2}(24)\right]^2 + h^2 = 13^2$$
$$12^2 + h^2 = 13^2$$
$$144 + h^2 = 169$$
$$h^2 = 25$$
$$h = 5$$
So the area of the given triangle is $\dfrac{1}{2}(24)(5) = 60$. Now any other isosceles triangle with equal sides of 13 must have a height and a base such that $\left(\dfrac{1}{2}(\text{base}), \text{height}, 13\right)$ is a Pythagorean triple. The only triple of this form is (5, 12, 13). The given triangle has base $2(12) = 24$, so the other possible base is

$2(5) = 10$. The height of this triangle would be 12, and the area would then be
$$\frac{1}{2}(10)(12) = 60 \text{ as required. So the base}$$
must be 10.

89. $\overset{\bullet\ \ \bullet}{AB}$ is a diameter of the circle, so $\angle ACB$ is a right angle. Then
$$(AC)^2 + (BC)^2 = (AB)^2$$
$$6^2 + 8^2 = (AB)^2$$
$$36 + 64 = (AB)^2$$
$$100 = (AB)^2$$
$$10 = AB$$
The radius is half the length of the diameter $\overset{\bullet\ \ \bullet}{AB}$, so the radius is $\dfrac{1}{2}(10) = 5$ in.

9.5 Exercises

1. True; if the volume is 64 cubic inches, one side of the cube is 4 inches because $4 \cdot 4 \cdot 4 = 64$. Then the area of one face is $4 \cdot 4 = 16$. A cube has six faces so that $6 \cdot 16 = 96$ square inches is the total surface area.

3. True; a dodecahedron has 12 faces.

5. False; the new cube will have eight times the volume of the original cube.

7. (a) $V = lwh$
$$= 2 \cdot 1\frac{1}{2} \cdot 1\frac{1}{4}$$
$$= 2 \cdot \frac{3}{2} \cdot \frac{5}{4}$$
$$= \frac{15}{4}$$
$$= 3\frac{3}{4} \text{ m}^3$$

(b) $S = 2lh + 2hw + 2lw$
$$= 2 \cdot 2 \cdot \frac{5}{4} + 2 \cdot \frac{5}{4} \cdot \frac{3}{2} + 2 \cdot 2 \cdot \frac{3}{2}$$
$$= 5 + \frac{15}{4} + 6$$
$$= 5 + 3\frac{3}{4} + 6$$
$$= 14\frac{3}{4} \text{ m}^2$$

9. It may be helpful to use parentheses to indicate multiplication when working with decimal values. Otherwise, it is possible to confuse the multiplication dot and the decimal points.

(a) $V = lwh = (6)(5)(3.2) = 96 \text{ in}^3$

(b) $S = 2lh + 2hw + 2lw$
$= 2(6)(3.2) + 2(3.2)(5) + 2(6)(5)$
$= 38.4 + 32 + 60$
$= 130.4 \text{ in}^2$

11. (a) $V = \frac{4}{3}\pi r^3$
$= \frac{4}{3}(3.14)(40)^3$
$= \frac{4}{3}(3.14)(64000)$
$\approx 267,946.67 \text{ ft}^3$

(b) $S = 4\pi r^2$
$= 4(3.14)(40)^2$
$= 4(3.14)(1600)$
$= 20,096 \text{ ft}^2$

13. (a) $V = \pi r^2 h$
$= (3.14)(5)^2(7)$
$= (3.14)(25)(7)$
$= 549.5 \text{ cm}^3$

(b) $S = 2\pi r^2 + 2\pi rh$
$= 2(3.14)(5)^2 + 2(3.14)(5)(7)$
$= 2(3.14)(25) + 2(3.14)(5)(7)$
$= 157 + 219.8$
$= 376.8 \text{ cm}^2$

15. (a) $V = \frac{1}{3}\pi r^2 h$
$= \frac{1}{3}(3.14)(3)^2(7)$
$= \frac{1}{3}(3.14)(9)(7)$
$= 65.94 \text{ m}^3$

(b) $S = \pi r\sqrt{r^2 + h^2} + \pi r^2$
$= (3.14)(3)\sqrt{3^2 + 7^2} + (3.14)(3)^2$
$= (3.14)(3)\sqrt{9 + 49} + (3.14)(9)$
$= (3.14)(3)\sqrt{58} + 28.6$
$= 9.42\sqrt{58} + 28.6$
$\approx 100.00 \text{ m}^2$

17. Remember that B represents the area of the base.
$$V = \frac{1}{3}Bh = \frac{1}{3}(8 \cdot 9) \cdot 7 = \frac{504}{3} = 168 \text{ in}^3$$

19. $V = \pi r^2 h$
$= (3.14)(6.3)^2(15.8)$
$= (3.14)(36.69)(15.8)$
$= 1969.10 \text{ cm}^3$

21. First find the radius by taking half of the diameter: $r = \frac{1}{2}(7.2) = 3.6$. Then use the formula for volume of a right circular cylinder.
$V = \pi r^2 h$
$= (3.14)(3.6)^2(10.5)$
$= (3.14)(12.96)(10.5)$
$\approx 427.29 \text{ cm}^3$

23. First find the radius by taking half of the diameter: $r = \frac{1}{2}(9) = 4.5$. Then use the formula for volume of a right circular cylinder.
$V = \pi r^2 h$
$= (3.14)(4.5)^2(8)$
$= (3.14)(20.25)(8)$
$= 508.68 \text{ cm}^3$

25. Remember that B represents the area of the base.
$V = \frac{1}{3}Bh$
$= \frac{1}{3}(230)^2 \cdot 137$
$= \frac{1}{3}(52900) \cdot 137$
$= \frac{7,247,300}{3}$
$\approx 2,415,766.67 \text{ m}^3$

27. Change $\dfrac{1}{2}$ to the decimal value 0.5 for ease of computation.

$$V = \frac{1}{3}\pi r^2 h$$

$$= \frac{1}{3}(3.14)(0.5)^2(2)$$

$$= \frac{1}{3}(3.14)(0.25)(2)$$

$$= \frac{1.57}{3}$$

$$= 0.52 \text{ m}^3$$

Table for Exercises 29–35

	r	d	V	S
29.	6 in.	12 in.	$288\pi \text{ in}^3$	$144\pi \text{ in}^2$
31.	5 ft	10 ft	$\frac{500}{3}\pi \text{ ft}^3$	$100\pi \text{ ft}^2$
33.	2 cm	4 cm	$\frac{32}{3}\pi \text{ cm}^3$	$16\pi \text{ cm}^2$
35.	1 m	2 m	$\frac{4}{3}\pi \text{ m}^3$	$4\pi \text{ m}^2$

29. $d = 2r = 2 \cdot 6 = 12$

$$V = \frac{4}{3}\pi r^3 = \frac{4}{3}\pi(6)^3 = \frac{4}{3}\pi(216) = 288\pi$$

$$S = 4\pi r^2 = 4\pi(6)^2 = 4\pi(36) = 144\pi$$

31. $r = \dfrac{1}{2}d = \dfrac{1}{2} \cdot 10 = 5$

$$V = \frac{4}{3}\pi r^3 = \frac{4}{3}\pi(5)^3 = \frac{4}{3}\pi(125) = \frac{500}{3}\pi$$

$$S = 4\pi r^2 = 4\pi(5)^2 = 4\pi(25) = 100\pi$$

33. Use the formula for the volume of a sphere to solve for r, by replacing V with the given value, $\dfrac{32}{3}\pi$.

$$V = \frac{4}{3}\pi r^3$$

$$\frac{32}{3}\pi = \frac{4}{3}\pi r^3$$

$$\frac{32}{3}\pi \div \frac{4}{3}\pi = \frac{4}{3}\pi r^3 \div \frac{4}{3}\pi$$

$$\frac{32\pi}{3} \cdot \frac{3}{4\pi} = r^3$$

$$8 = r^3$$

$$2 = r$$

$$d = 2r = 2 \cdot 2 = 4$$

$$S = 4\pi r^2 = 4\pi r^2 = 4\pi(2)^2 = 4\pi(4) = 16\pi$$

35. Use the formula for the surface area of a sphere to solve for r, by replacing S with the given value, 4π.

$$S = 4\pi r^2$$

$$4\pi = 4\pi r^2$$

$$\frac{4\pi}{4\pi} = r^2$$

$$1 = r^2$$

$$1 = r$$

$$d = 2r = 2 \cdot 1 = 2$$

$$V = \frac{4}{3}\pi r^3 = \frac{4}{3}\pi(1)^3 = \frac{4}{3}\pi(1) = \frac{4}{3}\pi$$

37. Volume is a measure of capacity.

39. The volume of the original cube is x^3. Let the length of the side of the new cube be represented by y. Then $y^3 = 2x^3$. Solve for y by taking the cube root of both sides of the equation.

$$y^3 = 2x^3$$

$$\sqrt[3]{y^3} = \sqrt[3]{2x^3}$$

$$y = \sqrt[3]{2x^3}$$

$$y = x\sqrt[3]{2}$$

41. If the new diameter is 3 times the old diameter, then the new volume will be 3^3 or 27 times greater. Therefore, the cost will also be 27 times greater, or $27 \cdot 300 = \$8100$.

43. If the new diameter is 5 times the old diameter, then the new volume will be 5^3 or 125 times greater. Therefore, the cost will also be 125 times greater, or $125 \cdot 300 = \$37{,}500$.

45. $V = lwh$

$60 = 6 \cdot 4 \cdot x$

$60 = 24x$

$2.5 = x$

47. In this exercise $x =$ the diameter of the sphere. Therefore $r = \dfrac{x}{2}$.

$$V = \frac{4}{3}\pi r^3$$

$$36\pi = \frac{4}{3}\pi\left(\frac{x}{2}\right)^3$$

$$36\pi = \frac{4}{3}\pi \cdot \frac{x^3}{8}$$

$$36 = \frac{4}{3}\frac{x^3}{8}$$

$$36 = \frac{x^3}{6}$$

$$216 = x^3$$

$$6 = x$$

49. Look at the figure and try some values for the edges of each side that will create the given areas. One side has edges 6 in. and 5 in.; the adjacent side has edges 5 in. and 7 in.; and the third side has edges 7 in. and 6 in. Write these values on the edges of the rectangular box to verify that it can be done to create the given areas. The three dimensions of the box, then, are 6, 7, and 5.

The volume of the box is $6 \cdot 7 \cdot 5 = 210$ in³.

51. The formula for the volume of a sphere is $V = \dfrac{4}{3}\pi r^3$, so the radius of the sphere must be found. To find the radius of the circle that is formed by the intersection, set 576π equal to πr^2, the formula for the area of a circle, and solve for r.

$$576\pi = \pi r^2$$

$$\frac{576\pi}{\pi} = \frac{\pi r^2}{\pi}$$

$$576 = r^2$$

$$24 = r$$

Now use the Pythagorean theorem to find the length of the hypotenuse, which is also the radius of the sphere.

$$7^2 + 24^2 = c^2$$

$$49 + 576 = c^2$$

$$625 = c^2$$

$$25 = c$$

Now compute the volume of the sphere.

$$V = \frac{4}{3}\pi r^3$$

$$= \frac{4}{3}\pi(25)^3$$

$$= \frac{4}{3}\pi(15625)$$

$$= \frac{62500}{3}\pi \text{ in}^3$$

53. Rotate the inscribed square 45° so that one of its diagonals is horizontal and the other vertical. Notice that the length of the diagonal is the same length as the side of the circumscribed square. This length is $2r$. That means that the area of the circumscribed square is $A = 4r^2$. Returning to the inscribed square, the length $2r$ is the length of the hypotenuse of a right triangle. Use the Pythagorean theorem to find the length of a side of this square.

Let $x =$ the length of each leg.

$$x^2 + x^2 = (2r)^2$$

$$2x^2 = 4r^2$$

$$x^2 = 2r^2$$

Because the area of this square is equal to x^2, the ratio of the two areas can be determined.

$$\frac{\text{area of the circumscribed square}}{\text{area of the inscribed square}} = \frac{4r^2}{2r^2} = \frac{2}{1}$$

The ratio is 2 to 1.

55. Draw a line connecting one diameter RT; draw a line connecting another diameter QS. Recall from section 9.2 that any angle inscribed in a semicircle is a right angle, which means that $\angle RPT$ and S are both right angles. From the Pythagorean theorem, $PR^2 + PT^2$ equals the square of the diameter, 12^2. Also, $PQ^2 + PS^2$ equals the square of the diameter. Finally,

$$PR^2 + PT^2 + PQ^2 + PS^2 = 12^2 + 12^2$$
$$= 288.$$

	Polyhedron	Faces (F)	Vertices (V)	Edges (E)	Value of F + V − E
57.	Tetrahedron	4	4	6	2
59.	Octahedron	8	6	12	2
61.	Icosahedron	20	12	30	2

9.6 Exercises

Exercises 1–7 represent reflection transformations. There is a one-to-one correspondence between each point in the original figure and each corresponding point in the image figure. The original figure and image figure are congruent hence preserving collinearity and distance.

The answers are given in gray for this section.

1.

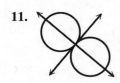

3.

5.

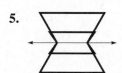

7. The figure is its own reflection image.

Exercises 9–11 represent figures that are their own reflections across the lines of symmetry.

9.

11.

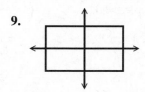

Exercises 13–19 represent the composition or product transformations of r_m followed by r_n, or $r_n \cdot r_m$.

13.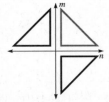

15.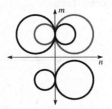

17.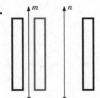

19.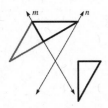

21. r_m

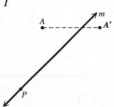

23. T

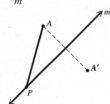

25. $T \cdot T$

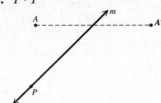

27. $T \cdot R_p$

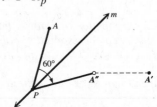

29. $r_m \cdot T$

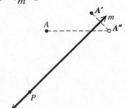

31. $r_m \cdot R_p$

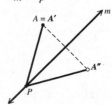

33. No, $T \cdot r_m$ is not a glide reflection since a reflection is not preserved.

35.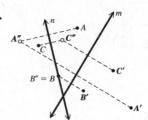

37.

39.

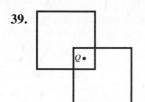

41.

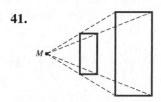

43.

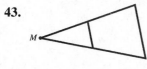

45.

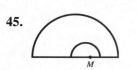

9.7 Exercises

The chart in the text characterizes certain properties of Euclidean and non-Euclidean geometries. Study it and use it to respond to Exercises 1–10.

1. Euclidean

3. Lobachevskian

5. greater than

7. Riemannian

9. Euclidean

11. No, both have no holes. They are of genus 0.

13. Yes, the slice of American cheese is of genus 0, and the slice of Swiss cheese is of genus 1 or more.

15. C; both are of genus 2, meaning they have two holes.

17. A and E; all are of genus 0, having no holes.

19. A and E; all three are of genus 0, having no holes.

21. None of them

23. A compact disc has one hole, so it is of genus 1.

25. A sheet of loose-leaf paper made for a three-ring binder has three holes, so it is of genus 3.

27. A wedding band has one hole, so it is of genus 1.

29. *A*, *C*, *D*, and *F* are even vertices because each has two paths leading to or from the vertex; *B* and *E* are odd because each has three paths leading to or from the vertex.

31. *A*, *B*, *C*, and *F* are odd because each has three paths leading to or from the vertex; *D*, *E*, and *G* are even. *D* and *E* each have two paths leading to or from the vertex; *G* has four.

33. *A*, *B*, *C*, and *D* are odd vertices because each has three paths leading to or from the vertex; *E* is even because it has four.

35. There are two odd vertices at the extremities and the rest are even. Therefore, the network is traversable.

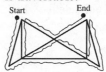

37. Not traversable; it has more than two odd vertices.

39. The network has exactly 2 odd vertices. It is traversable.

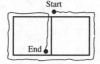

41. Yes; there are no rooms (vertices) with an odd number of doors (paths). Therefore, the house (network) is traversable.

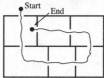

43. No; there are more than two rooms with an odd number of doors (paths).

45. (a)–(g)

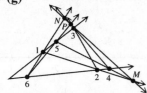

(h) Suppose that a hexagon is inscribed in an angle. Let each pair of opposite sides be extended so as to intersect. Then the three points of intersection thus obtained will lie in a straight line.

9.8 Exercises

1. The least number of these squares that can be put together edge to edge to form a larger square is 4.

3. The length of each edge of the new square is 2.

5. $\dfrac{\text{new size}}{\text{old size}} = \dfrac{4}{1} = 4$

7. The scale factor is $\dfrac{\text{new length}}{\text{old length}} = \dfrac{4}{1} = 4$.

The ratio of $\dfrac{\text{new size}}{\text{old size}} = \dfrac{16}{1} = 16$.

9. Each ratio in the bottom row is the square of the scale factor in the top row.

11.

Scale factor	Ratio of new size to old size
2	4
3	9
4	16
5	25
6	36
10	100

13. Some examples are: $3^d = 9$ and $d = 2$; $4^d = 16$ and $d = 2$; $5^d = 25$ and $d = 2$.

15. The scale factor between these two cubes is $\dfrac{\text{new length}}{\text{old length}} = \dfrac{2}{1} = 2$.

The ratio of $\dfrac{\text{new size}}{\text{old size}} = \dfrac{8}{1} = 8$.

17. Each ratio in the bottom row is the cube of the scale factor in the top row.

19. The scale factor between stage 1 and staage 2 is $\dfrac{3}{1} = \underline{3}$.

21. $3^d = 4$
Use trial and error:
$3^{1.5} = 5.196...$
$3^{1.25} = 3.948...$
$3^{1.26} = 3.992...$
$3^{1.27} = 4.036...$
$3^{1.261} = 3.996...$
$3^{1.262} = 4.001...$
$3^{1.263} = 4.005...$
$d = 1.262$ to three decimal places, or solve using logarithms.
$$3^d = 4$$
$$\ln 3^d = \ln 4$$
$$d \ln 3 = \ln 4$$
$$d = \frac{\ln 4}{\ln 3}$$
$d = 1.262$ to three decimal places.

23. Old size = 1, new size = $\underline{3}$

25. $2^d = 3$
Use trial and error:
$2^{1.5} = 2.828...$
$2^{1.6} = 3.031...$
$2^{1.55} = 2.928...$
$2^{1.58} = 2.990...$
$2^{1.59} = 3.010...$
$2^{1.584} = 2.998...$
$2^{1.585} = 3.000...$
$d = 1.585$ to three decimal places

Or solve using logarithms.

$$2^d = 3$$
$$\ln 2^d = \ln 3$$
$$d \ln 2 = \ln 3$$
$$d = \frac{\ln 3}{\ln 2}$$
$$d = 1.585 \text{ to three decimal places}$$

27. Given $k = 3.4$, $x = 0.8$ and formula
$y = kx(1 - x)$.
Note that rounded values are used below.
$$y = 3.4(0.8)(1 - 0.8) = 3.4(0.8)(0.2) = 0.544$$
$$y = 3.4(0.544)(1 - 0.544)$$
$$= 3.4(0.544)(0.456)$$
$$\approx 0.843$$
$$y = 3.4(0.843)(1 - 0.843)$$
$$= 3.4(0.843)(0.157)$$
$$\approx 0.450$$
$$y = 3.4(0.450)(1 - 0.450)$$
$$= 3.4(0.450)(0.550)$$
$$\approx 0.842$$
$$y = 3.4(0.842)(1 - 0.842)$$
$$= 3.4(0.842)(0.158)$$
$$\approx 0.452$$
$$y = 3.4(0.452)(1 - 0.452)$$
$$= 3.4(0.452)(0.548)$$
$$\approx 0.842$$
$$y = 3.4(0.842)(1 - 0.842)$$
$$= 3.4(0.842)(0.158)$$
$$\approx 0.452$$
The attractors are evidently 0.842 and 0.452.

Chapter 9 Test

1. (a) The measure of its complement is
$90 - 38 = 52°$.

(b) The measure of its supplement is
$180 - 38 = 142°$.

(c) It is an acute angle because it is less
than 90°.

2. The designated angles are supplementary;
their sum is 180°.
$$(2x + 16) + (5x + 80) = 180$$
$$7x + 96 = 180$$
$$7x = 84$$
$$x = 12$$
Then one angle measure is $2 \cdot 12 + 16 = 40°$
and the other angle measure is
$5 \cdot 12 + 80 = 140°$. A check is that their sum
is indeed 180°.

3. The angles are vertical so they have the
same measurement. Set the algebraic
expressions equal to each other.
$$7x - 25 = 4x + 5$$
$$3x = 30$$
$$x = 10$$
Then one angle measure is $7 \cdot 10 - 25 = 45°$
and the other angle measure is
$4 \cdot 10 + 5 = 45°$.

4. The designated angles are complementary;
their sum is 90°.
$$(4x + 6) + 10x = 90$$
$$14x + 6 = 90$$
$$14x = 84$$
$$x = 6$$
Then one angle measure is $4 \cdot 6 + 6 = 30°$
and the other angle measure is $10 \cdot 6 = 60°$.

5. The designated angles are supplementary.
$$(7x + 11) + (3x - 1) = 180$$
$$10x + 10 = 180$$
$$10x = 170$$
$$x = 17$$
Then one angle measure is $7 \cdot 17 + 11 = 30°$
and the other angle measure is
$3 \cdot 17 - 1 = 50°$. A check is that their sum is
indeed 180°.

6. These are alternate interior angles, which are
equal to each other.
$$13y - 26 = 10y + 7$$
$$3y = 33$$
$$y = 11$$
Then one angle measure is
$13 \cdot 11 - 26 = 117°$ and the other angle
measure is $10 \cdot 11 + 7 = 117°$.

7. Writing exercise; answers will vary.

8. Letter C is false because a triangle cannot
have both a right angle and an obtuse angle.
A right angle measures 90° and an obtuse
angle measures greater than 90°; however,
the sum of all three angles of any triangle is
exactly 180°.

9. The curve is simple and closed.

10. The curve is neither simple nor closed.

11. The sum of the three angle measures is 180°.
$$(3x+9)+(6x+3)+(21x-42)=180$$
$$30x-30=180$$
$$30x=210$$
$$x=7$$
Then one angle measure is $3 \cdot 7 + 9 = 30°$, a second angle measure is $6 \cdot 7 + 3 = 45°$, and the third angle measure is $21 \cdot 7 - 42 = 147 - 42 = 105°$. A check is that their sum is indeed 180°.

12. $A = lw = 6 \cdot 12 = 72 \text{ cm}^2$

13. $A = bh = 12 \cdot 5 = 60 \text{ in}^2$

14. $A = \dfrac{1}{2}bh = \dfrac{1}{2} \cdot 17 \cdot 8 = 68 \text{ m}^2$

15. $A = \dfrac{1}{2}h(b+B)$
$$= \dfrac{1}{2} \cdot 9(16+24)$$
$$= \dfrac{9}{2}(40)$$
$$= 180 \text{ m}^2$$

16. Replace A in the formula for the area of a circle and solve for r.
$$A = \pi r^2$$
$$144\pi = \pi r^2$$
$$\dfrac{144\pi}{\pi} = \dfrac{\pi r^2}{\pi}$$
$$144 = r^2$$
$$12 = r$$
Now replace r in the formula for the circumference.
$$C = 2\pi r = 2\pi \cdot 12 = 24\pi \text{ in.}$$

17. Use the formula for the circumference $C = \pi d$.
$$C = (3.14) \cdot 630 \approx 1978 \text{ ft}$$

18. Subtract the area of the triangle from the area of the semicircle. First, the area of the semicircle is:
$$A = \dfrac{1}{2}\pi r^2$$
$$= \dfrac{1}{2}(3.14)(10)^2$$
$$= \dfrac{1}{2}(3.14)(100)$$
$$= 157$$
Now find the area of the triangle.
$$A = \dfrac{1}{2}bh = \dfrac{1}{2} \cdot 20 \cdot 10 = 100$$
Finally, $157 - 100 = 57 \text{ cm}^2$.

19.

STATEMENTS	REASONS
1. $\angle CAB = \angle DBA$	1. Given
2. $DB = CA$	2. Given
3. $AB = AB$	3. Reflexive property
4. $\triangle ABD \cong \triangle BAC$	4. SAS Congruence Property

20. Let h = height of the pole.
$$\dfrac{h}{30} = \dfrac{30}{45}$$
$$45 \cdot h = 30 \cdot 30$$
$$\dfrac{45h}{45} = \dfrac{900}{45}$$
$$h = 20 \text{ feet}$$

21. Use the Pythagorean theorem to find c.
$$a^2 + b^2 = c^2$$
$$20^2 + 21^2 = c^2$$
$$400 + 441 = c^2$$
$$841 = c^2$$
$$\sqrt{841} = \sqrt{c^2}$$
$$29 = c$$
The length of the diagonal is 29 m.

22.

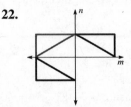

23.

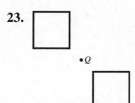

24. (a) $V = \frac{4}{3}\pi r^3$

$= \frac{4}{3}(3.14)(6)^3$

$= \frac{4}{3}(3.14)(216)$

$\approx 904.32 \text{ in}^3$

(b) $S = 4\pi r^2$

$= 4(3.14)(6)^2$

$= 4(3.14)(36)$

$= 452.16 \text{ in}^2$

25. (a) $V = lwh = 12 \cdot 9 \cdot 8 = 864 \text{ ft}^3$

(b) $S = 2lh + 2hw + 2lw$

$= 2 \cdot 12 \cdot 8 + 2 \cdot 8 \cdot 9 + 2 \cdot 12 \cdot 9$

$= 192 + 144 + 216$

$= 552 \text{ ft}^2$

26. (a) $V = \pi r^2 h$

$= (3.14)(6)^2(14)$

$= (3.14)(36)(14)$

$= 1582.56 \text{ m}^3$

(b) $S = 2\pi r^2 + 2\pi rh$

$= 2(3.14)(6)^2 + 2(3.14)(6)(14)$

$= 2(3.14)(36) + 2(3.14)(6)(14)$

$= 226.08 + 527.52$

$= 753.60 \text{ m}^2$

27. Writing exercise; answers will vary

28. (a) A page of a book and the cover of the same book are topologically equivalent because they both have no holes; they are of genus 0.

(b) A pair of glasses with the lenses removed and the Mona Lisa are not topologically equivalent. The glasses have two holes, but the Mona Lisa has none.

29. (a) Yes.

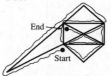

(b) No, because the network has more than two odd vertices.

30. $y = 2.1x(1 - x)$: Begin with $x = 0.6$ and iterate with a calculator to produce a sequence of numbers. Eventually, the number 0.5238095238 will repeat on the calculator. Thus the only attractor is 0.5238095238.

Chapter 10

10.1 Exercises

*In Exercises 1–7 consider the set
N = {A, B, C, D, E} for
{Alan, Bill, Cathy, David, and Evelyn}. List and
count the different ways of electing each of the
following slates of officers.*

1. A president and a treasurer
   Agreeing that the first letter represents the
   president and that the second represents the
   treasurer, we can generate systematically the
   following symbolic list and count the
   resulting possibilities: AB, AC, AD, AE;
   BA, BC, BD, BE; CA, CB, CD, CE; DA,
   DB, DC, DE; EA, EB, EC, ED. By
   counting, there are 20 ways to elect a
   president and a treasurer.

3. A president and a treasurer if the two
   officers must be the same sex.
   Since the men include A, B, and D, and the
   women are C and E, we are only interested
   in doubles that include combinations of just
   A, B, and D, or just C and E. The results are
   AB, AD, BA, BD, CE, DA, DB, and EC.
   Thus, there are
   8 ways the officers can be elected.

5. A president, a secretary, and a treasurer, if
   the president must be a man and the other
   two must be women.
   Generating a new symbolic list where the
   first member must be a man and the second
   and third, women, we get ACE, AEC, BCE,
   BEC, DCE, and DEC. The officers may be
   elected in 6 different ways.

*List and count the ways club N could appoint a
committee of three members under the following
conditions.*

7. There are no restrictions.
   One method would be to list all triples.
   Remembering, however, that ABC is the
   same committee as BAC or CAB, cross out
   all triples with the same three letters. We are
   left with: ABC, ABD, ABE, ACD, ACE,
   ADE, BCD, BCE, BDE, CDE. Therefore,
   there are 10 ways to select the 3-member
   committees with no restrictions.

*For Exercises 9–25, refer to Table 2 (the product
table for rolling two dice) in the text.*

9. Only 1 member of the product table (1, 1)
   represents an outcome where the sum of the
   dice is two.

11. Only 3 members of the product table (3, 1),
    (2, 2), (1, 3) represent outcomes where the
    sum of the dice is four.

13. There are 5 members of the product table,
    (5, 1), (4, 2), (3, 3), (2, 4), and (1, 5), which
    represent outcomes where the sum is six.

15. There are 5 members of the product table,
    (6, 2), (5, 3), (4, 4), (3, 5), and (2, 6), which
    represent outcomes where the sum is 8.

17. There are only 3 members of the product
    table, (6, 4), (5, 5), and (4, 6), which
    represent outcomes where the sum of the
    dice is ten.

19. Only 1 member, (6, 6), of the product table
    yields an outcome where the sum is twelve.

21. Half of all 36 outcomes suggested by the
    product table should represent a sum which
    is even; the other half, odd. Thus, they are
    18 outcomes which will be even. They are:
    (1, 1), (3, 1), (2, 2), (1, 3), (5, 1), (4, 2),
    (3, 3), (2, 4), (1, 5), (6, 2), (5, 3), (4, 4),
    (3, 5), (2, 6), (6, 4), (5, 5), (4, 6), (6, 6).

23. To find the sums between 6 and 10, we must
    count pairs in which the sum is 7, 8, or 9.
    Sum is 7: (1, 6), (2, 5), (3, 4), (4, 3), (5, 2),
    (6, 1)
    Sum is 8: (2, 6), (3, 5), (4, 4), (5, 3), (6, 2)
    Sum is 9: (6, 3), (5, 4), (4, 5), (3, 6)
    Since there are six pairs with a sum of 7,
    five pairs with a sum of 8, and 4 pairs with a
    sum of 9, there are 6 + 5 + 4 = 15 pairs with
    a sum between 6 and 10.

25. Construct a product table showing all
    possible two-digit numbers using digits
    from the set {2, 3, 5, 7}.

	2	3	5	7
2	22	23	25	27
3	32	33	35	37
5	52	53	55	57
7	72	73	75	77

27. The following numbers in the table are numbers with repeating digits: 22, 33, 55, and 77.

29. A counting number larger than 1 is prime if it is divisible by itself and 1 only. The following numbers in the table are prime numbers: 23, 37, 53, and 73.

31. Extend the tree diagram of Exercise 30 for four fair coins. Then list the ways of getting the following results.

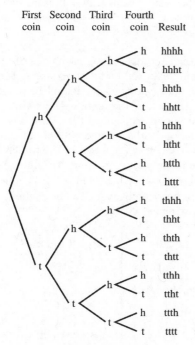

First Second Third Fourth
coin   coin   coin   coin   Result

hhhh
hhht
hhth
hhtt
hthh
htht
htth
httt
thhh
thht
thth
thtt
tthh
ttht
ttth
tttt

(a) More than three tails
There is only one such outcome: tttt.

(b) Fewer than three tails
List those outcomes with 0, 1 or 2 tails:
hhhh, hhht, hhth, hhtt, hthh, htht, htth, thhh, thht, thth, tthh.

(c) At least three tails
List those outcomes with 3 or 4 tails:
httt, thtt, ttht, ttth, tttt.

(d) No more than three tails
List those outcomes with 0, 1, 2, or 3 tails:
hhhh, hhht, hhth, hhtt, hthh, htht, htth, httt, thhh, thht, thth, thtt, tthh, ttht, ttth.

33. Begin with the largest triangles which have the long diagonals as their bases. There is 1 triangle on each side of the (2) diagonals. This gives 4 large triangles. Count the

intermediate sized triangles, each with a base along the outside edge of the large square. There are 4 of these. Furthermore, each of these intermediate sized triangles contain two right triangles within. There are a total of 8 of these. Thus, the total number of triangles is 4 + 4 + 8 = 16.

35. Begin with the larger right triangle at the center square. There are 4. Pairing two of these triangles with each other forms 4 isosceles triangles within the center box. Within each of the four right triangles in the square are two smaller right triangles for a total of 8. Associated with each exterior side of the octagon are 8 triangles, each containing two other right triangles (one of which has already been counted) for a total of 16. There are 4 more isosceles triangles which have their two equal side lengths as exterior edges of the octagon. Thus, the number of triangles contained in the figure is 4 + 4 + 8 + 16 + 4 = 36.

37. Label the figure as shown below, so that we can refer to the small squares by number.

	1	2	
3	4	5	6
7	8	9	10
	11	12	

Find the number of squares of each size and add the results.
There are twelve 1×1 squares, which are labeled 1 through 12.
Name the 2×2 squares by listing the small squares they contain:

1, 2, 4, 5	5, 6, 9, 10
8, 9, 11, 12	3, 4, 7, 8
4, 5, 8, 9	

There are five 2×2 squares. There are no squares larger than 2×2.
Thus, there are a total of 12 + 5 = 17 squares contained in the figure.

39. Examine carefully the figure in the text.
There are sixteen 1×1 squares with horizontal bases.
There are three 2×2 squares in each of the first and second rows, the second and third rows, as well as the third and fourth rows with horizontal bases. Thus, there are a total

of nine 2×2 squares with horizontal bases. There are two 3×3 squares with horizontal bases found in the first, second, and third rows as well as two 3×3 squares with horizontal base found in the second, third and fourth rows. Thus, there are a total of four 3×3 squares with horizontal bases. There is one 4×4 square (the large square itself).

Visualize the squares along the diagonals (at a slant).

There are twenty-four 1×1 squares with bases along diagonals.

There are thirteen 2×2 squares with bases along diagonals.

There are four 3×3 squares with bases along diagonals.

There is only one 4×4 square with bases along diagonals. Add the results.

Size	Number of squares
1×1 (horizontal)	16
1×1 (slant)	24
2×2 (horizontal)	9
2×2 (slant)	13
3×3 (horizontal)	4
3×3 (slant)	4
4×4 (horizontal)	1
4×4 (horizontal)	1
	72

There are 72 squares in the figure.

41. There are $3 \times 3 = 9$ cubes in each of the bottom two layers. This gives a total of 18 in the bottom two layers.

There are $3 \times 2 = 6$ cubes in each of the middle two layers. This gives a total of 12 in the middle two layers.

There are $3 \times 1 = 3$ cubes in the top two layers. This gives a total of 6 in the top two layers. Altogether, there are $18 + 12 + 6 = 36$ ($1 \times 1 \times 1$) cubes.

The visible cubes are:

Location	Number of cubes
Top two layers	6
Middle two layers	8
Bottom two layers	10
(exclude cubes in corners which have already been counted).	24

Thus, the number of cubes in the stack that are not visible is $36 - 24 = 12$. One could ignore the top two levels since each cube is visible.

43. There are 4 cubes along each edge of the bottom layer for a total of 10 cubes. There are 3 cubes along each edge of the second layer for a total of 6 cubes. There are 2 cubes along each edge of the third layer for a total of 3 cubes. Remember not to count the back corner cube twice. The top layer cube is visible, so ignore it. Thus, there are a total of $10 + 6 + 3 = 19$ ($1 \times 1 \times 1$) cubes in the bottom three layers. Of these, the following are visible.

Location	Number of cubes
Bottom layer	4
Second layer	3
Third layer	2
	9

Thus, the number of cubes in the stack that are not visible is $19 - 9 = 10$.

45. Label the figure as shown below.

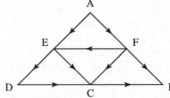

List all the paths in a systematic way. AFB, AFCB, AECB, AEDCB, AFECB, and AFEDCB represent all paths with the indicated restrictions. Thus, there are 6 paths.

47. To determine the number of ways in which 40 can be written as the sum of two primes, use trial and error in a systematic manner. Test each prime, starting with 2, as a possibility for the smaller prime.

(Since $40 - 2 = 38$, and 38 is not a prime, 2 will not work.) We obtain the following list:
$40 = 3 + 37$
$40 = 11 + 29$
$40 = 17 + 23$.
Thus, 40 can be written as the sum of two primes in 3 different ways.

49. Make a table to determine all possible sums, then determine the number of unique sums.

	2	2	3	3	5	8
2	4	4	5	5	7	10
2	4	4	5	5	7	10
3	5	5	6	6	8	11
3	5	5	6	6	8	11
5	7	7	8	8	10	13
8	10	10	11	11	13	16

The unique sums are 4, 5, 6, 7, 8, 10, 11, 13, and 16. Thus, 9 sums are possible.

51. If there are no ties, each time a game is played, the loser is eliminated. In order to determine the champion, 49 people (all but the champion) must be eliminated. Thus, it will take 49 games to determine the champion.

53. Make a systematic list or table of all three-digit numbers that have the sum of their digits equal to 22. Notice that since the largest possible sum of two digits is $9 + 9 = 18$, the smallest possible third digit in any of these number is 4.

499					
589	598				
679	688	697			
769	778	787	796		
859	868	877	886	895	
949	958	967	976	985	994

This table shows that there are
$1 + 2 + 3 + 4 + 5 + 6 = 21$ three-digit numbers that have the sum of their digits equal to 22.

55. The problem is essentially to find the number of ways to get a sum of 15 using the numbers 1, 3, and 4. There is one way using

all 1's, and there is one way using all 3's.
There are 4 ways using 1's and 3's:
$1 + 1 + 1 + 1 + 1 + 1 + 1 + 1 + 1 + 1 + 1 + 1$
$\quad + 3$
$1 + 1 + 1 + 1 + 1 + 1 + 1 + 1 + 1 + 3 + 3$
$1 + 1 + 1 + 1 + 1 + 1 + 3 + 3 + 3$
$1 + 1 + 1 + 3 + 3 + 3 + 3$
There are 3 ways using 1's and 4's:
$1 + 1 + 1 + 1 + 1 + 1 + 1 + 1 + 1 + 1 + 1 + 4$
$1 + 1 + 1 + 1 + 1 + 1 + 1 + 4 + 4$
$1 + 1 + 1 + 4 + 4 + 4$
There is only 1 way using 3's and 4's:
$3 + 4 + 4 + 4$
There are 5 ways using 1's, 3's and 4's:
$1 + 1 + 1 + 1 + 1 + 1 + 1 + 1 + 3 + 4$
$1 + 1 + 1 + 1 + 1 + 3 + 3 + 4$
$1 + 1 + 3 + 3 + 3 + 4$
$1 + 3 + 3 + 4 + 4$
$1 + 1 + 1 + 1 + 3 + 4 + 4$
Thus, there are 15 ways the order can be filled.

57. Draw a tree diagram showing all possible switch settings.

First Switch	Second Switch	Third Switch	Fourth Switch	Switch Settings

Thus, Pamela can choose 16 different switch settings.

59. There are five switches rather than four, and no two adjacent switches can be on. If no two adjacent switches can be on, the tree diagram that is constructed will not have two "1"s in succession.

First Switch	Second Switch	Third Switch	Fourth Switch	Fifth Switch	Switch Settings

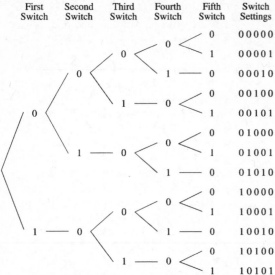

thus, Pamela can choose 13 different switch settings.

61. A line segment joins the points (8, 12) and (53, 234) in the Cartesian plane. Including its endpoints, how many lattice points does this line segment contain? (A lattice point is a point with integer coordinates.)

Any point (x, y) on the line segment, when used with either endpoint, must yield the same slope as that of the segment using both endpoints. Therefore, find the slope of the segment.

$$m = \frac{y_2 - y_1}{x_2 - x_1} = \frac{234 - 12}{53 - 8} = \frac{222}{45} = \frac{74}{15}$$

Set up the slope using the unknown point (x, y) and the known endpoint (8, 12).

$$m = \frac{y_2 - y_1}{x_2 - x_1} = \frac{y - 12}{x - 8}$$

Since the slope is the same for all points on the line segment, set these equal to each other and solve for y (in terms of x).

$$\frac{y - 12}{x - 8} = \frac{74}{15}$$
$$y - 12 = \frac{74}{15}(x - 8)$$
$$y = \frac{74}{15}(x - 8) + 12$$

All points on the line segment must be solutions for this equation. For the solutions to be integers (with x between 8 and 53), the denominator 15 will have to divide the value $(x - 8)$ evenly so that y remains an integer. That is, the number "$x - 8$" must be a multiple of 15 for values of x. Systematically trying integers for x from and including 8 to 53 will yield the following

results. All other values for x between 8 and 53 would not.

x	$x - 8$	Divisible by 15?
8	$8 - 8 = 0$	Yes
23	$23 - 8 = 15$	Yes
38	$38 - 8 = 30$	Yes
53	$53 - 8 = 45$	Yes

Thus, including the endpoints, there are 4 lattice points.

Note that using a graphing calculator with a table feature would provide a quicker solution. Set up the function

$$y = \frac{74}{15}(x - 8) + 12 \text{ in the calculator. Adjust}$$

the "table set" feature to begin at $x = 8$ with $\triangle x$ set to increase by 1 and create the table (set of ordered pairs) for x and y. Scanning the table for those y-values which are whole numbers will yield the same answers as above.

63. Each row will contain 25 matchsticks. If the grid is 12 matchsticks high there will be 13 rows of matchsticks (including the top and bottom rows). Therefore, there are $13 \times 25 = 325$ matchsticks in all of the rows. Each column contains 12 matchsticks. If the grid is 25 matchsticks wide, then there will be 26 columns of matchsticks counting the first and last columns. Thus, there are $26 \times 12 = 312$ matchsticks in the columns. Altogether, the number of matchsticks are $325 + 312 = 637$.

65. There are 8 people. Let C represent Chris and 0–6 the other 7 people who meet 0–6 others respectively. One of these is Chris' son. When we find how many people Chris had to meet, we will know who his son is since the son met the same number of people as Chris. One method to solve the problem is to use sets. Each set contains the people that the designated person met.
0: { }; he met nobody.
6: {1, 2, 3, 4, 5, C}; 6 met all but 0.
1: {6}; 6 met 1 (from 6's set) so 6 is the only person in the set.
5: {2, 3, 4, 6, C}; 5 met everyone but 0 and 1.
2: {5, 6}; 5 and 6 met 2 (from their sets) so 5 and 6 are the only people 2 met.
4: {3, 5, 6, C}: 4 met everyone by 0, 1, and

2.

3: $\{4, 5, 6\}$; 4, 5, and 6 met 3 (from their sets) so 4, 5, and 6 are the only people that 3 met.

C: $\{4, 5, 6\}$; same reasoning as 3, above.
Therefore, Chris met three people and shook three hands, and "3" is his son.

Wording may vary in answers for Exercise 67.

67. (a) Find the number of ways to select an ordered pair of letters from the set$\{A, B, C, D, E\}$ if repetition of letters is not allowed.

(b) Find the number of ways to select an ordered pair of letters from the set $\{A, B, C, D, E\}$ if the selection is done without replacement.

10.2 Exercises

1. Writing exercise; answers will vary.

3. (a) No, $(n + m)! \neq n! + m!$.

(b) Writing exercise; answers will vary.

5. (a) No, $(n - m)! \neq n! - m!$.

(b) Writing exercise; answers will vary.

Evaluate each expression without using a calculator.

7. $4! = 4 \cdot 3 \cdot 2 \cdot 1 = 24$

9. $\dfrac{9!}{7!} = \dfrac{9 \cdot 8 \cdot \cancel{7} \cdot \cancel{6} \cdot \cancel{5} \cdot \cancel{4} \cdot \cancel{3} \cdot \cancel{2} \cdot \cancel{1}}{\cancel{7} \cdot \cancel{6} \cdot \cancel{5} \cdot \cancel{4} \cdot \cancel{3} \cdot \cancel{2} \cdot \cancel{1}} = 72$

11. $\dfrac{5!}{(5-2)!} = \dfrac{5!}{3!} = \dfrac{5 \cdot 4 \cdot \cancel{3} \cdot \cancel{2} \cdot \cancel{1}}{\cancel{3} \cdot \cancel{2} \cdot \cancel{1}} = 20$

13. $\dfrac{8!}{6!(8-6)!} = \dfrac{8!}{6!2!}$

$= \dfrac{8 \cdot 7 \cdot \cancel{6} \cdot \cancel{5} \cdot \cancel{4} \cdot \cancel{3} \cdot \cancel{2} \cdot \cancel{1}}{\cancel{6} \cdot \cancel{5} \cdot \cancel{4} \cdot \cancel{3} \cdot 2 \cdot 1 \cdot \cancel{2} \cdot \cancel{1}}$

$= \dfrac{8 \cdot 7}{2 \cdot 1}$

$= \dfrac{56}{2}$

$= 28$

15. Evaluate $\dfrac{n!}{(n-r)!}$, where $n = 7$ and $r = 4$.

$\dfrac{7!}{(7-4)!} = \dfrac{7!}{3!} = \dfrac{7 \cdot 6 \cdot 5 \cdot 4 \cdot \cancel{3} \cdot \cancel{2} \cdot \cancel{1}}{\cancel{3} \cdot \cancel{2} \cdot \cancel{1}} = 840$

Evaluate each expression using a calculator. (Some answers may not be exact.) For Exercises 15–23, use the factorial key on a calculator, which is labeled $\boxed{x!}$ or $\boxed{n!}$ or, if using a graphing calculator, find "!" in the "Math" menu.

17. $10! = 3{,}628{,}800$

19. $\dfrac{12!}{5!} = 3{,}991{,}680$

21. $\dfrac{20!}{10! \cdot 10!} = 184{,}756$

23. Evaluate $\dfrac{n!}{(n-r)!}$, where $n = 17$ and $r = 8$.

$\dfrac{17!}{(17-8)!} = \dfrac{17!}{9!} = 980{,}179{,}200$

25. $\dfrac{6!}{2! \cdot 3!} = 60$

27. Since there are two possible outcomes for each switch (on/off), we have

$2 \cdot 2 \cdot 2 = 2^3 = 8.$

29. Writing exercise; answers will vary.

31. Using the fundamental counting principle, this may be considered as a three-part task. There would be $\underline{6} \cdot \underline{6} \cdot \underline{6} = 6^3 = 216$.

33. For each of the 10 pins, there are two possibilities: either the pin is up or the pin is down. So the total number of configurations is $2^{10} = 1024$.

Recall the club
$N = \{Alan, Bill, Cathy, David, Evelyn\}.$

35. This is a 5-part task. Use the fundamental counting principle to count the number of ways of lining up all five members for a photograph. There would be $\underline{5} \cdot \underline{4} \cdot \underline{3} \cdot \underline{2} \cdot \underline{1} = 120$ possibilities.

Similarly, one could use $n!$, where $n = 5$, to arrive at the total number or arrangements of a set of n objects.

37. This is a 2-part task. Since there are three males to choose from and two females to choose from, the number of possibilities is $\underline{3} \cdot \underline{2} = 6$.

In the following exercises, counting numbers are to be formed using only the digits 3, 4, and 5.

39. Choosing two-digit numbers may be considered a 2-part task. Since we can use any of the three given digits for each choice, there are $\underline{3} \cdot \underline{3} = 9$ different numbers that can be obtained.

41. Using the textbook hint, this may be considered a 3-part task. (1) Since there are only 3 positions that the two adjacent 4's can take (1st and 2nd, 2nd and 3rd, and 3rd and 4th positions), (2) two remaining positions that the 3 can take, and (3) the one last remaining digit must filled by the 5, there are $\underline{3} \cdot \underline{2} \cdot \underline{1} = 6$ different numbers that may be created.

43. Choosing from each of the three food categories is a 3-part task. There are five choices from the soup and salad category, two from the bread category, and four from the entree category. Applying the fundamental counting principle gives $5 \cdot 2 \cdot 4 = 40$ different dinners that may be selected.

45. Since there are 2 choices (T or F) for each question, we have $2^6 = 64$ possible ways.

For each situation in Exercises 47–51, use the table in the text to determine the number of different sets of classes Jessica can take.

47. All classes shown are available. Choose the number of possible courses from each category and apply the fundamental theorem: $\underline{2} \cdot \underline{3} \cdot \underline{4} \cdot \underline{5} = 120$

49. All sections of Minorities in America and Women in American Culture are filled already. The filled classes reduce the options in the Sociology category by 2. Thus, there are $\underline{2} \cdot \underline{3} \cdot \underline{4} \cdot \underline{3} = 72$ possible class schedules.

51. Funding has been withdrawn for three of the Education courses and for two of the Sociology courses. The reductions to the Education and Sociology categories leave only 1 Education course and 3 Sociology courses to choose from. Thus, there are $\underline{2} \cdot \underline{3} \cdot \underline{1} \cdot \underline{3} = 18$ possible class schedules.

53. This is a 3-part task. Applying the fundamental counting principle, there are $\underline{2} \cdot \underline{4} \cdot \underline{6} = 48$ different outfits that Don may wear.

55. The number of different ZIP codes that can be formed using all of those same five digits, 86726, would be the number of arrangements of the 5 digits. This is given by $\underline{5} \cdot \underline{4} \cdot \underline{3} \cdot \underline{2} \cdot \underline{1} = 5! = 120$.

57. There are 8 possibilities for the first digit. There are 10 possibilities for the second digit. There are 10 possibilities for the third digits. The total number of possible area codes is $8 \cdot 10 \cdot 10 = 800$.

Arne (A), Bobbette (B), Chuck (C), Deirdre (D), Ed (E), and Fran (F) have reserved six seats in a row at the theater, starting at an aisle seat.

59. Using the textbook hint, think of the problem as a six-part task, with the six parts described in (a)–(f) below.

  (a) From the sketch in the textbook, we see that there are 5 pairs of adjacent seats that Arne and Bobbette can occupy.

  (b) Given the two seats for A and B, they can be seated in 2 orders (A to the left of B, or B to the left of A).

  (c) C may occupy any of the 4 seats which are not taken by A or B.

  (d) Once A, B, and C have been seated, there are 3 remaining seats available for D.

  (e) Once A, B, C, and D have been seated, there are 2 remaining seats available for E.

  (f) Once all of the others have been seated, there is only 1 seat left for F.

Applying the fundamental counting principle, we conclude that the number of ways in which the six people can arrange themselves so that Aaron and Bobbette will be next to each other is $5 \cdot 2 \cdot 4 \cdot 3 \cdot 2 \cdot 1 = 240$.

61. Using the textbook hint, think of the problem as a six-part task, with the six parts described in (a)–(f) below.

(a) Any of the 6 people may sit on the aisle, so there are 6 choices.

(b) There are 3 men and 3 women. For whoever sits on the aisle, any of the 3 people of the opposite sex may occupy the second seat.

(c) For the third seat, we may choose either of the 2 remaining people of the first sex chosen.

(d) For the fourth seat, we may choose either of the 2 remaining people of the second sex chosen.

(e) For the fifth seat, there is only 1 choice, the remaining person of the first sex chosen.

(f) For the sixth seat, there is only one person left, so there is just 1 choice.

Multiplying the answers from (a)–(f), we conclude that the number of ways the people can arrange themselves if the men and women are to alternate with either a man or woman on the aisle is
$6 \cdot 3 \cdot 2 \cdot 2 \cdot 1 \cdot 1 = 72.$

63. Writing exercise; answers will vary.

65. The list will begin with the smallest number possible, 123,456. The list will continue with numbers that begin with the digit 1. There will be a total of
$1 \cdot 5 \cdot 4 \cdot 3 \cdot 2 \cdot 1 = 120$ numbers whose first digit is 1. Then the next numbers in the list will be the numbers beginning with the digit 2. There are 120 such numbers. Similarly, there are 120 numbers that begin with the digit 3 and 120 numbers that begin with the digit 4. The 481st number is the first number in the list that begins with the digit 5. There are $1 \cdot 1 \cdot 4 \cdot 3 \cdot 2 \cdot 1 = 24$ numbers whose first digit is 5 and whose second digit is 1, so the 500th number in the list must be of this form, since the 504th number is the number 516,432 (the largest possible number whose first digit is 5 and remaining digits are 1, 2, 3, 4, and 6). The next largest number (the 503rd number) is 516,423. The 502nd number is 516,342, the 501st number is 516,324. Finally, the 500th number is 516,243.

67. There are only 3 letters in the word INDIANA that repeat, and in a palindrome the letter D must always be the middle letter. There are 3 choices for the first letter, then 2 choices for the second letter, leaving only 1 possible letter as the third letter, for a total of $3 \cdot 2 \cdot 1 = 6$ possible palindromes.

10.3 Exercises

1. $\quad {}_9P_3 = \dfrac{9!}{(9-3)!}$

$\quad = \dfrac{9!}{6!}$

$\quad = \dfrac{9 \cdot 8 \cdot 7 \cdot 6!}{6!}$

$\quad = 9 \cdot 8 \cdot 7$

$\quad = 504$

3. $\quad {}_{11}C_7 = \dfrac{11!}{7!(11-7)!}$

$\quad = \dfrac{11!}{7!4!}$

$\quad = \dfrac{11 \cdot 10 \cdot 9 \cdot 8 \cdot 7!}{7!4 \cdot 3 \cdot 2 \cdot 1}$

$\quad = \dfrac{11 \cdot 10 \cdot 9 \cdot 8}{4 \cdot 3 \cdot 2 \cdot 1}$

$\quad = \dfrac{7920}{24}$

$\quad = 330$

5. Evaluate ${}_{20}P_4$.

$\quad {}_{20}P_4 = \dfrac{20!}{(20-4)!}$

$\quad = \dfrac{20!}{16!}$

$\quad = \dfrac{20 \cdot 19 \cdot 18 \cdot 17 \cdot 16!}{16!}$

$\quad = 20 \cdot 19 \cdot 18 \cdot 17$

$\quad = 116,280$

7. Evaluate $_9C_4$.

$$_9C_4 = \frac{9!}{4!(9-4)!}$$
$$= \frac{9!}{4!(5!)}$$
$$= \frac{9 \cdot 8 \cdot 7 \cdot 6 \cdot 5!}{4 \cdot 3 \cdot 2 \cdot 1 \cdot 5!}$$
$$= \frac{9 \cdot 8 \cdot 7 \cdot 6}{4 \cdot 3 \cdot 2 \cdot 1}$$
$$= \frac{3024}{24}$$
$$= 126$$

Use a calculator to evaluate each expression.

9. Use the *nPr* or *P(n, r)* button on a scientific calculator in the following order: 22 $\boxed{nPr}$ 9. Or, with a graphing calculator, find *nPr*. It is usually found in the probability menu. Insert, in the same order, 22 $\boxed{nPr}$ 9.

$$_{22}P_9 = 1.805037696 \times 10^{11}$$

11. Writing exercise; answers will vary.

13. Writing exercise; answers will vary.

15. **(a)** Permutation, since the order of the digits is important.

   (b) Permutation, since the order of the digits is important.

   (c) Combination, since the order is unimportant.

   (d) Combination, since the order is unimportant.

   (e) Permutation, since the order of the digits is important.

   (f) Combination, since the order is unimportant.

   (g) Permutation, since the order of the digits is important.

   (h) Permutation, since the order of the letters, digits, and symbols is important.

17. Since 5 different models will be built, items cannot be repeated. Also, order is important. Therefore, we use permutations. The number of ways in which Tyler can place

the homes on the lots is given by

$$_8P_5 = \frac{8!}{3!} = 6720.$$

19. Since no repetitions are allowed (one person will not be both) and the order of selection is important, the number of ways to choose a president and vice president is

$$_{12}P_3 = 12 \cdot 11 = 132.$$

21. Since no repetitions are allowed and the order of selection is important, the number of ways for the teacher to give the five different prizes to her students is

$$_{25}P_5 = 25 \cdot 24 \cdot 23 \cdot 22 \cdot 21 = 6,375,600.$$

23. **(a)** To get a sum of 10 we must use the digits $\{1, 2, 3, 4\}$ since $1 + 2 + 3 + 4 = 10$. They are not repeated and order is important. Thus, there are $_4P_4 = 4! = 24$ such numbers.

   (b) To get a sum of 11 we must use the digits $\{1, 2, 3, 5\}$ since $1 + 2 + 3 + 5 = 11$. They are not repeated and order is important. Thus, there are $_4P_4 = 4! = 24$ such numbers.

25. Samples are subsets, so use combinations. There are $24 - 6 = 18$ non-defective players, so we are to select 5 players from a set of 18. The number of samples which contain no defective players is

$$_{18}C_5 = \frac{18!}{5!13!} = 8568.$$

27. **(a)** Any hand represents a combination (or subset) of cards. Here we are choosing 5 cards from the 13 diamonds available. Thus, there are

$$_{13}C_5 = \frac{13!}{5!8!} = 1287 \text{ five-card hands.}$$

   (b) Since there are 26 black cards in the deck (13 spades and 13 clubs), the number of hands containing all black cards is $_{26}C_5 = \frac{26!}{5!21!} = 65,780.$

   (c) There are only 4 aces making it impossible to draw such a hand. there are 0 ways to do so.

29. (a) He has 6 lots to choose from. From the six, he can choose any three to build his standard homes on. Since the standard homes are all the same, the order is not important, and we have $_6C_3 = 20$ possible combinations or choices. Once these have been chosen, the remaining 3 lots will contain the deluxe models.

(b) Tyler may choose the positions for the two deluxe models. Once this has been done, the four standard models must go in the four remaining positions, which can only be done in one way. The number of different positions is therefore $_6C_2 = \dfrac{6!}{2!4!} = 15$.

Notice that the result will be the same if the four standard homes are positioned first. In this case, we obtain
$$_6C_4 = \dfrac{6!}{4!2!} = 15.$$

31. Assuming that the streets form a complete grid with 4 vertical lines and 7 horizontal lines, start at the bottom left corner of the grid. The shortest possible length of any path is 9 blocks long, and 3 of those must be walked to the north. Thus, there are $_9C_3 = 84$ different paths that may be followed to get to the northeast corner.

33. (a) The worst-case scenario is that the first four cards selected will be of different suits, then the 5th card chosen will be the same as one of the earlier choices. Any other scenario will require fewer cards to be drawn. Thus, there is a minimum of 5 cards to be drawn to obtain two cards of the same suit.

(b) By the 5th drawing we are guaranteed 2 cards are of the same suit. The 6th, 7th, and 8th drawings may give results that represent just 2 cards of the same suit for all 4 suits. But the 9th card must then be of one of the 4 suits, adding a 3rd card of that same suit. There must be a minimum of 9 cards drawn to guarantee three cards of the same suit.

35. For each of the first and third groups, there are $_{26}P_3$ possible arrangements of letters. For the second group, there are $_{10}P_3$ possible arrangements for the digits. Thus, by the fundamental counting principle, there are $_{26}P_3 \cdot _{10}P_3 \cdot _{26}P_3 = 175,219,200,000$ identification numbers.

37. Consider choosing the call letters for a station as a two-part task. The first part consists of choosing the first letter. Since the first letter must be K or W, this may be done in 2 ways. For the remaining three letters, we use permutations because order is important and repetition is not allowed. These three letters may be chosen from any of the 25 letters which were not used as the first letter, so the number of possibilities is $_{25}P_3$. Use the fundamental counting principle to combine the results from the two parts of the task. The number of possible call letters is $2 \cdot _{25}P_3 = 27,600$.

39. This is a two-part task. First choose the pitcher. This can be done in 7 ways. Now choose the players and batting order for the rest of the team. Since order is important and repetition is not allowed, use permutations. The number of choices is $_{12}P_8$. Use the fundamental counting principle to combine the results from the two parts of the task. The number of different batting orders is
$7 \cdot _{12}P_8 = 139,708,800$.

41. (a) The number of ways she can arrange her reading with replacement is found by using the fundamental counting principle.
$$7 \cdot 7 \cdot 7 \cdot 7 \cdot 7 \cdot 7 \cdot 7 = 7^7 = 823,543$$

(b) The number of ways she can arrange her reading without replacement can be found by permutations.
$_7P_7 = 7! = 5040$
Alternatively, one can apply the fundamental counting principle as well to get $7 \cdot 6 \cdot 5 \cdot 4 \cdot 3 \cdot 2 \cdot 1 = 7! = 5040$.

43. The number of ways each of the five groupings can be chosen is given by $_nC_r$ since the order in each grouping is not important. Apply the fundamental counting principle to find the total number of ways to

create the 5 groupings altogether.
$$_{15}C_1 \cdot {}_{14}C_2 \cdot {}_{12}C_3 \cdot {}_{9}C_4 \cdot {}_{5}C_5$$
$$= 37{,}837{,}800$$

45. Similar to Exercise 52, but adjusting for unwanted ordering of the three committees, the number of ways to distribute the people among the committees is

$$\frac{{}_{8}C_3 \cdot {}_{5}C_3 \cdot {}_{2}C_2}{2!} = 280.$$

47. Since each group of three non collinear points determines a triangle, we are looking for the number of 3-element subsets of a set of 20 elements. Since subsets are combinations, the number of triangles determined by 20 points in a plane, no three of which are collinear, is

$$_{20}C_3 = \frac{20!}{3!17!} = 1140.$$

49. (a) Since any pair of the 7 people may drive and the order of selection is important, the number of choices is

$$_7P_2 = \frac{7!}{5!} = 42.$$

(b) Consider choosing the drivers as a two-part task. There are only 3 choices for the driver of the sports car. Once the driver of the first sports car has been chosen, any of the remaining 6 people can be chosen to drive the second car. By the fundamental counting principle, the number of choices for drivers is $3 \cdot 6 = 18$.

(c) Choose the drivers first. Of the two, the first could be the driver of the sports car; and the second, the driver of the other vehicle. There are $_7P_2$ choices.

The second task is to pick the passenger for the sports car. There are five to choose from. By the fundamental counting principle, the total number of choices is $_7P_2 \cdot 5 = 210$.

51. The number of ways for three of the eight horses running to be 1st, 2nd, and 3rd place winners is given by $_8P_3 = 336$.

Thus, buying 336 different tickets will assure a winner.

53. Similar to Exercise 52, but adjusting for unwanted ordering of the three committees, the number of ways to distribute the people among the committees is

$$\frac{{}_{9}C_3 \cdot {}_{6}C_3 \cdot {}_{3}C_3}{3!} \cdot 3^3 = 7560.$$

55. (a) How many numbers can be formed using all six digits 4, 5, 6, 7, 8, and 9? Since the order or arrangement of these digits is important (each being a different number), we can consider the answer to the question to be
$_6P_6 = 6! = 720$ different numbers.

(b) The first number is 456,789; the second number is 456,798; the third number is 456,879 and so forth.
The number of arrangements (permutations) of the last five digits is given by 5! = 120. Thus, the 121st number is 546,789 (where we have moved to the sixth digit from right and interchanged the fifth and sixth digit—4 and 5). There are another 120 permutations (with this change on the fifth and sixth digits) bringing us to the 241st number: 645,789. In a similar manner, numbers beginning with the new first digit—6, we have another 120 permutations using the new set of numbers. For the 361st number, we change the first digit to seven, giving us 745,689. The 362nd number is 745,698. The 363rd number is 745,869 and the 364th number is 745,896.

57. (a) If any order will do, the number of possible arrangements is 6! = 720.

(b) If the bride and groom must be the last two in line (but in one of two possible orders) then the four attendants can be arranged in 4! ways. So the total number of arrangements is $2 \cdot 4! = 48$.

(c) If the groom must be last in line with the bride next to him, then the only people left to arrange are the four attendants, so the number of possible arrangements is 4! = 24.

59. (a) There are only two sets of distinct digits that add to 12. They are {1, 2, 3, 6} and {1, 2, 4, 5}. Try to find others. Therefore are 4! distinct permutations (which lead to a different counting number) for each set of digits. Thus, using the fundamental counting principle, the total number of counting numbers whose sum of digits is 12 is $2 \cdot 4! = 48$.

(b) There are only three sets of distinct digits that add to 13. They are $\{1, 2, 3, 7\}$, $\{1, 2, 4, 6\}$ and $\{1, 3, 4, 5\}$. Try to find others. There are 4! distinct permutations (which lead to a different counting number) for each set of digits. Therefore, using the fundamental counting principle, the total number of counting numbers whose sum of digits is 13 is $3 \cdot 4! = 72$.

61. $_{12}C_9 = \dfrac{12!}{9!(12-9)!} = \dfrac{12!}{9!3!} = 220$

$_{12}C_3 = \dfrac{12!}{3!(12-3)!} = \dfrac{12!}{3!9!} = 220$

Thus, $_{12}C_9 = {}_{12}C_3$.

10.4 Exercises

Read the following combination values directly from Pascal's triangle. For exercises 1–8, refer to Table 5 in the text.

1. To find the value of $_4C_2$ from Pascal's triangle, read entry number 2 in row 4 (remember that the top row is row "0" and that in row 4 the "1" is entry "0").
$_4C_2 = 6$

3. To find the value of $_6C_3$ from Pascal's triangle, read entry number 3 in row 6.
$_6C_3 = 20$

5. To find the value of $_8C_5$ from Pascal's triangle, read entry number 5 in row 8.
$_8C_5 = 56$

7. To find the value of $_9C_2$ from Pascal's triangle, read entry number 2 in row 9.
$_9C_2 = 36$

9. Selecting the committee is a two-part task. There are $_7C_1$ ways of choosing the one Democrat and $_3C_3$ way of choosing the remaining 3 Republicans. The combination values can be read from Pascal's triangle. By the fundamental counting principle, the total number of ways is $_7C_1 \cdot {}_3C_3 = 7 \cdot 1 = 7$.

11. A committee with exactly three Democrats will consist of three Democrats and one Republican. Selecting the committee is a two-part task. There are $_7C_3$ ways of choosing three Democrats and $_3C_1$ ways to choose the one remaining Republican. Hence, there are
$_7C_3 \cdot {}_3C_1 = 35 \cdot 3 = 105$ ways in total.

13. There are $_8C_3 = 56$ ways to choose three different positions for heads. Using Pascal's triangle, find row 8 entry 3. Remember to count first row and first entry as 0. The remaining positions will automatically be tails.

15. There are $_8C_5 = 56$ ways to choose exactly five different positions for heads. Using Pascal's triangle, this would be found in row 8, entry 5.

17. The number of selections for four rooms is given by $_9C_4 = 126$. Using Pascal's triangle, this would be found in row 9, entry 4.

19. The number of selections that succeed in locating the class is given by total number of selections (Exercise 17) minus the number of ways which will fail to locate the classroom (Exercise 18), or
$_9C_4 - {}_8C_4 = 126 - 70 = 56$ ways

21. The number of 0-element subsets for a set of five elements is entry 0 (the first entry) in row 5 of Pascal's triangle. This number is 1.

23. The number of 2-element subsets for a set of five elements is entry 2 (the third entry) in row 5 of Pascal's triangle. This number is 10.

25. The number of 4-element subsets for a set of five elements is entry 4 (the fifth entry) in row 5. This number is 5.

27. The total number of subsets is given by
$_5C_0 + {}_5C_1 + {}_5C_2 + {}_5C_3 + {}_5C_4 + {}_5C_5$
$= 1 + 5 + 10 + 10 + 5 + 1$
$= 32$.
This is the sum of elements in the fifth row of Pascal's triangle.

29. The even-numbered rows have a single entry in the middle of the row that is greater than all other entries in the row.

31. (a) All are multiplies of the row number.

(b) The same pattern holds.

(c) Row 11:

1 11 55 165 330 462 462 330 165 55 11 1
All are multiples of 11. Thus, the same pattern holds.

33. Following the indicated sums 1, 1, 2, 3, 5, the sequence continues 8, 13, 21, 34, ... A number in this sequence comes from the sum of the two preceding terms. This is the Fibonacci sequence.

35. Row 8 would be the next row to begin and end with 1, with all other entries 0 (each internal entry in row 8 of Pascal's triangle is even).

37. The sum of the squares of the entries across the top row equals the entry at the bottom vertex. Choose, for example, the second triangle from the bottom.

$$1^2 + 3^2 + 3^2 + 1^2 = 1 + 9 + 9 + 1$$
$$= 20 \text{ (the vertex value)}$$

39. Prove $_nC_r = {_{n-1}C_{r-1}} + {_{n-1}C_r}$.

$$_{n-1}C_{r-1} + {_{n-1}C_r}$$
$$= \frac{(n-1)!}{(r-1)![(n-1)-(r-1)]!} + \frac{(n-1)!}{r![(n-1)-r]!}$$
$$= \frac{(n-1)!}{(r-1)!(n-r)!} + \frac{(n-1)!}{r!(n-r-1)!}$$
$$= \frac{n}{n} \cdot \frac{(n-1)!}{(r-1)!(n-r)!} \cdot \frac{r}{r}$$
$$\qquad + \frac{n}{n} \cdot \frac{(n-1)!}{r!(n-r-1)!} \cdot \frac{(n-r)}{(n-r)}$$
$$= \frac{n! \cdot r}{n \cdot r! \cdot (n-r)!} + \frac{n! \cdot (n-r)}{n \cdot r! \cdot (n-r)!}$$
$$= \frac{n! \cdot r + n! \cdot (n-r)}{n \cdot r! \cdot (n-r)!}$$
$$= \frac{n! \cdot [r + (n-r)]}{n \cdot r! \cdot (n-r)!}$$
$$= \frac{n! \cdot \cancel{n}}{\cancel{n} \cdot r! \cdot (n-r)!}$$
$$= \frac{n!}{r!(n-r)!}$$
$$= {_nC_r}$$

41. The sum = N; any entry in the array equals the sum of the two entries immediately above it and immediately to its left.

43. The sum = N; any entry in the array equals the sum of the row of entries from the cell immediately above it to the left boundary of the array.

***EXTENSION:* MAGIC SQUARES**

1. 180° in a clockwise direction
Imagine a straight line (180°) from the top left corner to the bottom right corner of the magic square in Figure 9. That moves the 8 from its original position to the bottom right corner, and the other numbers follow.

2	7	6
9	5	1
4	3	8

3. 90° in a clockwise direction
Imagine the top left corner of the box that contains the number 17 as an origin. Rotate the entire square 90° clockwise. Then the 17 will be in the top right box and the 11 will be in the top left. All the other numbers follow.

11	10	4	23	17
18	12	6	5	24
25	19	13	7	1
2	21	20	14	8
9	3	22	16	15

5. 90° in a counterclockwise direction
Use the upper right corner of the box containing 15 as the pivot point. Rotate 90° in a counterclockwise direction. Then the 9 will be in the upper right box, and 15 will be in the upper left.

15	16	22	3	9
8	14	20	21	2
1	7	13	19	25
24	5	6	12	18
17	23	4	10	11

7. Figure 9, multiply by 3

24	9	12
3	15	27
18	21	6

$$MS = \frac{n(n^2 + 1)}{2} \cdot 3$$
$$= \frac{3(3^2 + 1)}{2} \cdot 3$$
$$= \frac{3(10)}{2} \cdot 3$$
$$= 45$$

9. Figure 11, divide by 2

$\frac{17}{2}$	12	$\frac{1}{2}$	4	$\frac{15}{2}$
$\frac{23}{2}$	$\frac{5}{2}$	$\frac{7}{2}$	7	8
2	3	$\frac{13}{2}$	10	11
5	6	$\frac{19}{2}$	$\frac{21}{2}$	$\frac{3}{2}$
$\frac{11}{2}$	9	$\frac{25}{2}$	1	$\frac{9}{2}$

$$MS = \frac{n(n^2 + 1)}{2} \div 2$$
$$= \frac{5(5^2 + 1)}{2} \div 2$$
$$= \frac{5(26)}{2} \div 2$$
$$= \frac{65}{2} = 32\frac{1}{2}$$

11. Using the third row, the magic sum is
281 + 467 + 59 = 807.
Then 807 − (71 + 257) = 479.
The missing entry is 479.

13. Using the first column, the magic sum is
389 + 71 + 347 = 807.
Then, 807 − (191 + 149) = 467.
The missing entry is 467.

15. Using the first column, the magic sum is
401 + 17 + 389 = 807.
Then, 807 − (257 + 281) = 269.
The missing entry is 269.

17. Using the second column, the magic sum is
68 + 72 + 76 = 216.

(a) 216 − (75 + 68) = 73

(b) 216 − (75 + 71) = 70

(c) Use the answer from (b) to find
216 − (72 + 70) = 74.

(d) 216 − (71 + 76) = 69

19. Using the second column to obtain the
magic sum, 20 + 14 + 21 + 8 + 2 = 65.

(a) 65 − (3 + 20 + 24 + 11) = 7

(b) 65 − (14 + 1 + 18 + 10) = 22

(c) 65 − (9 + 21 + 13 + 17) = 5

(d) 65 − (16 + 8 + 25 + 12) = 4

(e) Use the first column and (b):
65 − (3 + 9 + 16 + 22) = 15

(f) Use the third column and (a):
65 − (1 + 13 + 25 + 7) = 19

(g) Use the fourth column and (c):
65 − (24 + 18 + 12 + 5) = 6

(h) Use the fifth column and (d):
65 − (11 + 10 + 17 + 4) = 23

21. Use the "staircase method" to construct a
magic square of order 7, containing the
entries 1, 2, 3, ...

	31	40	49	2	11	20	
30	39	48	1	10	19	28	30
38	47	7	9	18	27	29	38
46	6	8	17	26	35	37	46
5	14	16	25	34	36	45	5
13	15	24	33	42	44	4	13
21	23	32	41	43	3	12	21
22	31	40	49	2	11	20	

23. The sum of the entries in the four corners is
16 + 13 + 4 + 1 = 34.

25. The entries in the diagonals are
16 + 10 + 7 + 1 + 13 + 11 + 6 + 4 = 68.
The entries not in the diagonals are
3 + 2 + 5 + 8 + 9 + 12 + 15 + 14 = 68.

27. Sum of cubes of diagonal entries:
$$16^3 + 10^3 + 7^3 + 1^3 + 13^3 + 11^3 + 6^3 + 4^3$$
$$= 4096 + 1000 + 343 + 1 + 2197 + 1331 + 216$$
$$+ 64$$
$$= 9248$$

Sum of cubes of entries not in the diagonals:
$3^2 + 2^3 + 5^3 + 8^3 + 9^3 + 12^3 + 15^3 + 14^3$
$= 27 + 8 + 125 + 512 + 729 + 1728 + 3375$
$$+ 2744$$
$= 9248$

29. $16^2 + 3^2 + 2^2 + 13^2 + 9^2 + 6^2 + 7^2 + 12^2$
$= 256 + 9 + 4 + 169 + 81 + 36 + 49 + 144$
$= 748;$

$5^2 + 10^2 + 11^2 + 8^2 + 4^2 + 15^2 + 14^2 + 1^2$
$= 25 + 100 + 121 + 64 + 16 + 225 + 196 + 1$
$= 748$

31.

$\rightarrow$	2	3	$\rightarrow$
5	$\rightarrow$	$\rightarrow$	8
9	$\rightarrow$	$\rightarrow$	12
$\rightarrow$	14	15	$\rightarrow$

16	2	3	13
5	11	10	8
9	7	6	12
4	14	15	1

The second and third columns are interchanged.

33. $a = 16, b = 2, c = -6$
Replace a, b, and c with these numbers to find the entries in the magic square.
$a + b = 16 + 2 = 18$
$a - b - c = 16 - 2 - (-6) = 20$
$a + c = 16 + (-6) = 10$
$a - b + c = 16 - 2 + (-6) = 8$
$a = 16$
$a + b - c = 16 + 2 - (-6) = 24$
$a - c = 16 - (-6) = 22$
$a + b + c = 16 + 2 + (-6) = 12$
$a - b = 16 - 2 = 14$
The magic square is then as follows.

18	20	10
8	16	24
22	12	14

35.

39	48	57	10	19	28	37
47	56	16	18	27	36	38
55	15	17	26	35	44	46
14	23	25	34	43	45	54
22	24	33	42	51	53	13
30	32	41	50	52	12	21
31	40	49	58	11	20	29

$$MS = \frac{7(2 \cdot 10 + 7^2 - 1)}{2}$$
$$= \frac{7(20 + 49 - 1)}{2}$$
$$= \frac{7(68)}{2}$$
$$= 238$$

37. There are many ways to find the magic sum. One way is by adding the top row:
$52 + 61 + 4 + 13 + 20 + 29 + 36 + 45 = 260$

39. $52 + 45 + 16 + 17 + 54 + 43 + 10 + 23$
$= 260$

41. Start by placing 1 in the fourth row, second column. Move up two, right one and place 2 in the second row, third column. If we now move up two, right one again, we will go outside the square. Drop down 5 cells (one complete column) to the bottom of the fourth column to place 3. Move up two, right one, to place 4 in the third row, fifth column. Moving up two, right one, we again go outside the square; move 5 cells to the left to place 5 in the first row, first column. Moving up two, right one, takes us two rows outside the square. Counting downward 5 cells, find that the fourth row, second column is blocked with a 1 already there. The number 6 is then placed just below the entry 5. Continue in this manner until all 24 numbers have been placed. Notice that in trying to enter the number 21, it is blocked by 16 which is already in the cell. Because 20 is already in the bottom cell of the last row, dropping a cell just "below" this one moves it to the top row, third column. The completed magic square is shown here.

	14	22	10	18	
	20	3	16	24	
5	13	21	9	17	5
6	19	2	15	23	11
12	25	8	16	4	12
18	1	14	22	10	
24	7	20	3	11	

43. First find the magic sum. Consider the known entries on the front of the cube. Let x = the unknown number in the column between the entries 4 and 27. The sum of this column is $x + 4 + 27 = x + 31$. The sum of the entries in the top row must be the same, but the known entries have a sum of 30. So the unknown entry in that row must have the value $x + 1$. Similarly, if y is the unknown entry in the bottom row, then $27 + y + 10 = 37 + y$. Also the sum of the known entries in the far right column is 36, so the unknown entry in that column must be $y + 1$. Then we know the following:
$4 + x + 27 = 27 + y + 10$ (1)
and the sum in the middle row is
$x + 25 + y + 1$, and
$x + 25 + y + 1 = 27 + y + 10$ (2)
Solve the first equation for x.
$x = y + 6$
Substitute the value into the second equation.
$$y + 6 + 25 + y + 1 = 27 + y + 10$$
$$2y + 32 = y + 37$$
$$y = 5$$
Thus, the magic sum is
$27 + y + 10 = 27 + 5 + 10 = 42$. Because every other number in the "magic cube" is given, now that the magic sum is known the unknown entries are easy to find. The completed cube is shown.

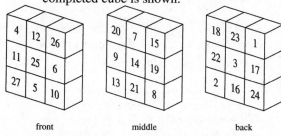

front   middle   back

1. The total number of subsets is 2^4 for a set with 4 elements. The only subset which is not a proper subset is the given set itself. Thus, by the complements principle, the number of proper subsets is
$$2^4 - 1 = 16 - 1 = 15.$$

3. By the fundamental counting principle, there are 2^7 different outcomes if seven coins are tossed. There is only one way to get no heads (all tails); thus, by the complements principle, there are
$2^7 - 1 = 128 - 1 = 127$ outcomes with at least one head.

5. By the fundamental counting principle, there are 2^7 different outcomes if seven coins are tossed. Of these, there is 1 way to get zero tails, and there are 7 ways to get one tail. Since "at least two" is the complement of "zero or one," the number of ways to get at least two tails is $2^7 - (1 + 7) = 128 - 8 = 120$.

Refer to Table 2 in the first section of the textbook chapter.

7. In Table 2 all columns except the first represent "at least 2 on all the green die." Thus, there are $36 - 6 = 30$ ways to achieve this outcome.

9. Counting the number of outcomes in row 4 (4 on red die) + those in column 4 (4 on green die) and subtracting the outcome counted twice, the number of outcomes with "a 4 on at least one of the dice" is
$6 + 6 - 1 = 11$.

11. Since there is only one "ace of spades" in the deck, there must be 51 other cards. Thus, there are 51 ways to draw "a card other than the ace of spades."

13. There are nine two-digit multiples of 10 (10, 20, 30, ..., 90). Altogether there are $9 \cdot 10 = 90$ two-digit numbers by the fundamental counting principle. Thus, by the complements principle, the number of "two-digit numbers which are not multiples of ten" is $90 - 9 = 81$.

15. (a) The number of different sets of three albums she could choose is $_{10}C_3 = 120$.

(b) The number which would not include *Southern Voice* is $_9C_3 = 84$.

(c) The number that would contain *Southern Voice* is $120 - 84 = 36$.

17. The total number of ways of choosing any three days of the week is $_7C_3$. The number of ways of choosing three days of the week that do not begin with S is $_5C_3$. Thus, the number of ways of choosing any three days of the week such that "at least one of them begin with S" is $_7C_3 - {_5C_3} = 25$.

19. If the order of selection is important, the number of choices of restaurants is $_8P_4$. The number of choices of restaurants that would not include seafood is $_5P_4$. Thus, the number of choices such that at least one of the four will serve seafood is $_8P_4 - {_5P_4} = 1560$.

21. The total number of ways to arrange 3 people among ten seats is $_{10}P_3$. The number of ways to arrange three people among the seven (non-aisle) seats is $_7P_3$. Therefore, by the complements principle, the number of arrangements with at least one aisle seat is $_{10}P_3 - {_7P_3} = 510$.

23. For the stations with 3 call letters, there are 2 choices for the first letter, 26 choices for the second letter, and 26 choices for the third letter, or $2 \cdot 26^2$ possibilities. Similarly, for stations with 4 call letters, there are $2 \cdot 26^3$ possibilities. So the total number of different call letter combinations is $2 \cdot 26^2 + 2 \cdot 26^3 = 36,504$.

25. "At least one officer" is the complement of "no officers." The number of ways of choosing 4-member search teams is $_{12}C_4$. The number of ways to choose 4-member search teams with no officers included is $_8C_4$. The total number of ways to choose the search team with at least one officer included is $_{12}C_4 - {_8C_4} = 425$.

27. Let C = the set of clubs and J = the set of jacks. Then, $C \cup J$ is the set of cards which are face cards or jacks, and $C \cap J$ is the set of cards which are both clubs and jacks, that is, the jack of clubs. Using the general additive counting principle, we obtain
$$n(C \cup J) = n(C) + n(J) - n(C \cap J)$$
$$= 13 + 4 - 1$$
$$= 16.$$

29. Let M = the set of students who enjoy music and C = the set of students who enjoy cinema. Then $M \cup C$ is the set of students who enjoy music or cinema, and $M \cap C$ is the set of students who enjoy both music and cinema. Using the additive principle, we obtain
$$n(M \cup C) = n(M) + n(C) - n(M \cap C)$$
$$= 25 + 22 - 18$$
$$= 29.$$

31. There are $_{13}C_5$ 5-card hands of hearts. Thus, by the complements principle, the number of hands containing "at least one card that is not a heart" is $2,598,960 - {_{13}C_5} = 2,597,673$.

33. The number of 5-card hands drawn from the 40 non-face cards in the deck is given by $_{40}C_5$. Thus, by the complements principle the number of 5-card hands with "at least one face card" is $2,598,960 - {_{40}C_5} = 1,940,952$.

35. If there are no duplicate flavors chosen, then the number of possible selections is $_6C_3 = 20$. But if we allow duplicate flavors, then after the first flavor is selected there are 6 choices for the second flavor and 6 choices for the third flavor, or 36 possible selections. So the total number of possible selections is $20 + 36 = 56$.

37. "At most two elements" is the same as 0, 1, or 2 elements. Thus, the number of subsets is $_{10}C_0 + {_{10}C_1} + {_{10}C_2} = 56$.

39. "More than two elements" is the complement of "at most two elements." Find the number of subsets with "at most two elements" by adding the number of 0-element subsets, 1-element subsets, and 2-element subsets.
$$_{10}C_0 + {_{10}C_1} + {_{10}C_2} = 1 + 10 + 45 = 56$$

There are 2^{10} subsets altogether. Thus, by the complements principle, the number of subsets of more than two elements is
$$2^{10} - 56 = 968.$$

41. The complement of "at least one letter or digit repeated" is "no letters or digits repeated." There are $_{26}P_2 \cdot {}_{10}P_3$ license plates with no digits repeated, and using the fundamental counting principle we have
$$26 \cdot 26 \cdot 10 \cdot 10 \cdot 10 = 26^2 \cdot 10^3$$ license plates where any letter of digit can be repeated. By the complements principle, the number of different license plates with at least one letter or digit repeated is
$$26^2 \cdot 10^3 - {}_{26}P_2 \cdot {}_{10}P_3 = 208,000.$$

43. Writing exercise; answers will vary.

45. To choose sites in only one state works as follows: The number of ways to choose three monuments in New Mexico is $_4C_3$, The number of ways to choose three monuments in Arizona is $_3C_3$, and the number of ways to pick three monuments in California is $_5C_3$. Since these components are disjoint, we may use the special additive principle. The number of ways to choose the monuments is
$$_4C_3 + {}_3C_3 + {}_5C_3 = 4 + 1 + 10 = 15.$$

47. "Sites in fewer than all three states" is the complement of choosing "sites in all three states." Since there are 12 monuments altogether, the total number of ways to select three monuments (with no restrictions) is $_{12}C_3$. Choosing sites in all three states requires choosing one site in each state, which can be done in $4 \cdot 3 \cdot 5$ ways. Using the complements principles, the number of ways to choose sites in fewer than all three states is
$$_{12}C_3 - 4 \cdot 3 \cdot 5 = 220 - 60 = 160.$$

49. Writing exercise; answers will vary.

51. Writing exercise; answers will vary.

53. Writing exercise; answers will vary.

Chapter 10 Test

1. To find three-digit numbers from the set $\{0, 1, 2, 3, 4, 5, 6\}$, use the fundamental counting principle: $\underline{6} \cdot \underline{7} \cdot \underline{7} = 294$.

2. To find odd three-digit numbers from the set $\{0, 1, 2, 3, 4, 5, 6\}$, use the fundamental counting principle: $\underline{6} \cdot \underline{7} \cdot \underline{3} = 126$.

3. To find three-digit numbers without repeated digits from the set $\{0, 1, 2, 3, 4, 5, 6\}$, use the fundamental counting principle: $\underline{6} \cdot \underline{6} \cdot \underline{5} = 180$.

4. To find three-digit multiples of five without repeated digits from the set $\{0, 1, 2, 3, 4, 5, 6\}$, use the fundamental counting principle and the special additive principle: Multiples of 5 end in "0" or "5." There are $\underline{6} \cdot \underline{5} \cdot \underline{1} = 30$ multiples that end in 0 and $\underline{5} \cdot \underline{5} \cdot \underline{1} = 25$ that end in 5. Thus, the number of three-digit multiples of five without repeated digits is $30 + 25 = 55$.

5. Make a systematic listing of triangles. Beginning with the smaller inside triangle, there are 4 right triangles off the horizontal bisector of the larger triangle. These triangles may be combined to create 4 larger isosceles triangles. There are 2 isosceles triangles—inside the upper left and lower left corners of the largest triangle and 1 larger right triangle above and 1 below the horizontal bisector of the larger triangle. Of course, count the largest isosceles triangle itself. The total number of triangles is $4 + 4 + 2 + 1 + 1 + 1 = 13$.

6.

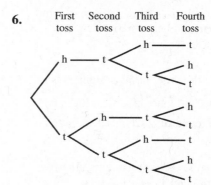

7. There is only one set of 4 digits that add to 30 $(9 + 8 + 7 + 6 = 30)$. The number of arrangements of the digits 9876 is given by $_4P_4 = 4! = 24$.

8. Counting clockwise, Tia and Jo can sit either 1 seat, 2 seats, or 3 seats apart, leaving 4 empty seats for their four friends. There are 3 arrangements for Tia and Jo and 4! arrangements for their four friends, so the total number of ways they can be arranged is $3 \cdot 4! = 72$.

9. $6! = 6 \cdot 5 \cdot 4 \cdot 3 \cdot 2 \cdot 1 = 720$

10. $\dfrac{8!}{6!} = \dfrac{8 \cdot 7 \cdot 6!}{6!} = 8 \cdot 7 = 56$

11. $\begin{aligned} _{12}P_3 &= \frac{12!}{(12-3)!} \\ &= \frac{12!}{9!} \\ &= \frac{12 \cdot 11 \cdot 10 \cdot 9!}{9!} \\ &= 12 \cdot 11 \cdot 10 \\ &= 1320 \end{aligned}$

12. $\begin{aligned} _8C_5 &= \frac{8!}{5!(8-5)!} \\ &= \frac{8!}{5! \cdot 3!} \\ &= \frac{8 \cdot 7 \cdot 6 \cdot 5!}{5! \cdot 3 \cdot 2 \cdot 1} \\ &= \frac{8 \cdot 7 \cdot 6}{3 \cdot 2 \cdot 1} \\ &= 56 \end{aligned}$

13. Since the arrangement of the letters is important and no repetitions are allowed, use $_{26}P_5$.
$$_{26}P_5 = 26 \cdot 25 \cdot 24 \cdot 23 \cdot 22 = 7,893,600$$

14. Since repetitions are allowed, sue the fundamental counting principle.
$$32^5 = 33,554,432$$

15. Since the order of the assignments is important, use $_7P_2$.
$$_7P_2 = \frac{7!}{(7-2)!} = \frac{7!}{5!} = \frac{7 \cdot 6 \cdot 5!}{5!} = 7 \cdot 6 = 42$$

16. Since the order of the remaining five assignments is important, use $_5P_5$ or 5!.
$$_5P_5 = 5! = 5 \cdot 4 \cdot 3 \cdot 2 \cdot 1 = 120$$

17. There are 6 letters to be arranged, but two letters are repeated: G appears 2 times and O appears 3 times. To account for this, we must divide out combinations that would imply order mattered. Thus, the number of arrangements is
$$\frac{6!}{2! \cdot 3!} = \frac{6 \cdot 5 \cdot 4 \cdot 3!}{2 \cdot 1 \cdot 3!} = \frac{6 \cdot 5 \cdot 4}{2} = 60.$$

18. Since order is not important, use $_{10}C_4$.
$$\begin{aligned} _{10}C_4 &= \frac{10!}{4!(10-4)!} \\ &= \frac{10!}{4!6!} \\ &= \frac{10 \cdot 9 \cdot 8 \cdot 7 \cdot 6!}{4 \cdot 3 \cdot 2 \cdot 1 \cdot 6!} \\ &= 210 \end{aligned}$$

19. Use the fundamental counting principle and account for any unwanted ordering.
$$\frac{_{10}C_2 \cdot \, _8C_2}{2!} = 630$$

20. Use the fundamental counting principle and account for any unwanted ordering.
$$\frac{_{10}C_5 \cdot \, _5C_5}{2!} = 126$$

21. Use the fundamental counting principle and account for any unwanted ordering.
$$\frac{_{10}C_4 \cdot \, _6C_4}{2!} = 1575$$

22. The complement of "a group of three or more of the players" is "zero, one, or two of the players." The total number of subsets of the 10 players is $2^{10} = 1024$.
The total number of 0-member subsets, 1-member subsets, and 2-member subsets is
$$_{10}C_0 + {_{10}C_1} + {_{10}C_2} = 1 + 10 + 45 = 56.$$
By the complements principle, the number of ways to select a group of three or more of the players is
$$\begin{aligned} 2^{10} - ({_{10}C_0} + {_{10}C_1} + {_{10}C_2}) &= 1024 - 56 \\ &= 968. \end{aligned}$$

23. With no restrictions, use the fundamental counting principle to determine the number of positions that a row of five switches may be set.
$$2 \cdot 2 \cdot 2 \cdot 2 \cdot 2 = 2^5 = 32$$

24. Use the fundamental counting principle to determine the number of positions that a row of five switches may be set if the first and fifth switches must be on.
$$1 \cdot 2 \cdot 2 \cdot 2 \cdot 1 = 2^3 = 8$$

25. Use the fundamental counting principle to determine the number of positions that a row of five switches may be set if the first and fifth switches must be set the same.
$$2 \cdot 2 \cdot 2 \cdot 2 \cdot 1 = 2^4 = 16$$

26. The following represent switch settings where "no two adjacent switches can both be off." Remember that "0" represents an "off" switch.

01010	10101	10111
01011	10110	11101
01101	11010	11110
01110	11011	
01111	11111	

There are 13 switch settings that satisfy the restriction.

27. There are only 2 switch settings in which no two adjacent switches can be set the same:
01010   10101

28. The complement of "at least two switches must be on" is "zero or one switch is on." Without restrictions there are

$2^5 = 32$ different switch settings. There is only 1 way for zero switches to be on and 5 ways for only 1 switch to be on. Thus, by the complements principle, the number of ways for at least two switches to be on is $32 - (1 + 5) = 26$.

29. Since the letter B must be a member, all that is necessary is to choose two members from the remaining six members. The number of three-element subsets is $_6C_2 = 15$.

30. Since the letters A and E must be members, all that is necessary is to choose one member from the five remaining members. The number of three-element subsets is $_5C_1 = 5$.

31. $_5C_2$ represents the number of three-element subsets that contain exactly one of the letters but not the other. There will be twice this number if we do the same for the 2nd letter. Thus, the number of three-element subsets with either A or E but not both is $2 \cdot {_5C_2} = 20$.

32. Since the letters A and D must be members, all that is necessary is to choose one member from the remaining five members. The number of three-element subsets is then $_5C_1 = 5$.

33. "More consonants than vowels" can happen two ways. One, with 3 consonants and no vowels, can be found by $_5C_3$. The second, with 2 consonants and 1 vowel, can be found by choosing the 1 vowel $_2C_1$ ways. Since one or the other will satisfy the restrictions (and they cannot both happen at the same time) apply the special addition principle to find the total number of subset choices for "more consonants than vowels."
$_5C_3 + {_5C_2} \cdot {_2C_1} = 30$

34. All possible paths are 9 blocks long, and 4 of those must be in an eastward direction. Thus, there are $_9C_4 = 126$ different paths to the garage.

35. Writing exercise; answers will vary.

36. Because $_nC_r$ and $_nC_{r+1}$ are the two entries just above $_{n+1}C_{r+1}$, evaluate $_{n+1}C_{r+1}$ by adding their values.
$_{n+1}C_{r+1} = {_nC_r} + {_nC_{r+1}} = 495 + 220 = 715$

37. The sequence of numbers obtained is 1, 2, 3, 4, 5, ...

38. Writing exercise; answers will vary.

Chapter 11

Exercise Set 11.1

The sample space is {red, yellow, blue}.

1. The number of regions in the sample space is $n(S) = 3$. Each region has the same area and thus, has the same likelihood of occurring. Therefore,

   (a) $P(\text{red}) = \dfrac{n(\text{red regions})}{n(S)} = \dfrac{1}{3}$.

   (b) $P(\text{yellow}) = \dfrac{n(\text{yellow regions})}{n(S)} = \dfrac{1}{3}$.

   (c) $P(\text{blue}) = \dfrac{n(\text{blue regions})}{n(S)} = \dfrac{1}{3}$.

3. The number of regions in the sample space is $n(S) = 6$. Each region (piece of the pie) has the same area and thus, has the same likelihood of occurring. The probability of landing on any one of the six regions is $\dfrac{1}{6}$, but we must account for the fact that some colors shade more than one region. Therefore,

   (a) $P(\text{red}) = \dfrac{n(\text{red regions})}{n(S)} = \dfrac{3}{6} = \dfrac{1}{2}$.

   (b) $P(\text{yellow}) = \dfrac{n(\text{yellow regions})}{n(S)}$
   $= \dfrac{2}{6}$
   $= \dfrac{1}{3}$.

   (c) $P(\text{blue}) = \dfrac{n(\text{blue regions})}{n(S)} = \dfrac{1}{6}$.

5. (a) The sample space is {1, 2, 3}.

   (b) The number of favorable outcomes is 2.

   (c) The number of unfavorable outcomes is 1.

   (d) The total number of possible outcomes is 3.

   (e) The probability of an odd number is given by
   $P(\text{odd number})$
   $= P(E)$
   $= \dfrac{\text{number of favorable outcomes}}{\text{total number of outcomes}}$
   $= \dfrac{2}{3}$.

   (f) The odds in favor of an odd number is given by
   Odds in favor
   $= \dfrac{\text{number of favorable outcomes}}{\text{number of unfavorable outcomes}}$
   $= \dfrac{2}{1}$, or 2 to 1

7. (a) The sample space is {11, 12, 13, 21, 22, 23, 31, 32, 33}.

   (b) The probability of an odd number is given by
   $P(\text{odd number})$
   $= P(E)$
   $= \dfrac{\text{number of favorable outcomes}}{\text{total number of outcomes}}$
   $= \dfrac{6}{9}$
   $= \dfrac{2}{3}$.

   (c) The probability of a number with repeated digit is given by
   $P(\text{number with repeated digits}) = \dfrac{3}{9}$
   $= \dfrac{1}{3}$.

   (d) The probability of a number greater than 30 is given by
   $P(\text{number greater than 30}) = \dfrac{3}{9} = \dfrac{1}{3}$.

   (e) The primes are {11, 13, 23, 31}. Thus, the probability for a prime number is given by $P(\text{prime number}) = \dfrac{4}{9}$.

9. (a) The odds against selecting a red ball is given by

Odds against

$$= \frac{\text{number of unfavorable outcomes}}{\text{number of favorable outcomes}}$$

$$= \frac{7}{4} \text{ or 7 to 4.}$$

(b) The odds against selecting a yellow ball is given by

Odds against

$$= \frac{\text{number of unfavorable outcomes}}{\text{number of favorable outcomes}}$$

$$= \frac{6}{5} \text{ or 6 to 5.}$$

(c) The odds against selecting a blue ball is given by

Odds against

$$= \frac{\text{number of unfavorable outcomes}}{\text{number of favorable outcomes}}$$

$$= \frac{9}{2} \text{ or 9 to 2.}$$

11. (a) $P(\text{Smiley Lewis}) = \dfrac{1}{50}$

(b) $P(\text{The Drifters}) = \dfrac{2}{50} = \dfrac{1}{25}$

(c) $P(\text{Bobby Darin}) = \dfrac{3}{50}$

(d) $P(\text{The Coasters}) = \dfrac{4}{50} = \dfrac{2}{25}$

(e) $P(\text{Fats Domino}) = \dfrac{5}{50} = \dfrac{1}{10}$

13. Product table for "sum"

		2nd die					
	+	1	2	3	4	5	6
	1	2	3	4	5	6	7
	2	3	4	5	6	7	8
1st	3	4	5	6	7	8	9
die	4	5	6	7	8	9	10
	5	6	7	8	9	10	11
	6	7	8	9	10	11	12

(a) Of the 36 possible outcomes, one gives a sum of 2, so $P(\text{sum is 2}) = \dfrac{1}{36}$.

(b) The sum of 3 appears 2 times in the body of the table. Thus,

$$P(\text{sum of 3}) = \frac{2}{36} = \frac{1}{18}.$$

(c) The sum of 4 appears 3 times in the body of the table. Thus,

$$P(\text{sum of 4}) = \frac{3}{36} = \frac{1}{12}.$$

(d) The sum of 5 appears 4 times in the table. Thus, $P(\text{sum of 5}) = \dfrac{4}{36} = \dfrac{1}{9}$.

(e) The sum of 6 appears 5 times in the table. Thus, $P(\text{sum of 6}) = \dfrac{5}{36}$.

(f) The sum of 7 appears 6 times in the table. Thus, $P(\text{sum of 7}) = \dfrac{6}{36} = \dfrac{1}{6}$.

(g) The sum of 8 appears 5 times in the table. Thus, $P(\text{sum of 8}) = \dfrac{5}{36}$.

(h) The sum of 9 appears 4 times in the table. Thus, $P(\text{sum of 9}) = \dfrac{4}{36} = \dfrac{1}{9}$.

(i) The sum of 10 appears 3 times in the table. Thus, $P(\text{sum of 10}) = \dfrac{3}{36} = \dfrac{1}{12}$.

(j) The sum of 11 appears 2 times in the table. Thus, $P(\text{sum of 11}) = \dfrac{2}{36} = \dfrac{1}{18}$.

(k) The sum of 12 appears 1 time in the table. Thus, $P(\text{sum of 12}) = \dfrac{1}{36}$.

In Exercise 15, answers are computed to three decimal places.

15. The probability that a randomly selected location in California will be forested is $\dfrac{51,250}{155,959} \approx 0.329$.

17. The possible ways to obtain a sum of 7 using regular 6-sided dice are 1 + 6, 2 + 5, 3 + 4, 4 + 3, 5 + 2, and 6 + 1, where the first number in each sum comes from the first die and the second number in each sum comes from the second die. Since these dice have been altered, the only possibilities for obtaining a sum of 7 are 2 + 5, 4 + 3, 5 + 2, and 6 + 1. So there are 4 possible ways out of a total of 36 possible sums, so the probability is $\dfrac{4}{36} = \dfrac{1}{9}$.

19. Since there is no dominance, only RR will result in red flowers. Thus, $P(\text{red}) = \dfrac{1}{4}$.

21. Since only rr will result in white flowers $P(\text{white}) = \dfrac{1}{4}$.

23. (a) Since round peas are dominant over wrinkled peas, the combinations RR, Rr, and rR will all result in round peas. Thus, $P(\text{round}) = \dfrac{3}{4}$.

    (b) Since wrinkled peas are recessive, only rr will result in wrinkled peas. Thus, $P(\text{wrinkled}) = \dfrac{1}{4}$.

25. Cystic fibrosis occurs in 1 of every 250,000 non-Caucasian births, so the empirical probability that cystic fibrosis will occur in a randomly selected non-Caucasian birth is $P = \dfrac{1}{250,000} = 0.000004$.

Construct a chart similar to Table 2 in the textbook and determine the probability of each of the following events.

		Second Parent	
	+	C	c
First Parent	C	CC	Cc
	c	cC	cc

27. C represents the normal (disease-free gene) and c represents the cystic fibrosis gene. Since c is a recessive gene, only the combination cc results in a child with the disease. Thus, the probability that their first child will have the disease is given by $P = \dfrac{1}{4}$.

29. Only the combination CC results in a child who neither has nor carries the disease, so the required probability is given by $P = \dfrac{1}{4}$.

Create a table that gives the possibilities when one parent is a carrier and the other is a non-carrier to answer Exercises 30–32.

		Second Parent	
	+	C	c
First Parent	C	CC	Cc
	C	CC	Cc

31. The combination Cc results in a child who is a healthy cystic fibrosis carrier. This combination occurs twice in the table (while the other combination that gives a carrier, cC, does not occur). Thus, the required probability is given by $P = \dfrac{2}{4} = \dfrac{1}{2}$.

33. Sickle-cell anemia occurs in about 1 of every 500 black baby births, so the empirical probability that a randomly selected black baby will have sickle-cell anemia is $P = \dfrac{1}{500} = 0.002$.

35. From Exercise 33, the probability that a particular black baby will have sickle-cell anemia is 0.002. Therefore, among 80,000 black baby births, about 0.002(80,000) = 160 occurrences of sickle-cell anemia would be expected.

For Exercise 37 let S represent the normal gene and s represent the sickle-cell gene. The possibilities for a child with parents who both have sickle-cell trait are given in the following table that gives the possibilities when one parent is a carrier and the other is a non-carrier.

		Second Parent	
+		S	s
First Parent	S	SS	Ss
	s	sS	ss

37. Since the combinations Ss and sS result in sickle-cell trait, the probability that the child will have sickle-cell trait is given by

$$P = \frac{1}{4} = \frac{1}{2}.$$

39. (a) The empirical probability formula is

$$P(E) = \frac{\text{number of times event occurred}}{\text{number of times experiment performed}}.$$

Since the number of times that the event described in the exercise has occurred is 0, the probability fraction has a numerator of 0. The denominator is some natural number n. Thus,

$$P(E) = \frac{0}{n} = 0.$$

(b) There is no basis for establishing a theoretical probability for this event.

(c) A woman may break the 10-second barrier at any time in the future, so it is possible that this event will occur.

41. Writing exercise; answers will vary.

One approach to Exercise 43 is to consider the following: Odds in favor $= \dfrac{a}{b}$; Odds against $= \dfrac{b}{a}$;

Probability of same event $= \dfrac{a}{a+b}$.

43. The odds in favor of event E are 12 to 19, where $a = 12$ and $b = 19$. Since

$$P(E) = \frac{a}{a+b}, \text{ we have}$$

$$P(E) = \frac{12}{12+19} = \frac{12}{31}.$$

45. From Table 1 in the text, there are 2,598,960 5-card poker hands. Of these 36 are straight flushes. Thus,

$$P(\text{straight flush}) = \frac{36}{2,598,960}$$
$$\approx 0.00001385.$$

Refer to Table 1 in the text. Answers are given to eight decimal places.

47. From Table 1 in the text, there are 2,598,960 5-card poker hands. Of these 624 are four of a kind. Thus,

$$P(\text{four of a kind}) = \frac{624}{2,598,960}$$
$$\approx 0.00024010.$$

49. Since there are 4 different suits, a hearts flush is $\dfrac{1}{4}$ of all possible non-royal and non-straight flushes. Thus, the probability is

$$\left(\frac{1}{4}\right) \cdot \left(\frac{5108}{2,598,960}\right) \approx 0.00049135.$$

51. Compare the area of the colored regions to the total area of the target. Using $\dfrac{22}{7}$ to approximate π, the areas of the colored regions are given by

blue: $\dfrac{22}{7}(2 \text{ ft})^2 - \dfrac{22}{7}\left(\dfrac{1}{2}\text{ ft}\right)^2 = \dfrac{165}{14}\text{ ft}^2$;

white: $\dfrac{22}{7}(4 \text{ ft})^2 - \dfrac{165}{14}\text{ ft}^2 = \dfrac{77}{2}\text{ ft}^2$;

red: $\dfrac{22}{7}(6 \text{ ft})^2 - \dfrac{22}{7}(4 \text{ ft})^2 = \dfrac{440}{7}\text{ ft}^2$.

The total area is given by

$$\pi(6 \text{ ft})^2 \approx \frac{22}{7} \cdot 36 \text{ ft}^2 = \frac{792}{7}\text{ ft}^2.$$

Thus, the probability of hitting a colored region is given by

$$P(\text{red}) = \left(\frac{440}{7}\right) \div \left(\frac{792}{7}\right)$$
$$= \left(\frac{440}{7}\right) \cdot \left(\frac{7}{792}\right)$$
$$= \frac{5}{9};$$

$$P(\text{white}) = \left(\frac{77}{2}\right) \div \left(\frac{792}{7}\right)$$
$$= \left(\frac{77}{2}\right) \cdot \left(\frac{7}{792}\right)$$
$$= \frac{49}{144};$$
$$P(\text{blue}) = \left(\frac{165}{14}\right) \div \left(\frac{792}{7}\right)$$
$$= \left(\frac{165}{14}\right) \cdot \left(\frac{7}{792}\right)$$
$$= \frac{5}{48}.$$

53. Use the fundamental counting principle to determine the number of favorable seating arrangements where each man will sit immediately to the left of his wife.
The first seat can be occupied by one of the three men, the second by his wife, the third by one of the two remaining men, etc. or, $3 \cdot 1 \cdot 2 \cdot 1 \cdot 1 \cdot 1 = 6$; since there are $6! = 720$ possible arrangements of the six people in the six seats, we have
$$P = \frac{6}{720} = \frac{1}{120} \approx 0.0083.$$

55. Use the fundamental counting principle to determine the number of ways the women can sit in three adjacent seats.
The first task is to decide in which seat the first woman is to sit. There are 4 choices (seats 1, 2, 3 or 4). Once this is decided, a second task would be to decide how many arrangements the three women can make sitting together (3!). The last task is to decide how many arrangements the three men could make sitting in the remaining three seats (3!). Thus, there are $4 \cdot 3! \cdot 3! = 144$ ways to accommodate the three women sitting together.
The probability of this occurring is given by
$$P = \frac{144}{720} = \frac{1}{5} = 0.2, \text{ where } 6! = 720 \text{ is the}$$
total number of seating arrangements possible.

57. Two distinct numbers are chosen randomly from the set $\left\{-2, -\frac{4}{3}, -\frac{1}{2}, 0, \frac{1}{2}, \frac{3}{4}, 3\right\}$.
To evaluate the probability that they will be the slopes of two perpendicular lines, find the size of the sample space and the size of the event of interest. The size of the sample space, $n(S)$, is given by $_7C_2 = 21$, since we

are choosing 2 items from a set of 7. The size of the event of interest, $n(E)$, is 2 (remember that perpendicular lines must have slopes that are the negative reciprocals of each other). These are either -2 and $\frac{1}{2}$ (either order) or $-\frac{4}{3}$ and $\frac{3}{4}$. Thus,
$$P = \frac{n(E)}{n(S)} = \frac{2}{21} \approx 0.095.$$

59. Since repetitions are not allowed and order is not important in selecting courses, use combinations. The number of ways of choosing any three courses from the list of twelve is $C(12, 3)$. Let F be the event of interest "all three courses selected are science courses." Then,
$$P(F) = \frac{\text{number of favorable outcomes}}{\text{total number of outcomes}}$$
$$= \frac{_5C_3}{_{12}C_3}$$
$$= \frac{10}{220}$$
$$= \frac{1}{22} \approx 0.045.$$

61. The total number of ways to make the three selections, in order, is given by
$$_{26}P_3 = 15,600.$$
Only 1 of these ways represents a success.
Thus, $P = \dfrac{1}{15,600} \approx 0.000064.$

63. The first eight primes are 2, 3, 5, 7, 11, 13, 17, 19. The sample space consists of all combinations of the set of 8 elements taken 2 at a time, the size of which is given by $_8C_2 = 28.$ Thus,
$E = \{19 + 5, 17 + 7, 13 + 11\}$ and
$$P = \frac{n(E)}{n(S)} = \frac{3}{28} \approx 0.107.$$

65. Two integers are randomly selected from the set $\{1, 2, 3, 4, 5, 6, 7, 8, 9\}$ and are added together. Find the probability that their sum is 11 if they are selected as follows:

(a) With replacement, the event of interest is $E = \{2 + 9, 9 + 2, 3 + 8, 8 + 3, 4 + 7, 7 + 4, 5 + 6, 6 + 5\}.$
Thus, $n(E) = 8$. Since there are

$9 \cdot 9 = 9^2 = 81$ ways of selecting the two digits to add together, we have $n(S) = 81$ and

$$P(\text{sum is eleven}) = \frac{8}{81} \approx 0.099.$$

(b) Without replacement, the event of interest is $E = \{2 + 9, 3 + 8, 4 + 7, 5 + 6\}$ and $n(E) = 4$. However (without replacement), $n(S) = C(9, 2) = 36$. Thus,

$$P(\text{sum is eleven}) = \frac{4}{36} = \frac{1}{9} \approx 0.111.$$

67. Only one number cannot be seen. If that number is six, the product of the remaining five sides is divisible by 6 because the factors 2 and 3 are part of the product. If the number that cannot be seen is not 6, then the product has a factor of 6 and is divisible by 6.

69. Let S be the sample space, which is the set of all two-digit numbers. Since the tens digit can be any of the nine digits 1 through 9 and the units digit can be any of the ten digits 0 through 9, $n(S) = 9 \cdot 10 = 90$.
Let E be the event "a palindromic two-digit number is chosen." A two-digit number will be palindromic only if both digits are the same. This repeated digit can be any of the nine digits 1 through 9, that is, $E = \{11, 22, 33, ..., 99\}$.
Thus, $n(E) = 9$, and

$$P(E) = \frac{n(E)}{n(S)} = \frac{9}{90} = \frac{1}{10}.$$

71. The number of possible groupings is $5 \cdot 3 = 15$. If A and B are in the same group and C and D are in the same group, then E and F must be in the same group, and there is only one possible grouping. Thus the probability of this event is $\frac{1}{15}$.

11.2 Exercises

1. Yes, since event A and event B cannot happen at the same time.

3. Writing exercise; answers will vary.

Use the sample space $S = \{1, 2, 3, 4, 5, 6\}$ for Exercises 5–9.

5. Let E be the event "not prime." Then, $E = \{1, 4, 6\}$, the non-prime numbers in S.
Thus, $P(E) = \frac{3}{6} = \frac{1}{2}.$

7. Let $E = \{2, 4, 6\}$ and $F = \{2, 3, 5\}$.
$$P(E \text{ or } F) = P(E) + P(F) - P(E \text{ and } F)$$
$$= \frac{3}{6} + \frac{3}{6} - \frac{1}{6}$$
$$= \frac{5}{6},$$
by the general addition rule of probability.

9. Let A be the event "less than 3" and B be the event "greater than 4." Thus, $A = \{1, 2\}$ and $B = \{5, 6\}$.
Since A and B are mutually exclusive events, use the special addition rule of probability:

$$P(A \text{ or } B) = P(A) + P(B) = \frac{2}{6} + \frac{2}{6} = \frac{4}{6} = \frac{2}{3}.$$

11. (a) Since the two events, drawing a king (K) and drawing a queen (Q) are mutually exclusive, use the special addition rule:
$$P(K \text{ or } Q) = P(K) + P(Q)$$
$$= \frac{4}{52} + \frac{4}{52}$$
$$= \frac{8}{52}$$
$$= \frac{2}{13}.$$

(b) Use the formula for finding odds in favor of an event E:

Odds in favor of $E = \dfrac{P(E)}{P(E')}$. Thus,

$$\frac{P(E)}{P(E')} = \frac{P(K \text{ or } Q)}{P(\text{not } (K \text{ or } Q))}$$
$$= \frac{P(K \text{ or } Q)}{1 - P(K \text{ or } Q)}$$
$$= \frac{\frac{2}{13}}{1 - \left(\frac{2}{13}\right)}$$
$$= \frac{\frac{2}{13}}{\frac{11}{13}}$$
$$= \frac{2}{13} \cdot \frac{13}{11}$$
$$= \frac{2}{11}, \text{ or 2 to 11.}$$

13. (a) Let S be the event of "drawing a spade" and F be the event of "drawing a face card." Then, $n(S) = 13$ and $n(F) = 12$. (There are 3 face cards in each of the 4 suits.) There are 3 face cards that are also spades so that $n(S$ and $F) = 3$. Thus, by the general additive rule for probability,

$$P(S \text{ or } F) = P(S) + P(F) - P(S \text{ and } F)$$
$$= \frac{13}{52} + \frac{12}{52} - \frac{3}{52}$$
$$= \frac{11}{26}.$$

(b) Use the formula for finding odds in favor of an event E:

Odds in favor of $E = \dfrac{P(E)}{P(E')}$. Thus,

$$\frac{P(E)}{P(E')} = \frac{P(S \text{ or } F)}{P(\text{not }(S \text{ or } F))}$$
$$= \frac{P(S \text{ or } F)}{1 - P(S \text{ or } F)}$$
$$= \frac{\frac{11}{26}}{1 - \left(\frac{11}{26}\right)}$$
$$= \frac{\frac{11}{26}}{\frac{15}{26}}$$
$$= \frac{11}{26} \cdot \frac{26}{15}$$
$$= \frac{11}{15}, \text{ or } 11 \text{ to } 15.$$

15. (a) Let H be the event "a heart is drawn" and S be the event "a seven is drawn." We want to find $P(H \text{ or } S)'$, or $P(H \cup S)'$. Since there are 13 hearts in the deck and 3 other sevens which are not hearts, $n(H \cup S) = 16$ and

$$P(H \cup S) = \frac{16}{52} = \frac{4}{13}.$$

Thus, $P(H \cup S)' = 1 - P(H \cup S)$
$$= 1 - \frac{4}{13}$$
$$= \frac{9}{13}.$$

(b) The number of cards which are hearts or are sevens totals 16. Thus the number of cards which are not hearts nor sevens is $52 - 16 = 36$. The odds in favor of not hearts nor sevens are 36 to 16, or 9 to 4.

Construct a table showing the sum for each of the 36 equally likely outcomes.

		2nd die				
+	1	2	3	4	5	6
1	2	3	4	5	6	7
2	3	4	5	6	7	8
1st die 3	4	5	6	7	8	9
4	5	6	7	8	9	10
5	6	7	8	9	10	11
6	7	8	9	10	11	12

17. Let E be the event of getting a sum which is an even number. Counting the number of occurrences in the sum table for these even outcomes represents the numerator of the probability fraction. Then, $P(E) = \dfrac{18}{36}$.

Let M be the event of getting sums which are multiples of three, $\{3, 6, 9, 12\}$. Counting the number of occurrences in the sum table for these outcomes represents the numerator of the probability fraction.

Then, $P(M) = \dfrac{12}{36}$ and $P(E \text{ and } M) = \dfrac{6}{36}$.

Thus, by the general addition rule,
$$P(E \cup M) = P(E) + P(M) - P(E \text{ and } M)$$
$$= \frac{18}{36} + \frac{12}{36} - \frac{6}{36}$$
$$= \frac{24}{36}$$
$$= \frac{2}{3}.$$

19. Since these are mutually exclusive events, use the special additive rule. Since there is only one sum less than 3 (the sum of 2),

$P(\text{sum less than } 3) = \dfrac{1}{36}$, and since there are six sums greater than 9,

$P(\text{sum greater than } 9) = \dfrac{6}{36}$.

Thus, $P(\text{sum less than 3 or greater than 9})$
$$= \frac{1}{36} + \frac{6}{36}$$
$$= \frac{7}{36}.$$

21. $P(S) = P(A \cup B \cup C \cup D)$
$$= P(A) + P(B) + P(C) + P(D)$$
$$= 1$$

Refer to Table 1 (11.1 in the textbook) and give answers to six decimal places.

23. Let F be the event of drawing a full house and S be the event of drawing a straight. Using Table 1 in the text to determine $n(F) = 3744$ and $n(S) = 10,200$, we have

$$P(F) = \frac{3744}{2,598,960} \text{ and } P(S) = \frac{10,200}{2,598,960}.$$

Since the events are mutually exclusive,
$P(F \text{ or } S) = P(F) + P(S)$

$$= \frac{3,744}{2,598,960} + \frac{10,200}{2,598,960}$$

$$= \frac{13,944}{2,598,960} \approx 0.005365.$$

25. The events are mutually exclusive, so use the special addition rule:
$P(\text{nothing any better than two pairs})$
$= P(\text{no pair}) + P(\text{one pair}) + P(\text{two pairs})$

$$= \frac{1,302,540}{2,598,960} + \frac{1,098,240}{2,598,960} + \frac{123,552}{2,598,960}$$

$$= \frac{2,524,332}{2,598,960} \approx 0.971285$$

27. "Par or above" is represented by all categories from 70 up. Since these are mutually exclusive, use the special addition rule:
$P(\text{Par or above})$
$= 0.30 + 0.23 + 0.09 + 0.06 + 0.04 + 0.03$
  $+ 0.01$
$= 0.76$

29. "Less than 90" is represented by all categories under the 90–94 category. Since these are mutually exclusive, use the special addition rule:
$P(\text{Less than 90})$
$= 0.04 + 0.06 + 0.14 + 0.30 + 0.23 + 0.09$
  $+ 0.06$
$= 0.92$

31. Odds of Brian's shooting below par

$$= \frac{P(\text{below par})}{P(\text{par or above})}$$

$P(\text{par or above}) = 0.76$ (Exercise 27), and
$P(\text{below par}) = 1 - P(\text{par or above})$
$$= 1 - 0.76$$
$$= 0.24,$$

by the complements rule of probability.

Thus, odds in favor $= \dfrac{0.24}{0.76} = \dfrac{6}{19}$, or 6 to 19.

33. Let x denote the sum of two distinct numbers selected randomly from the set $\{1, 2, 3, 4, 5\}$. Construct the probability distribution for the random variable x. Create a "sum table" to list the elements in the sample space. Note: can't use $1 + 1 = 2$, etc. Why?

+	1	2	3	4	5
1	–	3	4	5	6
2	3	–	5	6	7
3	4	5	–	7	8
4	5	6	7	–	9
5	6	7	8	9	–

Thus, the probability distribution is as follows.

x	$P(x)$
3	$\frac{2}{20} = 0.1$
4	$\frac{2}{20} = 0.1$
5	$\frac{4}{20} = 0.2$
6	$\frac{4}{20} = 0.2$
7	$\frac{4}{20} = 0.2$
8	$\frac{2}{20} = 0.1$
9	$\frac{2}{20} = 0.1$

For Exercises 35–37, let A be an event within the sample space S, and let $n(A) = a$ and $n(S) = s$.

35. $n(A') + n(A) = n(S)$
  $n(A') + a = s$
Thus, $n(A') = s - a.$

37. $P(A) + P(A') = \dfrac{a}{s} + \dfrac{s-a}{s} = \dfrac{a+s-a}{s} = \dfrac{s}{s} = 1$

We want to form three-digit numbers using the set of digits $\{0, 1, 2, 3, 4, 5\}$. For example, 501 and 224 are such numbers but 035 is not.

39. The number of three-digit numbers is, by the fundamental counting principle, $5 \cdot 6 \cdot 6 = 180$. Remember that we can't choose "0" for the first digit.

41. If one three-digit number is chosen at random from all those that can be made from the above set of digits, find the probability that the one chosen is not a multiple of 5.

The number of three-digit numbers is, by the fundamental counting principle, $5 \cdot 6 \cdot 6 = 180$ (Exercise 39).
There are $5 \cdot 6 \cdot 2 = 60$ three-digit numbers that are multiples of 5 (Exercise 40). Thus,

$$P(\text{multiple of 5}) = \frac{60}{180} = \frac{1}{3}.$$

By the complements rule,

$$P(\text{not a multiple of 5}) = 1 - P(\text{multiple of 5})$$
$$= 1 - \frac{60}{180}$$
$$= 1 - \frac{1}{3}$$
$$= \frac{2}{3}.$$

43. Since box C contains the same number of green marbles as blue and the number of green and blue marbles was the same to begin with, then box A and box B must contain exactly the same number of marbles after all the marbles are drawn. Because this is certain, the probability is 1.

11.3 Exercises

1. The events are independent since the outcome on the first toss has no effect on the outcome of the second toss.

3. The two planets are selected, without replacement, from the list in Table 7. The events "first is closer than Jupiter" and "second is farther than Neptune" are not independent. This is because the first selection was not replaced, which may affect the outcome of the second choice.

5. The answers are all guessed on a twenty-question multiple choice test. Let A be the event "first answer correct" and let B be the event "last answer correct." The events A and B are independent since the first answer choice does not affect the last answer choice.

7. The probability that the student selected is "female" is $\frac{52}{100} = \frac{13}{25}.$

9. Since a total of $100 - 31 = 69$ of the 100 students are "not motivated primarily by money," the probability is $\frac{69}{100}.$

11. Given that the student selected is motivated primarily by "sense of giving to society," the sample space is reduced to 31 students. Of these 14 are male, so the probability is $\frac{14}{31}.$

In Exercises 13–15 the first puppy chosen is replaced before the second is chosen. Note that "with replacement" means that the events may be considered independent and we can apply the special multiplication rule of probability.

13. The probability that "both select a poodle" is $P(P_1 \text{ and } P_2) = P(P_1) \cdot P(P_2)$

$$= \left(\frac{4}{7}\right) \cdot \left(\frac{4}{7}\right)$$
$$= \frac{16}{49}.$$

15. The probability that "Rebecka selects a terrier and Aaron selects a retriever" is given by $P(T_1 \text{ and } R_2) = P(T_1) \cdot P(R_2)$

$$= \left(\frac{2}{7}\right) \cdot \left(\frac{1}{7}\right)$$
$$= \frac{2}{49}.$$

In Exercises 17–21, the first puppy chosen is not replaced before the second is chosen. Thus, the events are not independent. Therefore, apply the general multiplication rule of probability.

17. The probability that "both select a poodle" is given by $P(P_1 \text{ and } P_2) = P(P_1) \cdot P(P_2|P_1)$

$$= \left(\frac{4}{7}\right) \cdot \left(\frac{3}{6}\right)$$
$$= \frac{2}{7}.$$

Remember that the second probability is conditional to the first event as having occurred and hence, the sample space is reduced.

19. The probability that "Aaron selects a retriever," given "Rebecka selects a poodle" is found by $P(R_2|P_1) = \frac{1}{6}.$

Note that Rebecka's choice decreased the sample space by one dog.

21. The probability that "Aaron selects a retriever," given "Rebecka selects a retriever" is found by $P(R_2|R_1) = \dfrac{0}{6} = 0$.

Note that after Rebecka's selection, there are no remaining retrievers for Aaron to select.

23. Since the cards are dealt without replacement, when the second card is drawn, there will be 51 cards left, of which 12 are spades. Thus, $P(S_2|S_1) = \dfrac{12}{51} = \dfrac{4}{17}$.

Note that both the event of interest (numerator) and the sample space (denominator) were reduced by the selection of the first card, a spade.

25. Since the cards are dealt without replacement, the events "first is face card" and "second is face card" are not independent, so be sure to use the general multiplication rule of probability. Let F_1 be the event "first is a face card" and F_2 be the event "second is a face card." Then,

$$
\begin{aligned}
P(\text{two face cards}) &= P(F_1 \text{ and } F_2) \\
&= P(F_1) \cdot P(F_2|F_1) \\
&= \frac{12}{52} \cdot \frac{11}{51} \\
&= \frac{3}{13} \cdot \frac{11}{51} \\
&= \frac{11}{221}.
\end{aligned}
$$

27. The probability that the "first card dealt is a jack and the second is a face card" is found by $P(J_1 \text{ and } F_2) = P(J_1) \cdot (F_2|J_1)$

$$
\begin{aligned}
&= \frac{4}{52} \cdot \frac{11}{51} \\
&= \frac{11}{663}.
\end{aligned}
$$

Remember that there are only 11 face cards left once the jack is drawn.

Use the results of Exercise 28 to find each of the following probabilities when a single card is drawn from a standard 52-card deck.

29. $P(\text{queen}|\text{face card}) = \dfrac{n(F \text{ and } Q)}{n(F)} = \dfrac{4}{12} = \dfrac{1}{3}$

31. $P(\text{red}|\text{diamond}) = \dfrac{n(D \text{ and } R)}{n(D)} = \dfrac{13}{13} = 1$

33. From the integers 1 through 10, the set of primes are $\{2, 3, 5, 7\}$. The set of odds are $\{1, 3, 5, 7, 9\}$. Since there are only three odd numbers in the set of 4 primes, the second probability fraction becomes $\dfrac{3}{4}$ and

$$P(\text{prime}) \cdot P(\text{odd}|\text{prime}) = \frac{4}{10} \cdot \frac{3}{4} = \frac{3}{10}.$$

This is the same value as computed in the text for $P(\text{odd}) \cdot P(\text{prime}|\text{odd})$, that is, the probability of selecting an integer from the set which is "odd and prime."

35. Since the birth of a boy (or girl) is independent of previous births, the probability of both authors having three boys successively is

$$\left(\frac{1}{2}\right) \cdot \left(\frac{1}{2}\right) \cdot \left(\frac{1}{2}\right) \cdot \left(\frac{1}{2}\right) \cdot \left(\frac{1}{2}\right) \cdot \left(\frac{1}{2}\right) = \frac{1}{64}.$$

37. Let S represent a sale purchase for more than \$100. Then,
$P(\text{both sales more than \$100})$
$= P(S_1 \text{ and } S_2)$
$= P(S_1) \cdot P(S_2)$
$= (0.70) \cdot (0.70)$
$= 0.490$

39. Let S represent a sale purchase for more than \$100. Then, S' represents a sale purchase that is not more than \$100. Since $P(S) = 0.70$, $P(S') = 1 - 0.70 = 0.30$, by the complements principle.
$P(\text{none of the first 3 sales more than \$100})$
$= P(\text{not } S_1 \text{ and not } S_2 \text{ and not } S_3)$
$= P(S_1') \cdot P(S_2') \cdot P(S_3')$
$= (0.30) \cdot (0.30) \cdot (0.30)$
$= 0.027$

41. Since the probability the cloud will move in the critical direction is 0.05, the probability that it will not move in the critical direction is 1 0.05 = 0.95, by the complements formula.

43. The probability that the cloud would not move in the critical direction for each launch is $1 - 0.05 = 0.95$. The probability that the cloud would not move in the critical direction for any 5 launches is $(0.95)^5$ by the special multiplication rule. The probability that any 5 launches will result in at least one cloud movement in the critical direction is the complement of the

probability that a cloud would not move in the critical direction for any 5 launches, or

$1-(0.95)^5 \approx 0.23$.

Three men and three women are waiting to be interviewed for jobs. If they are all selected in random order, find the probability of each of the following events.

45. P(all women first)

$= P(W_1) \cdot P(W_2|W_1) \cdot P(W_3|W_1 \text{ and } W_2)$

$= \dfrac{3}{6} \cdot \dfrac{2}{5} \cdot \dfrac{1}{4}$

$= \dfrac{1}{20}$

47. The probability that no man will be interviewed until at least two women have been interviewed is the same as the probability that the first two people interviewed are women. That is,

P(two women first) $= P(W_1) \cdot P\left(W_2|W_1\right)$

$= \dfrac{3}{6} \cdot \dfrac{2}{5}$

$= \dfrac{1}{5}$

49. The probability of selecting the fair coin out of the two coins is $\dfrac{1}{2}$. The probability of getting two heads using either coin is $\dfrac{2}{5}$.

So, the probability that the gambler selected the fair coin knowing he got two heads is

$\dfrac{1}{2} \cdot \dfrac{2}{5} = \dfrac{1}{5}$.

51. (a) "At least three" is the complement of "one or two" and we can't get two girls from one birth. Thus, find only the probability of having 2 girls with two births.

$P(\text{gg}) = P(\text{g}) \cdot P(\text{g}) = \dfrac{1}{2} \cdot \dfrac{1}{2} = \dfrac{1}{4}$.

Therefore, by the complements principle.

$P(\text{at least three births}) = 1 - \dfrac{1}{4} = \dfrac{3}{4}$.

(b) "At least four births" is the complement of "two or three births." From (a) above

$P(\text{two births to get gg}) = \dfrac{1}{4}$.

To calculate P(three births to get gg) examine Figure 2 in the text. The three successes are ggb, gbg, and bgg. Don't count, however, the outcome ggb as a success since this has already been counted when computing the probability associated with two births.

$P(\text{three births to get gg}) = \dfrac{2}{8} = \dfrac{1}{4}$.

P(two or three births to get gg)

$= P(\text{two births to get gg})$

$\qquad + P(\text{three births to get gg})$

$= \dfrac{1}{4} + \dfrac{1}{4}$

$= \dfrac{2}{4}$

$= \dfrac{1}{2}$

Finally, the probability of "at least four births" may be calculated by the complements rule:

$P(\text{at least four births to get gg}) = 1 - \dfrac{1}{2}$

$= \dfrac{1}{2}$.

(c) "At least five births" is the complement of "two or three or four births."

P(two or three or four births to obtain gg)

$= P(\text{two or three births to get gg})$

$\qquad + P(\text{four births to get gg})$

$P(\text{two or three births to get gg}) = \dfrac{1}{2}$

was calculated in (b) above. To calculate P(four births to get gg) extend the tree diagram (Figure 2 in the text). Count all of the outcomes with gg (two girls) as successes except bggb and ggbb since they have already been used in calculating the earlier probability. There are then 3 outcomes which may be considered as successes. Thus,

$P(\text{four births to get gg}) = \dfrac{3}{16}$.

P(two or three or four births to
      obtain gg)
$= P$(two or three births to get gg)
          $+ P$(four births to get gg)

$$= \frac{1}{2} + \frac{3}{16}$$
$$= \frac{8}{16} + \frac{3}{16}$$
$$= \frac{11}{16}$$

Finally, the probability of "at least five births to obtain gg" may be calculated by the complements rule:
P(at least five births to obtain gg)

$$= 1 - \frac{11}{16}$$
$$= \frac{5}{16}.$$

A coin, biased so that P(h) = 0.5200 and P(t) = 0.4800, it tossed twice. Give answers to four decimal places.

53. Since the two tosses are independent events, use the special multiplication rule:
$P(hh) = P(h) \cdot P(h)$
      $= (0.5200) \cdot (0.5200)$
      $= 0.2704.$

55. Since the two tosses are independent events, use the special multiplication rule:
$P(th) = P(t) \cdot P(h)$
      $= (0.4800) \cdot (0.5200)$
      $= 0.2496.$

57. Writing exercise; answers will vary.

59. Using the fundamental counting principle, there are $2 \cdot 2 \cdot 2 \cdot 2 \cdot 2 \cdot 2 = 2^6 = 64$ different switch settings. The probability of randomly getting 1 of the 64 possible settings is
$$\frac{1}{64} \approx 0.0156.$$

61. P(at least one duplication of switch settings)
$= 1 - P$(no duplication)
$$= 1 - \frac{63}{64}$$
≈ 0.016 for two neighbors;
$$1 - \frac{63}{64} \cdot \frac{62}{64} = 1 - \frac{63 \cdot 62}{(64^2)}$$
      ≈ 0.046 for three neighbors;

$$1 - \frac{63 \cdot 62 \cdot 61 \cdot 60 \cdot 59 \cdot 58 \cdot 57 \cdot 56}{(64)^8}$$
≈ 0.445 for nine neighbors;
and
$$1 - \frac{63 \cdot 62 \cdot 61 \cdot 60 \cdot 59 \cdot 58 \cdot 57 \cdot 56 \cdot 55}{(64)^9}$$
$\approx 0.523 > \frac{1}{2}$ for ten neighbors.

Thus, the minimum number of neighbors who must use this brand of opener before the probability of at least one duplication of settings is greater than $\frac{1}{2}$ is ten.

63. Since the events are not independent, use the general multiplication rule. Let R_1 represent rain on the first day, R_2 represent rain on the second day.
P(rain on two consecutive days in
    November)
$= P(R_1 \cap R_2)$
$= P(R_1) \cdot P(R_2|R_1)$
$= (0.500) \cdot (0.800)$
$= 0.400$

65. Since the events are not independent, use the general multiplication rule. Let R_1 represent rain on November 1, R_2 represent rain on November 2, and R_3 represent rain on November 3.
P(rain on November 1st and 2nd, but not on the 3rd)
$= P(R_1 \cap R_2 \cap \text{not } R_3)$
$= P(R_1) \cdot P(R_2|R_1) \cdot P(\text{not } R_3|R_2)$
$= (0.500) \cdot (0.800) \cdot (1 - 0.800)$
$= 0.080$
Note that the probability of not raining after a rainy day is the complement of the probability that it does rain.

67. To find the probability of "no engine failures," begin by letting F represent a failed engine. The probability that a given engine will fail, $P(F)$, is 0.10. This means that, by the complements principle, the probability that an engine will not fail is given by $P(\text{not } F) = 1 - 0.10 = 0.90$. Since engines "not failing" are independent events, use the special product rule.

P(no engine failures)

$= P(\text{not } F_1 \cap \text{not } F_2 \cap \text{not } F_3 \cap \text{not } F_4)$

$= P(\text{not } F_1) \cdot P(\text{not } F_2) \cdot P(\text{not } F_3)$
$\qquad \cdot P(\text{not } F_4)$

$= (0.90)^4$

$= 0.6561$

69. The probability of "exactly two engine failures" can be found by applying the fundamental counting principle, where the first task is to decide which two engines fail $_4C_2$; the second task, find the probability that one of these engines fails (0.10); followed by the second engine failing (0.20); followed by finding the probability the third engine does not fail $(1 - 0.30 = 0.70)$; followed by the probability that last engine does not fail $(1 - 0.30 = 0.70)$. Thus,

$P = {}_4C_2 \cdot (0.10) \cdot (0.20) \cdot (0.70)^2$

$\quad = 0.0588.$

Refer to text discussion of the rules for "one-and-one" basketball. Christine Ellington, a basketball player, has a 70% foul shot record. (She makes 70% of her foul shots.) Find the probability that, on a given one-and-one foul shooting opportunity, Christine will score the following number of points.

71. A one-and-one foul shooting opportunity means that Christine gets a second shot only if she makes her first shot. The probability of scoring "no points" means that she missed her first shot. Thus,

$P(\text{scoring no points})$

$= 1 - P(\text{scoring at least one point})$

$= 1 - 0.70$

$= 0.30.$

73. The probability of "scoring two points" is given by

$P(\text{scoring two points})$

$= P(\text{scoring the 1st shot and scoring the}$
$\qquad \text{2nd shot})$

$= P(\text{scoring on 1st shot}) \cdot P(\text{scoring on}$
$\qquad \text{2nd shot})$

$= (0.70) \cdot (0.70)$

$= 0.49.$

75. Writing exercise; answers will vary.

11.4 Exercises

1. Let heads be "success."

Then $n = 3$, $p = q = \dfrac{1}{2}$, and $x = 0$.

By the binomial probability formula,

$P(0) = {}_3C_0 \cdot \left(\dfrac{1}{2}\right)^0 \left(\dfrac{1}{2}\right)^{3-0}$

$\qquad = \dfrac{3!}{0!(3-0)!} \cdot 1 \cdot \left(\dfrac{1}{2}\right)^3$

$\qquad = \dfrac{3!}{1 \cdot 3!} \cdot \dfrac{1}{2^3}$

$\qquad = 1 \cdot \dfrac{1}{8}$

$\qquad = \dfrac{1}{8}.$

Note $_3C_0 = 1$, and we could easily reason this result without using the combination formula since there is only one way to choose 0 things from a set of 3 things—take none out.

3. Let heads be "success."

Then $n = 3$, $p = q = \dfrac{1}{2}$, and $x = 2$.

By the binomial probability formula,

$P(2 \text{ heads}) = {}_3C_2 \cdot \left(\dfrac{1}{2}\right)^2 \cdot \left(\dfrac{1}{2}\right)^{3-2}$

$\qquad = \dfrac{3!}{2!(3-2)!} \cdot \dfrac{1}{4} \cdot \dfrac{1}{2}$

$\qquad = \dfrac{3 \cdot 2 \cdot 1}{2 \cdot 1 \cdot 4 \cdot 2}$

$\qquad = \dfrac{3}{8}.$

5. Use the special addition rule for calculating the probability of "1 or 2 heads."

$P(1 \text{ or } 2 \text{ heads})$

$= P(1) + P(2)$

$= {}_3C_1 \cdot \left(\dfrac{1}{2}\right)^1 \cdot \left(\dfrac{1}{2}\right)^{3-1} + {}_3C_2 + \left(\dfrac{1}{2}\right)^2 \cdot \left(\dfrac{1}{2}\right)^{3-2}$

$= 3 \cdot \dfrac{1}{2} \cdot \dfrac{1}{4} + 3 \cdot \dfrac{1}{4} \cdot \dfrac{1}{2}$

$= \dfrac{6}{8}$

$= \dfrac{3}{4}$

7. "No more than 1" is the same as "0 or 1."
$P(0 \text{ or } 1)$
$= P(0) + P(1)$
$$= {}_3C_0 \cdot \left(\frac{1}{2}\right)^0 \left(\frac{1}{2}\right)^3 + {}_3C_1 \cdot \left(\frac{1}{2}\right)^1 \left(\frac{1}{2}\right)^2$$
$$= 1 \cdot 1 \cdot \frac{1}{8} + 3 \cdot \frac{1}{2} \cdot \frac{1}{4}$$
$$= \frac{1}{8} + \frac{3}{8}$$
$$= \frac{1}{2}$$

9. Assuming boy and girl babies are equally likely, find the probability that a family with three children will have exactly two boys.
$$P(2 \text{ boys}) = {}_3C_2 \cdot \left(\frac{1}{2}\right)^2 \left(\frac{1}{2}\right)^1 = 3 \cdot \frac{1}{4} \cdot \frac{1}{2} = \frac{3}{8}$$

11. If n fair coins are tossed, the probability of exactly x heads is the fraction whose numerator is entry number $\underline{x}$ of row number $\underline{n}$ in Pascal's triangle, and whose denominator is the sum of the entries in row number $\underline{n}$. That is x; n; n.

For Exercises 13–19, refer to Pascal's triangle. Since seven coins are tossed, we will use row number 7 of the triangle. (Recall that the first row is row number 0 and that the first entry in each row is entry number 0.) The sum of the numbers in row 7 is $2^7 = 128$, which will be the denominator in each of the probability fractions.

```
            1
          1   1
        1   2   1
      1   3   3   1
    1   4   6   4   1
  1   5  10  10   5   1
1   6  15  20  15   6   1
1  7  21  35  35  21   7  1
```

13. For the probability "1 head," the numerator of the probability fraction is entry 1 of row 7 of Pascal's triangle, or 7, and the denominator is the sum of the elements in row 7, or
$1 + 7 + 21 + 35 + 35 + 21 + 7 + 1 = 128.$
Thus, $P(1 \text{ head}) = \dfrac{7}{128}$.

15. For he probability "3 heads," the numerator of the probability fraction is entry number 3 of row 7 of Pascal's triangle, or 35, and the denominator is the sum of the elements in row 7, or 128. Thus, $P(3 \text{ heads}) = \dfrac{35}{128}$.

17. For the probability "5 heads," the numerator of the probability fraction is entry number 5 of row 7 of Pascal's triangle, or 21, and the denominator is the sum of the elements in row 7, or 128. Thus, $P(5 \text{ heads}) = \dfrac{21}{128}$.

19. For the probability "7 heads," the numerator of the probability fraction is entry 7 of row 7 of Pascal's triangle, or 1, and the denominator is the sum of the elements in row 7, or 128. Thus, $P(7 \text{ heads}) = \dfrac{1}{128}$.

For Exercises 21–23, a fair die is rolled three times and a 4 is considered "success," while all other outcomes are "failures."

21. Here $n = 3$, $p = \dfrac{1}{6}$, $q = \dfrac{5}{6}$, and $x = 1$.
$$P(1) = {}_3C_1 \cdot \left(\frac{1}{6}\right)^1 \left(\frac{5}{6}\right)^2 = 3 \cdot \frac{1}{6} \cdot \frac{25}{36} = \frac{25}{72}$$

23. Here $n = 3$, $p = \dfrac{1}{6}$, $q = \dfrac{5}{6}$, and $x = 3$.
$$P(3) = {}_3C_3 \cdot \left(\frac{1}{6}\right)^3 \left(\frac{5}{6}\right)^0 = 1 \cdot \frac{1}{216} \cdot 1 = \frac{1}{216}$$

Answers are rounded to three decimal places.

25. Here $n = 5$, $p = \dfrac{1}{3}$, and $x = 4$. Since $p = \dfrac{1}{3}$,
$$q = 1 - p = 1 - \frac{1}{3} = \frac{2}{3}.$$
Substitute these values into the binomial probability formula:
$$P(4) = {}_5C_4 \cdot \left(\frac{1}{3}\right)^4 \left(\frac{2}{3}\right)^1 = 5 \cdot \frac{2}{3^5} \approx 0.041$$

27. Here $n = 20$, $p = \dfrac{1}{8}$, and $x = 2$. Since
$$p = \frac{1}{8}, \quad q = 1 - p = 1 - \frac{1}{8} = \frac{7}{8}.$$
Substitute these values into the binomial

probability formula:

$$P(2) = {}_{20}C_2 \cdot \left(\frac{1}{8}\right)^2 \left(\frac{7}{8}\right)^{18}$$

$$= \frac{20!}{2!18!} \cdot \frac{7^{18}}{8^2 8^{18}}$$

$$= 190 \cdot \frac{7^{18}}{8^{20}}$$

$$\approx 0.268$$

29. Writing exercise; answers will vary.

31. Writing exercise; answers will vary.

For Exercises 33–35, let a correct answer be a "success." Then $n = 10$, $p = \frac{2}{6} = \frac{1}{3}$, and $q = 1 - p = \frac{2}{3}$.

33. The probability of getting "exactly 7 correct answers" is given by

$$P(7 \text{ correct answers }) = {}_{10}C_7 \cdot \left(\frac{1}{3}\right)^7 \left(\frac{2}{3}\right)^3$$

$$= 120 \cdot \frac{2^3}{3^{10}}$$

$$\approx 0.016.$$

35. "At least seven" means seven, eight, nine, or ten correct answers.
$P(7 \text{ or } 8 \text{ or } 9 \text{ or } 10)$
$= P(7) + P(8) + P(9) + P(10)$

$$= {}_{10}C_7 \cdot \left(\frac{1}{3}\right)^7 \left(\frac{2}{3}\right)^3 + {}_{10}C_8 \cdot \left(\frac{1}{3}\right)^8 \left(\frac{2}{3}\right)^2$$

$$+ {}_{10}C_9 \cdot \left(\frac{1}{3}\right)^9 \left(\frac{2}{3}\right)^1 + {}_{10}C_{10} \cdot \left(\frac{1}{3}\right)^{10} \left(\frac{2}{3}\right)^0$$

$$\approx 0.01626 + 0.00305 + 0.00034 + 0.00002$$

$$\approx 0.01967$$

$$\approx 0.020$$

37. For "exactly 1 to have undesirable side effects," $x = 1$ and $n = 8$, $p = 0.35$, $q = 1 - p = 0.65$. Thus,

$$P(1) = {}_8C_1 \cdot (0.35)^1 (1 - 0.35)^{8-1}$$

$$= 8(0.35)(0.65)^7$$

$$\approx 0.137$$

39. "More than two" is the complement of 0, 1, or 2. Thus,
$P(\text{more than two})$
$= 1 - P(0, 1, \text{ or } 2)$
$= 1 - [P(0) + P(1) + P(2)]$
$= 1 - [{}_8C_0 \cdot (0.35)^0 (0.65)^8$
$\qquad + {}_8C_1 \cdot (0.35)^1 (0.65)^7$
$\qquad + {}_8C_2 \cdot (0.35)^2 (0.65)^6]$
$\approx 1 - [1 \cdot 1 \cdot (0.03186448)$
$\qquad + 8(0.35)(0.04902228)$
$\qquad + (28)(0.1225)(0.07541889)]$
≈ 0.572

41. For the probability that "from 4 through 6" will attend college, $n = 9$, $p = 0.60$, and $q = 1 - p = 0.40$. Thus,
$P(4 \text{ or } 5 \text{ or } 6) = P(4) + P(5) + P(6)$

$$= {}_9C_4 \cdot (0.60)^4 (0.40)^5$$

$$+ {}_9C_5 \cdot (0.60)^5 (0.40)^4$$

$$+ {}_9C_6 \cdot (0.60)^6 (0.40)^3$$

$$\approx 0.167 + 0.251 + 0.251$$

$$\approx 0.669$$

43. "At least 3" means three, four, five, six, seven, eight or nine enroll in college.
$P(3 \text{ or } 4 \text{ or } 5 \text{ or } 6 \text{ or } 7 \text{ or } 8 \text{ or } 9)$
$= P(3) + P(4) + P(5) + P(6) + P(7)$
$\qquad + P(8) + P(9)$

$$= {}_9C_3 \cdot (0.60)^3 (0.40)^6$$

$$+ {}_9C_4 \cdot (0.60)^4 (0.40)^5$$

$$+ {}_9C_5 \cdot (0.60)^5 (0.40)^4$$

$$+ {}_9C_6 \cdot (0.60)^6 (0.40)^3$$

$$+ {}_9C_7 \cdot (0.60)^7 (0.40)^2$$

$$+ {}_9C_8 \cdot (0.60)^8 (0.40)^1$$

$$+ {}_9C_9 \cdot (0.60)^9 (0.40)^0$$

$$\approx 0.074 + 0.167 + 0.251 + 0.251 + 0.161$$

$$+ 0.060 + 0.010$$

$$\approx 0.974$$

45. "At least half of the 6 trees" is the complement of "0, 1, or 2 trees." Here $p = 0.65$, $q = 1 - 0.65 + 0.35$ and $n = 6$. Using the complements rule, the probability is found by

P(at least half)
$= 1 - P(0 \text{ or } 1 \text{ or } 2)$
$= 1 - [P(0) + P(1) + P(2)]$
$= 1 - [_6C_0 \cdot (0.65)^0 (0.35)^6$
$\qquad + _6C_1 \cdot (0.65)^1 (0.35)^5$
$\qquad + _6C_2 \cdot (0.65)^2 (0.35)^4]$
$\approx 1 - [0.0018 + 0.0205 + 0.0951]$
$\approx 1 - 0.1174$
$\approx 0.883.$

47. If p is the probability that a ball selected at random is black, $x = 2$, and $n = 4$, then
P(two of four balls are black)
$= _4C_2 \cdot p^2 (1-p)^2$
$= 6p^2 (1-p)^2$

49. To end up 10 miles east of the starting point, Abby must go 10 miles east and 0 miles west, so she must toss 10 heads and no tails. Use the binomial probability formula with
$n = 10$, $p = \dfrac{1}{2}$, $q = \dfrac{1}{2}$, and $x = 10$.
Then
P(10 heads, 0 tails) $= P(10)$
$$= _{10}C_{10} \cdot \left(\frac{1}{2}\right)^{10} \left(\frac{1}{2}\right)^0$$
$$= 1 \cdot \frac{1}{1024} \cdot 1$$
$$= \frac{1}{1024} \approx 0.001.$$

51. To end up 6 miles west of the starting point, Abby must go 8 miles west and 2 miles east, so she must toss 8 tails and 2 heads. Use the binomial probability formula with $n = 10$,
$p = \dfrac{1}{2}$, $q = \dfrac{1}{2}$, and $x = 8$. Then,
P(8 tails, 2 heads) $= P(8)$
$$= _{10}C_8 \cdot \left(\frac{1}{2}\right)^8 \left(\frac{1}{2}\right)^2$$
$$= 45 \cdot \left(\frac{1}{2}\right)^{10}$$
$$= \frac{45}{1024} \approx 0.044.$$

53. To end up 2 miles east of the starting point, Abby must go 6 blocks east and 4 blocks west (since $6 + 4 = 10$ and $6 - 4 = 2$), so she must toss 6 heads and 4 tails. Use the binomial probability formula with $n = 10$,

$p = \dfrac{1}{2}$, $q = \dfrac{1}{2}$, and $x = 6$. Then,
P(6 heads, 4 tails) $= P(6)$
$$= _{10}C_6 \cdot \left(\frac{1}{2}\right)^6 \left(\frac{1}{2}\right)^4$$
$$= 210\left(\frac{1}{1024}\right)$$
$$= \frac{210}{1024}$$
$$= \frac{105}{512} \approx 0.205.$$

55. To find the probability that Abby will end up at least 2 miles from the starting point, use the complements rule. The complement of "at least 2 miles from the start" is "less than 2 miles from the start" or "1 mile from the start" or "exactly at the start." However, it is impossible for Abby to end up 1 mile from the starting point (either east or west). In order for two integers to have a sum of 10, they must be both even or both odd. In either case, their difference will be even. Thus, Abby can never end up an odd number of miles from the starting point. In order to end up at the starting point, Abby must go 5 miles east and 5 miles west. Let
$p = \dfrac{1}{2}$, $q = \dfrac{1}{2}$, and $x = 5$:
P(5 heads, 5 tails) $= P(5)$
$$= _{10}C_5 \cdot \left(\frac{1}{2}\right)^5 \left(\frac{1}{2}\right)^5$$
$$= 252\left(\frac{1}{2}\right)^{10}$$
$$= 252\left(\frac{1}{1024}\right)$$
$$= \frac{252}{1024}$$
$$= \frac{63}{256}$$
Then, the probability that Abby ends up at least 2 miles from the starting point is
$$1 - P(5) = 1 - \frac{63}{256} = \frac{193}{256} \approx 0.754.$$

11.5 Exercises

1. Writing exercise; answers will vary.

3. Five fair coins are tossed. A tree diagram may be helpful to create the following sample space.

Use the following sample space to create the individual probabilities.

hhhhh	hhhht	hhhth	hhhtt
hhthh	hhtht	hhtth	hhttt
hthhh	hthht	hthth	hthtt
htthh	httht	httth	htttt
thhhh	thhht	thhth	thhtt
ththh	ththt	ththth	thttt
tthhh	tthht	tthth	tthtt
ttthh	tttht	tttth	ttttt

Number of heads, x	Probability $P(x)$	Product $x \cdot P(x)$
0	$\frac{1}{32}$	0
1	$\frac{5}{32}$	$\frac{5}{32}$
2	$\frac{10}{32}$	$\frac{20}{32}$
3	$\frac{10}{32}$	$\frac{30}{32}$
4	$\frac{5}{32}$	$\frac{20}{32}$
5	$\frac{1}{32}$	$\frac{5}{32}$

Thus, the expected value is given by:
Expected number of heads

$$= 0 + \frac{5}{32} + \frac{20}{32} + \frac{30}{32} + \frac{20}{32} + \frac{5}{32}$$
$$= \frac{80}{32}$$
$$= \frac{5}{2}.$$

5. List the given information in a table. Then calculate $P(x)$, the product $x \cdot P(x)$, and their total.

Number Rolled	Payoff	Probability	Product
6	$3	$\frac{1}{6}$	$\$\left(\frac{3}{6}\right)$
5	$2	$\frac{1}{6}$	$\$\left(\frac{2}{6}\right)$
4	$1	$\frac{1}{6}$	$\$\left(\frac{1}{6}\right)$
1–3	$0	$\frac{3}{6}$	$0

Expected value: $\$\left(\frac{6}{6}\right) = \$1$

7. List the given information in a table. Then complete the table as follows.

Number Rolled	Payoff	Probability	Product
1	–$1	$\frac{1}{6}$	$-\$\left(\frac{1}{6}\right)$
2	$2	$\frac{1}{6}$	$\$\left(\frac{2}{6}\right)$
3	–$3	$\frac{1}{6}$	$-\$\left(\frac{3}{6}\right)$
4	$4	$\frac{1}{6}$	$\$\left(\frac{4}{6}\right)$
5	–$5	$\frac{1}{6}$	$-\$\left(\frac{5}{6}\right)$
6	$6	$\frac{1}{6}$	$\$\left(\frac{6}{6}\right)$

Expected value: $\$\left(\frac{3}{6}\right) = 50¢$

The expected net winnings for this game are 50¢.

9. List the given information in a table, and complete the probability and product columns. Remember that the expected value is the sum of the product column.

Number of heads	Payoff	Probability $P(x)$	Product $x \cdot P(x)$
3	10¢	$\frac{1}{8}$	$\left(\frac{10}{8}\right)¢$
2	5¢	$\frac{3}{8}$	$\left(\frac{15}{8}\right)¢$
1	3¢	$\frac{3}{8}$	$\left(\frac{9}{8}\right)¢$
0	0¢	$\frac{1}{8}$	0¢

Expected value: $\left(\frac{34}{8}\right)¢ = \left(\frac{17}{4}\right)¢$

Since it costs 5¢ to play, the expected net winnings are $\frac{17}{4}¢ - 5¢ = \frac{17}{4}¢ - \frac{20}{4}¢ = -\frac{3}{4}¢.$

Because the expected net winnings are not zero, 5¢ is not a fair price to pay to play this game.

11. The expected number of absences on a given day is
$$x_1 \cdot P(x_1) + x_2 \cdot P(x_2) + x_3 \cdot P(x_3)$$
$$+ x_4 \cdot P(x_4) + x_5 \cdot P(x_5)$$
$$= 0(0.18) + 1(0.26) + 2(0.29) + 3(0.23)$$
$$+ 4(0.04)$$
$$= 1.69$$

A college foundation raises funds by selling raffle tickets for a new car worth $36,000.

13. (a) Since 600 tickets are sold, a person who buys one ticket will have a probability of $\frac{1}{600} \approx 0.00167$ of winning the car and a $1 - 0.0017 = 0.9983$ probability of not winning anything. For this person, the expected value is $36,000(0.00167) + $0(0.9983) \approx 60, and the expected *net* winnings (since the ticket costs $120) are
$$60 - $120 = -$60.$$

(b) By selling 600 tickets at $120 each, the foundation takes in
$$600($120) = $72,000.$$
Since they had to spend $36,000 for the car, the total profit for the foundation is
$$\text{revenue} - \text{cost} = \text{profit}$$
$$72,000 - $36,000 = $36,000.$$

(c) Without having to pay for the car, the foundation's total profit will be all of the revenue from the ticket sales, which is $72,000.

Five thousand raffle tickets are sold. One first prize of $1000, two second prizes of $500 each, and three third prizes of $100 each will be awarded, with all winners selected randomly.

15. The associated probabilities are
$$P(\text{1st prize}) = \frac{1}{5000}, \quad P(\text{2nd prize}) = \frac{2}{5000},$$
and $P(\text{3rd prize}) = \frac{3}{5000}$.

The expected winnings, ignoring the cost of the raffle ticket, are given by
$$1000\left(\frac{1}{5000}\right) + $500\left(\frac{2}{5000}\right) + $100\left(\frac{3}{5000}\right)$$
$$= $.20 + $.20 + $.06$$
$$= $.46, \text{ or } 46¢.$$

17. Since 5000 tickets were sold for $1 each, the sponsor's revenue was $5000($1) = 5000. The sponsor's cost was the sum of all the prizes:
$$1($1000) + 2($500) + 3($100) = $2300.$$
Therefore, the sponsor's profit is
$$\text{revenue} - \text{cost} = \text{profit}$$
$$5000 - $2300 = $2700.$$

19. List the given information in a table, and complete the probability and product columns. Remember that the expected value is the sum of the product column.

Number of families	Probability $P(x)$	Product $x \cdot P(x)$
1020	$\frac{1020}{10000}$	$1 \cdot \left(\frac{1020}{10000}\right) = \frac{1020}{10000}$
3370	$\frac{3370}{10000}$	$2 \cdot \left(\frac{3370}{10000}\right) = \frac{6740}{10000}$
3510	$\frac{3510}{10000}$	$3 \cdot \left(\frac{3510}{10000}\right) = \frac{10530}{10000}$
1340	$\frac{1340}{10000}$	$4 \cdot \left(\frac{1340}{10000}\right) = \frac{5360}{10000}$
510	$\frac{510}{10000}$	$5 \cdot \left(\frac{510}{10000}\right) = \frac{2550}{10000}$
80	$\frac{80}{10000}$	$6 \cdot \left(\frac{80}{10000}\right) = \frac{480}{10000}$
170	$\frac{170}{10000}$	$0 \cdot \left(\frac{170}{10000}\right) = 0$

Expected value: $\frac{26680}{10000} \approx 2.7$

21. The expected value is
$$200(0.2) + 300(0.5) + (-800)(0.3)$$
$$= 40 + 150 - 240$$
$$= -50$$
Since this expected value is negative, the expected change in the number of electronics jobs is a decrease of 50.

23. Writing exercise; answers will vary.

25. The optimist viewpoint would ignore the probabilities and hope for the best possible outcome, which is Project *C* since it may return up to $340,000.

27. If the contestant takes a chance on the other two prizes, the expected winnings will be
$$5000(0.20) + $8000(0.15) = $1000 + $1200$$
$$= $2200.$$

29. Expectation for Insuring

	Net Profit x	Probability $P(x)$	Product $x \cdot P(x)$
Rain	$100,000 + $30,000 − $25,000 = $105,000	0.20	$21,000
No rain	$100,000 − $25,000 = $75,000	0.80	$60,000

Expected Profit: $81,000

31. Do not purchase the insurance because $86,000 > $81,000.

33. Compute the remaining values in Column 5 (Expected Value).
Row 5: 5000(0.30) = 1500
Row 6: 30,000(0.10) = 3000
Row 7: 25,000(0.70) = 17,500
Row 8: 45,000(0.60) = 27,000

35. Jessica's total "expected" additional volume is
$4000 + $3000 + $1500 + $3000 + $17,500 + $27,000
= $56,000.

37. Her total existing volume is
$10,000 + $30,000 + $25,000 + $35,000 + $15,000
= $115,000.
The increase in volume would be
$171,000 − $115,000 = $56,000.
That would be an increase of
$$\frac{56,000}{115,000} \approx 0.487, \text{ or } 48.7\%.$$

EXTENSION: ESTIMATING PROBABILITIES BY SIMULATION

1. Writing exercise; answers will vary.

3. No, since the probability of an individual girl's birth is (nearly) the same as that for a boy.

5. Let each of the 50 numbers correspond to one family. For example, the first number, 51592, with middle digits—1(boy), 5(girl), 9(girl)—represents a family with 2 girls and 1 boy. The last number whose middle digits are 800 represents the 50th family which has 1 girl and 2 boys—a success, and so on.

Examining each number, we count (tally) 18 successes. Therefore,
$$P(2 \text{ boys and 1 girl}) = \frac{18}{50} = 0.36.$$
Observe that this is quite close to the 0.375 predicted by the theoretical value.

Refer to discussion in text regarding foul shooting in basketball. After completing the indicated tally, find the empirical probability that, on a given opportunity, Christine will score as follows.

To construct the tally for Exercise 7, begin as follows: Since the first number in the table of random digits is 5, which represents a hit, Christine will get a second shot. The second digit is 7, representing a miss on the second shot. Record the results of the first two shots (the first one-and-one opportunity) as "one point." The second and third opportunities correspond to the pair 3, 4 and 0, 5. Record each of these results as "two points." For the fourth opportunity, the digit 9 indicates that the first shot was missed, so Christine does not get a second shot. In this case only one digit is used. Record this result as "zero points." Continue in this manner until 50 one-and-one opportunities are obtained. The results of the tally are as follows. Note that this, in effect, is a frequency distribution as discussed in Chapter 11.

Number of Points	Tally frequency
0	15
1	6
2	29
Total	50

7. From the tally, we see that 1 point shots occur 6 times. Thus,
$$P(1 \text{ point}) = \frac{6}{50} = 0.12.$$

9. Answers will vary.

11. Writing exercise; answers will vary.

Chapter 11 Test

1. Writing exercise; answers will vary.

2. Writing exercise; answers will vary.

3. There are 39 non-hearts and 13 hearts, so the odds against getting a heart are 39 to 13, or 3 to 1 (when reduced).

4. There are 2 red queens and 50 other cards, so the odds against getting a red queen are 50 to 2, or 25 to 1.

5. There are 12 face cards altogether. Of these, there are 6 black face cards and 2 more non-black kings for a total of 8 cards. There are $52 - 8 = 44$ other cards in the deck. Thus, the odds against getting a black face card or king are 44 to 8, or 11 to 2.

6.

		Second Parent	
		C	c
First Parent	C	CC	Cc
	c	cC	cc

7. There are two outcomes (cC or Cc) indicating that the next child will be a carrier. Thus, $P(\text{carrier}) = \dfrac{2}{4} = \dfrac{1}{2}$.

8. There is one 'favorable' outcome (cc) and three 'unfavorable' outcomes (CC, Cc, and cC). Thus, the odds that a child will have the disease (cc) are 1 to 3.

9. Use the fundamental counting principle where the first task is to calculate the probability of the initial employee choosing any day of the week $\left(\dfrac{7}{7}\right)$, the second task is the probability for the second employee to choose any other day of the seek $\left(\dfrac{6}{7}\right)$. In a similar manner, the third employee's probability must involve a choice from one of the five remaining days with a resulting probability of $\left(\dfrac{5}{7}\right)$. Thus,

$$P = \frac{7}{7} \cdot \frac{6}{7} \cdot \frac{5}{7} = \frac{30}{49}.$$

10. Let E_1, E_2, and E_3 represent the 3 employees. The following seven outcomes would be considered a success, or favorable outcome, for our probability:
{(M: E_1, E_2, E_3), (Tue: E_1, E_2, E_3), (W: E_1, E_2, E_3), ..., (Sun: E_1, E_2, E_3)}.
The remainder of the sample space outcomes include those where employees are split between Tuesday and Thursday or Saturday and Sunday. Listing the Tuesday,

Thursday possibilities we have:
{(Tue: E_1; Th: E_2, E_3),
(Tue: E_2; Th: E_1, E_3),
(Tue: E_3; Th: E_1, E_2),
(Tue: E_1, E_2; Th: E_3),
(Tue: E_1, E_3; Th: E_2),
(Tue: E_2, E_3; Th: E_1)} for a total of six more outcomes. Similarly, if all employees choose a day beginning with "S" i.e. Saturday or Sunday there would be an additional six outcomes for a total of nineteen outcomes in our sample space, i.e. $n(S) = 19$. Thus the probability that all employees choose the same day to work given that all three select a day beginning with the same letter is:

$$P = \frac{\text{number of favorable outcomes}}{\text{total number of outcomes, } n(S)} = \frac{7}{19}.$$

11. The complement of "exactly two choose the same day" is "all three choose different days (Exercise 9) or all three choose the same day." The probability that all three choose the same day is $\dfrac{7}{7} \cdot \dfrac{1}{7} \cdot \dfrac{1}{7} = \dfrac{1}{49}$. Thus, the probability of "exactly two choosing the same day" is given by

$$P = 1 - \left(\frac{30}{49} + \frac{1}{49}\right) = \frac{18}{49}.$$

Observe that the calculation involves both the complements rule and the special addition rule.

Two numbers are randomly selected without replacement from the set $\{1, 2, 3, 4, 5\}$.

12. To find the probability that "both numbers are even," use combinations to select the number of successes—ways of selecting the two even numbers, $_2C_2$, and to calculate the total number of ways of selecting two of the numbers from the 5, $_5C_2$. Thus,

$$P(\text{selecting two even numbers}) = \frac{_2C_2}{_5C_2}$$
$$= \frac{1}{10}.$$

As in many of the exercises an alternate solution may be considered here: Let E_1 represent the event of selecting an even number as the first selection and E_2, selecting an even number as the second selection. Using the general multiplication

rule the probability is given by
$$P(E_1 \text{ and } E_2) = P(E_1) \cdot P(E_2|E_1)$$
$$= \frac{2}{5} \cdot \frac{1}{4}$$
$$= \frac{1}{10}.$$

13. To find the probability that "both numbers are prime" use combinations. Since there are three prime numbers {2, 3, 5}, use $_3C_2$ to calculate the number of successes. To calculate the total number of ways of selecting two of the numbers from the 5, use $_5C_2$. Thus,

$$P(\text{selecting two prime numbers}) = \frac{_3C_2}{_5C_2}$$
$$= \frac{3}{10}.$$

14. Create a "product (sum) table" to list the elements in the sample space and the successes (event of interest). Note that "without replacement," one can only use a selected number once. Hence, there are no diagonal values in the table.

2nd number

+	1	2	3	4	5
1	–	3	4	5	6
2	3	–	5	6	7
3	4	5	–	7	8
4	5	6	7	–	9
5	6	7	8	9	–

2nd number

There are 12 odd sums in the table and a total of 20 sums in the sample space. Thus,
$$P(\text{sum is odd}) = \frac{12}{20} = \frac{3}{5}.$$

15. Similar to Exercise 14, create a "product table" to list the elements in the sample space and the successes (event of interest). Note that "without replacement," one can only use a selected number once. Hence, there are no diagonal values in the table.

2nd number

×	1	2	3	4	5
1	–	2	3	4	5
2	2	–	6	8	10
3	3	6	–	12	15
4	4	8	12	–	20
5	5	10	15	20	–

2nd number

There are 6 odd products in the table and a total of 20 products in the sample space.

Thus, $P(\text{product is odd}) = \frac{6}{20} = \frac{3}{10}.$

A three-member committee is selected randomly from a group consisting of three men and two women.

16. Let x represent the number of men on the committee. Then,

x	$P(x)$
0	0
1	$\frac{3}{10}$
2	$\frac{6}{10}$
3	$\frac{1}{10}$

Where,
$$P(1) = \frac{C(3, 1) \cdot C(2, 2)}{C(5, 3)}$$
$$P(2) = \frac{C(3, 2) \cdot C(2, 1)}{C(5, 3)}$$
$$P(3) = \frac{C(3, 3) \cdot C(2, 0)}{C(5, 3)}$$
Why is $P(0) = 0$?

17. The probability that the "committee members are not all men" is the complement of the "committee are all men," $P(3)$. Hence, use the complements rule,
$P(\text{committee members are not all men})$
$$= 1 - P(3)$$
$$= 1 - \frac{1}{10}$$
$$= \frac{9}{10}.$$

18. Complete the table begun in Exercise 16 as an aid to calculating the expected number of men (sum of product column).

x	$P(x)$	$x \cdot P(x)$
0	0	0
1	$\frac{3}{10}$	$\frac{3}{10}$
2	$\frac{6}{10}$	$\frac{12}{10}$
3	$\frac{1}{10}$	$\frac{3}{10}$

Expected number: $\frac{18}{10} = \frac{9}{5}$

Create a "product (sum) table" such as below for the "sum" of rolling two dice.

2nd die

+	1	2	3	4	5	6
1	2	3	4	5	6	7
2	3	4	5	6	7	8
3	4	5	6	7	8	9
4	5	6	7	8	9	10
5	6	7	8	9	10	11
6	7	8	9	10	11	12

1st die

Use for Exercises 19–22.

19. There are 6 doubles values and 36 possible values. The probability of doubles is therefore, $\frac{6}{36} = \frac{1}{6}$.

20. There are 35 outcomes with a "sum greater than 2" and 36 possible outcomes. Thus the odds in favor of a "sum greater than 2" are 35 to 1.

21. To find the odds against a "sum of 7 or 11" count the sums that satisfy the condition "sum of 7 or 11." There are 8 such sums. It follows that there are $36 - 8 = 28$ sums that are not 7 or 11. Thus, the odds against a "sum of 7 or 11" are 28 to 8, or 7 to 2.

22. Since there are 4 sums that are even and less than 5,
$$P(\text{sum that is even and less than 5}) = \frac{4}{36} = \frac{1}{9}.$$

For Exercises 23–26, the chance of making par on any one hole is 0.78.

23. By the special multiplication rule,
$$P(\text{making par on all three holes}) = 0.78^3 \approx 0.475$$

24. Use the fundamental counting principle, where the first task is to find the number of ways to choose the two holes he scores par on followed by the tasks of assigning a probability for each hole. Note that since 0.78 is the probability of scoring par, $1 - 0.78 = 0.22$ is the probability of not scoring par on a hole.
$P(\text{makes par on exactly 2 holes})$
$= {}_3C_2 \cdot (0.78)^2 \cdot (0.22)$
$= (3)(0.6084)(0.22)$
≈ 0.402

25. "At least one of the three holes" is the complement of "none of the three holes." Since the probability of not making par on any of the three holes is 0.22^3,
$P(\text{at least one of the three holes})$
$= 1 - (0.22)^3$
$= 1 - 0.010648$
$\approx 0.989.$

26. Use the special multiplication rule since these probabilities are independent. The probability that he makes par on the first and third holes but not on the second is found by $P = (0.78)(0.22)(0.78) \approx 0.134.$

Two cards are drawn, without replacement, from a standard 52-card deck for Exercises 27–30.

27. Let R_1 and R_2 represent the two red cards. Since the cards are not replaced, the events are not independent. Use the general multiplication rule:
$$P(R_1 \text{ and } R_2) = P(R_1) \cdot P(R_2|R_1)$$
$$= \frac{26}{52} \cdot \frac{25}{51}$$
$$= \frac{1}{2} \cdot \frac{25}{51}$$
$$= \frac{25}{102}$$

28. Let C_1 and C_2 represent two cards of the same color. Since the cards are not replaced, the events are not independent. Use the general multiplication rule. Note that since it doesn't matter what color the first card is, its probability is 1.

Thus, $P(C_1 \text{ and } C_2) = P(C_1) \cdot P(C_2|C_1)$

$$= \frac{52}{52} \cdot \frac{25}{51}$$

$$= \frac{25}{51}.$$

29. The first card drawn limits the sample space to 51 cards where all 4 queens are still in the deck. The probability then is given by

$P(\text{queen given the first card is an ace})$

$= P(Q_2|A_1)$

$$= \frac{4}{51}.$$

30. In the event "the first card is a face card and the second is black," the first (face) card may be red or black. Since this affects the probability associated with the second card, look at both cases and use the special addition rule to add (since we are "or-ing") the results.

Case 1 (first card is a red face card):

$P(F_1 \text{ and } B_2) = P(F_1) \cdot P(B_2|F_1)$

$$= \frac{6}{52} \cdot \frac{26}{51}$$

$$= \frac{3}{51}$$

$$= \frac{1}{17}.$$

Case 2 (first card is a black face card):

$P(F_1 \text{ and } B_2) = P(F_1) \cdot P(B_2|F_1)$

$$= \frac{6}{52} \cdot \frac{25}{51}$$

$$= \frac{3}{26} \cdot \frac{25}{51}$$

$$= \frac{1}{26} \cdot \frac{25}{17}$$

$$= \frac{25}{26 \cdot 17}.$$

The probability is $P(\text{Case 1 or Case 2})$

$= P(\text{Case 1}) + P(\text{Case 2})$

$$= \frac{1}{17} + \frac{25}{26 \cdot 17}$$

$$= \frac{26 \cdot 1}{26 \cdot 17} + \frac{25}{26 \cdot 17}$$

$$= \frac{26 + 25}{26 \cdot 17}$$

$$= \frac{51}{26 \cdot 17}$$

$$= \frac{3 \cdot 17}{26 \cdot 17}$$

$$= \frac{3}{26}.$$

Chapter 12

12.1 Exercises

1. (a) Remember that f represents the frequency of each data value, that $\dfrac{f}{n}$ is a comparison of each frequency to the frequency to the overall number of data values.

x	f	$\dfrac{f}{n}$
0	10	$\dfrac{10}{30} \approx 33\%$
1	7	$\dfrac{7}{30} \approx 23\%$
2	6	$\dfrac{6}{30} = 20\%$
3	4	$\dfrac{4}{30} \approx 13\%$
4	2	$\dfrac{2}{30} \approx 7\%$
5	1	$\dfrac{1}{30} \approx 3\%$

(b)

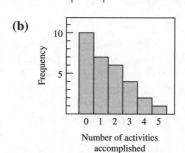

(c)

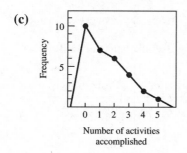

3. (a)

Class Limits	Tally	Frequency f	Relative Frequency $\dfrac{f}{n}$
45–49	|||	3	$\dfrac{3}{54} \approx 5.6\%$
50–54	ᵀᴴᴸ ᵀᴴᴸ ||||	14	$\dfrac{14}{54} \approx 25.9\%$
55–59	ᵀᴴᴸ ᵀᴴᴸ ᵀᴴᴸ		16	$\dfrac{16}{54} \approx 29.6\%$
60–64	ᵀᴴᴸ ᵀᴴᴸ ᵀᴴᴸ ||	17	$\dfrac{17}{54} \approx 31.5\%$
65–69	||||	4	$\dfrac{4}{54} \approx 7.4\%$

Total: $n = 54$

(b)

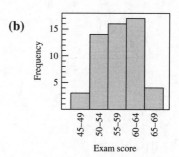

(c)

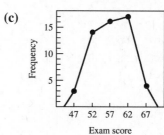

5. (a)

Class Limits	Tally	Frequency f	Relative Frequency $\dfrac{f}{n}$
70–74	||	2	$\dfrac{2}{30} \approx 6.7\%$
75–79			1	$\dfrac{1}{30} \approx 3.3\%$
80–84	|||	3	$\dfrac{3}{30} = 10.0\%$
85–89	||	2	$\dfrac{2}{30} \approx 6.7\%$
90–94	ᵀᴴᴸ	5	$\dfrac{5}{30} \approx 16.7\%$
95–99	ᵀᴴᴸ	5	$\dfrac{5}{30} \approx 16.7\%$
100–104	ᵀᴴᴸ		6	$\dfrac{6}{30} = 20.0\%$
105–109	||||	4	$\dfrac{4}{30} \approx 13.3\%$
110–114	||	2	$\dfrac{2}{30} \approx 6.7\%$

Total: $n = 30$

(b)

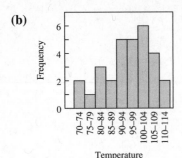

(c)

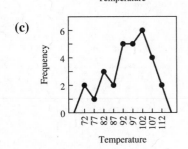

7.

0	7 9 8
1	1 1 2 8 9 4 3 1 0 5 0 5 5
2	7 0 9 6 6 2 2 5 2 3 4 4
3	8 1

9.

0	8 5 4 9 6 9 4 8
1	6 0 1 8 8 2 4 0 2 8 6 3
2	6 1 2 5 1 3
3	0 4 6
4	4

11. From the graph, receipts exceeded outlays in 2001.

13. From the graph, receipts appear to have climbed faster than outlays in 2005, 2006, 2007, and 2011.

15.

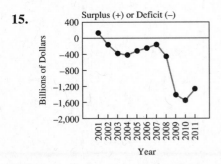

17. From the graph for consumer price index, the highest index occurred in 2008 and was about 3.8%.

19. Writing exercise; answers will vary.

21. The greatest single expense category is represented by the largest portion of the circle graph, and that is Medicare & Medicaid spending. The portion of the graph for this category represents 23% of the spending, and thus also 23% of the area of the circle. To find the central angle of the sector, find 23% of 360°.
$(0.23)(360) \approx 83°$

23. To calculate the number of degrees in each sector of the circle, multiply each percentage in decimal form times 360°. Here are a few examples:
$0.33(360) \approx 119°$
$0.25(360) = 90°$
$0.12(360) \approx 43°$

Trained in military, or courses

Trained in school

Formal training from employers

Informal on-the-job training

No particular training

25. Examine the graph to see that she would be about 79.

27. (a) Examine the 6% curve to see that Claire's money will run out at age 76. If she reaches age 70, her money would last for about $76 - 70 = 6$ years.

(b) Writing exercise; answers will vary.

29. Writing exercise; answers will vary.

31. Writing exercise; answers will vary.

33. Writing exercise; answers will vary.

35. (a)

Letter	Probability
A	$\frac{0.08}{0.385} \approx 0.208$
E	$\frac{0.13}{0.385} \approx 0.338$
I	$\frac{0.065}{0.385} \approx 0.169$
O	$\frac{0.08}{0.385} \approx 0.208$
U	$\frac{0.03}{0.385} \approx 0.078$

(b)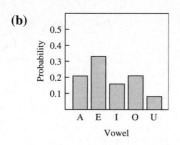

37. Writing exercise; answers will vary.

39. (a) Read the table in Exercise 38 to see that the probability is 0.225 that a given student studied 30–39 hours.

(b) The student would fall into either the 40–49 hour range or the 50–59 hour range. Add the probabilities: 0.175 + 0.100 = 0.275.

(c) Fewer than 30 hours means either 20–29 hours or 10–19 hours. Add the probabilities: 0.275 + 0.150 = 0.425.

(d) At least 50 hours means 50 or more; the categories included are 50–59, 60–69, and 70–79. Add the probabilities: 0.100 + 0.050 + 0.025 = 0.175.

41. (a)

Class Limits	Probability
Sailing	$\frac{9}{40} = 0.225$
Hang gliding	$\frac{5}{40} = 0.125$
Snowboarding	$\frac{7}{40} = 0.175$
Bicycling	$\frac{3}{40} = 0.075$
Canoeing	$\frac{12}{40} = 0.300$
Rafting	$\frac{4}{40} = 0.100$

(b) Empirical

(c) Writing exercise; answers will vary.

12.2 Exercises

1. (a) $\bar{x} = \dfrac{7 + 9 + 12 + 14 + 34}{5} = \dfrac{76}{5} = 15.2$

(b) The data is given in order from smallest to largest; the median is 12.

(c) There is no mode.

3. (a) $\bar{x} = \dfrac{218 + 230 + 196 + 224 + 196 + 233}{6}$

$= \dfrac{1297}{6}$

≈ 216.2

(b) Arrange the values from smallest to largest: 196, 196, 218, 224, 230, 233. Find the mean of the two middle numbers: $\dfrac{218 + 224}{2} = \dfrac{442}{2} = 221$.

(c) The mode is 196.

5. (a) $\bar{x} = \dfrac{\begin{array}{c}3.1 + 4.5 + 6.2 + 7.1 + 4.5 + 3.8\\ + 6.2 + 6.3\end{array}}{8}$

$= \dfrac{41.7}{8}$

≈ 5.2

(b) Arrange the values from smallest to largest or vice versa: 3.1, 3.8, 4.5, 4.5, 6.2, 6.2, 6.3, 7.1. Find the mean of the two middle numbers: $\dfrac{4.5 + 6.2}{2} = 5.35$.

(c) The set of values is bimodal: 4.5 and 6.2.

7. (a) $\bar{x} = \dfrac{\begin{array}{c}.78 + .93 + .66 + .94 + .87 + .62\\ + .74 + .81\end{array}}{8}$

$= \dfrac{6.35}{8}$

≈ 0.8

(b) Arrange the values from smallest to largest or vice versa: 0.62, 0.66, 0.74, 0.78, 0.81, 0.87, 0.93, 0.94. Find the mean of the two middle numbers: $\dfrac{0.78 + 0.81}{2} = 0.795$.

(c) There is no mode.

9. (a) The sum of the data is 1032.

$\bar{x} = \dfrac{1032}{8} = 129$

(b) Arrange the values from smallest to largest or vice versa: 125, 125, 127, 128, 128, 131, 132, 136. The two middle numbers are both 128.

The median is $\dfrac{128+128}{2} = 128$.

(c) There are two modes: 125 and 128.

11. (a) The sum of the number of departures (in millions) is 106.7.

$\bar{x} = \dfrac{106.7}{10} \approx 10.7$ million

(b) Arrange the data from smallest to largest. The two middle numbers are 10.6 and 10.7. The median is

$\dfrac{10.6+10.7}{2} = 10.65$ million.

(c) The mode is 10.6 million.

13. (a) The sum of the number of fatalities is 739.

$\bar{x} = \dfrac{739}{10} = 73.9$

(b) Arrange the data from smallest to largest. The two middle numbers are 13 and 22. The median is $\dfrac{13+22}{2} = 17.5$.

(c) The mode is 0.

15. The mean using the new data is

$\bar{x} = \dfrac{474}{10} = 47.4$. The median and mode remain the same.

17. The sum of the spending is $298.3 billion.

The mean is $\bar{x} = \dfrac{298.3}{5} \approx \$59.7$ billion.

19. The sum of the data is 36.92. Then

$\bar{x} = \dfrac{36.92}{7} = 5.27$ seconds.

21. The sum for the new list is 16.94. Then

$\bar{x} = \dfrac{16.94}{7} = 2.42$ seconds.

23. The mean was affected more.

25. $\bar{x} = \dfrac{79+81+44+89+79+90}{6} = \dfrac{462}{6} = 77$

Arrange in order from smallest to largest; 44, 79, 79, 81, 89, 90. The median is

$\dfrac{79+81}{2} = 80$. The mode is 79.

27. Let x = the score he must make. Replace the score of 44 with x.

$$\dfrac{x+79+79+81+89+90}{6} = 85$$

$$\dfrac{6}{1} \cdot \dfrac{x+79+79+81+89+90}{6} = 85 \cdot 6$$

$$x+418 = 510$$

$$x = 92$$

29. (a)

Value	Frequency	Value · Frequency
615	17	10,455
590	7	4130
605	9	5445
579	14	8106
586	6	3516
600	5	3000
Totals	58	34,652

Then $\bar{x} = \dfrac{34,652}{58} \approx 597.4$.

(b) From part (a), there are 58 items. The formula for the position of the median is $\dfrac{\sum f + 1}{2} = \dfrac{58+1}{2} = 29.5$.

This means that the median is halfway between the 29th and 30th item. A chart showing cumulative frequency shows the value. Note that the values must be arranged in order.

Value	Frequency	Cumulative Frequency
579	14	14
586	6	20
590	7	27
600	5	32
605	9	41
615	17	58

The value 600 is the median.

(c) Examine the frequency column to see that the mode is 615.

31.

Units	Grade	Units · Grade Value
4	C	$4 \cdot 2 = 8$
7	B	$7 \cdot 3 = 21$
3	A	$3 \cdot 4 = 12$
3	F	$3 \cdot 0 = 0$
17		41

Then $\bar{x} = \dfrac{41}{17} \approx 2.41$, to the nearest hundredth.

33. The sum of the populations (in millions) is 3242. Then $\bar{x} = \dfrac{3242}{5} \approx 648$ million.

35. China: $\dfrac{1,339,000,000 \text{ people}}{3,601,000 \text{ sq mi}}$
≈ 372 persons per square mile

India: $\dfrac{1,157,000,000 \text{ people}}{1,148,000 \text{ sq mi}}$
≈ 1008 persons per square mile

United States: $\dfrac{307,000,000 \text{ people}}{3,537,000 \text{ sq mi}}$
≈ 87 persons per square mile

Indonesia: $\dfrac{240,000,000 \text{ people}}{741,000 \text{ sq mi}}$
≈ 324 persons per square mile

Brazil: $\dfrac{199,000,000 \text{ people}}{3,265,000 \text{ sq mi}}$
≈ 61 persons per square mile

37. The sum of the number of PCs in use in 2008 (in millions) is 1969.1. Then
$\bar{x} = \dfrac{1969.1}{6} \approx 328.2$ million.

39. (a) The mean number of crew members per plane is $\bar{x} = \dfrac{11+9+6+7}{4} = \dfrac{33}{4} = 8.25$.

(b) The median number of crew members per plane is $\dfrac{7+9}{2} = 8$.

41. (a)

Flight	Total Persons per Plane
American #111	$11 + 76 + 5 = 92$
United #175	$9 + 51 + 5 = 65$
American #77	$6 + 53 + 5 = 64$
United #93	$7 + 33 + 4 = 44$

The mean number of persons per plane is $\bar{x} = \dfrac{92+65+64+44}{4} = \dfrac{265}{4} = 66.25$.

(b) The median number of persons per plane is $\dfrac{64+65}{2} = 64.5$.

43. The median number of silver medals is $\dfrac{6+6}{2} = 6$.

45. (a) The sum of the "Total" column is 194.
$\bar{x} = \dfrac{194}{10} = 19.4$

(b) The median is $\dfrac{15+16}{2} = 15.5$.

(c) The mode is 11.

47. (a) $x = \dfrac{47+51+53+56+\cdots+96}{34}$
$= \dfrac{2544}{34}$
≈ 74.8

(b) The scores are listed from smallest to the largest. Because there is an even number of scores, the median is the mean of the two middle numbers, the 17th and 18th scores: $\dfrac{77+78}{2} = 77.5$.

49. Writing exercise; answers will vary.

51. (a)

Value	Frequency	Value · Frequency
0	1	0
1	1	1
3	1	3
14	2	28
15	1	15
16	2	32
17	2	34
18	3	54
19	1	19
20	1	20
Totals	15	206

$$\bar{x} = \frac{206}{15} = 13.7$$

Position of the median is $\frac{15+1}{2} = 8$.
The position is the 8th piece of data.

Value	Frequency	Cumulative Frequency
0	1	1
1	1	2
3	1	3
14	2	5
15	1	6
16	2	8
17	2	10
18	3	13
19	1	14
20	1	15

Examine the table to see that the median is located in the row for the value of 16. The mode is 18 with a frequency of 3.

(b) The median, 16, is most representative of the data.

53. 2, 3, 5, 7; the median is $\frac{3+5}{2} = 4$.

(a) There are 4 numbers listed.

(b) $\bar{x} = \frac{2+3+5+7}{4} = \frac{17}{4} = 4.25$

55. 1, 3, 6, 10, 15, 21; the median is $\frac{6+10}{2} = 8$.

(a) There are 6 numbers listed.

(b) $\bar{x} = \frac{1+3+6+10+15+21}{6} = \frac{56}{6} \approx 9.33$

57. Arrange the numbers from smallest to largest. At this point it is uncertain where x will lie. However, if a single number must be the mean, median, and mode, then one of the given numbers must be that number. Because the median is the middle number and five values are given, the median must be 70 or 80. Try each of these as a mean to see which one works.

$$\frac{60+70+80+110+x}{5} = 70$$
$$\frac{5}{1} \cdot \frac{320+x}{5} = 70 \cdot 5$$
$$320+x = 350$$
$$x = 30$$

This value does not work.

$$\frac{60+70+80+110+x}{5} = 80$$
$$\frac{5}{1} \cdot \frac{320+x}{5} = 80 \cdot 5$$
$$320+x = 400$$
$$x = 80$$

This value works. The set of numbers is {60, 70, 80, 80, 110}. The value 80 is the mean, median, and mode.

59. No

61. Writing exercise; answers will vary.

12.3 Exercises

1. The sample standard deviation will be larger because the denominator is $n-1$ instead of n.

3. (a) The range is $15 - 2 = 13$.

(b) To find the standard deviation:

1. First find the mean, $\bar{x} = \dfrac{56}{7} = 8$.

2 and 3. Find each deviation from the mean $(x - \bar{x})$ and square each deviation $(x - \bar{x})^2$.

Data	Deviations	Squared Deviations
2	$2 - 8 = -6$	$(-6)^2 = 36$
5	$5 - 8 = -3$	$(-3)^2 = 9$
6	$6 - 8 = -2$	$(-2)^2 = 4$
8	$8 - 8 = 0$	$(0)^2 = 0$
9	$9 - 8 = 1$	$(1)^2 = 1$
11	$11 - 8 = 3$	$(3)^2 = 9$
15	$15 - 8 = 7$	$(7)^2 = 49$
Total		108

4. Sum the squared deviations. The sum is 108.
5. Divide by $n - 1$.

$$\frac{108}{7-1} = 18$$

6. Take the square root.

$$\sqrt{18} \approx 4.24$$

5. (a) The range is $41 - 22 = 19$.

(b) To find the standard deviation:

1. First find the mean, $\bar{x} = \dfrac{210}{7} \approx 30$.

2 and 3. Find each deviation from the mean $(x - \bar{x})$ and square each deviation $(x - \bar{x})^2$.

Data	Deviations	Squared Deviations
27	$27 - 30 = -3$	$(-3)^2 = 9$
34	$34 - 30 = 4$	$(4)^2 = 16$
22	$22 - 30 = -8$	$(-8)^2 = 64$
41	$41 - 30 = 11$	$(11)^2 = 121$
30	$30 - 30 = 0$	$(0)^2 = 0$
25	$25 - 30 = -5$	$(-5)^2 = 25$
31	$31 - 30 = 1$	$(1)^2 = 1$
Total		236

4. Sum the squared deviations. The sum is 236.
5. Divide by $n - 1$.

$$\frac{236}{7-1} \approx 39.3$$

6. Take the square root.

$$\sqrt{39.3} \approx 6.27$$

Some of the details are omitted in the following exercises. See Exercises 3–6 for details of computing standard deviation. A spreadsheet is a very useful tool in obtaining the intermediate calculations.

7. (a) The range is $348 - 308 = 40$.

(b) To find the standard deviation:
1. First find the mean.

$$\bar{x} = \frac{3233}{10} = 323.3$$

2 and 3. Find each deviation from the mean $(x - \bar{x})$ and square each deviation $(x - \bar{x})^2$.

4. Sum the squared deviations. The sum is 1192.1.
5. Divide by $n - 1$.

$$\frac{1192.1}{10-1} \approx 132.456$$

6. Take the square root.

$$\sqrt{132.456} \approx 11.51$$

9. (a) The range is $85.62 - 84.48 = 1.14$.

(b) To find the standard deviation:
1. First find the mean.

$$\bar{x} = \frac{763.62}{9} \approx 84.85$$

2 and 3. Find each deviation from the mean $(x - \bar{x})$ and square each deviation $(x - \bar{x})^2$.
4. Sum the squared deviations. The sum is 1.0915.
5. Divide by $n - 1$.

$$\frac{1.0915}{9-1} \approx 0.13644$$

6. Take the square root.

$$\sqrt{0.13644} \approx 0.37$$

11. (a) The range is $13 - 1 = 12$.

(b) To find the standard deviation:
1. First find the mean.

Value	Frequency	Value · Frequency
13	3	39
10	4	40
7	7	49
4	5	20
1	2	2
Totals	21	150

Then $\bar{x} = \frac{150}{21} \approx 7.143$.

2 and 3. Find each deviation from the mean $(x - \bar{x})$ and square each deviation $(x - \bar{x})^2$. These steps are shown in the table.

Value	Deviations	Squared Deviations	Freq · $(x - \bar{x})^2$
13	5.857	34.304	3(34.304)
10	2.857	8.162	4(8.162)
7	−0.143	0.020	7(0.020)
4	−3.143	9.878	5(9.878)
1	−6.143	37.736	2(37.736)
Total			260.562

4. The fourth column shows the frequency of each value multiplied by the squared deviation. After multiplying each of these, find the sum: 260.562.
5. Divide by $n - 1$. Remember that the total number of values is 21.

$$\frac{260.562}{21-1} = 13.0281$$

6. Take the square root.

$$\sqrt{13.0281} \approx 3.61$$

13. According to Chebyshev's theorem, the fraction of scores that lie within 2 standard deviations of the mean is at least

$$1 - \frac{1}{2^2} = 1 - \frac{1}{4} = \frac{3}{4}.$$

15. According to Chebyshev's theorem, the fraction off scores that lie within $\frac{5}{2}$ standard deviations of the mean is at least

$$1 - \frac{1}{\left(\frac{5}{2}\right)^2} = 1 - \frac{1}{\frac{25}{4}} = 1 - \frac{4}{25} = \frac{21}{25}.$$

17. According to Chebyshev's theorem, the fraction of scores that lie within 3 standard deviations of the mean is at least

$$1 - \frac{1}{(3)^2} = 1 - \frac{1}{9} = \frac{8}{9}.$$ Divide 8 by 9 and change the decimal $0.\overline{8}$ to 88.9%.

19. According to Chebyshev's theorem, the fraction of scores that lie within $\frac{5}{3}$ standard deviations of the mean is at least

$$1 - \frac{1}{\left(\frac{5}{3}\right)^2} = 1 - \frac{1}{\frac{25}{9}} = 1 - \frac{9}{25} = \frac{16}{25}.$$

Divide 16 by 25 and change the decimal 0.64 to 64%.

21. Since 64 is 2 standard deviations below the mean $(80 - 2 \cdot 8 = 64)$ and 96 is 2 standard deviations above the mean $(80 + 2 \cdot 8 = 96)$, find the minimum fraction of values that lie within 2 standard deviations of the mean.

See Exercise 13 for the answer $\frac{3}{4}$.

23. Since 48 is 4 standard deviations below the mean $(80 - 4 \cdot 8 = 48)$ and 112 is 4 standard deviations above the mean $(80 + 4 \cdot 8 = 112)$, find the minimum fraction of values that lie within 4 standard deviations of the mean. See Exercise 14 for the answer $\frac{15}{16}$.

25. This is equivalent to finding the largest fraction of values that lie outside 2 standard deviations from the mean. There are at least $1 - \dfrac{1}{2^2} = 1 - \dfrac{1}{4} = \dfrac{3}{4}$ of the values within 2 standard deviations of the mean. Thus, the largest fraction of values that lie outside this range would be: $1 - \dfrac{3}{4} = \dfrac{1}{4}$.

27. To find how many standard deviations below the mean 52 is, use $\dfrac{52 - 80}{8} = -3.5$.

Also, the value 108 is 3.5 standard deviations above the mean. Then we must find the largest fraction of values that lie outside 3.5 or $\dfrac{7}{2}$ standard deviations from the mean. There are at least

$1 - \dfrac{1}{\left(\frac{7}{2}\right)^2} = 1 - \dfrac{1}{\left(\frac{49}{4}\right)} = 1 - \dfrac{4}{49} = \dfrac{45}{49}$ of the

values within 3.5 standard deviations of the mean. Thus, the largest fraction of values that lie outside this range of values would be

$1 - \dfrac{45}{49} = \dfrac{4}{49}$.

29. The sum of the values is $2430. Then

$x = \dfrac{2430}{12} = \$202.50$.

31. The standard deviation is $80.38. Then $202.50 − 80.38 = \$122.12$, and $202.50 + 80.38 = \$282.88$. There are six bonus amounts that fall within these two boundaries: $175, $185, $190, $205, $210, and $215.

33. $1 - \dfrac{1}{2^2} = 1 - \dfrac{1}{4} = \dfrac{3}{4}$

Then $\dfrac{3}{4}$ of the data is $\dfrac{3}{4} \cdot 12 = 9$. There should be at least 9 amounts.

35. (a) Using the six steps to find the standard deviation of the samples:
Sample A
1. Find the mean.

Value, x	Frequency, f	$x \cdot f$
3	2	6
4	1	4
7	1	7
8	1	8
Totals	5	25

Then $\overline{x}_A = \dfrac{25}{5} = 5$.

2 and 3. find each deviation from the mean $(x - \overline{x})$ and square each deviation, $(x - \overline{x})^2$. These steps are shown in the table.

x	$x - \overline{x}$	$(x - \overline{x})^2$	$f \cdot (x - \overline{x})^2$
3	−2	4	8
4	−1	1	1
7	2	4	4
8	3	9	9
Totals			22

4. The fourth column shows the frequency of each value multiplied by the squared deviation. After multiplying each of these, find the sum: 22.
5. Divide by $n − 1$. Remember that the total number of values is 5.

$\dfrac{22}{5 - 1} = 5.5$

6. Take the square root.

$S_A = \sqrt{5.5} \approx 2.35$

Sample B
1. Find the value.

Value, x	Frequency, f	$x \cdot f$
3	1	3
5	1	5
6	1	6
7	1	7
8	1	8
10	2	20
Totals	7	49

Then, $\overline{x}_B = \dfrac{49}{7} = 7$.

2 and 3. Find each deviation from the mean $(x - \overline{x})$ and square each deviation, $(x - \overline{x})^2$. These steps are shown in the table.

x	$x - \overline{x}$	$(x - \overline{x})^2$	$f \cdot (x - \overline{x})^2$
3	–4	16	16
5	–2	4	4
6	–1	1	1
7	0	0	0
8	1	1	1
10	3	9	18
Total			40

4. The fourth column shows the frequency of each value multiplied by the squared deviation. After multiplying each of these, find the sum: 40.
5. Divide by $n - 1$. Remember that the total number of values is 7.

$$\dfrac{40}{7 - 1} = 6.67$$

6. Take the square root.

$$S_B = \sqrt{6.67} \approx 2.58$$

(b) To find the sample coefficients of variation:

$$V_A = \dfrac{S_A}{\overline{x}} \cdot 100 \approx \dfrac{2.35}{5} \cdot 100 \approx 46.9$$

Note: Answers may vary in accuracy. Above answer made use of most accurate value of S_A which was carried over in calculator from the initial calculated value of S.

$$V_B = \dfrac{S_B}{\overline{x}} \cdot 100 \approx \dfrac{2.58}{7} \cdot 100 \approx 36.9$$

(c) Sample B has the higher dispersion as indicated by the larger standard deviation.

(d) Sample A has the higher relative dispersion as indicated by the larger coefficient of variation.

37. (a) Calculate sample means:
Sample A

Value, x	Frequency, f
64	1
65	1
70	1
71	1
74	1
Totals 344	5

Thus, $\overline{x}_A = \dfrac{344}{5} = 68.8$.

Sample B

Value, x	Frequency, f
60	1
62	1
69	1
70	1
72	1
Totals 333	5

Thus, $\overline{x}_B = \dfrac{333}{5} = 66.6$.

(b) Calculate sample standard deviations:
Sample A

x	$x - \overline{x}$	$(x - \overline{x})^2$
64	–4.8	23.04
65	–3.8	14.44
70	1.2	1.44
71	2.2	4.84
74	5.2	27.04
Total		70.8

$$\dfrac{70.8}{5 - 1} = 17.1$$

$$S_A = \sqrt{17.7} \approx 4.21$$

Sample B

x	$x - \overline{x}$	$(x - \overline{x})^2$
60	−6.6	43.56
62	−4.6	21.16
69	2.4	5.76
70	3.4	11.56
72	5.4	29.16
Total		111.2

$$\frac{111.2}{5-1} = 27.8$$

$$S_B = \sqrt{27.8} \approx 5.27$$

(c) Brand A should last longer, since $\overline{x}_A > \overline{x}_B$.

(d) Brand A should have the more consistent lifetime, since $S_A < S_B$.

39. To answer this question, the standard deviation values must be compared. The standard deviation for Brand A is given as $S_A = 2116$ miles. To calculate the standard deviation for Brand B modules, follow the procedure given in the text or use a calculator or spreadsheet. The value of the standard deviation is $S_B = 3539$ miles.

Brand A is more consistent because the value of the standard deviation is smaller.

41. $\overline{x} = \dfrac{13 + 14 + 17 + 19 + 21 + 22 + 25}{7}$

$= \dfrac{131}{7}$

≈ 18.71

To find the standard deviation:
1. The mean is 18.71.
2 and 3. Find each deviation from the mean $(x - \overline{x})$ and square each deviation $(x - \overline{x})^2$. These steps are shown in the table.

Data	Deviations	Squared Deviations
13	$13 - 18.71$ $= -5.71$	$(-5.71)^2 = 32.6041$
14	$14 - 18.71$ $= -4.71$	$(-4.71)^2 = 22.1841$
17	$17 - 18.71$ $= -1.71$	$(-1.71)^2 = 2.9241$
19	$19 - 18.71$ $= 0.29$	$(0.29)^2 = 0.0841$
21	$21 - 18.71$ $= 2.29$	$(2.29)^2 = 5.2441$
22	$22 - 18.71$ $= 3.29$	$(3.29)^2 = 10.8241$
25	$25 - 18.71$ $= 6.29$	$(6.29)^2 = 39.5641$
Total		113.4287

4. Sum the squared deviations. The sum is 113.4287.
5. Divide by $n - 1$.

$$\frac{113.4287}{7-1} \approx 18.90478$$

6. Take the square root. $\sqrt{18.90478} \approx 4.35$

43. $\bar{x} = \dfrac{3+4+7+9+11+12+15}{7}$

$\qquad = \dfrac{61}{7}$

$\qquad \approx 8.71$

To find the standard deviation:
1. The mean is 8.71.
2 and 3. Find each deviation from the mean $(x-\bar{x})$ and square each deviation $(x-\bar{x})^2$. These steps are shown in the table.

Data	Deviations	Squared Deviations
3	$3-8.71$ $=-5.71$	$(-5.71)^2 = 32.6041$
4	$4-8.71$ $=-4.71$	$(-4.71)^2 = 22.1841$
7	$7-8.71$ $=-1.71$	$(-1.71)^2 = 2.9241$
9	$9-8.71$ $=0.29$	$(0.29)^2 = 0.0841$
11	$11-8.71$ $=2.29$	$(2.29)^2 = 5.2441$
12	$12-8.71$ $=3.29$	$(3.29)^2 = 10.8241$
15	$15-8.71$ $=6.29$	$(6.29)^2 = 39.5641$
Total		113.4287

4. Sum the squared deviations. The sum is 113.4287.
5. Divide by $n-1$.

$\dfrac{113.4287}{7-1} \approx 18.90478$

6. Take the square root. $\sqrt{18.90478} \approx 4.35$

45. $\bar{x} = \dfrac{39+42+51+57+63+66+75}{7}$

$\qquad = \dfrac{393}{7}$

$\qquad \approx 56.14$

To find the standard deviation:
1. The mean is 56.14.
2 and 3. Find each deviation from the mean $(x-\bar{x})$ and square each deviation $(x-\bar{x})^2$. These steps are shown in the table.

Data	Deviations	Squared Deviations
39	$39-56.14$ $=-17.14$	$(-17.14)^2 = 293.7796$
42	$42-56.14$ $=-14.14$	$(-14.14)^2 = 199.9396$
51	$51-56.14$ $=-5.14$	$(-5.14)^2 = 26.4196$
57	$57-56.14$ $=0.86$	$(0.86)^2 = 0.7396$
63	$63-56.14$ $=6.86$	$(6.86)^2 = 47.0596$
66	$66-56.14$ $=9.86$	$(9.86)^2 = 97.2196$
75	$75-56.14$ $=18.86$	$(18.86)^2 = 355.6996$
Total		1020.8572

4. Sum the squared deviations. The sum is 1020.8572.
5. Divide by $n-1$.

$\dfrac{1020.8572}{7-1} \approx 170.14287$

6. Take the square root.
$\sqrt{170.14287} \approx 13.04$

47. Writing exercise; answers will vary.

49. Control cities:

$x = \dfrac{+1+(-8)+(-5)+0}{4} = \dfrac{-12}{4} = -3.0$

51. To find the standard deviation for the control cities:

Steps 1, 2, and 3. Find each deviation from the mean for the control cities, $(x-\bar{x})$ and square each deviation, $(x-\bar{x})^2$. These steps are shown in the table.

Data	Deviations	Squared Deviations
+1	$1 - (-3.0)$ $= 4.0$	$(4.0)^2 = 16$
-8	$-8 - (-3.0)$ $= -5.0$	$(-5.0)^2 = 25$
-5	$-5 - (-3.0)$ $= -2.0$	$(-2.0)^2 = 4$
0	$0 - (-3.0)$ $= 3.0$	$(3.0)^2 = 9$
Total		54

4. Sum the squared deviations. The sum is 54.

5. Divide by $n - 1$.
$$\frac{54}{4-1} \approx 18$$

6. Take the square root. $\sqrt{18} \approx 4.2$

53. $15.5 - 7.95 = 7.55$; $15.5 + 7.95 = 23.45$

55. No, because the individual data items cannot be identified.

57. Writing exercise; answers will vary.

12.4 Exercises

For each of Exercises 1–4, make use of z-scores.

1. Find 15% of 40 items: $0.15(40) = 6$. Select the 7th item in the data set, which is 58.

3. The third decile is the same as 30%. Find 30% of 40 items: $0.30(40) = 12$. Select the 13th item in the data set, which is 62.

5. The z-score for Neil: $x = \dfrac{5-4.6}{2.1} = 0.19$

   The z-score for Janet: $z = \dfrac{6-4.9}{2.3} = 0.48$

   Janet's score is greater than Chris's score, so her score is better.

7. The z-score for Nicole's Brand A tires:
$$z = \frac{37,000 - 45,000}{4500} \approx -1.78$$
   The z-score for Yvette's Brand B tires:
$$z = \frac{35,000 - 38,000}{2080} \approx -1.44$$
   Yvette's score is greater than Nicole's score, so that score is better.

9. Find the mean by dividing the sum of values in the Population column (1918 million) by 10, the number of trading countries.
$$\overline{x} = \frac{1918}{10} = 191.8$$
   Find s by using the six-step process described in Section 12.3 or by entering the data into a calculator or spreadsheet that contains the function to calculate s. The standard deviation is approximately 405. Use the formula $z = \dfrac{x - \overline{x}}{s}$, where x is Japan's population.
$$z = \frac{127 - 191.8}{405} \approx -0.2$$

11. The sum of the Exports column is 745 (billion U.S.\$).
$$\overline{x} = \frac{745}{10} = 74.5$$
   Find s by using the six-step process described in Section 12.3 or by entering the data into a calculator or spreadsheet that contains the function to calculate s. The standard deviation is approximately 77. Use the formula $z = \dfrac{x - \overline{x}}{s}$, where x is exports to Mexico.
$$z = \frac{151 - 74.5}{77} \approx 1.0$$

13. The fifteenth percentile in population is found by taking 15% of the 10 scores or $(0.15)(10) = 1.5$. Take the fifteenth percentile to be the second item up, Saudi Arabia, when the countries are arranged from smallest (bottom) to largest population (top).

15. The fourth decile in imports is found by taking 40% of the 10 scores or $(0.40)(10) = 4$. After arranging the import values from lowest to largest, take the fifth score up, United Kingdom, to be the fourth decile.

17. Compare the corresponding z-scores to determine which country was relatively higher. From Exercise 10, China's imports z-score is 1.7. Calculating Canada's exports z-score (by calculator) we arrive at 2.4. Canada's exports z-score is relatively higher since $2.4 > 1.7$.

19. (a) The median is $54.5 billion.

(b) The range is $261 - 12 = 249$.

(c) The middle half of the items extend from $29 billion to $70 billion.

21. Writing exercise; answers will vary.

23. Writing exercise; answers will vary.

25. Writing exercise; answers will vary.

27. The "skewness coefficient" is a measure of the overall distribution.

29. Both are skewed to the right, imports about twice as much as exports.

31. (a) No; this would only be true if Q_1 and Q_3 are symmetric about Q_2.

(b) Writing exercise; answers will vary.

33. Writing exercise; answers will vary.

35. Writing exercise; answers will vary.

37. The second, Q_2, which is the same value as the median average, can be found by calculating the mean average of the fifth and sixth (two middle) scores: $\dfrac{99.9 + 100.5}{2} = 100.2$.

The first quartile is the same value as the median of all scores below the Q_2. Since there are five scores below Q_2, choose the middle or third score for $Q_1 \cdot Q_1$ is thus 97.6.

The third quartile is the median of all values above Q_2. Since there are five scores, we choose the middle, or third score, which is 104.4 for Q_3.

39. The sixty-fifth percentile, P_{65}, is found by taking 65% of the 10 scores of $(0.65)(10) = 6.5$. Take the seventh score up, 103.2, to be P_{65}.

41. The midquartile $= \dfrac{Q_1 + Q_3}{2}$
$= \dfrac{97.6 + 104.4}{2}$
$= 101.0$

43.

45. If Eli Manning had completed one more pass in 2009, then
$A = 509$
$C = 318$
$T = 27$
$Y = 4021$
$I = 14$
and Rating = 93.3.

47. Solve the rating formula for I if
$A = 497$
$C = 336$
$T = 49$
$Y = 4557$
and Rating = 121.1.

$$121.1 = \frac{\left(250 \cdot \frac{336}{497}\right) + \left(1000 \cdot \frac{49}{497}\right) + \left(12.5 \cdot \frac{4557}{497}\right) + 6.25 - \left(1250 \cdot \frac{I}{497}\right)}{3}$$

$363.3 \approx 169.0 + 98.6 + 114.6 + 6.25 - 2.5I$
$363.3 = 388.45 - 2.5I$
$-25.15 = -2.5I$
$10 \approx I$

Peyton Manning was intercepted 10 times.

12.5 Exercises

1. Discrete because the variable can take on only fixed number values such as 1, 2, 3, etc., up to and including 50.

3. Continuous because the variable is not limited to fixed values. It is measurable rather than countable.

5. Discrete because the variable is limited to fixed values.

7. This represents all values to the right of the mean, which is 50% of the total number of values or 50% of 100 students: 0.50(100) = 50 students.

9. These values are 1 standard deviation above and 1 standard deviation below the mean, respectively; (86 ± 1). By the empirical rule, 68% of all scores lie within 1 standard deviation of the mean. Then $0.68(100) = 68$.

11. Less than 100 represents all values below the mean. This is 50% of the area under the curve or 50% of all the data.

13. The score 70 lies 2 standard deviations below the mean, and 130 lies 2 standard deviations above the mean. According to the empirical rule, 95% of the data lies within 2 standard deviations of the mean.

15. To find the percent of area between the mean and 1.5 standard deviations, use Table 10 to locate a z-score of 1.5. Read the value of A in the next column.
This is the area from the mean to the corresponding value of z. The area is 0.433, which is 43.3%.

17. Because of the symmetry of the normal curve, the area between the mean and a z value of -1.08 is the same as that from the mean to a z value of $+1.08$. Use Table 10 to locate a z-score of 1.08. Read the value of A in the next column. this is the area from the mean to the corresponding value of z. The area is 0.360, which is 36.0%.

19. It is helpful to sketch the area under the normal curve.

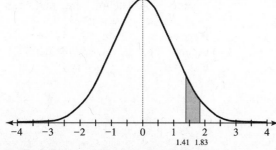

To find the area between $z = 1.41$ and

$z = 1.83$, use the z-table to find the area between the mean and $z = 1.83$. The area is 0.466. Find the area between the mean and $z = 1.41$, which is 0.421. Subtracting, $0.466 - 0.421 = 0.045$, which is 4.5%.

21. It is helpful to sketch the area under the normal curve.

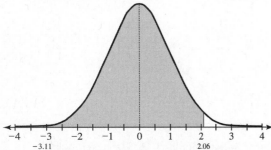

Add the areas under the curve from the mean to $z = -3.11$ to that from the mean to $z = 2.06$. The first area is the same as that from the mean to $z = +3.11$, which is 0.499. The area from the mean to $z = 2.06$ is 0.480. Find the total area: $0.499 + 0.480 = 0.979$ or 97.9%.

23. If 10% of the total area is to the right of the z-score, there is $50\% - 10\% = 40\%$ of the area between the mean and the value of z. From the z-table, find 0.40 in the A column. This area under the curve of 40% or 0.40 yields a z-score of 1.28.

25. If 9% of the total area is to the left of z, the z-score is below the mean, so the answer will be negative. (A sketch helps to see this.) Subtract 9% from 50% to find the amount of area between the mean and the value of z: $50\% - 9\% = 41\%$. Now find 0.41 in the A column to locate the appropriate value for z: 1.34. The z-score is -1.34.

27. Since the mean is 600 hr, we can expect half of the bulbs or 5000 bulbs to last at least 600 hr.

29. Find the z-score for 675:
$$z = \frac{675 - 600}{50} = 1.5.$$
Find the z-score for 740:
$$z = \frac{740 - 600}{50} = 2.8.$$
It is helpful to sketch the area under the normal curve.

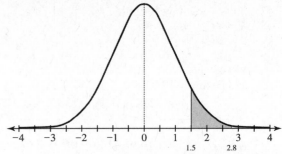

Find the amount of area under the normal curve between 1.5 and 2.8 by finding the corresponding values for A and then subtracting the smaller from the larger:
$0.497 - 0.433 = 0.064$
Finally, find 6.4% of 10,000:
$0.064(10000) = 640$

31. Find the z-score for 740:
$$z = \frac{740 - 600}{50} = 2.8.$$
It is helpful to sketch the area under the normal curve.

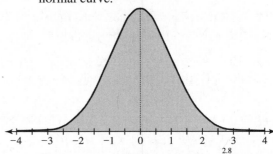

Find the amount of area under the normal curve between the mean and 2.8 by finding the corresponding value for A: 0.497. Add this area value to 0.5: $0.497 + 0.5 = 0.997$. Finally, find 99.7% of 10,000: $0.997(10000) = 9970$.

33. Because the mean is 1850 and the standard deviation is 150, the value 1700 corresponds to $z = -1$. Use the empirical rule to evaluate the corresponding area under the curve. The area between $z = -1$ and the mean is half of 68% or 34 %. The area to the right of the mean is 50%. The total area to the right of $z = -1$ is $34 + 50 = 84\%$.
If the z-table is used, the answer is 84.1%, because the area value in the table for $z = 1$ is 0.341.

35. Find the z-score for 1750:
$$z = \frac{1750 - 1850}{150} \approx -0.67.$$
Find the z-score for 1900:
$$z = \frac{1900 - 1850}{150} = 0.33.$$
It is helpful to sketch the area under the normal curve.

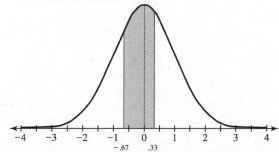

Find the amount of area under the normal curve between −0.67 and 0.33 by finding the corresponding values from area and then adding the values: $0.249 + 0.129 = 0.378$. This is 37.8%.

37. The z-score corresponding to 24 oz when $s = 0.5$ is found by $z = \dfrac{24 - 24.5}{0.5} = -1$. The fraction of boxes that are underweight is equivalent to the area under the curve to the left of −1. It is helpful to sketch the area under the normal curve.

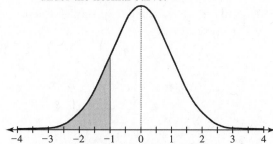

From the z-table, the area under the curve between the mean and +1 is 0.341. Subtract 0.341 from 0.5 to obtain the area under the curve to the right of +1: $0.5 - 0.341 = 0.159$. Because of symmetry, this is also the amount of area under the curve to the left of −1. The answer is 0.159 or 15.9%.

39. The z-score corresponding to 24 oz when $s = 0.3$ is found by $z = \dfrac{24 - 24.5}{0.3} \approx -1.67$. The fraction of boxes that are underweight is equivalent to the area under the curve to the left of −1.67.

It is helpful to sketch the area under the normal curve.

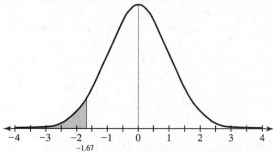

From the z-table, the area under the curve between the mean and +1.67 is 0.453. Subtract 0.453 from 0.5 to obtain the area under the curve to the right of +1.67: $0.5 - 0.453 = 0.047$. Because of symmetry, this is also the amount of area under the curve to the left of −1.67. The answer is 0.047 or 4.7%.

41. The mean plus 2.5 times the standard deviation, $\overline{x} + 2.5s$, corresponds to $z = 2.5$ no matter what the values of $\overline{x}$ and s are:
$$z = \frac{value - mean}{standard\ deviation}$$
$$= \frac{(\overline{x} + 2.5s) - \overline{x}}{s}$$
$$= \frac{2.5s}{s}$$
$$= 2.5.$$
The fraction of the population between the mean and $z = 2.5$ is 0.494, by the z-table. The fraction below the mean is 0.5. The sum of these is $0.5 + 0.494 = 0.994$ or 99.4%.

43. The RDA is the value corresponding to $z = 2.5$. Use the z-score formula to find x:
$$z = \frac{x - \overline{x}}{s}$$
$$2.5 = \frac{x - 159}{12}$$
$$(12) \cdot 2.5 = \frac{x - 159}{12} \cdot \frac{12}{1}$$
$$30 = x - 159$$
$$189 = x$$
The RDA is 189 units.

45. Find the z-score for 7: $z = \dfrac{7 - 7.47}{3.6} \approx -0.13$.

Find the z-score for 8: $z = \dfrac{8 - 7.47}{3.6} \approx 0.15$.

The area under the curve to the left of −0.13 and the area under the curve to the right of 0.15 must be added to answer the question.

Use the z-table to obtain 0.052 for the area between the mean and +0.13. The area to the right of 0.13 is then $0.5 - 0.052 = 0.448$. Because of symmetry, this is the amount of area to the left of −0.13.

Use the z-table to obtain 0.060 for the area between the mean and 0.15. The area to the right of 0.15 is then $0.5 - 0.060 = 0.440$. The sum of these two areas is $0.448 + 0.440 = 0.888$.

47. Find the z-score for 2.2:

$$z = \frac{2.2 - 1.5}{0.4} = 1.75.$$

Find the area under the normal curve between the mean and 1.75, which is 0.460. Subtract 0.460 from 0.5 to obtain the area under the curve to the right of $z = 1.75$: $0.5 - 0.460 = 0.040$ or 4.0%. Find 4.0% of five dozen: $(0.04)(5 \cdot 12) = 2.4$. The answer is about 2 eggs.

49. Find the area as a percent between $\overline{x} + \left(\frac{1}{2}\right)s$ and $\overline{x} + \left(\frac{3}{2}\right)s$. This is the area between $z = \frac{1}{2}$ or 0.5 and $z = \frac{3}{2}$ or 1.5. The area under the curve from the mean to $z = 0.5$ is 0.191; the area under the curve from the mean to $z = 1.5$ is 0.433. Subtract: $0.433 - 0.191 = 0.242$ or 24.2%.

51. Writing exercise; answers will vary.

53. To find the bottom cutoff grade, $0.15 + 0.08 = 0.23$ must be the amount of area under the normal curve to the right of the grade. That means that the amount of area between the mean and this cutoff grade is $0.5 - 0.23 = 0.27$. Locate this area value in the A column; it corresponds to a z-score of 0.74. Then use the z-score formula to find x:

$$z = \frac{x - \overline{x}}{s}$$
$$0.74 = \frac{x - 75}{5}$$
$$(5) \cdot (0.74) = \frac{x - 75}{5} \cdot \frac{5}{1}$$
$$3.7 = x - 75$$
$$78.7 = x$$

Rounded to the nearest whole number, the cutoff score should be 79.

55. To find the bottom cutoff grade, 0.08 must be the amount of area under the normal curve to the left of the grade. That means that the amount of area between the mean and this cutoff grade is $0.5 - 0.08 = 0.42$. Locate this area value in the A column; it corresponds to a z-score of 1.40 or 1.41. Use 1.405. The negative z-score, however, must be used because the grade is below the mean.

$$z = \frac{x - \overline{x}}{s}$$
$$-1.405 = \frac{x - 75}{5}$$
$$(5) \cdot (-1.405) = \frac{x - 75}{5} \cdot \frac{5}{1}$$
$$-7.025 = x - 75$$
$$68 \approx x$$

The cutoff grade should be 68.

57. Replace $\overline{x}$, s and z in the z-score formula and solve for x:

$$z = \frac{x - \overline{x}}{s}$$
$$1.44 = \frac{x - 76.8}{9.42}$$
$$(9.42) \cdot (1.44) = \frac{x - 76.8}{9.42} \cdot \frac{9.42}{1}$$
$$13.5648 = x - 76.8$$
$$90.4 \approx x$$

59. Replace $\overline{x}$, s and z in the z-score formula and solve for x:

$$z = \frac{x - \overline{x}}{s}$$
$$-3.87 = \frac{x - 76.8}{9.42}$$
$$(9.42) \cdot (-3.87) = \frac{x - 76.8}{9.42} \cdot \frac{9.42}{1}$$
$$-36.4554 = x - 76.8$$
$$40.3 \approx x$$

61. Writing exercise; answers will vary.

***EXTENSION:* REGRESSION AND CORRELATION**

1. The equation of the least squares line is $y' = ax + b$ where a and b are found as follows:

$$a = \frac{n(\Sigma xy) - (\Sigma x)(\Sigma y)}{n(\Sigma x^2) - (\Sigma x)^2}$$

$$= \frac{10(75) - (30)(24)}{10(100) - (30)^2}$$

$$= \frac{750 - 720}{1000 - 900}$$

$$= \frac{30}{100}$$

$$= 0.3, \text{ and}$$

$$b = \frac{\Sigma y - a(\Sigma x)}{n}$$

$$= \frac{24 - 0.3(30)}{10}$$

$$= \frac{24 - 9}{10}$$

$$= 1.5.$$

Then the equation for the least squares line is $y' = 0.3x + 1.5$.

3. The regression equation is $y' = 0.3x + 1.5$. Find y' when $x = 3$ tons.

$y = 0.3(3) + 1.5 = 0.9 + 1.5 = 2.4$ decimeters

5. The regression equation is $y' = 0.556x - 17.8$. Find y' when $x = 120°$.

$y' = 0.556(120) - 17.8$

$\quad = 66.76 - 17.8$

$\quad = 48.92°$

7. The table shows how to calculate all the sums that are needed in the formula for the least squares line.

x	y	x^2	y^2	xy
62	120	3844	14400	7440
62	140	3844	19600	8680
63	130	3969	16900	8190
65	150	4225	22500	9750
66	142	4356	20164	9372
67	130	4489	16900	8710
68	135	4624	18225	9180
68	175	4624	30625	11900
70	149	4900	22201	10430
72	168	51840	28224	12096
$\Sigma x = 663$	$\Sigma y = 1439$	$\Sigma x^2 = 44059$	$\Sigma y^2 = 209739$	$\Sigma xy = 95748$

The equation of the least squares line is $y' = ax + b$ where a and b are found as follows.

$$a = \frac{n(\Sigma xy) - (\Sigma x)(\Sigma y)}{n(\Sigma x^2) - (\Sigma x)^2}$$

$$= \frac{10(95748) - (663)(1439)}{10(44059) - (663)^2}$$

$$= \frac{957480 - 954057}{440590 - 439569}$$

$$= \frac{3423}{1021}$$

$$\approx 3.35, \text{ and}$$

$$b = \frac{\Sigma y - a(\Sigma x)}{n}$$

$$= \frac{1439 - 3.35(663)}{10}$$

$$= \frac{1439 - 2221.05}{10}$$

$$= -78.2$$

The value for b calculated here differs slightly from the one in the text because of rounding error.

It is highly recommended that a scientific calculator or spreadsheet be used, because then rounding is only done at the very end of all the calculations. There the value of a was rounded before using it in the calculation for b. Then the equation for the least squares line is $y' = 3.35x - 78.2$.

The value for b in the text is -78.4.

9. The regression equation is
    $y' = 3.35x - 78.4$. Find y' when $x = 70$ in.
    $y' = 3.35(70) - 78.4$
    $\quad = 234.5 - 78.4$
    $\quad \approx 156$ pounds

11.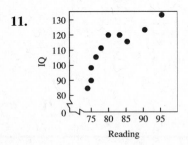

13. The regression equation is $y' = 2x - 51$.
    Find y' when $x = 65$.
    $y' = 2(65) - 51 = 130 - 51 = 79$

15. Use a scientific calculator or a spreadsheet to calculate the various sums needed in the formula, or see Exercise 7 for the detailed process. Actually, a calculator or spreadsheet can calculate the value of r.

    $\Sigma x = 15$; $\Sigma y = 418$; $\Sigma x^2 = 55$; $\Sigma y^2 = 30266$; $\Sigma xy = 1186$; and $n = 6$.

    Substitute the sums into the formula for the coefficient of correlation.

    $$r = \frac{n(\Sigma xy) - (\Sigma x)(\Sigma y)}{\sqrt{n(\Sigma x^2) - (\Sigma x)^2} \cdot \sqrt{n(\Sigma y^2) - (\Sigma y)^2}}$$

    $$= \frac{6(1186) - (15)(418)}{\sqrt{6(55) - (15)^2} \cdot \sqrt{6(30266) - (418)^2}}$$

    $$= \frac{7116 - 6270}{\sqrt{330 - 225} \cdot \sqrt{181596 - 174724}}$$

    $$= \frac{846}{\sqrt{105} \cdot \sqrt{6872}}$$

    $$\approx \frac{846}{10.25 \cdot 82.90}$$

    $$\approx 0.996$$

17.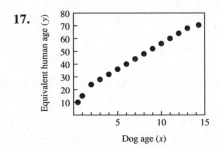

19. Writing exercise; answers will vary.

21. Use a scientific calculator or a spreadsheet to calculate the various sums needed in the formula. Actually, a calculator or spreadsheet can calculate the appropriate values for a and b needed for the regression line equation. See Exercise 7 for the details of calculating the following sums.

    $\Sigma x = 56$; $\Sigma y = 77.7$; $\Sigma x_2 = 560$;

    $\Sigma y^2 = 1110.43$; $\Sigma xy = 786.4$ and with

    $n = 8$.

    The equation of the least squares line is $y' = ax + b$ where

$$a = \frac{n(\Sigma xy)-(\Sigma x)(\Sigma y)}{n(\Sigma x^2)-(\Sigma x)^2}$$
$$= \frac{8(786.4)-(56)(77.7)}{8(560)-(56)^2}$$
$$= \frac{6291.2-4351.2}{4480-3136}$$
$$= \frac{1940}{1344}$$
$$\approx 1.44, \text{ and}$$
$$b = \frac{\Sigma y - a(\Sigma x)}{n}$$
$$= \frac{77.7-1.44(56)}{8}$$
$$= \frac{77.7-80.64}{8}$$
$$\approx -0.37$$

Again, it is recommended that a scientific calculator or spreadsheet be used to calculate these values in order to avoid rounding error. (The value of a has been rounded here.) The value for b that is calculated in the text is -0.39. Then the equation for the least squares line $y' = 1.44x - 0.39$.

23. The linear correlation of 0.99 is strong.

25. Use a scientific calculator or a spreadsheet to calculate the various sums needed in the formulas. Actually, a calculator or spreadsheet can calculate the appropriate values for a and b needed for the regression line equation. See Exercise 7 for the details of calculating the following sums.
$\Sigma x = 163; \Sigma y = 1583$
$\Sigma x^2 = 3345; \Sigma y^2 = 254,085$
$\Sigma xy = 26,162$ and $n = 10$
The equation of the least squares line is $y' = ax + b$ where
$$a = \frac{n(\Sigma xy)-(\Sigma x)(\Sigma y)}{n(\Sigma x^2)-(\Sigma x)^2}$$
$$= \frac{10(26,162)-(163)(1583)}{10(3345)-(163)^2}$$
$$\approx 0.5219$$
and
$$b = \frac{\Sigma y - a(\Sigma x)}{n}$$
$$= \frac{1583-0.5219(163)}{10}$$
$$\approx 149.8$$
Then the equation for the regression line is $y' = 0.5219x + 149.8$.

27. The linear correlation is weak to moderate.

Chapter 12 Test

1. Examine the scores in the table. Students who copied less than 10% generally improved their exam performance over the course of the semester.

2. Students with copy rates from 10% to 30% or greater than 50% did better on exam 3 than exam 2.

3. Students with copy rates from 30% to 50% consistently had lower scores from one exam to the next throughout the semester.

4. Writing exercise; answers will vary.

5. (a) $\bar{x} = \frac{398+250+215+73+64}{5}$
$$= \frac{1000}{5}$$
$$= 200 \text{ million barrels}$$

(b) The range is $398 - 64 = 334$ million barrels.

(c) Calculate the deviations from the mean $(x - \bar{x})$ and the squared deviations $(x - \bar{x})^2$.

x	$x - \bar{x}$	$(x - \bar{x})^2$
398	198	39,204
250	50	2500
215	15	225
73	−127	16,129
64	−136	18,496
Total		76,554

$$\frac{\text{Total}}{n-1} = \frac{76,554}{5-1} = 19,138.5$$
$$s = \sqrt{19,138.5} \approx 138 \text{ million barrels}$$

(d) $V = \frac{s}{\bar{x}} \cdot 100 = \frac{138}{200} \cdot 100 = 69$

(e) $\bar{x} = \frac{398+250+215+73+64+14\cdot 26}{5+26}$
$$= \frac{1364}{31}$$
$$= 44 \text{ million barrels}$$

6. (a) Arrange the data for height in order from smallest to largest. The middle value is 281 feet.

(b) Arrange the data for girth in order from smallest to largest. The median is 761 inches. The first quartile is the median of the values below 761 inches or

$$\frac{512+522}{2}=517 \text{ inches.}$$

(c) The eighth decile is found by taking 80% of the 9 data entries, or $(0.80)(9) = 7.2$. Take the 8th score up, 1290, to be the eighth decile.

7. (a) $T = G + H + 0.25C$
$T = 647 + 96 + 0.25(74)$
$= 743 + 18.5$
≈ 762

(b) Based on girth alone, the Common Bald Cypress would have ranked 7th.

(c) The ninth ranked tree had 773 total points. If the Common Bald Cypress had been 12 feet taller, it would have had $762 + 12 = 774$ total points and would have displaced the ninth ranked tree.

(d) With a circumference (girth) of 647 inches, the diameter would have been about $\frac{647}{\pi} \approx 205.9$ inches, or about 17 feet.

8.

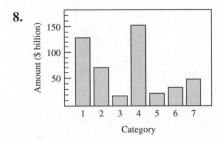

9. Of the $787 billion in the whole stimulus bill, $152.0 billion was assigned to health, or about $\frac{152.0}{787} \approx 0.193$ or 19.3%.

10. $152.0 billion − $20 billion = $132.0 billion went to other health categories, or about $\frac{132.0}{787} \approx 0.168$ or 16.8%.

11. The three greatest categories combined were $128.2 + 152.0 + 70.3 = 350.5 billion of the stimulus bill, or $\frac{350.5}{787} \approx 0.445$ or 44.5%.

12. A tally column is not shown here, but it is useful when creating a frequency distribution by hand.

Class Limits	Frequency f	Relative frequency $\frac{f}{n}$
6–10	3	$\frac{3}{22} \approx 0.14$
11-15	6	$\frac{6}{22} \approx 0.27$
16–20	7	$\frac{7}{22} \approx 0.32$
21–25	4	$\frac{4}{22} \approx 0.18$
26–30	2	$\frac{2}{22} \approx 0.09$

13. (a)

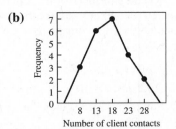

(b)

14. If the first class had limits of 7–9, this would be a width of 3. The smallest value is 8, and the largest value is 30. All the data must be included within the classes that are constructed. The classes would be:
7–9
10–12
13–15
16–18
19–21
22–24
25–27
28–30.
this is a total of 8 classes.

15.

Value	Frequency	Value · Frequency
8	3	24
10	8	80
12	10	120
14	8	112
16	5	80
18	1	18
Totals	35	434

Then $\bar{x} = \dfrac{434}{35} = 12.4$.

16. There are 35 items. The formula for the position of the median is
$$\frac{\Sigma f + 1}{2} = \frac{35 + 1}{2} = 18.$$
This means that the median is 18th item when the values are arranged from smallest to largest. A chart showing cumulative frequency shows that the value is in the row for a cumulative frequency of 21.

Value	Frequency	Cumulative Frequency
8	3	3
10	8	11
12	10	21
14	8	29
16	5	34
18	1	35

The value 12 is the median.

17. Examine the frequency column to see that the mode is 12, with a frequency of 10.

18. The range of the data is $18 - 8 = 10$.

19. The leaves are arranged in rank order (from smallest to largest).

```
3 | 3 8
4 | 3 5 8 9
5 | 0 2 5
6 | 1 1 4 5 6 7 7 8
7 | 0 1 2 3 7 7 8 9 9
8 | 0 4 4
9 | 1
```

20. There are 33 items. The formula for the position of the median is
$$\frac{\Sigma f + 1}{2} = \frac{33 + 1}{2} = 17.$$
This means that the median is the 17th item when the values are arranged from smallest to largest. Count the values in the stem-and-leaf display in the text to see that the 17th value is 35.

21. The mode is 33; it occurs 3 times.

22. The range of the data is $60 - 23 = 37$.

23. The third decile is the same as the 30th percentile. First, $0.30(33) = 9.9$. Then the 10th value is the location of the third decile. This value is 31.

24. The eighty-fifth percentile is located by first taking 85% of the 33 data items: $0.85(33) = 28.05$. Then the 29th value is located at the 85th percentile. This value is 49.

25. The values shown in the box plot are: Minimum value, 23; Maximum value, 60; Q_1, 29.5; Q_2, 35; and Q_3, 43.

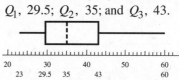

26. (a) The scores 70 and 90 are each two standard deviations away from the mean, because the mean is 80 and the value of one standard deviation is 5. According to the empirical rule, approximately 95% of the data lie within two standard deviations of the mean.

(b) The score of 95 is three standard deviations above the mean; the score of 65 is three standard deviations below the mean. According to the empirical rule, approximately 99.7% of the data lie within this interval. Therefore, $100 - 99.7 = 0.3\%$ lie outside this interval.

(c) The score of 75 is one standard deviation below the mean. Using the empirical rule, if 68% of the scores lie within one standard deviation of the mean, then 34% of the scores lie between 75 and 80. Subtract 34% from 50% to find the percentage of scores that are less than 75: $50 - 34 = 16\%$.

(d) The score of 85 is one standard deviation above the mean; the score of 90 is two standard deviations above the mean. From Exercise 21 we know that about 34% of the scores lie between 80 and 85. Also from the empirical rule, approximately 95% of the scores lie within two standard deviations. Half of 95% is 47.5%. Subtract 34% from 47.5% to find the percentage of scores between 85 and 90: 47.5 – 34 = 13.5%.

27. First find the z-score for 6.5:
$$z = \frac{6.5 - 5.5}{2.1} \approx 0.48.$$

It is always helpful to sketch the area under the normal curve that is being sought. Then find the amount of area between the mean and $z = 0.48$ from Table 10 in Section 13.5. The amount is 0.184. This area must be added to 0.5, the amount of area under the normal curve below the mean:
0.184 + 0.5 = 0.684

28. First find the z-scores for 6.2 and 9.4:
$$z = \frac{6.2 - 5.5}{2.1} \approx 0.33$$
$$z = \frac{9.4 - 5.5}{2.1} \approx 1.86$$

It is always helpful to sketch the area under the normal curve that is being sought.

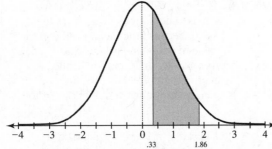

We are trying to find the amount of area under the curve between a z-score of 0.33 and a z-score of 1.86. Find the amount of area between the mean and $z = 1.86$ from Table 10 in Section 13.5. The amount is 0.469. Find the amount of area between the mean and $z = 0.33$; it is 0.129. the smaller value must be subtracted from the larger.
0.469 – 0.129 = 0.340

29. Find the means of the three groups.

Eastern teams: $\dfrac{84.2 + 79.0}{2} = 81.6$

Central teams: $\dfrac{5(76.4) + 6(78.0)}{11} \approx 77.3$

Western teams: $\dfrac{4(86.0) + 5(84.0)}{9} \approx 84.9$

The teams in the West had the greatest winning average. (Note that we must use a weighting factor, frequency, in the mean calculations for the Central and Western teams since there are a different number of teams in each of these leagues.)

30. Find the means of the standard deviations.

Eastern teams: $\dfrac{15.5 + 14.1}{2} = 14.8$

Central teams: $\dfrac{5(10.9) + 6(9.7)}{11} \approx 10.25$

Western teams: $\dfrac{4(9.0) + 5(10.9)}{9} \approx 10.06$

The Eastern teams were the least "consistent" because their standard deviation (variation) is the largest. (Note that again we must use a weighting factor, frequency, in the average calculations for the Central and Western teams since there are a different number of teams in each of these leagues.)

31. The average number of games won for all West Division teams is 84.9. See the Western Division teams winning average in Exercise 29.

32. Boston Red Sox:
$$z = \frac{x - \bar{x}}{5} = \frac{95 - 84.2}{15.5} = \frac{10.8}{15.5} = 0.70$$
Los Angeles Dodgers:
$$z = \frac{x - \bar{x}}{5} = \frac{95 - 84.0}{10.9} = \frac{11.0}{10.9} = 1.01$$
Thus, the Dodgers did relatively better, since $z = 1.01 > 0.70$.

Chapter 13

Note: Except where it is stated that a table has been used, exercises in this chapter have been completed with a calculator or spreadsheet. To ensure as much accuracy as possible, rounded values have been avoided in intermediate steps as much as possible. When rounded values have been necessary, several decimal places have been carried throughout the exercise; only the final answer has been rounded to 1 or 2 decimal places. In most cases, answers involving money are rounded to the nearest cent.

1. Use the formula $I = Prt$ with $P = \$800$; $r = 0.06$; and $t = 1$ yr.
$$I = Prt = (\$800)(0.06)(1) = \$48$$

3. Use the formula $I = Prt$ with $P = \$920$; $r = 0.07$; and $t = \dfrac{9}{12}$ yr. Remember that t must be expressed in years.
$$I = Prt = (\$920)(0.07)\left(\frac{9}{12}\right) = \$48.30$$

5. Use the formula $I = Prt$ with $P = \$2675$; $r = 0.073$; and $t = 2.5$ yr.
$$I = Prt = (\$2675)(0.073)(2.5) \approx \$488.19$$

7. (a) Use the formula $A = P(1 + rt)$, with $P = \$700$; $r = 0.03$; and $t = 6$.
$$\begin{aligned} A &= P(1 + rt) \\ &= \$700(1 + 0.03 \cdot 6) \\ &= \$700(1 + 0.18) \\ &= \$700(1.18) \\ &= \$826 \end{aligned}$$

   (b) Use the formula $A = P\left(1 + \dfrac{r}{m}\right)^n$, with m = number of periods per year and n = total number of periods.
$$\begin{aligned} A &= P\left(1 + \frac{r}{m}\right)^n \\ &= \$700\left(1 + \frac{0.03}{1}\right)^6 \\ &= \$700(1.03)^6 \\ &= \$700(1.194052297) \\ &\approx \$835.84 \end{aligned}$$

9. (a) Use the formula $A = P(1 + rt)$, with $P = \$2500$; $r = 0.02$; and $t = 3$.
$$\begin{aligned} A &= P(1 + rt) \\ &= \$2500(1 + 0.02 \cdot 3) \\ &= \$2500(1 + 0.06) \\ &= \$2500(1.06) \\ &= \$2650 \end{aligned}$$

   (b) Use the formula $A = P\left(1 + \dfrac{r}{m}\right)^n$, with m = number of periods per year and n = total number of periods.
$$\begin{aligned} A &= P\left(1 + \frac{r}{m}\right)^n \\ &= \$2500\left(1 + \frac{0.02}{1}\right)^3 \\ &= \$2500(1.02)^3 \\ &= \$2500(1.061208) \\ &\approx \$2653.02 \end{aligned}$$

11. Use the formula $I = Prt$ with $P = \$7500$; $r = 0.05$; and $t = \dfrac{4}{12}$ yr. Remember that t must be expressed in years.
$$\begin{aligned} I &= Prt \\ &= \$(7500)(0.05)\left(\frac{4}{12}\right) \\ &= \$125 \end{aligned}$$

13. First use the formula $I = Prt$ with $P = \$12{,}800$; $r = 0.07$; and $t = \dfrac{10}{12}$. Remember that t must be expressed in years.
$$\begin{aligned} I &= Prt \\ &= \$(12800)(0.07)\left(\frac{10}{12}\right) \\ &= \$746.67 \end{aligned}$$
Now the interest amount must be added to the loan amount to find the total amount Chris must repay.
$$\$746.67 + \$12800 = \$13{,}546.67$$

15. To find the compound interest that was earned, subtract the principal from the final amount.
$$\$1143.26 - \$975 = \$168.26$$

17. To find the final amount add the principal to the compound interest.
$$\$480 + \$337.17 = \$817.17$$

19. To find the final amount, use the formula for

future value $A = P\left(1+\dfrac{r}{m}\right)^n$, with

$P = \$7500$, $r = 0.035$, $m = 1$ and
$n = 1 \cdot 25 = 25$.

$A = P\left(1+\dfrac{r}{m}\right)^n$

$= \$7500\left(1+\dfrac{0.035}{1}\right)^{25}$

$= \$7500(1.035)^{25}$

$= \$7500(2.363244984)$

$\approx \$17,724.34$

Then to find the compound interest, subtract
the principal from this final amount.
$\$17724.34 - \$7500 = \$10224.34$

21. To find the final amount, use the formula for

future value $A = P\left(1+\dfrac{r}{m}\right)^n$.

(a) $A = P\left(1+\dfrac{r}{m}\right)^n$

$= \$2000\left(1+\dfrac{0.04}{1}\right)^{3}$

$= \$2000(1.04)^{3}$

$= \$2000(1.124864)$

$\approx \$2249.73$

(b) $A = P\left(1+\dfrac{r}{m}\right)^n$

$= \$2000\left(1+\dfrac{0.04}{2}\right)^{2\cdot3}$

$= \$2000(1.02)^{6}$

$= \$2000(1.126162419)$

$\approx \$2252.32$

(c) $A = P\left(1+\dfrac{r}{m}\right)^n$

$= \$2000\left(1+\dfrac{0.04}{4}\right)^{4\cdot3}$

$= \$2000(1.01)^{12}$

$= \$2000(1.12682503)$

$\approx \$2253.65$

23. To find the final amount, use the formula for

future value $A = P\left(1+\dfrac{r}{m}\right)^n$.

(a) $A = P\left(1+\dfrac{r}{m}\right)^n$

$= \$18,000\left(1+\dfrac{0.01}{1}\right)^{5}$

$= \$18,000(1.01)^{5}$

$= \$18,000(1.05101005)$

$\approx \$18,918.18$

(b) $A = P\left(1+\dfrac{r}{m}\right)^n$

$= \$18,000\left(1+\dfrac{0.01}{2}\right)^{2\cdot5}$

$= \$18,000(1.005)^{10}$

$= \$18,000(1.051140132)$

$\approx \$18,920.52$

(c) $A = P\left(1+\dfrac{r}{m}\right)^n$

$= \$18,000\left(1+\dfrac{0.01}{4}\right)^{4\cdot5}$

$= \$18,000(1.0025)^{20}$

$= \$18,000(1.051205503)$

$\approx \$18,921.70$

25. (a) First find the future value by using the

formula $A = P\left(1+\dfrac{r}{m}\right)^n$ with $P = \$850$,

$r = 0.016$, $m = 2$, and $n = 2 \cdot 4 = 8$.

$A = P\left(1+\dfrac{r}{m}\right)^n$

$= \$850\left(1+\dfrac{0.016}{2}\right)^{8}$

$= \$850(1.008)^{8}$

$= \$850(1.065820961)$

$\approx \$905.95$

Then subtract the principal from this
amount.
$\$905.95 - \$850 = \$55.95$

(b) Find the future value, changing to $m = 4$ and $n = 4 \cdot 4 = 16$.

$$A = P\left(1 + \frac{r}{m}\right)^n$$
$$= \$850\left(1 + \frac{0.016}{4}\right)^{16}$$
$$= \$850(1.004)^{16}$$
$$= \$850(1.06595631)$$
$$\approx \$906.06$$

Then subtract the principal from this amount.
$\$06.06 - \$850 = \$56.06$

(c) Find the future value, changing to $m = 12$ and $n = 12 \cdot 4 = 48$.

$$A = P\left(1 + \frac{r}{m}\right)^n$$
$$= \$850\left(1 + \frac{0.016}{12}\right)^{48}$$
$$= \$850(1.001\overline{3})^{48}$$
$$= \$850(1.066046954)$$
$$\approx \$906.14$$

Then subtract the principal from this amount.
$\$906.14 - \$850 = \$56.14$

(d) Find the future value, changing to $m = 365$ and $n = 365 \cdot 4 = 1460$.

$$A = P\left(1 + \frac{r}{m}\right)^n$$
$$= \$850\left(1 + \frac{0.016}{365}\right)^{1460}$$
$$= \$850(1.000043836)^{1460}$$
$$= \$850(1.066090903)$$
$$\approx \$906.18$$

Then subtract the principal from this amount.
$\$906.18 - \$850 = \$56.18$

(e) Use the formula for continuous growth $A = Pe^{rt}$ with $P = \$850$, $r = 0.016$, and $t = 4$ yr.

$$A = Pe^{rt}$$
$$= \$850 \cdot e^{0.016 \cdot 4}$$
$$= \$850 \cdot e^{0.064}$$
$$= \$850(1.066092399)$$
$$\approx \$906.18$$

Then subtract the principal from this amount.
$\$906.18 - \$850 = \$56.18$

27. Writing exercise; answers will vary.

29. Use the present value formula for simple interest with $A = \$1500$, $r = 0.08$, and $t = \frac{4}{12} = 0.\overline{3}$.

$$P = \frac{A}{1 + rt} = \frac{\$1500}{1 + (0.08)(0.\overline{3})} \approx \$1461.04$$

31. Use the present value formula for compound interest with $A = \$1000$, $r = 0.06$, $m = 1$, and $n = 1 \cdot 5$.

$$P = \frac{A}{\left(1 + \frac{r}{m}\right)^n}$$
$$= \frac{\$1000}{\left(1 + \frac{0.06}{1}\right)^5}$$
$$= \frac{\$1000}{(1.06)^5}$$
$$= \frac{\$1000}{1.3382256}$$
$$\approx \$747.26$$

33. Use the present value formula for compound interest with $A = \$9860$, $r = 0.08$, $m = 2$, and $n = 2 \cdot 10 = 20$.

$$P = \frac{A}{\left(1 + \frac{r}{m}\right)^n}$$
$$= \frac{\$9860}{\left(1 + \frac{0.08}{2}\right)^{20}}$$
$$= \frac{\$9860}{(1.04)^{20}}$$
$$= \frac{\$9860}{2.1911231}$$
$$\approx \$4499.98$$

35. Use the present value formula for compound interest with $A = \$500,000$, $r = 0.05$, $m = 4$, and $n = 4 \cdot 30 = 120$.

$$P = \frac{A}{\left(1 + \frac{r}{m}\right)^n}$$
$$= \frac{\$500,000}{\left(1 + \frac{0.05}{4}\right)^{120}}$$
$$= \frac{\$500,000}{(1.0125)^{120}}$$
$$= \frac{\$500,000}{4.440213229}$$
$$\approx \$112,607.20$$

37. Use the present value formula for compound interest with $A = \$500,000$, $r = 0.05$, $m = 365$, and $n = 365 \cdot 30 = 10,950$.

$$P = \frac{A}{\left(1+\frac{r}{m}\right)^n}$$

$$= \frac{\$500,000}{\left(1+\frac{0.05}{365}\right)^{10,950}}$$

$$\approx \$111,576.54$$

39. Use the formula $Y = \left(1+\frac{r}{m}\right)^m - 1$, with $r = 0.02$ and $m = 1$.

$$Y = \left(1+\frac{0.02}{1}\right)^1 - 1$$

$$= (1.02) - 1$$

$$= 0.02 \text{ or } 2.000\%$$

41. Use the formula $Y = \left(1+\frac{r}{m}\right)^m - 1$, with $r = 0.02$ and $m = 4$.

$$Y = \left(1+\frac{0.02}{4}\right)^4 - 1$$

$$= (1.005)^4 - 1$$

$$= 1.020150501 - 1$$

$$= 0.020150501 \text{ or } 2.015\%$$

43. Use the formula $Y = \left(1+\frac{r}{m}\right)^m - 1$, with $r = 0.02$ and $m = 365$.

$$Y = \left(1+\frac{0.02}{365}\right)^{365} - 1$$

$$= (1.000054795)^{365} - 1$$

$$= 1.020200781 - 1$$

$$= 0.020200781 \text{ or } 2.020\%$$

45. Use the formula $Y = \left(1+\frac{r}{m}\right)^m - 1$, with $r = 0.02$ and $m = 10,000$.

$$Y = \left(1+\frac{0.02}{10,000}\right)^{10,000} - 1$$

$$= (1.000002)^{10,000} - 1$$

$$= 1.02020132 - 1$$

$$= 0.02020132 \text{ or } 2.020\%$$

47. Use the yield of 1.400% in the rate in the formula $A = P + I$.

$$A = P + I$$

$$= \$30,000 + (0.01400)(30,000)$$

$$= \$30,420$$

49. Use the formula $A = P\left(1+\frac{r}{m}\right)^n$ with A having the value $2P$ because we want the amount to double the principal.

$$2P = P\left(1+\frac{0.04}{4}\right)^{4n}$$

$$2 = (1+0.01)^{4n}$$

$$\ln 2 = 4n \cdot \ln(1.01)$$

$$\frac{\ln 2}{\ln(1.01)} = 4n$$

$$\frac{69.66071689}{4} = n$$

$$17.41517922 \approx n$$

This is 17 years and 0.41517922 of a year: $0.41517922(365) \approx 151.54$. The answer then is about 17 years and 152 days.

51. Solve $Y = \left(1+\frac{r}{m}\right)^m - 1$ for r.

$$Y = \left(1+\frac{r}{m}\right)^m - 1$$

$$Y+1 = \left(1+\frac{r}{m}\right)^m$$

$$(Y+1)^{1/m} = \left[\left(1+\frac{r}{m}\right)^m\right]^{1/m}$$

$$(Y+1)^{1/m} = 1+\frac{r}{m}$$

$$(Y+1)^{1/m} - 1 = \frac{r}{m}$$

$$\frac{r}{m} = (Y+1)^{1/m} - 1$$

$$r = m[(Y+1)^{1/m} - 1]$$

53. (a) Use the formula $Y = \left(1+\frac{r}{m}\right)^m - 1$ where $r = 0.038$ and $m = 365$ to find Y for the Bank A.

$$Y = \left(1+\frac{0.038}{365}\right)^{365} - 1$$

$$= (1.00010411)^{365} - 1$$

$$\approx 1.038729178 - 1$$

$$\approx 0.038729178$$

This is a yield of about 3.87%.
Now replace Y with this value in
decimal form, m with 4, and solve for r
for Bank B using the formula derived in
Exercise 51.

$r = m[(Y+1)^{1/m} - 1]$

$= 4(1.038729178)^{1/4} - 4$

$\approx 4(1.009544769) - 4$

≈ 0.038179076

$\approx 3.818\%$

The Bank B has a nominal rate of about
3.818%.

(b) *Bank A*

Use the formula $A = P\left(1+\dfrac{r}{m}\right)^n$ with

$P = \$2000$, $r = 0.038$, $m = 365$, and

$n = \dfrac{10}{12}(365)$.

$A = \$2000\left(1+\dfrac{0.038}{365}\right)^{\frac{10}{12}(365)}$

$\approx \$2064.34$

Bank B

Use the formula $A = P\left(1+\dfrac{r}{m}\right)^n$ with P

$= \$2000$, $r = 0.03818$, $m = 4$, and $n = 3$.
Because the time period is 10 months,
she earns interest for 3 quarters (or nine
months).

$A = \$2000\left(1+\dfrac{0.03818}{4}\right)^3$

$\approx \$2057.82$

She should choose Bank A. She earns
$6.52 more interest.

(c) *Bank A*

Use the formula $A = P\left(1+\dfrac{r}{m}\right)^n$ with

$P = \$6000$, $r = 0.038$, $m = 365$, and
$n = 365$.

$A = P\left(1+\dfrac{r}{m}\right)^n$

$A = \$6000\left(1+\dfrac{0.038}{365}\right)^{365}$

$\approx \$6232.38$

Bank B

$A = P\left(1+\dfrac{r}{m}\right)^n$

$A = \$6000\left(1+\dfrac{0.03818}{4}\right)^4$

$\approx \$6232.38$

There is no difference in the amount for
the banks. There interest for both is
$232.38.

55. Years to double $= \dfrac{70}{\text{Annual inflation rate}}$

Years to double $= \dfrac{70}{2}$

Years to double $= 35$ years

57. Years to double $= \dfrac{70}{\text{Annual inflation rate}}$

Years to double $= \dfrac{70}{9}$

Years to double $= 7.8$ or about 8 years

59. Let $r =$ the inflation rate as a percent.

Years to double $= \dfrac{70}{\text{Annual inflation rate}}$

$7 = \dfrac{70}{r}$

$7r = 70$

$r = 10.0\%$

61. Let $r =$ the inflation rate as a percent.

Years to double $= \dfrac{70}{\text{Annual inflation rate}}$

$22 = \dfrac{70}{r}$

$22r = 70$

$r \approx 3.2\%$

63. Use the formula for continuous
compounding $A = Pe^{rt}$, changing r and t as
appropriate.

$A = \$5.89e^{0.02(5)}$

$A \approx \$6.51$

$A = \$5.89e^{0.02(15)}$

$A \approx \$7.95$

$A = \$5.89e^{0.10(5)}$

$A \approx \$9.71$

$A = \$5.89e^{0.10(15)}$

$A \approx \$26.40$

65. Use the formula for continuous compounding $A = Pe^{rt}$, changing r and t as appropriate.

$A = \$18500e^{0.02(5)}$
$A \approx \$20,400$

$A = \$18500e^{0.02(15)}$
$A \approx \$25,000$

$A = \$18500e^{0.10(5)}$
$A \approx \$30,500$

$A = \$13800e^{0.10(15)}$
$A \approx \$82,900$

67. Use the inflation proportion and Table 4 to calculate the price of article in 2009.

$$\frac{\text{price in year 2009}}{\text{price in year 2000}} = \frac{\text{CPI in year 2009}}{\text{CPI in year 2000}}$$

$$\frac{P}{175} = \frac{214.5}{172.2}$$

$$175 \cdot \frac{P}{175} = 175 \cdot \frac{214.5}{172.2}$$

$$P \approx \$218$$

69. Use the inflation proportion and Table 4 to calculate the price of article in 2009.

$$\frac{\text{price in year 2009}}{\text{price in year 1996}} = \frac{\text{CPI in year 2009}}{\text{CPI in year 1996}}$$

$$\frac{P}{1099} = \frac{214.5}{156.9}$$

$$1099 \cdot \frac{P}{1099} = 1099 \cdot \frac{214.5}{156.9}$$

$$P \approx \$1502$$

71. Use the formula $P = \dfrac{A}{\left(1+\frac{r}{m}\right)^n}$ with

$A = 300 \cdot \$300,000 = \$90,000,000$; $r = 0.07$;

$m = 4$; $n = 4 \cdot \dfrac{18}{12} = 6$.

$$P = \frac{A}{\left(1+\frac{r}{m}\right)^n}$$

$$P = \frac{\$90000000}{\left(1+\frac{0.07}{4}\right)^6} = \$81,102,828.75$$

13.2 Exercises

1. Subtract the down payment amount from $3450.
$\$3450 - \$500 = \$2950$

3. The sum of the total amount financed and the total interest owed is
$\$2950 + \$472 = \$3422$.

5. Total cost = cost of appliances + interest
$$= \$3450 + \$472$$
$$= \$3922$$

7. Use the simple interest formula with $P = \$13,500$, $r = 0.09$, and $t = 3$.
$I = Prt = (\$13,500)(0.09)(3) = \$3645$

9. From Exercise 8, the total amount owed on the loan is $17,145. The number of monthly payments is $3 \cdot 12 = 36$, so the amount of each monthly payment is
$$\frac{\$17145}{36} = \$476.25.$$

11. The interest charge is:
$I = Prt = (\$4500)(0.09)(3) = \$1215$.
The total amount owed is:
$P + I = \$4500 + \$1215 = \$5715$.
There are $3 \cdot 12 = 36$ monthly payments, so the amount of each monthly payment is:
$$\frac{\$5715}{36} = \$158.75.$$

13. The interest charge is:
$$I = Prt = (\$750)(0.074)\left(\frac{18}{12}\right) = \$83.25.$$

The total amount owed is:
$P + I = \$750 + \$83.25 = \$833.25$.
There are 18 monthly payments, so the amount of each monthly payment is:
$$\frac{\$833.25}{18} = \$46.29.$$

15. The interest charge is:
$$I = Prt = (\$535)(0.111)\left(\frac{16}{12}\right) = \$79.18.$$

The total amount owed is:
$P + I = \$535 + \$79.18 = \$614.18$.
There are 16 monthly payments, so the amount of each monthly payment is:
$$\frac{\$614.18}{16} = \$38.39.$$

17. First subtract the down payment from the purchase price of the furniture to find the value of P.
$8500 - $3000 = $5500
Now use the formula for simple interest with $P = 5500, $r = 0.10$, and $t = 2.5$.
$I = Prt = ($5500)(0.10)(2.5) = 1375
The total amount owed is: $P + I = $5500 + $1375 = 6875.
There are $2.5(12) = 30$ monthly payments, so the amount of each monthly payment is:
$$\frac{$6875}{30} = $229.17.$$

19. First subtract the down payment from the purchase price of the furniture to find the value of P.
$14240 - $2900 = $11340

Now use the formula for simple interest with $P = 11340, $r = 0.10$, and $t = \dfrac{48}{12} = 4$.

$I = Prt = ($11340)(0.10)(4) = 4536
The total amount owed is: $P + I = $11340 + $4536 = 15876.

There are 48 monthly payments, so the amount of each monthly payment is: $\dfrac{$15876}{48} = 330.75.

21. First find the total amount to be paid.
$(48)(314.65) = $15,103.20$
Then the rate is 0.098 and the time is 4 years to correspond to the 48 months.
Let $x =$ amount borrowed.
$$\text{Simple interest} + \text{Amount borrowed} = $15103.20$$
$$(0.098)(4)x + x = $15103.20$$
$$0.392x + x = $15103.20$$
$$1.392x = $15103.20$$
$$x = $10,850$$

23. The total number of payments will be $12 \cdot t$, where $t =$ the time in years. The amount to be paid per year is equivalent to multiplying the monthly payment by 12. This yields: $($172.44) \cdot 12 = 2069.28.
Substitute $P = 8000, $r = 0.092$, into the formula $P + Prt = A$, and solve for t.
$$$8000 + ($8000)(0.092)t = $2069.28t$$
$$$8000 + $736t = $2069.28t$$
$$$8000 + $736t - $736t = $2069.28t - $736t$$
$$$8000 = $1333.28t$$
$$\frac{$8000}{$1333.28} = \frac{$1333.28t}{$1333.28}$$
$$6 \text{ years} = t$$

25. $($249.94)(0.014) = 3.50

27. $($419.95)(0.0138) = 5.80

29. $($1073.40)(0.01425) = 15.30

31. First, make a table showing all the transactions, and compute the balance due on each date.

Date	Balance due
May 9	$728.36
May 17	$728.36 - $200 = $528.36
May 30	$528.36 + $46.11 = $574.47
June 3	$574.47 + $64.50 = $638.97

Next, tabulate the balance due figures, along with the number of days until the balance changed. Use this data to calculate the sum of the daily balances.

Date	Balance due	Number of days until balance changed	$\left(\begin{array}{c}\text{Balance}\\\text{due}\end{array}\right)\times\left(\begin{array}{c}\text{Number}\\\text{of days}\end{array}\right)$
May 9	$728.36	8	$728.36 \times 8 = \$5826.88$
May 17	$528.36	13	$528.36 \times 13 = \$6868.68$
May 30	$574.47	4	$574.47 \times 4 = \$2297.88$
June 3	$638.97	6	$638.97 \times 6 = \$3833.82$
Totals		31	$18,827.26

(a) Average daily balance $= \dfrac{\text{Sum of daily balances}}{\text{Days in billing period}} = \dfrac{\$18827.26}{31} = \$607.33$

(b) Finance charge $= (0.013)(\$607.33) = \$7.90$

(c) Add the finance charge to the June 3 balance of $638.97 to get account balance for next billing:
$638.97 + \$7.90 = \$646.87$

33. First, make a table showing all the transactions, and compute the balance due on each date.

Date	Balance due
June 11	$462.42
June 15	$462.42 - \$106.45 = \$355.97$
June 20	$355.97 + \$115.73 = \$471.70$
June 24	$471.70 + \$74.19 = \$545.89$
July 3	$545.89 - \$115.00 = \$430.89$
July 6	$430.89 + \$68.49 = \$499.38$

Next, tabulate the balance due figures, along with the number of days until the balance changed. Use this data to calculate the sum of the daily balances.

Date	Balance due	Number of days until balance changed	$\left(\begin{array}{c}\text{Balance}\\\text{due}\end{array}\right)\times\left(\begin{array}{c}\text{Number}\\\text{of days}\end{array}\right)$
June 11	$462.42	4	$462.42 \times 4 = \$1849.68$
June 15	$355.97	5	$355.97 \times 5 = \$1779.85$
June 20	$471.70	4	$471.70 \times 4 = \$1886.80$
June 24	$545.89	9	$545.89 \times 9 = \$4913.01$
July 3	$430.89	3	$430.89 \times 3 = \$1292.67$
July 6	$499.38	5	$499.38 \times 5 = \$2496.9$
Totals		30	$14,218.91

(a) Average daily balance

$$= \frac{\text{Sum of daily balances}}{\text{Days in billing period}}$$

$$= \frac{\$14218.91}{30}$$

$$= \$473.96$$

(b) Finance charge $= (0.013)(\$473.96)$
$$= \$6.16$$

(c) Add the finance charge to the July 6 balance of $499.38 to get account balance for next billing:
$499.38 + $6.16 = $505.54

35. Using the average daily balance method, she has a balance of $720 for the first 28 days of the billing period. For the last 3 days of the billing period the balance drops to $120 because of her $600 payment. Then,

$$\text{Average daily balance} = \frac{\$720 \cdot 28 + \$120 \cdot 3}{31}$$
$$= \$661.94,$$

and her finance charge is
$(0.014)(\$661.94) = \$9.27$.

37. (a) $(\$2900)(0.013167) = \$38.18$

(b) At the end of the second month, the interest is first added to the balance before being multiplied by the interest rate.
$(\$2900 + \$38.18)(0.013167) = \$38.69$

(c) At the end of the third month, both interest amounts are added to the balance before being multiplied by the interest rate.
$(\$2900 + \$38.18 + \$38.69)(0.013167)$
$= \$39.20$

39. From Exercises 37 and 38, the total interest is $116.07. Use the formula $I = Prt$, with

$I = \$116.07$, $P = \$2900$, and $t = \frac{1}{4}$ or 0.25.

$\$116.07 = (\$2900)(r)(0.25)$
$116.07 = 725r$
$0.160 = r$, which is 16.0%

41. There is a 2% charge for each cash advance of $100. because she had six cash advances of $100 each, the total charge is
$6 \times 0.02(100) = \$12$.
Add this amount to the late payment fee of $15 and the over-the-credit-limit fee of $5:
$12 + $15 + $5 = $32.

43. Writing exercise; answers will vary.

45. Writing exercise; answers will vary.

47. Writing exercise; answers will vary.

49. (a) If she chooses Bank A, their estimated total yearly cost will be
$(\$900)(0.0118)(12) = \$127.44$.

(b) If she chooses Bank B, their estimated total yearly cost will be
$\$30 + (\$900)(0.0101)(12) = \$139.08$.

13.3 Exercises

1. Find the finance charge per $100 of the amount financed.
$$\frac{75}{1000} \times 100 = \$7.50$$
In Table 5, find the "12 payments" row and read across to find the number closest to 7.50, which is 7.46. Read up to find the APR, which is 13.5%.

3. Find the finance charge per $100 of the amount financed.
$$\frac{750}{6600} \times 100 = \$11.36$$
In table 5, find the "30 payments" row and read across to find the number closest to 11.36, which is 11.35. Read up to find the APR, which is 8.5%.

5. First find the amount financed by subtracting the down payment from the purchase price.
$3000 − $500 = $2500
Now find the total payments by adding the amount financed to the finance charge.
$2500 + $250 = $2750

The monthly payment is $\frac{\$2750}{24} = \$114.58$.

7. First find the amount financed by subtracting the down payment from the purchase price.
$3950 − $300 = $3650
Now find the total payments by adding the amount financed to the finance charge.
$3650 + $800 = $4450

The monthly payment is $\frac{\$4450}{48} = \$92.71$.

9. First find the amount financed by subtracting the down payment from the purchase price.
$4190 − $390 = $3800
Now find the finance charge. The interest rate of 6% will be charged on the amount financed for 1 year (12 payments).
$I = Prt = (3800)(0.06)(1) = \$228$
Next find the finance charge per $100 of the amount financed.
$\dfrac{228}{3800} \times 100 = \$6.00$
In Table 5, the number closest to 6.00 in the "12 payments" row is 6.06. Read up to find the APR, which is 11.0%.

11. First find the amount financed by subtracting the down payment from the purchase price.
$7480 − $2200 = $5280
Now find the finance charge. The interest rate of 5% will be charged on the amount financed for 1.5 years (18 payments).
$I = Prt = (5280)(0.05)(1.5) = \$396$
Next find the finance charge per $100 of the amount financed.
$\dfrac{396}{5280} \times 100 = \$7.50$
In Table 5, the number closest to 7.50 in the "18 payments" row is 7.69. Read up to find the APR, which is 9.5%.

13. (a) Find the intersection of the row for 18 payments and the column for 11.0% to find the value of h: $8.93.

  (b) Use the formula $U = kR\left(\dfrac{h}{100+h}\right)$ to find unearned interest, with
k = remaining number of payments,
R = regular monthly payment, and
h = finance charge per $100.
$$U = (18)(346.70)\left(\dfrac{8.93}{100+8.93}\right)$$
$$= (18)(346.70)\left(\dfrac{8.93}{108.93}\right)$$
$$= \$511.60$$

  (c) The payoff amount is equal to the current payment plus the sum of the scheduled remaining payments minus the unearned interest.
$346.70 + (18)($346.70) − 511.60
= $6075.70

15. (a) Find the intersection of the row for 6 payments and the column for 9.5% to find the value of h: $2.79.

  (b) Use the formula $U = kR\left(\dfrac{h}{100+h}\right)$ to find unearned interest, with
k = remaining number of payments,
R = regular monthly payment, and
h = finance charge per $100.
$$U = (6)(\$595.80)\left(\dfrac{\$2.79}{\$100+\$2.79}\right)$$
$$= (6)(\$595.80)\left(\dfrac{\$2.79}{\$102.79}\right)$$
$$= \$97.03$$

  (c) The payoff amount is equal to the current payment plus the sum of the scheduled remaining payments minus the unearned interest.
$595.80 + (6)($595.80) − $97.03
= $4073.57

17. (a) The amount of the total payments is
(24)($91.50) = $2196.
The finance charge is the difference between this amount and the purchase price.
$2196 − $1990 = $206

  (b) Find the finance charge per $100 of the amount financed.
$\dfrac{206}{1990} \times 100 = \$10.35$
In Table 5, the number closest to 10.35 in the "24 payments" row is 10.19. Read up to find the APR, which is 9.5%.

19. (a) To find the finance charge use the simple interest formula. The interest rate of 6% will be charged on the amount financed for 1.5 years.
$I = Prt = (\$2000)(0.06)(1.5) = \$180$

  (b) Find the finance charge per $100 of the amount financed.
$\dfrac{180}{2000} \times 100 = \$9.00$
In Table 5, the number closest to 9.00 in the "18 payments" row is 8.93. Read up to find the APR, which is 11.0%.

21. (a) Using the actuarial method, first find the APR. If the loan were not paid off early, the total payments would be $(18)(\$201.85) = \$3633.30$, and the finance charge would be $\$3633.30 - \$3310 = \$323.30$. The finance charge per $100 of the amount financed would be

$$\frac{\$323.30}{\$3310} \times 100 = \$9.77.$$

In Table 5, the number 9.77 appears in the "18 payments" row and the 12.0% column, so the APR is 12%. Because the total number of payments remaining is 6, move up to the "6 payments" row to find $h = \$3.53$, the value of h needed to the actuarial method. Use the formula

$$U = kR\left(\frac{h}{100+h}\right)$$ to find unearned

interest, with k = remaining number of payments, R = regular monthly payment, and h = finance charge per $100.

$$U = (6)(\$201.85)\left(\frac{\$3.53}{\$100+\$3.53}\right)$$
$$= (6)(\$201.85)\left(\frac{\$3.53}{\$103.53}\right)$$
$$= \$41.29$$

(b) From part (a), the original finance charge is $323.30. Use the rule of 78 to find unearned interest, with $k = 6$, $n = 18$, and $F = \$323.30$.

$$U = \frac{k(k+1)}{n(n+1)} \times F$$
$$= \frac{6(6+1)}{18(18+1)} \times \$323.30$$
$$= \$39.70$$

23. (a) Using the actuarial method, first find the APR. If the loan were not paid off early, the total payments would be $(60)(\$641.58) = \$38,494.80$, and the finance charge would be $\$38494.80 - \$29850 = \$8644.80$. The finance charge per $100 of the amount financed would be

$$\frac{\$8644.80}{\$29850} \times 100 = \$28.96.$$

In Table 5, the number 28.96 appears in the "60 payments" row and the 10.5% column, so the APR is 10.5%. Because the total number of payments remaining is 12, move up to the "12 payments"

row to find $h = \$5.78$, the value of h needed in the actuarial method.

Use the formula $U = kR\left(\frac{h}{100+h}\right)$ to find unearned interest, with k = remaining number of payments, R = regular monthly payment, and h = finance charge per $100.

$$U = (12)(\$641.58)\left(\frac{\$5.78}{\$100+\$5.78}\right)$$
$$= (12)(\$641.58)\left(\frac{\$5.78}{\$105.78}\right)$$
$$= \$420.68$$

(b) From part (a), the original finance charge is $8644.80. Use the rule of 78 to find unearned interest, with $k = 12$, $n = 60$, and $F = \$8644.80$.

$$U = \frac{k(k+1)}{n(n+1)} \times F$$
$$= \frac{12(12+1)}{60(60+1)} \times \$8644.80$$
$$= \$368.47$$

25. (a) Use the finance charge formula with $n = 4$ and APR = 0.086.

$$h = \frac{n \times \frac{APR}{12} \times \$100}{1 - \left(1 + \frac{APR}{12}\right)^{-n}} - \$100$$
$$= \frac{4 \times \frac{0.086}{12} \times \$100}{1 - \left(1 + \frac{0.086}{12}\right)^{-4}} - \$100$$
$$= \$1.80$$

(b) Use the actuarial formula for unearned interest with $k = 4$, $R = \$212$, and $h = \$1.80$.

$$U = kR\left(\frac{h}{\$100+h}\right)$$
$$= 4(\$212)\left(\frac{\$1.80}{\$100+\$1.80}\right)$$
$$= \$14.99$$

(c) The payoff amount is equal to the current payment added to the sum of the remaining payments minus the unearned interest.
$$\$212 + 4(\$212) - \$14.99 = \$1045.01$$

27. *Finance Company*
Find the sum of the amount borrowed and the interest that will be charged.
$$\$5000 + (\$5000)(0.065)(3) = \$5975$$

Then, the finance charge is
$5975 − $5000 = $975.
The finance charge per $100 is the amount
financed would be $\dfrac{\$975}{\$5000} \times 100 = \$19.50$.
In Table 5, the number closest to 19.50 in
the "36 payments" row is 19.57. Read up to
find the APR, which is 12.0%.

Credit Union
Find the total amount she would pay.
36($164.50) = $5922.00
Then, the finance charge is
$5922 − $5000 = $922.
The finance charge per $100 of the amount
financed would be $\dfrac{\$922}{\$5000} \times 100 = \$18.44$.
In Table 5, the number closest to 18.44 in
the "36 payments" row is 18.71. Read up to
find the APR, which is 11.5%.
The credit union offers the better choice.

29. Use the finance charge formula with $n = 6$
and APR = 0.115. (See Exercise 27.)

$$h = \dfrac{n \times \frac{\text{APR}}{12} \times \$100}{1 - \left(1 + \frac{\text{APR}}{12}\right)^{-n}} - \$100$$

$$= \dfrac{6 \times \frac{0.115}{12} \times \$100}{1 - \left(1 + \frac{0.115}{12}\right)^{-6}} - \$100$$

$$= \$3.38$$

Then, use the actuarial formula for unearned
interest with $k = 6$, $R = \$164.50$, and
$h = \$3.38$.

$$U = kR\left(\dfrac{h}{\$100 + h}\right)$$

$$= 6(\$164.50)\left(\dfrac{\$3.38}{\$100 + \$3.38}\right)$$

$$= \$32.27$$

31. Writing exercise; answers will vary.

33. Writing exercise; answers will vary.

35. Use the rule of 78 to set up an inequality
with $n = 36$.

$$\dfrac{k(k+1)}{n(n+1)} \times F \geq 0.10F$$

$$\dfrac{k(k+1)}{n(n+1)} \geq 0.10$$

$$k(k+1) \geq 0.10 \times n(n+1)$$

$$k(k+1) \geq 0.10 \times 36(36+1)$$

$$k^2 + k \geq 133.2$$

This could be solved as a quadratic
inequality; however, trial and error could
also be applied. Because k must be an
integer, try $k = 11$. Test the inequality:
$11^2 + 11 \geq 133.2$ Not true
Try $k = 12$. Test the inequality:
$12^2 + 12 \geq 133.2$ True
The value of k must be at least 12.

37. Writing exercise; answers will vary.

39. Writing exercise; answers will vary.

41. Writing exercise; answers will vary.

43. If the loan is paid off in 8 months then the
portion (fraction) of the finance charge that
should be saved can be calculated from
adding the remaining 4 months of the
'Fraction of Finance Charge Owed' column
in the table.

$$\dfrac{1}{78} + \dfrac{2}{78} + \dfrac{3}{78} + \dfrac{4}{78} = \dfrac{10}{78} = \dfrac{5}{39}$$

13.4 Exercises

1. In Table 6, find the 10.0% row and read
across to the column for 20 years to find
entry $9.65022. Since this is the monthly
payment amount needed to amortize a loan
of $1000 and this loan is for $70,000, the
required monthly payment is
$70 \times \$9.65022 = \$675.52$.

3. Because the interest rate of 8.7 is not in
Table 6, use the formula for regular monthly
payment with $P = \$57,300$, $r = 0.087$, and
$t = 25$.

$$R = \dfrac{P\left(\frac{r}{12}\right)\left(1 + \frac{r}{12}\right)^{12t}}{\left(1 + \frac{r}{12}\right)^{12t} - 1}$$

$$= \dfrac{\$57300\left(\frac{0.087}{12}\right)\left(1 + \frac{0.087}{12}\right)^{12(25)}}{\left(1 + \frac{0.087}{12}\right)^{12(25)} - 1}$$

$$= \$469.14$$

5. Because the interest rate of 12.5% is not in
Table 6, use the formula for regular monthly
payment with $P = \$227,750$, $r = 0.125$, and

$t = 25.$

$$R = \frac{P\left(\frac{r}{12}\right)\left(1+\frac{r}{12}\right)^{12t}}{\left(1+\frac{r}{12}\right)^{12t}-1}$$

$$= \frac{\$227750\left(\frac{0.125}{12}\right)\left(1+\frac{0.125}{12}\right)^{12(25)}}{\left(1+\frac{0.125}{12}\right)^{12(25)}-1}$$

$$= \$2483.28$$

7. Because the interest rate of 7.6 and the term of the loan, 22 years, are not in Table 6, use the formula for regular monthly payment with $P = \$132,500$, $r = 0.076$, and $t = 22$.

$$R = \frac{p\left(\frac{r}{12}\right)\left(1+\frac{r}{12}\right)^{12t}}{\left(1+\frac{r}{12}\right)^{12t}-1}$$

$$= \frac{\$132500\left(\frac{0.076}{12}\right)\left(1+\frac{0.076}{12}\right)^{12(22)}}{\left(1+\frac{0.076}{12}\right)^{12(22)}-1}$$

$$= \$1034.56$$

9. (a) To find the total payment, use Table 6. Find the row for 10.0% interest; read over to the 30 year column to find 8.77572. Since this is the monthly payment amount needed to amortize a loan of $1000 and this loan is for $58,500, the required monthly payment is $\dfrac{\$58500}{\$1000} \times \$8.77572 = \$513.38.$

(b) This total payment includes both principal and interest. For the first month, interest is charged on the full amount of the mortgage, so use the formula $I = Prt$, with $P = 58500$, $r = 0.10$, and $t = \dfrac{1}{12}.$

$$(\$58500)(0.10)\left(\frac{1}{12}\right) = \$487.50$$

(c) The remainder of the total payment is applied to the principal, so the principal payment is
$\$513.38 - \$487.50 = \$25.88.$

(d) The balance of the principal is
$\$58500 - \$25.88 = \$58,474.12.$

11. (a) To find the total payment, use Table 6. Find the row for 6.5% interest; read over to the 15 year column to find

8.71107. Since this is the monthly payment amount needed to amortize a loan of $1000, and this loan is for $143,200, the required monthly payment is
$$\frac{\$143,200}{\$1000} \times \$8.71107 = \$1247.43.$$

(b) This total payment includes both principal and interest. For the first month, interest is charged on the full amount of the mortgage, so use the formula $I = Prt$, with $P = \$143,200$, $r = 0.065$, and $t = \dfrac{1}{12}.$

$$(\$143200)(0.065)\left(\frac{1}{12}\right) = \$775.67$$

(c) The remainder of the total payment is applied to the principal, so the principal payment is
$\$1247.43 - \$775.67 = \$471.76.$

(d) The balance of the principal is
$\$143200 - \$471.76 = \$142,728.24.$

(e) Every monthly payment is the same, so the second monthly payment is
$\$1247.43.$

(f) The interest payment for the second month is

$$(\$142728.24)(0.065)\left(\frac{1}{12}\right) = \$773.11.$$

(g) The principal payment for the second month is
$\$1247.43 - \$773.11 = \$474.32.$

(h) The balance of principal after the second month is
$\$142,728.24 - \$474.32 = \$142,253.92.$

13. (a) Because the interest rate of 8.2 is not in Table 6, use the formula for regular monthly payment with $P = \$113,650$, $r = 0.082$, and $t = 10$.

$$R = \frac{P\left(\frac{r}{12}\right)\left(1+\frac{r}{12}\right)^{12t}}{\left(1+\frac{r}{12}\right)^{12t}-1}$$

$$= \frac{\$113650\left(\frac{0.082}{12}\right)\left(1+\frac{0.082}{12}\right)^{12(10)}}{\left(1+\frac{0.082}{12}\right)^{12(10)}-1}$$

$$= \$1390.93$$

(b) This total payment includes both principal and interest. For the first month, interest is charged on the full amount of the mortgage, so use the formula $I = Prt$, with $P = \$113{,}650$,

$r = 0.082$, and $t = \dfrac{1}{12}$.

$$(\$113650)(0.082)\left(\frac{1}{12}\right) = \$776.61$$

(c) The remainder of the total payment is applied to the principal, so the principal payment is
$\$1390.93 - \$776.61 = \$614.32$.

(d) The balance of the principal is
$\$113650 - \$614.32 = \$113{,}035.68$.

(e) Every monthly payment is the same, so the second monthly payment is $\$1390.93$.

(f) The interest payment for the second month is

$$(\$113035.68)(0.082)\left(\frac{1}{12}\right) = \$772.41$$

(g) The principal payment for the second month is
$\$1390.93 - \$772.41 = \$618.52$.

(h) The balance of principal after the second month is
$\$113{,}035.68 - \$618.52 = \$112{,}417.16$.

15. Use Table 6 to find the monthly amortization payment (principal and interest).

$\dfrac{\$62300}{\$1000} \times \$7.75299 = \$483.01$

The monthly tax and insurance payment is
$\dfrac{\$610 + \$220}{12} = \$69.17$.

The total monthly payment, including taxes and insurance, is
$\$483.01 + \$69.17 = \$552.18$.

17. Use Table 6 to find the monthly amortization payment (principal and interest).

$\dfrac{\$89560}{\$1000} \times \$11.35480 = \$1016.94$

The monthly tax and insurance payment is

$\dfrac{\$915 + \$409}{12} = \$110.33$.

The total monthly payment, including taxes and insurance is
$\$1016.94 + \$110.33 = \$1127.27$.

19. Because the interest rate is not in Table 6, use the formula for regular monthly payment with $P = \$115{,}400$, $r = 0.088$, and $t = 20$.

$$R = \frac{P\left(\frac{r}{12}\right)\left(1 + \frac{r}{12}\right)^{12t}}{\left(1 + \frac{r}{12}\right)^{12t} - 1}$$

$$= \frac{\$115400\left(\frac{0.088}{12}\right)\left(1 + \frac{0.088}{12}\right)^{12(20)}}{\left(1 + \frac{0.088}{12}\right)^{12(20)} - 1}$$

$= \$1023.49$

The monthly tax and insurance payment is
$\dfrac{\$1295.16 + \$444.22}{12} = \$144.95$.

The total monthly payment, including taxes and insurance, is
$\$1023.49 + \$144.95 = \$1168.44$.

21. $12 \times 30 = 360$

23. The total interest is
$\$387{,}532.80 - \$140{,}000 = \$247{,}532.80$.

25. (a) In Table 8 read the heading of the column on the left to see that the monthly payment is $\$304.01$.

(b) Read the heading of the column on the right to see that the monthly payment is $\$734.73$.

27. Payment number 12 is the last payment for the year.

(a) Balance of principal is $\$59{,}032.06$.

(b) Balance of principal is $\$59{,}875.11$.

29. Compare the Interest Payment column with the Principal Payment column.

(a) The first payment in which the principal payment is higher is number 176.

(b) The first payment which the principal payment is higher is number 304.

31. Use Table 6 and an annual rate of 7.5% interest. Read across to the 10-year column to find 11.87018; this is the monthly payment for a $1000 mortgage. Multiply by 60 to obtain the monthly payment for a $60,000 mortgage:
$60 \times \$11.87018 = \$712.21$.
There would be $10 \times 12 = 120$ monthly payments: $120 \times \$712.21 = \$85,465.20$. Then, the interest is the difference between this amount and the loan amount.
$\$85,465.20 - \$60000 = \$25,465.20$

33. Use Table 6 and an annual rate of 7.5% interest. Read across to the 30-year column to find 6.99215; this is the monthly payment for a $1000 mortgage. Multiply by 60 to obtain the monthly payment for a $60,000 mortgage: $60 \times \$6.99215 = \$419.53$.
There would be $30 \times 12 = 360$ monthly payments: $360 \times \$419.53 = \$151,030.80$. Then, the interest is the difference between this amount and the loan amount.
$\$151030.80 - \$60000 = \$91,030.80$

35. (a) Add the initial index rate and the margin to obtain the ARM interest rate.
$6.5 + 2.5 = 9.0\%$
In Table 6 find 9.0% in the annual rate column and read across to the column for a 20-year mortgage to find 8.99726. Multiply this figure by 75 to obtain the initial monthly payment.
$75 \times \$8.99726 = \$674.79$

(b) The interest rate for the second adjustment period is given by the ARM interest rate.
$8.0 + 2.5 = 10.5\%$
Use the formula for Regular monthly payment with $P = \$73,595.52$ (the Adjusted Balance), $r = 0.105$, and $t = 19$.

$$R = \frac{P\left(\frac{r}{12}\right)\left(1+\frac{r}{12}\right)^{12t}}{\left(1+\frac{r}{12}\right)^{12t} - 1}$$

$$= \frac{\$73595.52\left(\frac{0.105}{12}\right)\left(1+\frac{0.105}{12}\right)^{12(19)}}{\left(1+\frac{0.105}{12}\right)^{12(19)} - 1}$$

$$= \$746.36$$

(c) The change in monthly payment is the difference in the two amounts from parts (a) and (b).
$\$746.36 - \$674.79 = \$71.57$

37. (a) The ARM interest rate is 2% plus 7.5% or 9.5%. Then,
$$\frac{0.095 \times \$50000}{12} = \$395.83.$$

(b) To find the first monthly payment first add 2% and 7.5% to obtain 9.5% as the ARM interest rate. In Table 6 find 9.5% in the annual rate column and read across to the 20-year mortgage column to find 9.32131. Multiply this figure by 50 to obtain the monthly payment for the $50,000 mortgage.
$50 \times \$9.32131 = \$466.07$

39. From Exercise 38, the monthly payment for the first month of the second year is $531.09. the monthly adjustment at the end of the second year is the difference between this amount and the monthly payment amount at the end of the first year.
$\$531.09 - \$466.07 = \$65.02$

41. The down payment is 20% of the purchase price of the house.
$0.20 \times \$175,000 = \$35,000$
Then, the mortgage amount is the difference between the purchase price of the house and this figure.
$\$175,000 - \$35,000 = \$140,000$

43. From Exercise 42, the Loan fee is
$0.02 \times \$140,000 = \$2800$. Add this figure to the other closing costs listed in the text to obtain the total closing costs of $4275.

45. Regular monthly mortgage payment
$= 205 \cdot \$5.99551$
$\approx \$1229$
From Example 7 we know that the monthly income tax savings (for a 20 or 30 year mortgage) is $356.
Monthly property taxes = $160
Total monthly expense $= \$1229 + \$160$
$= \$1389$
Net monthly cost of home $= \$1389 - \$356$
$= \$1033$
Initial net monthly savings $= \$1273 - \$1033$
$= \$240$

47. (a) Monthly mortgage payment
$= 200 \cdot \$8.17083$
$\approx \$1634$

(b) Monthly property tax $= \dfrac{\$960}{12} \approx \$80$

Monthly insurace cost $= \dfrac{\$480}{12} \approx \$40$

Monthly house payment
$= \$1634 + \$80 + \$40$
$\approx \$1754$

(c) Monthly interest
$= \$200,000(0.055)\left(\dfrac{1}{12}\right)$
$\approx \$917$

(d) Monthly tax deductible expense
$= \$917 + \$80$
$= \$997$

(e) Monthly income tax savings
$= \$997 \cdot 20\%$
$\approx \$199$
Net monthly cost of home
$= \$1754 - \$199$
$= \$1555$

49. (a) Monthly mortgage payment
$= 200 \cdot \$11.35480$
$\approx \$2271$

(b) Monthly property tax $= \dfrac{\$1092}{12} = \$91$

Monthly insurance cost $= \dfrac{\$540}{12} = \$45$

Monthly house payment
$= \$2271 + \$91 + \$45$
$= \$2407$

(c) Monthly interest
$= \$200,000(0.065)\left(\dfrac{1}{12}\right)$
$\approx \$1083$

(d) Monthly tax deductible expense
$= \$1083 + \$91$
$= \$1174$

(e) Monthly income tax savings
$= \$1174 \cdot 30\%$
$\approx \$352$
Net monthly cost of home
$= \$2407 - \$352$
$\approx \$2055$

51. Writing exercise; answers will vary.

53. Writing exercise; answers will vary.

55. Writing exercise; answers will vary.

57. Writing exercise; answers will vary.

59. Writing exercise; answers will vary.

13.5 Exercises

1. $20.93

3. $1.88 per share lower

5. $16.37

7. $0.80 per share

9. 3,327,000 shares

11. 13.69% lower

13. 29.62

15. $600 \cdot \$71.28 = \$42,768.00$

17. $500 \cdot \$79.37 = \$39,685.00$

19. The basic cost of stock (Principal Amount) is given by
(Price per share)$\times$(Number of shares)
$= \$13.57 \cdot 60$
$= \$814.20$.
Since this principal amount falls in the first tier of the commission structure (see table in text) the broker's commission is $35 plus 1.7% of this amount:
Broker's commission
$= \$35 + 0.017(\$814.20)$
$= \$48.84$.
The total cost of the shares is
$\$814.20 + \$48.84 = \$863.04$.

21. Principal Amount $= \$15.35 \cdot 355$
$= \$5449.25$
Since commission in this exercise is automated and the number of shares is less than 1000, the broker's commission is $29.95 (see table in text).
Total Cost $= \$5449.25 + \$29.95 = \$5479.20$

23. Principal Amount $= \$27.35 \cdot 2500$
$= \$68,375.00$
Since this principal amount falls in the fifth tier of the commission structure (see table in text), the broker's commission is $155 plus 0.11% of this amount:

Broker's Commission
$$= \$155 + 0.0011(\$68,375.00)$$
$$= \$230.21$$
Total Cost $= \$68,375.00 + \$230.21$
$$= \$68,605.21$$

25. Principal Amount $= \$40.86 \cdot 2400$
$$= \$98,064.00$$
Commission $= \$0.03(2400) = \$72.00$
Total Cost $= \$98,064.00 + \$72.00$
$$= \$98,136.00$$

27. The basic cost (Principal Amount) of stock is given by $\$28.94 \cdot 400 = \$11,576.00$. Since this principal amount falls in the third tier of the commission structure (see table in text), the broker's commission is $76 plus 0.34% of the principal.
Broker's commission
$$= \$76 + 0.0034(\$11,576.00)$$
$$= \$115.36$$
The SEC fee is
$$\frac{\$11,576.00}{\$1000} \cdot \$0.0169 = \$0.19563 = \$0.20.$$
The seller receives
$\$11,576.00 - \$115.36 - \$0.20 = \$11,460.44$

29. The basic cost (Principal Amount) of stock is given by $\$71.28 \cdot 500 = \$35,640.00$. Since commission in this exercise is automated and the number of shares is less than 1000, the broker's commission is $29.95 (see table in text).
The SEC fee is
$$\frac{\$35,640.00}{\$1000} \cdot \$0.0169 = \$0.602316 = \$0.61.$$
The seller receives
$\$35,640.00 - \$29.95 - \$0.61 = \$35,609.44.$

31. The basic cost (Principal Amount) of stock is given by $\$79.37 \cdot 1350 = \$107,149.50$. Since commission in this exercise is automated and the number of shares is more than 1000, the broker's commission is $\$0.03(1350) = \$40.50$.
The SEC fee is
$$\frac{\$107,149.50}{\$1000} \cdot \$0.0169 = \$1.81082655$$
$$= \$1.82.$$
The seller receives
$\$107,149.50 - \$40.50 - \$1.82$
$= \$107,107.18.$

33. The basic cost (Principal Amount) of stock is given by $\$14.24 \cdot 1480 = \$21,075.20$. Since this principal amount falls in the fourth tier of the commission structure (see table in text), the broker's commission is $100 plus 0.22% of the principal.
Commission $= \$100 + 0.0022(\$21,075.20)$
$$= \$146.37.$$
The SEC fee is
$$\frac{\$21,075.20}{\$1000} \cdot \$0.0169 = \$0.35617088$$
$$= \$0.36.$$
The seller receives
$\$21,075.20 - \$146.37 - \$0.36$
$= \$20,928.47.$

35. Purchase cost
Principal Amount $= \$33.69 \cdot 100 = \$3369.00$
Commission $= \$65 + 0.0066(\$3369.00)$
$$= \$87.24$$
Total cost $= \$3369.00 + \$87.24 = \$3456.24$

Sales profit
Principal Amount $= \$73.30 \cdot 20 = \$1466.00$
Commission $= \$35 + 0.017(\$1466.00)$
$$= \$59.92$$
SEC fee $= \dfrac{\$1466.00}{\$1000} \cdot \$0.0169 = \$0.03$
He receives
$\$1466 - \$59.92 - \$0.03 = \$1406.05.$
This results in a
$\$3456.24 - \$1406.05 = \$2050.19$ net paid out.

37. (a) The total purchase price is
($20 per share) $\times$ (40 shares) = $800.

(b) The total dividend amount is
($2 per share) $\times$ (40 shares) = $80.

(c) The capital gain is found by multiplying the change in price per share times the number of shares.
($44 − $20) $\times$ (40 shares) = $960

(d) The total return is the sum of the dividends and the capital gain.
$80 + $960 = $1040

(e) The percentage return is the quotient of the total return and the total cost as a percent.
$$\frac{\$1040}{\$800} \times 100 = 130\%$$

39. (a) The total purchase price is
    ($12.50 per share) × (100 shares)
    = $1250.

    (b) The total dividend amount is
    ($1.08 per share) × (100 shares) = $108.

    (c) The capital gain is found by multiplying
    the change in price per share times the
    number of shares.
    ($10.15 − $12.50) × (100 shares)
    = −$235

    (d) The total return is the sum of the
    dividends and the capital gain.
    $108 + (−$235) = −$127

    (e) The percentage return is the quotient of
    the total return and the total cost as a
    percent.
    $$\frac{-\$127}{\$1250} \times 100 = -10.16\%$$

41. Using the formula for simple interest with
    $P = \$1000$, $r = 0.055$, and $t = 5$ the total
    return is:
    $I = Prt = \$1000 \times 0.055 \times 5 = \$275.$

43. Using the formula for simple interest with
    $P = \$10,000$, $r = 0.0711$, and $t = \dfrac{3}{12}$ the
    total return is:
    $$I = Prt = \$10000 \times 0.0711 \times \frac{3}{12} = \$177.75.$$

45. (a) Use the formula for net asset value with
    $A = \$875$ million, $L = \$36$ million, and
    $N = 80$ million.
    $$NAV = \frac{A-L}{N} = \frac{875-36}{80} = \$10.49$$

    (b) Find the number of shares purchased by
    dividing the amount invested by the net
    asset value. Round to the nearest share.
    $$\frac{\$3500}{\$10.49} = 334 \text{ shares}$$

47. (a) Use the formula for net asset value with
    $A = \$2.31$ billion ($2,310 million),
    $L = \$135$ million, and $N = 263$ million.
    $$NAV = \frac{A-L}{N} = \frac{2310-135}{263} = \$8.27$$

    (b) Find the number of shares purchased by
    dividing the amount invested by the net

asset value. Round to the nearest share.
$$\frac{\$25470}{\$8.27} = 3080 \text{ shares}$$

49. (a) Find the monthly return by multiplying
    the monthly percentage return, as a
    decimal, times the amount invested.
    $0.013 \times \$645 = \$8.39$

    (b) Find the annual return by multiplying
    the monthly return by 12.
    $12 \times \$8.39 = \$100.68$

    (c) The annual percentage return is the
    ratio of the annual return to the amount
    invested, expressed as a percent.
    $$\frac{\$100.68}{\$645} \times 100 = 15.6\%$$

51. (a) Find the monthly return by multiplying
    the monthly percentage return, as a
    decimal, times the amount invested.
    $0.023 \times \$2498 = \$57.45$

    (b) Find the annual return by multiplying
    the monthly return by 12.
    $12 \times \$57.45 = \$689.40$

    (c) The annual percentage return is the
    ratio of the annual return to the amount
    invested, expressed as a percent.
    $$\frac{\$689.40}{\$2498} \times 100 = 27.6\%$$

53. (a) The beginning value of the investment
    is the product of the beginning net asset
    value and the number of shares
    purchased.
    $\$9.63 \times 125 = \$1203.75$

    (b) To find the first monthly return,
    multiply the monthly percentage return,
    in decimal form, times the beginning
    value from part (a).
    $0.015 \times \$1203.75 = \$18.06$

    (c) Using Example 9 from the text as a
    guide, find the effective annual rate of
    return as follows. First find the return
    relative.
    $1 + 1.5\% = 1 + 0.015 = 1.015$
    Raise this value to the 12th power to get
    the annual return relative and subtract 1
    to get the percentage rate.
    $(1.015)^{12} - 1 \approx 0.1956$, which is
    19.56%.

55. (a) The beginning value of the investment is the product of the beginning net asset value and the number of shares purchased.
$11.94 \times 350 = \$4179$

(b) To find the first monthly return, multiply the monthly percentage return, in decimal form, times the beginning value from part (a).
$0.0183 \times \$4179 = \$76.48$

(c) Using Example 9 from the text as a guide, find the effective annual rate of return as follows. First find the return relative.
$1 + 1.83\% = 1 + 0.0183 = 1.0183$
Raise this value to the 12th power to get the annual return relative and subtract 1 to get the percentage rate.
$(1.0183)^{12} - 1 \approx 0.2431$, which is 24.31%

57. (a) Principal Amount
$10 \times \$1.00 = \$10.00$

(b) Broker-assisted Commission
$\$35 + 0.017 \times \$10.00 = \$35.17$

(c) Automated Commission = $29.95

59. (a) Principal Amount
$400 \times \$1.00 = \$400.00$

(b) Broker-assisted Commission
$\$35 + 0.017 \times \$400.00 = \$41.80$

(c) Automated Commission = $29.95

61. (a) Principal Amount
$4000 \times \$1.00 = \$4000.00$

(b) Broker-assisted Commission
$\$65 + 0.0066 \times \$4000.00 = \$91.40$

(c) Automated Commission
$\$0.03 \times 4000.00 = \$120.00$

63. A broker-assisted purchase is cheaper than an automated purchase only when a relatively _large_ number of shares are purchased for a relatively _low_ price per share.

65.
$$V = R\left(\frac{(1+r)^n - (1+i)^n}{r-i}\right)$$
$$= \$2000\left(\frac{(1+0.05)^{25} - (1+0.03)^{25}}{0.05-0.03}\right)$$
$$= \$129,258$$

67.
$$V = R\left(\frac{(1+r)^n - (1+i)^n}{r-i}\right)$$
$$= \$5000\left(\frac{(1+0.065)^{40} - (1+0.01)^{40}}{0.065-0.01}\right)$$
$$= \$993,383$$

69. (a)
$$V = \frac{(1-t)R((1+r)^n - 1)}{r}$$
$$= \frac{(1-0.15)\$1000((1+0.05)^{30} - 1)}{0.05}$$
$$= \$56,473.02$$

(b)
$$V = \frac{R((1+r(1-t))^n - 1)}{r}$$
$$= \frac{\$1000((1+0.05(1-0.15))^{30} - 1)}{0.05}$$
$$= \$49,712.70$$

71. (a)
$$V = \frac{(1-t)R((1+r)^n - 1)}{r}$$
$$= \frac{(1-0.35)\$1500((1+0.06)^{10} - 1)}{0.06}$$
$$= \$12,851.28$$

(b)
$$V = \frac{R((1+r(1-t))^n - 1)}{r}$$
$$= \frac{R((1+0.06(1-0.35))^{10} - 1)}{0.06}$$
$$= \$11,651.81$$

73. (a)
$$A = P\left(1+\frac{r}{m}\right)^n$$

$$\frac{A}{P} = \left(1+\frac{r}{m}\right)^n$$

$$\left(\frac{A}{P}\right)^{1/n} = \left(\left(1+\frac{r}{m}\right)^n\right)^{1/n}$$

$$\left(\frac{A}{P}\right)^{1/n} = \left(1+\frac{r}{m}\right)$$

$$\left(\frac{A}{P}\right)^{1/n} - 1 = \frac{r}{m}$$

$$r = m\left(\left(\frac{A}{P}\right)^{1/n} - 1\right)$$

(b)
$$r = m\left(\left(\frac{A}{P}\right)^{1/n} - 1\right)$$

$$= 4\left(\left(\frac{85.49}{50}\right)^{1/(4 \cdot 10)} - 1\right)$$

$$\approx 0.054$$

$$= 5.4\%$$

75. *Aggressive Growth*
7% of $20,000 = 0.07 × $20000 = $1400
Growth
43% of $20,000 = 0.43 × $20000 = $8600
Growth & Income
31% of $20,000 = 0.31 × $20000 = $6200
Income
14% of $20,000 = 0.14 × $20000 = $2800
Cash
5% of $20,000 = 0.05 × $20000 = $1000

77. *Aggressive Growth*
2% of $400,000 = 0.02 × $400000 = $8000
Growth
28% of $400,000 = 0.28 × $400000
$= $112,000$
Growth & Income
36% of $400,000 = 0.36 × $400000
$= $144,000$
Income
29% of $400,000 = 0.29 × $400000
$= $116,000$
Cash
5% of $400,000 = 0.05 × $400000
$= $20,000$

79. First find the difference of 100% and 25%.
100% − 25% = 75%
Then find 75% of 5%
0.75 × 0.05 = 0.0375
The tax-exempt rate of return is 3.75%.

81. First find the difference of 100% and 35%.
100% − 35% = 65%
Then find 65% of 8%.
0.65 × 0.08 = 0.052
The tax-exempt rate of return is 5.2%.

83. Writing exercise; answers will vary.

85. Writing exercise; answers will vary.

87. Writing exercise; answers will vary.

89. Writing exercise; answers will vary.

***EXTENSION: PONZI SCHEMES AND
OTHER INVESTMENT FRAUDS***

1.

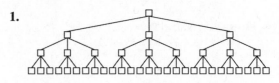

3. 3^{N-1}

5. Profit = $27 − $1 = $26

7. The total number of members in levels 1
through N, inclusive, is $1+3+9+\cdots+3^{N-1}$.
This is the sum of a <u>geometric</u> sequence
with $a = \underline{1}$, $r = \underline{3}$, and $n = \underline{N}$. So the total
number of members is $\dfrac{3^N-1}{\underline{2}}$.

9. If $N = 6$,
$13 \cdot 3^{N-3} = 13 \cdot 3^{6-3} = 13 \cdot 3^3 = 13 \cdot 27 = 351.$
351 members lose if the scheme runs
through level 6 and then fails.

11. Fraction who lose $= \dfrac{3^{N-1}+3^{N-2}+3^{N-3}}{\frac{3^N-1}{2}}$

$$= 3^{N-3}(3^2+3+1)\cdot\frac{2}{3^N-1}$$

$$= \frac{3^{N-3}(13\cdot 2)}{3^N-1}$$

$$= \frac{3^{N-3}\cdot 26}{3^N-1}$$

$$> \frac{3^{N-3}\cdot 26}{3^N}$$

$$= \frac{26}{3^{N-(N-3)}}$$

$$= \frac{26}{3^3}$$

$$= \frac{26}{27}$$

More than $\dfrac{26}{27}$ of all members will lose.

13. Profit $= \$1 - \$1 = \$0.00$

15. You are in the chart only from when you pay your entry fee until you move up three levels and get paid.

17. Writing exercise; answers will vary.

19. From Example 4, at the end of the first year eight investors are required.
In the second year, the following numbers of investors are required at the end of each quarter:
first quarter: 16 investors
second quarter: 32 investors
third quarter: 64 investors
fourth quarter: 128 investors

21. 10% of the 4000 investors is 400 investors. If these 400 investors invest another $1000 rather than taking out profit, another $400 \cdot \$1000 = \$400{,}000$ is invested in the scheme, for a total of $\$4{,}000{,}000 + \$400{,}000 = \$4{,}400{,}000$ that stays with the operator going into the second year.

23. In the 88 years from 1920 to 2008, with 3% inflation, $4 million would be equivalent to
$(\$4 \text{ million})(1.03)^{88} \approx \$54 \text{ million}$ in today's dollars.

$\dfrac{\$21 \text{ billion}}{\$54 \text{ million}} = \dfrac{\$21{,}000 \text{ million}}{\$54 \text{ million}} \approx 388.9$
Madoff's $21 billion is nearly 400 times as much as Ponzi's $4 million, after adjusting for inflation.

25. Writing exercise; answers will vary.

Chapter 13 Test

1. Use the formula $A = P(1 + rt)$, with $P = \$100$; $r = 0.06$; and $t = 5$.
$A = P(1+rt)$
$\quad = 100(1+0.06\cdot 5)$
$\quad = 100(1+0.30)$
$\quad = 100(1.30)$
$\quad = \$130$

2. To find the final amount, use the formula for future value $A = P\left(1+\dfrac{r}{m}\right)^n$, with $P = \$50$, $r = 0.08$, $m = 4$ and $n = 4 \cdot 2 = 8$ (compounded 4 times per year for 2 years).
$A = P\left(1+\dfrac{r}{m}\right)^n$
$\quad = 50\left(1+\dfrac{0.08}{4}\right)^8$
$\quad = 50(1.02)^8$
$\quad = 50(1.171659)$
$\quad = \$58.58$

3. Use the formula $Y = \left(1+\dfrac{r}{m}\right)^m - 1$, with $r = 0.03$ and $m = 12$.
$Y = \left(1+\dfrac{0.03}{12}\right)^{12} - 1$
$\quad = (1.0025)^{12} - 1$
$\quad \approx 0.0304159569$
This is 3.04%, to the nearest hundredth of a percent.

4. Years to double $= \dfrac{70}{\text{Annual inflation rate}}$
Years to double $= \dfrac{70}{5}$
Years to double $= 14$ years

5. Use the present value formula for compound interest with $A = \$100,000$, $r = 0.04$, $m = 2$, and $n = 2 \cdot 10 = 20$.

$$P = \frac{A}{\left(1 + \frac{r}{m}\right)^n}$$

$$= \frac{100000}{\left(1 + \frac{0.04}{2}\right)^{20}}$$

$$= \frac{100000}{(1.02)^{20}}$$

$$\approx \$67297.13$$

6. The interest due is 1.6% of $680.
$0.016 \times \$680 = \$10.88$

7. Use the simple interest formula with $P = \$3000$, $r = 0.075$, and $t = 2$. Because he made a down payment of $1000, he will pay interest on only $3000. In this formula, t must be expressed in years, so 24 months is equivalent to 2 years.
$I = Prt = (3000)(0.075)(2) = \$450$

8. The total amount owed is
$P + I = \$3000 + \$450 = \$3450$.
There are 24 monthly payments, so the amount of each monthly payment is
$$\frac{\$3450}{24} = \$143.75.$$

9. Find the finance charge per $100 of the amount financed.
$$\frac{\$450}{\$3000} \times 100 = \$15$$
In Table 6, find the "24 payments" row and read across to find the number closest to 15. Since 15.23 (associated with an APR of 14.0%) is the last entry in the table, choose 14.0%, for an approximate APR.

10. Use the Actuarial Method formula
$$U = kR\left(\frac{h}{100 + h}\right)$$ to find unearned interest, with k = remaining number of payments, R = regular monthly payment, and h = finance charge per $100. Use the results of Exercises 8 and 9 in formula. Use Table 6, with 6 payments and APR = 14%, to obtain $h = \$4.12$.
$$U = (6)(\$143.75)\left(\frac{\$4.12}{100 + \$4.12}\right)$$
$$= (6)(\$143.75)\left(\frac{\$4.12}{\$104.12}\right)$$
$$\approx \$34.13$$

11. First find the finance charge by multiplying $5 by 6.
$6 \times \$5 = \$30$
Find the finance charge per $100 of the amount financed.
$$\frac{\$5}{\$150} \times \$100 \approx \$3.33$$
In Table 6, the number closest to $3.33 in the "6 payments" row is $3.38. Read up to find the APR, which is 11.5%.

12. Writing exercise; answers will vary.

13. In Table 7, find the 7.0% row and read across to the column for 20 years to find entry $7.75299. Since this is the monthly payment amount needed to amortize a loan of $1000 and this loan is for $150,000, the required monthly payment is
$$\frac{\$150000}{\$1000} \times \$7.75299 = \$1162.95.$$

14. Subtract the down payment from the purchase price of the house to find the principal.
$\$218000 - (0.20 \times \$218000) = \$174,400$
In Table 7, find the 8.5% row and read across to the column for 30 years to find entry $7.68913. Since this is the monthly payment amount needed to amortize a loan of $1000, and this loan is for $174,400, the required monthly payment is
$$\frac{\$174400}{\$1000} \times \$7.68913 = \$1340.98.$$
The monthly tax and insurance payment is
$$\frac{\$1500 + \$750}{12} = \$187.50.$$
The total monthly payment, including taxes and insurance, is
$\$1340.98 + \$187.50 = \$1528.48.$

15. The "two points" charged by the lender mean 2%. This additional cost is the percentage taken of the amount borrowed, the principal from Exercise 14.
$0.02 \times \$174400 = \$3488$

16. Writing exercise; answers will vary.

17. If the index started at 7.85%, adding the margin of 2.25% gives
$7.85\% + 2.25\% = 10.1\%$.
Because the periodic rate cap is 2%, the interest rate during the second year cannot exceed $10.1\% + 2.00\% = 12.1\%$.

18. 15,235,100 shares

19. First find how much she paid for the stock by multiplying the price per share by 1000.
$12.75 × 1000 = $12,750
Calculate the amount of money she received from both dividends.
$1.38 × 1000 + $1.02 × 1000 = $2400
Subtract this amount from the total price she paid.
$12750 − $2400 = $10350
Finally subtract this amount from the price for which she sold the stock.
($10.36 × 1000) − $10350 = $10

20. Compare her amount of return to how much she originally paid for the stock.
$$\frac{\$10}{\$12750} = 0.00078$$
This is 0.078%. This percentage is not the annual rate of return, because the time period was for more than one year.

21. $V = \dfrac{(1-t)R((1+r)^n - 1)}{r}$

$ = \dfrac{(1-0.25)\$1800((1+0.06)^{30} - 1)}{0.06}$

$ = \$106,728.55$

22. Writing exercise; answers will vary.

Chapter 14

14.1 Exercises

1. (a) The complement of $30°$ is
$90° - 30° = 60°$.

   (b) The supplement of $30°$ is
$180° - 30° = 150°$.

3. (a) The complement of $45°$ is
$90° - 45° = 45°$.

   (b) The supplement of $45°$ is
$180° - 45° = 135°$.

5. (a) The complement of $89°$ is
$90° - 89° = 1°$.

   (b) The supplement of $89°$ is
$180° - 89° = 91°$.

7. Given an angle measures x degrees and two angles are complementary if their sum is $90°$, then the complement of an angle of $x°$ is $(90 - x)°$.

9. $62°18' + 21°41' = 83°59'$

11. $71°58' + 47°29' = 118°87'$
$= 118°(60 + 27)'$
$= 119°27'$

13. $90° - 51°28' = 89°60' - 51°28' = 38°32'$

15. $90° - 72°58'11'' = 89°59'60'' - 72°58'11''$
$= 17°1'49''$

Remember for Exercises 17–21:
$$1' = \frac{1°}{60} \text{ and } 1'' = \frac{1°}{3600}.$$

17. $20°54' = 20° + \dfrac{54°}{60'} = 20° + 0.900° = 20.900°$

19. $91°35'54'' = 91° + \dfrac{35°}{60} + \dfrac{54°}{3600} \approx 91.598°$

21. $274°18'59'' = 274° + \dfrac{18°}{60} + \dfrac{59°}{3600} \approx 274.316°$

23. $31.4296° = 31° + 0.4296°$
$= 31° + (0.4296)(60')$
$= 31° + 25.776'$
$= 31° + 25' + (0.776)(60'')$
$\approx 31° + 25' + 47''$
$= 31°25'47''$

25. $89.9004° = 89° + 0.9004°$
$= 89° + (0.9004)(60')$
$= 89° + 54.024'$
$= 89° + 54' + (0.024)(60'')$
$\approx 89° + 54' + 1''$
$= 89°54'1''$

27. $178.5994° = 178° + 0.5994°$
$= 178° + (0.5994)(60')$
$= 178° + 35.964'$
$= 178° + 35' + (0.964)(60'')$
$\approx 178° + 35' + 58''$
$= 178°35'58''$

29. $-40°$ is coterminal with $-40° + 360° = 320°$.

31. $-125°$ is coterminal with
$-125° + 360° = 235°$.

33. $539°$ is coterminal with $539° - 360° = 179°$.

35. $850°$ is coterminal with
$850° - 2(360°) = 130°$.

37. $30°$
A coterminal angle can be obtained by adding an integer multiple of $360°$ or $30° + n \cdot 360°$.

39. $60°$
A coterminal angle can be obtained by adding an integer multiple of $360°$:
$60° + n \cdot 360°$.

41. A positive angle coterminal with $75°$ is
$75° + 360° = 435°$.
A negative angle coterminal with $75°$ is
$75° - 360° = -285°$.
These angles are in quadrant I.

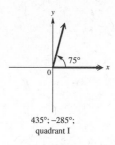

435°; −285°;
quadrant I

43. A positive angle coterminal with 174° is
174° + 360° = 534°.
A negative angle coeterminal with 174° is
174° − 360° = −186°.
These angles are in quadrant II.

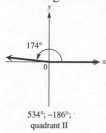

534°; −186°;
quadrant II

45. A positive angle coterminal with 300° is
300° + 360° = 660°.
A negative angle coterminal with 300° is
300° − 360° = −60°.
these angles are in quadrant IV.

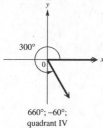

660°; −60°;
quadrant IV

47. A positive angle coterminal with −61° is
−61° + 360° = 299°.
A negative angle coterminal with −61° is
−61° − 360° = −421°.
These angles are in quadrant IV.

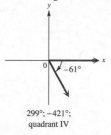

299°; −421°;
quadrant IV

14.2 Exercises

1.

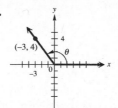

3.

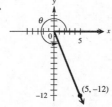

5. (−3, 4)

$x = -3,\ y = 4$

$r = \sqrt{(-3)^2 + 4^2} = \sqrt{9 + 16} = \sqrt{25} = 5$

$\sin\theta = \dfrac{y}{r} = \dfrac{4}{5}$ $\qquad \cos\theta = \dfrac{x}{r} = \dfrac{-3}{5} = -\dfrac{3}{5}$

$\tan\theta = \dfrac{y}{x} = \dfrac{4}{-3} = -\dfrac{4}{3}$ $\qquad \cot\theta = \dfrac{x}{y} = \dfrac{-3}{4} = -\dfrac{3}{4}$

$\sec\theta = \dfrac{r}{x} = \dfrac{5}{-3} = -\dfrac{5}{3}$ $\qquad \csc\theta = \dfrac{r}{y} = \dfrac{5}{4}$

7. (0, 2)

$x = 0,\ y = 2$

$r = \sqrt{(0)^2 + 2^2} = \sqrt{0 + 4} = \sqrt{4} = 2$

$\sin\theta = \dfrac{y}{r} = \dfrac{2}{2} = 1$ $\qquad \cos\theta = \dfrac{x}{r} = \dfrac{0}{2} = 0$

$\tan\theta = \dfrac{y}{x} = \dfrac{2}{0};$ $\qquad \cot\theta = \dfrac{x}{y} = \dfrac{0}{2} = 0$
undefined

$\sec\theta = \dfrac{r}{x} = \dfrac{2}{0};$ $\qquad \csc\theta = \dfrac{r}{y} = \dfrac{2}{2} = 1$
undefined

9. $\left(1, \sqrt{3}\right)$

$x = 1,\ y = \sqrt{3}$

$r = \sqrt{(1)^2 + \left(\sqrt{3}\right)^2} = \sqrt{1 + 3} = \sqrt{4} = 2$

$\sin\theta = \dfrac{y}{r} = \dfrac{\sqrt{3}}{2}$

$\cos\theta = \dfrac{x}{r} = \dfrac{1}{2}$

$\tan\theta = \dfrac{y}{x} = \dfrac{\sqrt{3}}{1} = \sqrt{3}$

$$\cot\theta = \frac{x}{y} = \frac{1}{\sqrt{3}} = \frac{\sqrt{3}}{3}$$

$$\sec\theta = \frac{r}{x} = \frac{2}{1} = 2$$

$$\csc\theta = \frac{r}{y} = \frac{2}{\sqrt{3}} = \frac{2\sqrt{3}}{3}$$

(Note that for cot θ and csc θ answers, the denominator is "rationalized.") For example,

$$\cot\theta = \frac{x}{y} = \frac{1}{\sqrt{3}} \cdot \frac{\sqrt{3}}{\sqrt{3}} = \frac{\sqrt{3}}{3}.$$

11. (3, 5)

$x = 3, y = 5$

$$r = \sqrt{3^2 + 5^2} = \sqrt{9 + 25} = \sqrt{34}$$

$$\sin\theta = \frac{y}{r} = \frac{5}{\sqrt{34}} = \frac{5\sqrt{34}}{4}$$

$$\cos\theta = \frac{x}{r} = \frac{3}{\sqrt{34}} = \frac{3\sqrt{34}}{34}$$

$$\tan\theta = \frac{y}{x} = \frac{5}{3}$$

$$\cot\theta = \frac{x}{y} = \frac{3}{5}$$

$$\sec\theta = \frac{r}{x} = \frac{\sqrt{34}}{3}$$

$$\csc\theta = \frac{r}{y} = \frac{\sqrt{34}}{5}$$

13. (−8, 0)

$x = -8, y = 0$

$$r = \sqrt{(-8)^2 + 0^2} = \sqrt{64 + 0} = \sqrt{64} = 8$$

$$\sin\theta = \frac{y}{r} = \frac{0}{8} = 0 \qquad \cos\theta = \frac{x}{r} = \frac{-8}{8} = -1$$

$$\tan\theta = \frac{y}{x} = \frac{0}{-8} = 0 \qquad \cot\theta = \frac{x}{y} = \frac{-8}{0};$$
$$\text{undefined}$$

$$\sec\theta = \frac{r}{x} = \frac{8}{-8} = -1 \qquad \csc\theta = \frac{r}{y} = \frac{8}{0};$$
$$\text{undefined}$$

15. Writing exercise, answers will vary.

17. The value r is the distance from a point (x, y) on the terminal side of the angle to the origin.

In Exercises 19–25, $r = \sqrt{x^2 + y^2}$, which is positive.

19. In quadrant II, y is positive, so $\frac{y}{r}$ is positive.

21. In quadrant III, y is negative, so $\frac{y}{r}$ is negative.

23. In quadrant IV, x is positive, so $\frac{x}{r}$ is positive.

25. In quadrant IV, x is positive and y is negative, so $\frac{y}{x}$ is negative.

For Exercises 27–31 choose any point on the terminal side of a 90° angle such as (0, 1). Then $r = \sqrt{x^2 + y^2} = \sqrt{0^2 + 1^2} = 1.$

27. Since $x = 0, r = 1,$ and $\cos\theta = \frac{x}{r}$,

$$\cos 90° = \frac{0}{1} = 0.$$

29. Since $x = 0, y = 1,$ and $\tan\theta = \frac{y}{x}$,

$$\tan 90° = \frac{1}{0}, \text{ which is undefined.}$$

31. Since $x = 0, r = 1,$ and $\sec\theta = \frac{r}{x}$,

$$\sec 90° = \frac{1}{0}, \text{ which is undefined.}$$

33. Selecting a point such as (−1, 0) on the terminal side of the angle 180°, we have $r = \sqrt{x^2 + y^2} = \sqrt{(-1)^2 + 0^2} = 1.$

Since $y = 0, r = 1,$ and $\sin\theta = \frac{y}{r}$,

$$\sin 180° = \frac{0}{1} = 0.$$

35. Selecting a point such as (−1, 0) on the terminal side of the angle 180°, we have $r = \sqrt{x^2 + y^2} = \sqrt{(-1)^2 + 0^2} = 1.$

Since $x = -1$, $y = 0$, and $\tan\theta = \dfrac{y}{x}$,

$$\tan 180° = \dfrac{0}{-1} = 0.$$

37. Selecting a point such as $(0, 1)$ on the terminal side of the angle $-270°$, we have

$$r = \sqrt{x^2 + y^2} = \sqrt{0^2 + (1)^2} = 1.$$

Since $y = 1$, $r = 1$, and $\sin\theta = \dfrac{y}{r}$,

$$\sin(-270°) = \dfrac{1}{1} = 1.$$

39. Selecting a point such as $(1, 0)$ on the terminal side of the angle $0°$, we have

$$r = \sqrt{x^2 + y^2} = \sqrt{1^2 + 0^2} = 1.$$

Since $x = 1$, $y = 0$, and $\tan\theta = \dfrac{y}{x}$,

$$\tan 0° = \dfrac{0}{1} = 0.$$

41. Selecting a point such as $(-1, 0)$ on the terminal side of the angle $180°$ we have

$$r = \sqrt{x^2 + y^2} = \sqrt{(-1)^2 + 0^2} = 1.$$

Since $x = -1$, $r = 1$, and $\cos\theta = \dfrac{x}{r}$,

$$\cos 180° = \dfrac{-1}{1} = -1.$$

14.3 Exercises

1. Since $\tan\theta = \dfrac{1}{\cot\theta}$ and $\cot\theta = -3$,

$$\tan\theta = \dfrac{1}{-3} = -\dfrac{1}{3}.$$

3. Since $\sin\theta = \dfrac{1}{\csc\theta}$ and $\csc\theta = 3$,

$$\sin\theta = \dfrac{1}{3}.$$

5. Since $\cot\beta = \dfrac{1}{\tan\beta}$ and $\tan\beta = -\dfrac{1}{5}$,

$$\cot\beta = \dfrac{1}{-\frac{1}{5}} = -5.$$

7. Since $\csc\alpha = \dfrac{1}{\sin\theta}$ and $\sin\alpha = \dfrac{\sqrt{2}}{4}$,

$$\csc\alpha = \dfrac{1}{\frac{\sqrt{2}}{4}} = \dfrac{4}{\sqrt{2}} = 2\sqrt{2}.$$

9. Since $\tan\theta = \dfrac{1}{\cot\theta}$ and $\cot\theta = -\dfrac{\sqrt{5}}{3}$,

$$\tan\theta = \dfrac{1}{-\frac{\sqrt{5}}{3}} = -\dfrac{3}{\sqrt{5}} = -\dfrac{3\sqrt{5}}{5}.$$

11. Since $\sin\theta = \dfrac{1}{\csc\theta}$ and

$$\csc\theta = 1.5 = 1\dfrac{1}{2} = \dfrac{3}{2}, \quad \sin\theta = \dfrac{1}{\frac{3}{2}} = \dfrac{2}{3}.$$

13. Since the sine function is positive and the cosine function is negative only for angles in quadrant II, α must be an angle in quadrant II.

15. Since the tangent function is positive and the sine function is positive only for angles in quadrant I, γ must be an angle in quadrant I.

17. Since the tangent function and the cosine function are negative only for angles in quadrant II, ω must be an angle in quadrant II.

19. Since the cosine function is negative for angles in quadrants II and III, β may be an angle in quadrant II or III.

21. Since the sine function is negative for angles in quadrants III and IV, θ may be an angle in quadrant III or IV.

For Exercises 23–29 each angle is assumed to be in standard position (initial ray along the positive x-axis).

23. Since $74°$ is an acute angle, its terminal ray lies in quadrant I. Since all trig functions are positive for angles in quadrant I, the trigonometric functions of $74°$ are all positive.

25. Since $183°$ is an angle in quadrant III, for any point (x, y) on the terminal ray, $x < 0$ and $y < 0$. Note that r is always positive. The signs of the trigonometric functions for $183°$ are

$$\sin 183°: \dfrac{y}{r} = \dfrac{-}{+} = -$$

$$\csc 183°: \dfrac{r}{y} = \dfrac{+}{-} = -$$

$\cos 183°$: $\dfrac{x}{r} = \dfrac{-}{+} = -$

$\sec 183°$: $\dfrac{r}{x} = \dfrac{+}{-} = -$

$\tan 183°$: $\dfrac{y}{x} = \dfrac{-}{-} = +$

$\cot 183°$: $\dfrac{x}{y} = \dfrac{-}{-} = +$

27. Since $302°$ is an angle in quadrant IV, for any point (x, y) on the terminal ray, $x > 0$ and $y < 0$. Note that r is always positive. The signs of the trigonometric functions for $302°$ are

$\sin 302°$: $\dfrac{y}{r} = \dfrac{-}{+} = -$

$\csc 302°$: $\dfrac{r}{y} = \dfrac{+}{-} = -$

$\cos 302°$: $\dfrac{x}{r} = \dfrac{+}{+} = +$

$\sec 302°$: $\dfrac{r}{x} = \dfrac{+}{+} = +$

$\tan 302°$: $\dfrac{y}{x} = \dfrac{-}{+} = -$

$\cot 302°$: $\dfrac{x}{y} = \dfrac{+}{-} = -$

29. Since $-700°$ is an angle whose terminal ray lies in quadrant I, all trigonometric functions of $-700°$ are positive.

31. Since $\tan^2 \alpha + 1 = \sec^2 \alpha$ and $\sec \alpha = 3$,

$\tan^2 \alpha + 1 = 3^2 = 9$

$\tan^2 \alpha = 9 - 1 = 8$

$\tan \alpha = \pm\sqrt{8} = \pm 2\sqrt{2}$

Since α is in quadrant IV, $\tan \alpha$ is negative and $\tan \theta = -2\sqrt{2}$.

33. Since $\sin^2 \theta + \cos^2 \alpha = 1$ and $\cos \alpha = -\dfrac{1}{4}$,

$\sin^2 \alpha + \left(-\dfrac{1}{4}\right)^2 = 1$

$\sin^2 \alpha + \dfrac{1}{16} = 1$

$\sin^2 \alpha = 1 - \dfrac{1}{16} = \dfrac{15}{16}$

$\sin \alpha = \pm\dfrac{\sqrt{15}}{4}$.

Since α is in quadrant II, $\sin \alpha$ is positive and $\sin \alpha = \dfrac{\sqrt{15}}{4}$.

35. Since $\sin^2 \theta + \cos^2 \theta = 1$ and $\cos \theta = \dfrac{1}{3}$,

$\sin^2 \theta + \left(\dfrac{1}{3}\right)^2 = 1$

$\sin^2 \theta + \dfrac{1}{9} = 1$

$\sin^2 \theta = 1 - \dfrac{1}{9} = \dfrac{8}{9}$

$\sin \theta = \pm\dfrac{\sqrt{8}}{3}$.

Since θ is in quadrant IV, $\sin \theta$ is negative and $\sin \theta = -\dfrac{\sqrt{8}}{3}$.

Since $\tan \theta = \dfrac{\sin \theta}{\cos \theta}$,

$\tan \theta = \dfrac{-\dfrac{\sqrt{8}}{3}}{\dfrac{1}{3}}$

$\tan \theta = -\dfrac{\sqrt{8}}{3} \cdot \dfrac{3}{1}$

$= -\sqrt{8}$

$= -2\sqrt{2}$.

37. Since $\sin \beta = \dfrac{1}{\csc \beta}$ and $\csc \beta = -4$,

$\sin \beta = \dfrac{1}{-4} = -\dfrac{1}{4}$.

Since $\sin^2 \beta + \cos^2 \beta = 1$ and $\sin \beta = -\dfrac{1}{4}$,

$\left(-\dfrac{1}{4}\right)^2 + \cos^2 \beta = 1$

$\dfrac{1}{16} + \cos^2 \beta = 1$

$\cos^2 \beta = 1 - \dfrac{1}{16} = \dfrac{15}{16}$

$\cos \beta = \pm\dfrac{\sqrt{15}}{4}$

Since β is in quadrant III, $\cos \beta$ is negative,

$\cos \beta = -\dfrac{\sqrt{15}}{4}$.

39. $\tan\alpha = -\dfrac{15}{8}$, with α in quadrant II. Since

$\tan\alpha = \dfrac{y}{x}$, let $x = -8$ and $y = 15$.

$r = \sqrt{x^2 + y^2}$
$= \sqrt{(-8)^2 + 15^2}$
$= \sqrt{64 + 225}$
$= \sqrt{289}$
$= 17$

$\sin\alpha = \dfrac{y}{r} = \dfrac{15}{17}$

$\cos\alpha = \dfrac{x}{r} = \dfrac{-8}{17} = -\dfrac{8}{17}$

$\tan\alpha = \dfrac{y}{x} = \dfrac{15}{-8} = -\dfrac{15}{8}$

$\cot\alpha = \dfrac{x}{y} = \dfrac{-8}{15} = -\dfrac{8}{15}$

$\sec\alpha = \dfrac{r}{x} = \dfrac{17}{-8} = -\dfrac{17}{8}$

$\csc\alpha = \dfrac{r}{y} = \dfrac{17}{15}$

41. $\cot\gamma = \dfrac{3}{4}$, with γ in quadrant III. Since

$\cot\gamma = \dfrac{x}{y}$, let $x = -3$ and $y = -4$.

$r = \sqrt{x^2 + y^2}$
$= \sqrt{(-3)^2 + (-4)^2}$
$= \sqrt{9 + 16}$
$= \sqrt{25}$
$= 5$

$\sin\gamma = \dfrac{y}{r} = \dfrac{-4}{5} = -\dfrac{4}{5}$

$\cos\gamma = \dfrac{x}{r} = \dfrac{-3}{5} = -\dfrac{3}{5}$

$\tan\gamma = \dfrac{y}{x} = \dfrac{-4}{-3} = \dfrac{4}{3}$

$\cot\gamma = \dfrac{x}{y} = \dfrac{-3}{-4} = \dfrac{3}{4}$

$\sec\gamma = \dfrac{r}{x} = \dfrac{5}{-3} = -\dfrac{5}{3}$

$\csc\gamma = \dfrac{r}{y} = \dfrac{5}{-4} = -\dfrac{5}{4}$

43. $\tan\beta = \sqrt{3}$, with β in quadrant III. Since

$\tan\beta = \dfrac{y}{x}$, let $y = -\sqrt{3}$ and $x = -1$.

$x^2 + y^2 = r^2$
$(-1) + (-\sqrt{3})^2 = r^2$
$\sqrt{1+3} = r$
$2 = r$

$\sin\beta = \dfrac{y}{r} = \dfrac{-\sqrt{3}}{2} = -\dfrac{\sqrt{3}}{2}$

$\cos\beta = \dfrac{x}{r} = \dfrac{-1}{2} = -\dfrac{1}{2}$

$\tan\beta = \dfrac{y}{x} = \dfrac{-\sqrt{3}}{-1} = \sqrt{3}$

$\cot\beta = \dfrac{x}{y} = \dfrac{-1}{-\sqrt{3}} = \dfrac{\sqrt{3}}{3}$

$\sec\beta = \dfrac{r}{x} = \dfrac{2}{-1} = -2$

$\csc\beta = \dfrac{r}{y} = \dfrac{2}{-\sqrt{3}} = -\dfrac{2\sqrt{3}}{3}$

45. $\sin\beta = \dfrac{\sqrt{5}}{7}$, with $\tan\beta > 0$. For sine and tangent to both be positive, β must lie in quadrant I. Since $\sin\beta = \dfrac{y}{r}$, let $y = \sqrt{5}$ and $r = 7$.

$x^2 + y^2 = r^2$
$x = \pm\sqrt{r^2 - y^2}$
$= \pm\sqrt{7^2 - (\sqrt{5})^2}$
$= \pm\sqrt{49 - 5}$
$= \sqrt{44}$
$= 2\sqrt{11}$

since β is in quadrant I.

$\sin\beta = \dfrac{y}{r} = \dfrac{\sqrt{5}}{7}$

$\cos\beta = \dfrac{x}{r} = \dfrac{2\sqrt{11}}{7}$

$\tan\beta = \dfrac{y}{x} = \dfrac{\sqrt{5}}{2\sqrt{11}} \cdot \dfrac{\sqrt{11}}{\sqrt{11}} = \dfrac{\sqrt{55}}{22}$

$\cot\beta = \dfrac{x}{y} = \dfrac{2\sqrt{11}}{\sqrt{5}} \cdot \dfrac{\sqrt{5}}{\sqrt{5}} = \dfrac{2\sqrt{55}}{5}$

$$\sec \beta = \frac{r}{x} = \frac{7}{2\sqrt{11}} \cdot \frac{\sqrt{11}}{\sqrt{11}} = \frac{7\sqrt{11}}{22}$$

$$\csc \beta = \frac{r}{y} = \frac{7}{\sqrt{5}} \cdot \frac{\sqrt{5}}{\sqrt{5}} = \frac{7\sqrt{5}}{5}$$

47. Writing exercise; answers will vary.

14.4 Exercises

1. $\sin A = \dfrac{\text{side opposite}}{\text{hypotenuse}} = \dfrac{3}{5}$

$\cos A = \dfrac{\text{side adjacent}}{\text{hypotenuse}} = \dfrac{4}{5}$

$\tan A = \dfrac{\text{side opposite}}{\text{side adjacent}} = \dfrac{3}{4}$

$\cot A = \dfrac{\text{side adjacent}}{\text{side opposite}} = \dfrac{4}{3}$

$\sec A = \dfrac{\text{hypotenuse}}{\text{side adjacent}} = \dfrac{5}{4}$

$\csc A = \dfrac{\text{hypotenuse}}{\text{side opposite}} = \dfrac{5}{3}$

3. $\sin A = \dfrac{\text{side opposite}}{\text{hypotenuse}} = \dfrac{21}{29}$

$\cos A = \dfrac{\text{side adjacent}}{\text{hypotenuse}} = \dfrac{20}{29}$

$\tan A = \dfrac{\text{side opposite}}{\text{side adjacent}} = \dfrac{21}{20}$

$\cot A = \dfrac{\text{side adjacent}}{\text{side opposite}} = \dfrac{20}{21}$

$\sec A = \dfrac{\text{hypotenuse}}{\text{side adjacent}} = \dfrac{29}{20}$

$\csc A = \dfrac{\text{hypotenuse}}{\text{side opposite}} = \dfrac{29}{21}$

5. $\sin A = \dfrac{\text{side opposite}}{\text{hypotenuse}} = \dfrac{n}{p}$

$\cos A = \dfrac{\text{side adjacent}}{\text{hypotenuse}} = \dfrac{m}{p}$

$\tan A = \dfrac{\text{side opposite}}{\text{side adjacent}} = \dfrac{n}{m}$

$\cot A = \dfrac{\text{side adjacent}}{\text{side opposite}} = \dfrac{m}{n}$

$\sec A = \dfrac{\text{hypotenuse}}{\text{side adjacent}} = \dfrac{p}{m}$

$\csc A = \dfrac{\text{hypotenuse}}{\text{side opposite}} = \dfrac{p}{n}$

7. $a = 5, b = 12$

$c^2 = a^2 + b^2$

$c^2 = 5^2 + 12^2$

$c^2 = 25 + 144$

$c^2 = 169$

$c = 13$

$\sin B = \dfrac{\text{side opposite}}{\text{hypotenuse}} = \dfrac{12}{13}$

$\cos B = \dfrac{\text{side adjacent}}{\text{hypotenuse}} = \dfrac{5}{13}$

$\tan B = \dfrac{\text{side opposite}}{\text{side adjacent}} = \dfrac{12}{5}$

$\cot B = \dfrac{\text{side adjacent}}{\text{side opposite}} = \dfrac{5}{12}$

$\sec B = \dfrac{\text{hypotenuse}}{\text{side adjacent}} = \dfrac{13}{5}$

$\csc B = \dfrac{\text{hypotenuse}}{\text{side opposite}} = \dfrac{13}{12}$

9. $a = 6, c = 7$

$c^2 = a^2 + b^2$

$7^2 = 6^2 + b^2$

$b^2 = 49 - 36$

$b^2 = 13$

$b = \sqrt{13}$

$\sin B = \dfrac{\text{side opposite}}{\text{hypotenuse}} = \dfrac{\sqrt{13}}{7}$

$\cos B = \dfrac{\text{side adjacent}}{\text{hypotenuse}} = \dfrac{6}{7}$

$\tan B = \dfrac{\text{side opposite}}{\text{side adjacent}} = \dfrac{\sqrt{13}}{6}$

$\cot B = \dfrac{\text{side adjacent}}{\text{side opposite}} = \dfrac{6}{\sqrt{13}} = \dfrac{6\sqrt{13}}{13}$

$\sec B = \dfrac{\text{hypotenuse}}{\text{side adjacent}} = \dfrac{7}{6}$

$\csc B = \dfrac{\text{hypotenuse}}{\text{side opposite}} = \dfrac{7}{\sqrt{13}} = \dfrac{7\sqrt{13}}{13}$

11. $\tan 50° = \cot(90° - 50°) = \cot 40°$

13. csc 47° = sec(90° − 47°) = sec 43°

15. tan 25.4° = cot(90° − 25.4°) = cot 64.6°

17. cos 13°30′ = sin(90° − 13°30′)
$\qquad$ = sin(89°60′ − 13°30′)
$\qquad$ = sin 76°30′

19. Sketch a 30°-60°-90° right triangle.

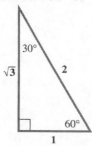

$$\tan 30° = \frac{\text{side opposite}}{\text{side adjacent}} = \frac{1}{\sqrt{3}} \cdot \frac{\sqrt{3}}{\sqrt{3}} = \frac{\sqrt{3}}{3}$$

21. Sketch a 30°-60°-90° right triangle.

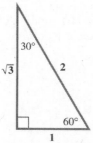

$$\sin 30° = \frac{\text{side opposite}}{\text{hypotenuse}} = \frac{1}{2}$$

23. Sketch a 45°-45°-90° right triangle.

$$\csc 45° = \frac{\text{hypotenuse}}{\text{side opposite}} = \frac{\sqrt{2}}{1} = \sqrt{2}$$

25. Sketch a 45°-45°-90° right triangle.

$$\cos 45° = \frac{\text{side adjacent}}{\text{hypotenuse}} = \frac{1}{\sqrt{2}} \cdot \frac{\sqrt{2}}{\sqrt{2}} = \frac{\sqrt{2}}{2}$$

27. Sketch a 30°-60°-90° right triangle.

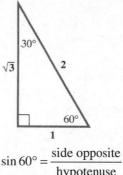

$$\sin 60° = \frac{\text{side opposite}}{\text{hypotenuse}} = \frac{\sqrt{3}}{2}$$

29. Sketch a 30°-60°-90° right triangle.

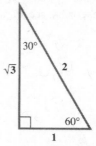

$$\tan 60° = \frac{\text{side opposite}}{\text{side adjacent}} = \frac{\sqrt{3}}{1} = \sqrt{3}$$

31. 180° − 98° = 82°

33. −135° + 360° = 225°
225° − 180° = 45°

35. 750° − 2(360°) = 30°

37. 120°
The reference angle is 60°. Since 120° is in quadrant II, the cosine, tangent, cotangent, and secant are negative.

$$\sin 120° = \sin 60° = \frac{\sqrt{3}}{2}$$

$$\cos 120° = -\cos 60° = -\frac{1}{2}$$

$$\tan 120° = -\tan 60° = -\sqrt{3}$$

$$\cot 120° = -\cot 60° = \frac{-\sqrt{3}}{3}$$

$$\sec 120° = -\sec 60° = -2$$

$$\csc 120° = \csc 60° = \frac{2\sqrt{3}}{3}$$

39. 150°
The reference angle is 30°. Since 150° is in quadrant II, the cosine, tangent, cotangent, and secant are negative.

$$\sin 150° = \sin 30° = \frac{1}{2}$$

$$\cos 150° = -\cos 30° = -\frac{\sqrt{3}}{2}$$

$$\tan 150° = -\tan 30° = -\frac{\sqrt{3}}{3}$$

$$\cot 150° = -\cot 30° = -\sqrt{3}$$

$$\sec 150° = -\sec 30° = -\frac{2\sqrt{3}}{3}$$

$$\csc 150° = \csc 30° = 2$$

41. 240°

The reference angle is 60°. Since 240° is in quadrant III, the sine, cosine, secant, and cosecant are negative.

$$\sin 240° = -\sin 60° = -\frac{\sqrt{3}}{2}$$

$$\sin 240° = -\cos 60° = -\frac{1}{2}$$

$$\tan 240° = \tan 60° = \sqrt{3}$$

$$\cot 240° = \cot 60° = \frac{\sqrt{3}}{3}$$

$$\sec 240° = -\sec 60° = -2$$

$$\csc 240° = -\csc 60° = -\frac{2\sqrt{3}}{3}$$

43. 315°

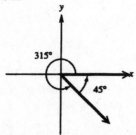

The reference angle is 45°. Since 315° is in quadrant IV, the sine tangent, cotangent, and cosecant are negative.

$$\sin 315° = -\sin 45° = -\frac{\sqrt{2}}{2}$$

$$\cos 315° = \cos 45° = \frac{\sqrt{2}}{2}$$

$$\tan 315° = -\tan 45° = -1$$

$$\cot 315° = -\cot 45° = -1$$

$$\sec 315° = \sec 45° = \sqrt{2}$$

$$\csc 315° = -\csc 45° = -\sqrt{2}$$

45. 420°

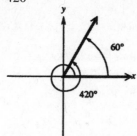

The reference angle is 60°. Since 420° is in quadrant I, all functions are positive.

$$\sin 420° = \sin 60° = \frac{\sqrt{3}}{2}$$

$$\cos 420° = \cos 60° = \frac{1}{2}$$

$$\tan 420° = \tan 60° = \sqrt{3}$$

$$\cot 420° = \cot 60° = \frac{\sqrt{3}}{3}$$

$$\sec 420° = \sec 60° = 2$$

$$\csc 420° = \csc 60° = \frac{2\sqrt{3}}{3}$$

47. 495°

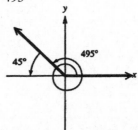

The reference angle is 45°. Since 495° is in quadrant II, the cosine, tangent, cotangent, and secant are negative.

$$\sin 495° = \sin 45° = \frac{\sqrt{2}}{2}$$

$$\cos 495° = -\cos 45° = -\frac{\sqrt{2}}{2}$$

$$\tan 495° = -\tan 45° = -1$$

$$\cot 495° = -\cot 45° = -1$$

$$\sec 495° = -\sec 45° = -\sqrt{2}$$

$$\csc 495° = \csc 45° = \sqrt{2}$$

49. 750° is coterminal with
750° − 2(360°) = 750° − 720° = 30°.

$$\sin 750° = \sin 30° = \frac{1}{2}$$

$$\cos 750° = \cos 30° = \frac{\sqrt{3}}{2}$$

$$\tan 750° = \tan 30° = \frac{\sqrt{3}}{3}$$

$$\cot 750° = \cot 30° = \sqrt{3}$$

$$\sec 750° = \sec 30° = \frac{2\sqrt{3}}{3}$$

$$\csc 750° = \csc 30° = 2$$

51. 1500° is coterminal with
$1500° - 4(360°) = 1500° - 1440° = 60°.$

$$\sin 1500° = \sin 60° = \frac{\sqrt{3}}{2}$$

$$\cos 1500° = \cos 60° = \frac{1}{2}$$

$$\tan 1500° = \tan 60° = \sqrt{3}$$

$$\cot 1500° = \cot 60° = \frac{\sqrt{3}}{3}$$

$$\sec 1500° = \sec 60° = 2$$

$$\csc 1500° = \csc 60° = \frac{2\sqrt{3}}{3}$$

53. −390° is coterminal with
$−390° + (2)(360°) = 330°.$
The reference angle is 30°. Since −390° is in quadrant IV, the sine, tangent, cotangent, and cosecant are negative.

$$\sin(−390)° = −\sin 30° = −\frac{1}{2}$$

$$\cos(−390)° = \cos 30° = \frac{\sqrt{3}}{2}$$

$$\tan(−390)° = −\tan 30° = −\frac{\sqrt{3}}{3}$$

$$\cot(−390)° = −\cot 30° = −\sqrt{3}$$

$$\sec(−390)° = \sec 30° = \frac{2\sqrt{3}}{3}$$

$$\csc(−390)° = −\csc 30° = −2$$

55. −1020° is coterminal with
$−1020° + (3)(360°) = −1020° + 1080° = 60°.$

$$\sin(−1020°) = \sin 60° = \frac{\sqrt{3}}{2}$$

$$\cos(−1020°) = \cos 60° = \frac{1}{2}$$

$$\tan(−1020°) = \tan 60° = \sqrt{3}$$

$$\cot(−1020°) = \cot 60° = \frac{\sqrt{3}}{3}$$

$$\sec(−1020°) = \sec 60° = 2$$

$$\csc(−1020°) = \csc 60° = \frac{2\sqrt{3}}{3}$$

57. 30°

$$\tan 30° = \frac{\sqrt{3}}{3}$$

$$\cot 30° = \sqrt{3}$$

59. 60°

$$\sin 60° = \frac{\sqrt{3}}{2}$$

$$\cot 60° = \frac{\sqrt{3}}{3}$$

$$\csc 60° = \frac{2\sqrt{3}}{3}$$

61. 135°
$\tan 135° = −\tan 45° = −1$
$\cot 135° = −\cot 45° = −1$

63. 210°

$$\cos(210°) = −\cos 30° = −\frac{\sqrt{3}}{2}$$

$$\sec(210°) = −\sec 30° = −\frac{2\sqrt{3}}{3}$$

65. Since the sine is negative, θ must lie in quadrant III or IV. Since the absolute value of $\sin \theta$ is $\frac{1}{2}$, the reference angle must be 30°. The quadrant III angle θ is $180° + 30° = 210°$, and the quadrant IV angle θ is $360° − 30° = 330°$.

67. Since the tangent is positive, θ must lie in quadrant I or III. Since $\tan \theta = 1$, the reference angle must be 45°. The quadrant I angle θ is 45°, and the quadrant III angle θ is $180° + 45° = 225°$.

69. Since the sine is positive, θ must lie in quadrant I or II. Since the value of $\sin \theta$ is $\frac{\sqrt{3}}{2}$, the reference angle must be 60°. The quadrant I angle θ is 60°, and the quadrant II angle θ is $180° − 60° = 120°$.

71. Since the secant is negative, θ must lie in quadrant II or III. Since the absolute value of sec θ is 2, the reference angle must be $60°$. The quadrant II angle θ is $180° - 60° = 120°$, and the quadrant III angle θ is $180° + 60° = 240°$.

73. Since the sine is negative, θ must lie in quadrant III or IV. Since the absolute value of sin θ is $\dfrac{\sqrt{2}}{2}$, the reference angle must be $45°$. The quadrant III angle θ is $180° + 45° = 225°$, and the quadrant IV angle θ is $360° - 45° = 315°$.

75. Since the tangent is negative, θ must lie in quadrant II or IV. Since the $\tan\theta = -\sqrt{3}$, the reference angle must be $60°$. The quadrant II angle θ is $180° - 60° = 120°$, and the quadrant IV angle θ is $360° - 60° = 300°$.

77. Since $\cos\theta = 0$, θ must lie on the y-axis. $\theta = 90°$ or $270°$.

14.5 Exercises

For the following exercises, be sure your calculator is degree mode. If your calculator accepts angles in degrees, minutes, and seconds, it is not necessary to change angles to decimal degrees. Keystroke sequences may vary on the type and/or model of calculator being used.

1. $\tan 29°30'$

$$\tan 29°30' = \tan\left(29 + \frac{30}{60}\right)$$
$$= \tan 29.5°$$
$$\approx 0.5657728$$

3. $\cot 41°24'$

$$\cot 41°24' = \frac{1}{\tan 41°24'}$$
$$= \frac{1}{\tan\left(41 + \frac{24}{60}\right)}$$
$$= \frac{1}{\tan 41.4°}$$
$$\approx \frac{1}{0.8816186}$$
$$\approx 1.1342773$$

5. $\sec 13°15'$

$$\sec 13°15' = \frac{1}{\cos\left(13 + \frac{15}{60}\right)}$$
$$= \frac{1}{\cos 13.25°}$$
$$= \frac{1}{0.9733793}$$
$$\approx 1.0273488$$

7. $\sin 39°40'$

$$\sin 39°40' = \sin\left(39 + \frac{40}{60}\right)$$
$$\approx \sin(39.6666667)°$$
$$\approx 0.6383201$$

9. $\csc 145°45'$

$$\csc 145°45' = \frac{1}{\sin\left(145 + \frac{45}{60}\right)}$$
$$= \frac{1}{\sin 145.75°}$$
$$\approx \frac{1}{0.5628049}$$
$$\approx 1.7768146$$

11. $\cos 421°30'$

$$\cos 421°30' = \cos\left(421 + \frac{30}{60}\right)$$
$$= \cos 421.5°$$
$$\approx 0.4771588$$

13. $\tan(-80°6')$

$$\tan(-80°6') = \tan\left(-\left(80 + \frac{6}{60}\right)\right)$$
$$= \tan(-80.1)°$$
$$\approx -5.7297416$$

15. $\cot(-512°20')$

$$\cot(-512°20') = \frac{1}{\tan(-512°20')}$$
$$= \frac{1}{\tan\left(-\left(512 + \frac{20}{60}\right)\right)}$$
$$= \frac{1}{\tan(-512.33333°)}$$
$$\approx \frac{1}{0.5242698}$$
$$\approx 1.9074147$$

Depending upon your calculator use "arc," or "INV" followed by trig function. For some calculators you may use "$\sin^{-1}$" for arc sin, etc.

17. $\sin \theta = 0.84802194$

$\theta = \arcsin(0.84802194)$ or

$\quad = \sin^{-1}(0.84802194)$ or

$\quad = \text{INV} \sin(0.84802194)$

$\quad \approx 57.997172°$

19. $\sec \theta = 1.1606249$

Since $\cos\theta = \dfrac{1}{\sec\theta}$,

$\cos\theta = \dfrac{1}{1.1606249} \approx 0.8616048$

$\theta = \arccos(0.8616048)$ or

$\quad = \cos^{-1}(0.8616048)$ or

$\quad = \text{INV}\cos(0.8616048)$

$\quad \approx 30.502748°.$

21. $\sin \theta = 0.72144101$

$\theta = \arcsin(0.72144101)$ or

$\quad = \sin^{-1}(0.72144101)$ or

$\quad = \text{INV}\sin(0.72144101)$

$\quad \approx 46.173581°.$

23. $\tan \theta = 6.4358841$

$\theta = \tan^{-1}(6.4358841) \approx 81.168073°$

25. $A = 36°20', \ c = 964$ m

$A + B = 90°$

$\quad B = 90° - A$

$\quad\quad = 90° - 36°20'$

$\quad\quad = 89°60' - 36°20'$

$\quad\quad = 53°40'$

$\sin A = \dfrac{a}{c}$

$\quad a = c \sin A$

$\quad a = 964 \sin 36°20'$

Use a calculator and round answer to three significant digits.

$a \approx 571$ m

$\cos A = \dfrac{b}{c}$

$\quad b = c \cos A$

$\quad b = 964 \cos 36°20'$

Use a calculator and round answer to three significant digits.

$b \approx 777$ m

27. $N = 51.2°, \ m = 124$ m

$M + N = 90°$

$\quad M = 90° - N$

$\quad\quad = 90° - 51.2°$

$\quad\quad = 38.8°$

$\tan N = \dfrac{n}{m}$

$\quad n = m \tan N = 124 \cdot \tan 51.2°$

Use a calculator and round answer to three significant digits.

$n \approx 154$ m

$\cos N = \dfrac{m}{p}$

$\quad p = \dfrac{m}{\cos N} = \dfrac{124}{\cos 51.2°}$

Use a calculator and round answer to three significant digits.

$p \approx 198$ m

29. $B = 42.0892°, \ b = 56.851$ cm

$A + B = 90°$

$\quad A = 90° - B$

$\quad\quad = 90° - 42.0892°$

$\quad\quad \approx 47.9108°$

$\sin B = \dfrac{b}{c}$

$\quad c = \dfrac{b}{\sin B} = \dfrac{56.851}{\sin 42.0892°} \approx 84.816$ cm

$\tan B = \dfrac{b}{a}$

$\quad a = \dfrac{b}{\tan B} = \dfrac{56.851}{\tan 42.0892°} \approx 62.942$ cm

31. $A = 28.00°, \ c = 17.4$ ft

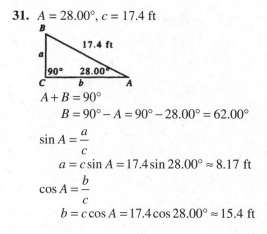

$A + B = 90°$

$\quad B = 90° - A = 90° - 28.00° = 62.00°$

$\sin A = \dfrac{a}{c}$

$\quad a = c \sin A = 17.4 \sin 28.00° \approx 8.17$ ft

$\cos A = \dfrac{b}{c}$

$\quad b = c \cos A = 17.4 \cos 28.00° \approx 15.4$ ft

33. $B = 73.00°, \ b = 128$ in.

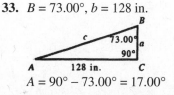

$A = 90° - 73.00° = 17.00°$

$$\tan 73.00° = \frac{128}{a}$$

$$a = \frac{128}{\tan 73.00°} \approx 39.1 \text{ in.}$$

$$\sin 73.00° = \frac{128}{c}$$

$$c = \frac{128}{\sin 73.00°} \approx 134 \text{ in.}$$

35. $a = 76.4$ yd, $b = 39.3$ yd

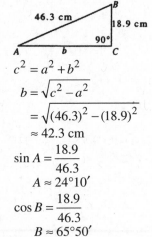

$$c^2 = a^2 + b^2$$

$$c = \sqrt{a^2 + b^2}$$

$$= \sqrt{(76.4)^2 + (39.3)^2}$$

$$\approx 85.9 \text{ yd}$$

$$\tan A = \frac{76.4}{39.3}$$

$$A \approx 62°50'$$

$$\tan B = \frac{39.3}{76.4}$$

$$B \approx 27°10'$$

37. $a = 18.9$ cm, $c = 46.3$ cm

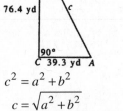

$$c^2 = a^2 + b^2$$

$$b = \sqrt{c^2 - a^2}$$

$$= \sqrt{(46.3)^2 - (18.9)^2}$$

$$\approx 42.3 \text{ cm}$$

$$\sin A = \frac{18.9}{46.3}$$

$$A \approx 24°10'$$

$$\cos B = \frac{18.9}{46.3}$$

$$B \approx 65°50'$$

39. $A = 53°24'$, $c = 387.1$ ft

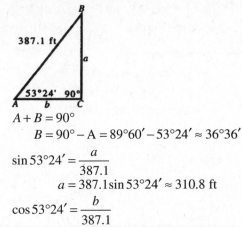

$$A + B = 90°$$

$$B = 90° - A = 89°60' - 53°24' \approx 36°36'$$

$$\sin 53°24' = \frac{a}{387.1}$$

$$a = 387.1 \sin 53°24' \approx 310.8 \text{ ft}$$

$$\cos 53°24' = \frac{b}{387.1}$$

$$b = 387.1 \cos 53°24' \approx 230.8 \text{ ft}$$

41. $B = 39°9'$, $c = 0.6231$ m

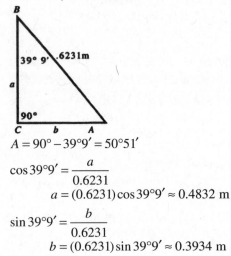

$$A = 90° - 39°9' = 50°51'$$

$$\cos 39°9' = \frac{a}{0.6231}$$

$$a = (0.6231) \cos 39°9' \approx 0.4832 \text{ m}$$

$$\sin 39°9' = \frac{b}{0.6231}$$

$$b = (0.6231) \sin 39°9' \approx 0.3934 \text{ m}$$

43. Let h = the distance the ladder goes up the wall.

$$\sin 43°50' = \frac{h}{13.5}$$

$$h = 13.5 \sin 43°50' \approx 9.3496000$$

The ladder goes up the wall 9.35 m.

45. Let x = the length of the guy wire.

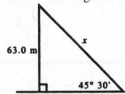

$$\sin 45°30' = \frac{63.0}{x}$$
$$x \sin 45°30' = 63.0$$
$$x = \frac{63.0}{\sin 45°30'} \approx 88.328020$$

The length of the guy wire is 88.3 m.

47. The two right triangles are congruent. So corresponding sides are congruent. Since the sides that compose the base of the isosceles triangle are congruent, each side is

$\frac{1}{2}(42.36)$, or 21.18 inches. Let $x =$ the

length of each of the two equal sides of the isosceles triangle.

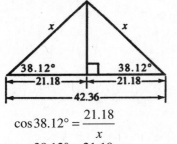

$$\cos 38.12° = \frac{21.18}{x}$$
$$x \cos 38.12° = 21.18$$
$$x = \frac{21.18}{\cos 38.12°} \approx 26.921918$$

The length of each of the two equal sides of the triangle is 26.92 inches.

49.
$$\tan 30.0° = \frac{y}{1000}$$
$$y = 1000 \tan 30.0° \approx 577$$

However, the observer's eye height is 6 feet from the ground, so the cloud ceiling is $577 + 6 = 583$ feet.

51. Let $h =$ height of the tower.

In triangle ABC, $\tan 34.6° = \dfrac{h}{40.6}$
$$h = 40.6 \tan 34.6°$$
$$h \approx 28.0.$$
The height of the tower is 28.0 m.

53. Let $d =$ the distance from the top B of the building to the point on the ground A.

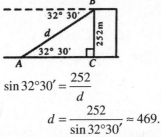

$$\sin 32°30' = \frac{252}{d}$$
$$d = \frac{252}{\sin 32°30'} \approx 469.$$

The distance from the top of the building to the point on the ground is 469 m.

55. Let $x =$ the height of the taller building, $h =$ the difference in height between the shorter and taller buildings, and $d =$ the distance between the buildings along the ground.

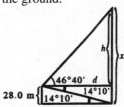

$$\frac{d}{28.0} = \cot 14°10'$$
$$d = (28.0) \cot 14°10'$$
$$= 110.9262493$$
$$\approx 111 \text{ m}$$

$$\frac{h}{d} = \tan 46°40'$$
$$h = d \tan 46°40'$$
$$h = (110.9262493) \tan 46°40' \approx 118 \text{ m}$$
$$x = h + 28.0 = 118 + 28.0 = 146 \text{ m}$$
The height of the taller building is 146 m.

57. Let $x =$ the distance from the assigned target.

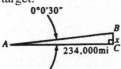

In triangle ABC,

$$\tan 0°0'30'' = \frac{x}{234000}$$
$$x = 234000 \tan 0°0'30''$$
$$x \approx 34.0$$
The distance from the assigned target is 34.0 mi.

59. $a = \dfrac{1}{2}(24)$, so $a = 12$.

$b = a\sqrt{3}$, so $b = 12\sqrt{3}$.

$d = b$, so $d = 12\sqrt{3}$.

$c = d\sqrt{2}$, so $c = \left(12\sqrt{3}\right)\left(\sqrt{2}\right) = 12\sqrt{6}$.

14.6 Exercises

1. $A = 37°$, $B = 48°$, $c = 18$ m

$C = 180° - A - B = 180° - 37° - 48° = 95°$

$\dfrac{b}{\sin B} = \dfrac{c}{\sin C}$

$\quad b = \dfrac{c \sin B}{\sin C} = \dfrac{18 \sin 48°}{\sin 95°} \approx 13$ m

$\dfrac{a}{\sin A} = \dfrac{c}{\sin C}$

$\quad a = \dfrac{c \sin A}{\sin C} = \dfrac{18 \sin 37°}{\sin 95°} \approx 11$ m

3. $A = 27.2°$, $C = 115.5°$, $c = 76.0$ ft

$B = 180° - A - C$

$\quad = 180° - 27.2° - 115.5°$

$\quad = 37.3°$

$\dfrac{a}{\sin A} = \dfrac{c}{\sin C}$

$\quad a = \dfrac{c \sin A}{\sin C} = \dfrac{76.0 \sin 27.2°}{\sin 115.5°} \approx 38.5$ ft

$\dfrac{b}{\sin B} = \dfrac{c}{\sin C}$

$\quad b = \dfrac{c \sin B}{\sin C} = \dfrac{76.0 \sin 37.3°}{\sin 115.5°} \approx 51.0$ ft

5. $A = 68.41°$, $B = 54.23°$, $a = 12.75$ ft

$C = 180° - A - B$

$\quad = 180° - 68.41° - 54.23°$

$\quad \approx 57.36°$

$\dfrac{a}{\sin A} = \dfrac{b}{\sin B}$

$\quad b = \dfrac{a \sin B}{\sin A} = \dfrac{12.75 \sin 54.23°}{\sin 68.41°} \approx 11.13$ ft

$\dfrac{a}{\sin A} = \dfrac{c}{\sin C}$

$\quad c = \dfrac{a \sin C}{\sin A}$

$\quad c = \dfrac{12.75 \sin 57.36°}{\sin 68.41°} \approx 11.55$ ft

7. $A = 87.2°$, $b = 75.9$ yd, $C = 74.3°$

$B = 180° - A - C$

$\quad = 180° - 87.2° - 74.3°$

$\quad \approx 18.5°$

$\dfrac{a}{\sin A} = \dfrac{b}{\sin B}$

$\quad a = \dfrac{b \sin A}{\sin B} = \dfrac{75.9 \sin 87.2°}{\sin 18.5°} \approx 239$ yd

$\dfrac{b}{\sin B} = \dfrac{c}{\sin C}$

$\quad c = \dfrac{b \sin C}{\sin B} = \dfrac{75.9 \sin 74.3°}{\sin 18.5°} \approx 230$ yd

9. $B = 20°50'$, $AC = 132$ ft, $C = 103°10'$

$A = 180° - B - C$

$\quad = 180° - 20°50' - 103°10'$

$\quad = 56°00'$

$\dfrac{AC}{\sin B} = \dfrac{AB}{\sin C}$

$\quad AB = \dfrac{AC \sin C}{\sin B} = \dfrac{132 \sin 103°10'}{\sin 20°50'} \approx 361$ ft

$\dfrac{BC}{\sin A} = \dfrac{AC}{\sin B}$

$\quad BC = \dfrac{AC \sin A}{\sin B} = \dfrac{132 \sin 56°00'}{\sin 20°50'} \approx 308$ ft

11. $A = 39.70°$, $C = 30.35°$, $b = 39.74$ m

$B = 180° - A - C$

$\quad = 180° - 39.70° - 30.35°$

$\quad = 109.95°$

$\quad \approx 110.0°$

$\dfrac{a}{\sin A} = \dfrac{b}{\sin B}$

$\quad a = \dfrac{b \sin A}{\sin B}$

$\quad = \dfrac{39.74 \sin 39.70°}{\sin 109.95°}$

$\quad \approx 27.01$ m

$\dfrac{b}{\sin B} = \dfrac{c}{\sin C}$

$\quad c = \dfrac{b \sin C}{\sin B}$

$\quad = \dfrac{39.74 \sin 30.35°}{\sin 109.95°}$

$\quad \approx 21.36$ m

13. $B = 42.88°$, $C = 102.40°$, $b = 3974$ ft

$A = 180° - B - C$
$\quad = 180° - 42.88° - 102.40°$
$\quad = 34.72°$

$\dfrac{a}{\sin A} = \dfrac{b}{\sin B}$

$\quad a = \dfrac{b \sin A}{\sin B} = \dfrac{3974 \sin 34.72°}{\sin 42.88°} \approx 3326$ ft

$\dfrac{b}{\sin B} = \dfrac{c}{\sin C}$

$\quad c = \dfrac{b \sin C}{\sin B} = \dfrac{3974 \sin 102.40°}{\sin 42.88°} \approx 5704$ ft

15. $A = 39°54'$, $a = 268.7$ m, $B = 42°32'$

$C = 180° - A - B$
$\quad = 180° - 39°54' - 42°32'$
$\quad = 97°34'$

$\dfrac{a}{\sin A} = \dfrac{b}{\sin B}$

$\quad b = \dfrac{a \sin B}{\sin A}$

$\quad\quad = \dfrac{268.7 \sin 42°32'}{\sin 39°54'}$

$\quad\quad \approx 283.2$ m

$\dfrac{a}{\sin A} = \dfrac{c}{\sin C}$

$\quad c = \dfrac{a \sin C}{\sin A}$

$\quad\quad = \dfrac{268.7 \sin 97°34'}{\sin 39°54'}$

$\quad\quad \approx 415.2$ m

17.

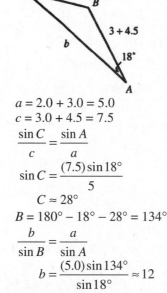

$A = 180° - B - C$
$\quad = 180° - 112°10' - 15°20'$
$\quad = 52°30'$

$\dfrac{BC}{\sin A} = \dfrac{AB}{\sin C}$

$\quad AB = \dfrac{BC \sin C}{\sin A} = \dfrac{354 \sin 15°20'}{\sin 52°30'} \approx 118$ m

19. $\dfrac{x}{\sin 54.8°} = \dfrac{12.0}{\sin 70.4°}$

$\quad x = \sin 54.8° \cdot \dfrac{12.0}{\sin 70.4°}$

$\quad x \approx 10.4$ in.

21. Label the centers of the atoms A, B, and C.

$a = 2.0 + 3.0 = 5.0$
$c = 3.0 + 4.5 = 7.5$

$\dfrac{\sin C}{c} = \dfrac{\sin A}{a}$

$\sin C = \dfrac{(7.5) \sin 18°}{5}$

$\quad C \approx 28°$

$B = 180° - 18° - 28° = 134°$

$\dfrac{b}{\sin B} = \dfrac{a}{\sin A}$

$\quad b = \dfrac{(5.0) \sin 134°}{\sin 18°} \approx 12$

The distance between the centers of atoms A and C is 12.

23. By the law of cosines,

$a^2 = 3^2 + 8^2 - 2(3)(8) \cos 60°$

$\quad = 9 + 64 - 48\left(\dfrac{1}{2}\right)$

$\quad = 73 - 24$

$a^2 = 49$

$\quad a = 7$.

25. $\cos \theta = \dfrac{b^2 + c^2 - a^2}{2bc}$

$\quad = \dfrac{1^2 + \left(\sqrt{3}\right)^2 - 1^2}{2(1)\left(\sqrt{3}\right)}$

$\quad = \dfrac{1 + 3 - 1}{2\sqrt{3}}$

$\quad = \dfrac{3}{2\sqrt{3}}$

$\quad = \dfrac{\sqrt{3}}{2}$

$\theta = 30°$

27. $C = 28.3°$, $b = 5.71$ in., $a = 4.21$ in.

$$c^2 = a^2 + b^2 - 2ab\cos C$$
$$= (4.21)^2 + (5.71)^2$$
$$\qquad - 2(4.21)(5.71)\cos 28.3°$$
$$c \approx \sqrt{7.9964337} \approx 2.83 \text{ in.}$$

Keep all digits of $\sqrt{c^2}$ in the calculator for use in the next calculation. If 2.83 is used, the answer will vary slightly due to round-off error. Find angle A next, since it is the smaller angle and must be acute.

$$\sin A = \frac{a\sin C}{c}$$
$$= \frac{(4.21)\sin 28.3°}{\sqrt{c^2}}$$
$$\approx 0.70581857$$
$$A = 44.9°$$
$$B = 180° - 44.9° - 28.3° = 106.8°$$

29. $C = 45.6°$, $b = 8.94$ m, $a = 7.23$ m

$$c^2 = a^2 + b^2 - 2ab\cos C$$
$$= (7.23)^2 + (8.94)^2$$
$$\qquad - 2(7.23)(8.94)\cos 45.6°$$
$$c \approx \sqrt{41.74934078} \approx 6.46 \text{ m}$$

Find angle A next, since it is the smaller angle and must be acute.

$$\sin A = \frac{a\sin C}{c}$$
$$= \frac{(7.23)\sin 45.6°}{\sqrt{c^2}}$$
$$\approx 0.79946437$$
$$A = 53.1°$$
$$B = 180° - 53.1° - 45.6° = 81.3°$$

31. $A = 80°40'$, $b = 143$ cm, $c = 89.6$ cm

$$a^2 = b^2 + c^2 - 2bc\cos A$$
$$= 143^2 + (89.6)^2$$
$$\qquad - 2(143)(89.6)\cos 80°40'$$
$$a \approx \sqrt{24321.25341} \approx 156 \text{ cm}$$

Find angle C next, since it is the smaller angle and must be acute.

$$\sin C = \frac{c\sin A}{a}$$
$$= \frac{(89.6)\sin 80°40'}{\sqrt{a^2}}$$
$$\approx 0.56692713$$
$$C = 34°30'$$
$$B = 180° - 80°40' - 34°30' = 64°50'$$

33. $B = 74.80°$, $a = 8.919$ in., $c = 6.427$ in.

$$b^2 = a^2 + c^2 - 2ac\cos B$$
$$= (8.919)^2 + (6.427)^2$$
$$\qquad - 2(8.919)(6.427)\cos 74.80°$$
$$b \approx 9.529 \text{ in.}$$

Find angle C next, since it is the smaller angle and must be acute.

$$\sin C = \frac{c\sin B}{b}$$
$$= \frac{(6.427)\sin 74.80°}{\sqrt{b^2}}$$
$$\approx 0.65089219$$
$$C \approx 40.61°$$
$$A = 180° - 74.80° - 40.61° = 64.59°$$

35. $A = 112.8°$, $b = 6.28$ m, $c = 12.2$ m

$$a^2 = b^2 + c^2 - 2bc\cos A$$
$$= (6.28)^2 + (12.2)^2$$
$$\qquad - 2(6.28)(12.2)\cos 112.8°$$
$$a \approx 15.7 \text{ m}$$

Angle A is obtuse, so both B and C are acute. Find either angle next.

$$\sin B = \frac{b\sin A}{a}$$
$$= \frac{(6.28)\sin 112.8°}{\sqrt{a^2}}$$
$$\approx 0.36787465$$
$$B = 21.6°$$
$$C = 180° - 112.8° - 21.6° = 45.6°$$

37. $a = 3.0$ ft, $b = 5.0$ ft, $c = 6.0$ ft

Angle C is the largest, so find it first.

$$c^2 = a^2 + b^2 - 2ab\cos C$$
$$\cos C = \frac{a^2 + b^2 - c^2}{2ab}$$
$$= \frac{3.0^2 + 5.0^2 - 6.0^2}{2(3.0)(5.0)}$$
$$\approx -0.06666667$$
$$C = 94°$$
$$\sin B = \frac{b\sin C}{c} = \frac{5.0\sin 94°}{6.0} \approx 0.83147942$$
$$B = 56°$$
$$A = 18° - 56° - 94° = 30°$$

39. $a = 9.3$ cm, $b = 5.7$ cm, $c = 8.2$ cm

Angle A is largest, so find it first.

$$a^2 = b^2 + c^2 - 2bc\cos A$$
$$\cos A = \frac{5.7^2 + 8.2^2 - 9.3^2}{2(5.7)(8.2)} \approx 0.14163457$$
$$A = 82°$$

$$\sin B = \frac{b \sin A}{a} = \frac{5.7 \sin 82°}{9.3} \approx 0.60672455$$
$$B = 37°$$
$$C = 180° - 82° - 37° = 61°$$

41. a = 42.9 m, b = 37.6 m, c = 62.7 m
Angle C is the largest, so find it first.

$$c^2 = a^2 + b^2 - 2ab \cos C$$
$$\cos C = \frac{a^2 + b^2 - c^2}{2ab}$$
$$= \frac{42.9^2 + 37.6^2 - 62.7^2}{2(42.9)(37.6)}$$
$$\approx -0.20988940$$
$$C = 102.1°$$
$$\sin B = \frac{b \sin C}{c}$$
$$= \frac{37.6 \sin 102.1°}{62.4}$$
$$\approx 0.58632321$$
$$B = 35.9°$$
$$A = 180° - 35.9° - 102.1° = 42.0°$$

43. Find AB, or c in the following triangle.

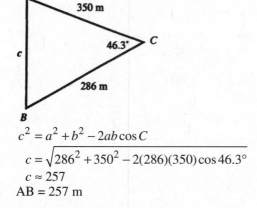

$$c^2 = a^2 + b^2 - 2ab \cos C$$
$$c = \sqrt{286^2 + 350^2 - 2(286)(350) \cos 46.3°}$$
$$c \approx 257$$
$$AB = 257 \text{ m}$$

45. Solve for x.
$$x^2 = 25^2 + 25^2 - 2(25)(25) \cos 52° \approx 480$$
$$x \approx 22 \text{ ft}$$

47. Let A = the angle between the beam and the 45-ft cable.
$$\cos A = \frac{45^2 + 90^2 - 60^2}{2(45)(90)}$$
$$A = 36°$$
Let B = the angle between the beam and the 60-ft cable.
$$\cos B = \frac{90^2 + 60^2 - 45^2}{2(90)(60)}$$
$$B = 26°$$

EXTENSION: AREA FORMULAS FOR TRIANGLES

1. Using $\mathcal{A} = \frac{1}{2}bh$, $\mathcal{A} = \frac{1}{2}(1)(\sqrt{3}) = \frac{\sqrt{3}}{2}$.

Using $\mathcal{A} = \frac{1}{2}ab \sin C$,

$$\mathcal{A} = \frac{1}{2}(\sqrt{3})(1) \sin 90° = \frac{1}{2}(\sqrt{3})(1)1 = \frac{\sqrt{3}}{2}.$$

3. Using $\mathcal{A} = \frac{1}{2}bh$, $\mathcal{A} = \frac{1}{2}(1)(\sqrt{2}) = \frac{\sqrt{2}}{2}$.

Using $\mathcal{A} = \frac{1}{2}ab \sin C$,

$$\mathcal{A} = \frac{1}{2}(2)(1) \sin 45° = \frac{1}{2}(2)(1)\left(\frac{\sqrt{2}}{2}\right) = \frac{\sqrt{2}}{2}.$$

5. A = 42.5°, b = 13.6 m, c = 10.1 m
Angle A is included between sides b and c.

$$\text{Area} = \frac{1}{2}bc \sin A$$
$$= \frac{1}{2}(13.6)(10.1) \sin 42.5°$$
$$\approx 46.4 \text{ m}^2$$

7. B = 124.5°, a = 30.4 cm, c = 28.4 cm
Angle B is included between sides a and c.

$$\text{Area} = \frac{1}{2}ac \sin B$$
$$= \frac{1}{2}(30.4)(28.4) \sin 124.5°$$
$$\approx 356 \text{ cm}^2$$

9. A = 56.80°, b = 32.67 in., c = 52.89 in.
Angle A is included between sides b and c.

$$\text{Area} = \frac{1}{2}bc \sin A$$
$$= \frac{1}{2}(32.67)(52.89) \sin 56.80°$$
$$\approx 722.9 \text{ in}^2$$

11. Using $A = \frac{1}{2}bh$,

$$A = \frac{1}{2}(16)\left(3\sqrt{3}\right) = 24\sqrt{3} \approx 41.57.$$

To use Heron's Formula, first find the semiperimeter.

$$s = \frac{1}{2}(a+b+c)$$
$$= \frac{1}{2}(6+14+16)$$
$$= \frac{1}{2}(36)$$
$$= 18$$

Now find the area of the triangle.

$$\text{Area} = \sqrt{s(s-a)(s-b)(s-c)}$$
$$= \sqrt{18(18-6)(18-14)(18-16)}$$
$$= \sqrt{18(12)(4)(2)}$$
$$= \sqrt{1728}$$
$$= \sqrt{576 \cdot 3}$$
$$= 24\sqrt{3}$$
$$\approx 41.57$$

Both formulas give the same area.

13. $a = 12$ m, $b = 16$ m, $c = 25$ m

$$s = \frac{1}{2}(a+b+c)$$
$$= \frac{1}{2}(12+16+25)$$
$$= \frac{1}{2}(53)$$
$$= 26.5$$
$$\text{Area} = \sqrt{s(s-a)(s-b)(s-c)}$$
$$= \sqrt{(26.5)(14.5)(10.5)(1.5)}$$
$$\approx 78 \text{ m}^2$$

15. $a = 154$ cm, $b = 179$ cm, $c = 183$ cm

$$s = \frac{1}{2}(a+b+c)$$
$$= \frac{1}{2}(154+179+183)$$
$$= \frac{1}{2}(516)$$
$$= 258$$
$$\text{Area} = \sqrt{s(s-a)(s-b)(s-c)}$$
$$= \sqrt{(258)(104)(79)(75)}$$
$$\approx 12,609 \text{ cm}^2$$

17. $a = 76.3$ ft, $b = 109$ ft, $c = 98.8$ ft

$$s = \frac{1}{2}(a+b+c)$$
$$= \frac{1}{2}(76.3+109+98.8)$$
$$= 142.05$$
$$\text{Area} = \sqrt{s(s-a)(s-b)(s-c)}$$
$$= \sqrt{(142.05)(65.75)(33.05)(43.25)}$$
$$\approx 3654 \text{ ft}^2$$

19. $\text{Area} = \frac{1}{2}ab\sin C$

$$= \frac{1}{2}(16.1)(15.2)\sin 125°$$
$$\approx 100 \text{ m}^2$$

21. Find the area of the region.

$$s = \frac{1}{2}(a+b+c) = \frac{1}{2}(75+68+85) = 114$$
$$\text{Area} = \sqrt{s(s-a)(s-b)(s-c)}$$
$$= \sqrt{(114)(39)(46)(29)}$$
$$\approx 2435.35706 \text{ m}^2$$

Let n = number of cans needed.

$$n = \frac{(\text{area in m}^2)}{(\text{m}^2 \text{ per can})} = \frac{2435.35761}{75} \approx 32.47$$

She will need to open 33 cans.

Chapter 14 Test

1. $74°17'54'' = 74° + \dfrac{17°}{60} + \dfrac{54°}{3600} \approx 74.2983°$

2. $360° + (-157°) = 203°$

3. $(2, -5)$, $x = 2$, $y = -5$

$$r = \sqrt{x^2 + y^2} = \sqrt{2^2 + (-5)^2} = \sqrt{29}$$
$$\sin\theta = \frac{y}{r} = \frac{-5}{\sqrt{29}} \cdot \frac{\sqrt{29}}{\sqrt{29}} = \frac{-5\sqrt{29}}{29}$$
$$\cos\theta = \frac{x}{r} = \frac{2}{\sqrt{29}} \cdot \frac{\sqrt{29}}{\sqrt{29}} = \frac{2\sqrt{29}}{29}$$
$$\tan\theta = \frac{y}{x} = \frac{-5}{2} = -\frac{5}{2}$$

4. If $\cos\theta < 0$, then θ is in quadrant II or III. If $\cot\theta > 0$, then θ is in quadrant I or III. Therefore, θ terminates in quadrant III.

5. If $\cos\theta = \dfrac{4}{5} = \dfrac{x}{r}$, let $x = 4$ and $r = 5$.

$$x^2 + y^2 = r^2$$
$$4^2 + y^2 = 5^2$$
$$y^2 = \pm\sqrt{9}$$
$$y = \pm 3$$

Since θ is in quadrant IV, y is negative, so $y = -3$.

$$\sin\theta = \frac{y}{r} = \frac{-3}{5} = -\frac{3}{5}$$

$$\tan\theta = \frac{y}{x} = \frac{-3}{4} = -\frac{3}{4}$$

$$\cot\theta = \frac{x}{y} = \frac{4}{-3} = -\frac{4}{3}$$

$$\sec\theta = \frac{r}{x} = \frac{5}{4}$$

$$\csc\theta = \frac{r}{y} = \frac{5}{-3} = -\frac{5}{3}$$

6. $\sin A = \dfrac{\text{side opposite}}{\text{hypotenuse}} = \dfrac{12}{13}$

$\cos A = \dfrac{\text{side adjacent}}{\text{hypotenuse}} = \dfrac{5}{13}$

$\tan A = \dfrac{\text{side opposite}}{\text{side adjacent}} = \dfrac{12}{5}$

$\cot A = \dfrac{\text{side adjacent}}{\text{side opposite}} = \dfrac{5}{12}$

$\sec\theta = \dfrac{\text{hypotenuse}}{\text{side adjacent}} = \dfrac{13}{5}$

$\csc\theta = \dfrac{\text{hypotenuse}}{\text{side opposite}} = \dfrac{13}{12}$

7. Use a 30°-60°-90° right triangle to find the exact values.

(a) $\cos 60° = \dfrac{1}{2}$

(b) $\tan 45° = 1$

(c) $\tan(-270°) = \tan(-270° + 360°)$
$$= \tan 90°$$
$$= \frac{1}{0}; \text{ undefined}$$

(d) $\sec 210° = \sec(210° - 180°)$
$$= -\sec 30°$$
$$= -\frac{2}{\sqrt{3}} \cdot \frac{\sqrt{3}}{\sqrt{3}}$$
$$= -\frac{2\sqrt{3}}{3}$$

(e) $\csc(-180°) = \dfrac{1}{\sin(-180°)}$
$$= \frac{1}{0}; \text{ undefined}$$

(f) $\sec 135° = -\sec(180° - 135°)$
$$= -\sec 45°$$
$$= -\frac{\sqrt{2}}{1}$$
$$= -\sqrt{2}$$

8. Check to see that your calculator is in degree mode.

(a) $\sin 78°21'$
$$\sin 78°21' = \sin\left(78 + \frac{21}{60}\right)°$$
$$= \sin 78.35°$$
$$\approx 0.97939940$$

(b) $\tan 11.7689°$
$\tan 11.7689° \approx 0.20834446$

(c) $\sec 58.9041°$
$$\sec 58.9041° = \frac{1}{\cos 58.9041°}$$
$$\approx \frac{1}{0.516472054}$$
$$\approx 1.9362132$$

(d) $\cot 13.5°$
$$\cot 13.5° = \frac{1}{\tan 13.5°}$$
$$\approx \frac{1}{0.2400787591}$$
$$\approx 4.16529977$$

9. $\sin\theta = 0.27843196$
$$\theta = \arcsin(0.27843196) \approx 16.16664145$$

10. $\cos\theta = \dfrac{-\sqrt{2}}{2}$

Cosine is negative in quadrants II and III. The reference angle θ' is 45°. In quadrant II θ is $180° - 45° = 135°$. In quadrant III θ is $180° + 45° = 225°$.

11. $A = 58°30'$, $c = 748$, $C = 90°$

$\cos A = \dfrac{b}{c}$

$\quad b = c\cos A = (748)\cos 58°30' \approx 391$

$\sin A = \dfrac{a}{c}$

$\quad a = c\sin A = (748)\cos 58°30' \approx 638$

$A + B = 90°$

$\quad B = 90° - A = 90° - 58°30' = 31°30'$

12.

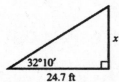

Let x be the height of the flag pole.

$\tan 32°10' = \dfrac{x}{24.7}$

$\quad x = (24.7)\tan 32°10' \approx 15.5 \text{ ft}$

13. Use the law of sines.

$\dfrac{\sin 25.2°}{6.92} = \dfrac{\sin B}{4.82}$

$\quad \sin B = \dfrac{(\sin 25.2°)(4.82)}{6.92}$

$\quad B = \sin^{-1}\!\left(\dfrac{(\sin 25.2°)(4.82)}{6.92}\right) \approx 17.3°$

Using the fact that the sum of the interior angles of a triangle add to 180°.

$C = 180° - (A + B)$

$\quad = 180° - (25.2° + 17.3°)$

$\quad = 137.5°$

14. Use the law of cosines.

$c^2 = a^2 + b^2 - 2ab\cos C$

$\quad = 75^2 + 130^2 - 2(75)(130)\cos 118°$

$\quad \approx 31680$

$c = \sqrt{31680} \approx 180 \text{ km}$

15. Use the law of cosines.

$b^2 = a^2 + c^2 - 2ac\cos B$

$2ac\cos B = a^2 + c^2 - b^2$

$\cos B = \dfrac{a^2 + c^2 - b^2}{2ac}$

$B = \cos^{-1}\!\left(\dfrac{a^2 + c^2 - b^2}{2ac}\right)$

$\quad = \cos^{-1}\!\left(\dfrac{(17.3)^2 + (29.8)^2 - (22.6)^2}{2(17.3)(29.8)}\right)$

$\quad \approx 49.0°$

16. $A = 111°$, $B = 41.0°$, $c = 326$ ft

$C = 180° - A - B$

$\quad = 180° - 111° - 41.0°$

$\quad = 28.0°$

$\dfrac{a}{\sin A} = \dfrac{c}{\sin C}$

$\quad a = \dfrac{c\sin A}{\sin C} = \dfrac{326\sin 111°}{\sin 28.0°} \approx 648 \text{ ft}$

$\dfrac{b}{\sin B} = \dfrac{c}{\sin C}$

$\quad b = \dfrac{c\sin B}{\sin C} = \dfrac{326\sin 41.0°}{\sin 28.0°} \approx 456 \text{ ft}$

17. $A = 60°$, $b = 30$ m, $c = 45$ m

$a^2 = b^2 + c^2 - 2bc\cos A$

$\quad = (30)^2 + (45)^2 - 2(30)(45)\cos 60°$

$a = \sqrt{1575} \approx 40 \text{ m}$

$\dfrac{\sin B}{30} = \dfrac{\sin 60°}{\sqrt{a^2}}$

$\quad B \approx 41°$

$C = 180° - 60° - 41° = 79°$

18. Find angle C.

$C = 180° - (47°20' + 24°50')$

$\quad = 180° - (72°10')$

$\quad = 107°50'$

Use this result and the law of sines to find AC.

$\dfrac{8.4}{\sin 107°50'} = \dfrac{AC}{\sin 47°20'}$

$\quad AC = \dfrac{(8.4)(\sin 47°20')}{\sin 107°50'}$

$\quad = \dfrac{(89.4)(\sin 47.333°)}{\sin 107.833°}$

$\quad \approx 6.5 \text{ mi}$

Drop a perpendicular line from C to segment AB.

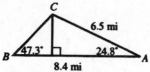

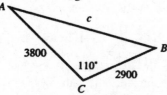

From the resulting triangle(s),

$$\sin 24.833° = \frac{h}{6.5}$$
$$h = (6.5)(\sin 24.833°) \approx 2.7 \text{ mi.}$$

The balloon is about 2.7 miles above the ground.

19. Let c = the length of the tunnel.

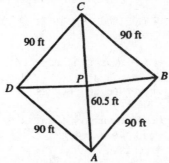

Use the law of cosines to find c.

$$\begin{aligned}
c^2 &= a^2 + b^2 - 2ab\cos C \\
&= 2900^2 + 3800^2 \\
&\quad - 2(2900)(3800)\cos 110° \\
&\approx 30,388,124
\end{aligned}$$

$c = \sqrt{30,388,124} \approx 5500$ m

The tunnel is about 5500 meters long.

20. Let A = home plate, B = first base, C = second base, D = third base, P = pitcher's rubber. Draw AC through P, draw PB and PD.

In triangle ABC, angle $B = 90°$, angle A = angle $C = 45°$.

$$AC = \sqrt{90^2 + 90^2} = 90\sqrt{2}$$
$$PC = 90\sqrt{2} - 60.5 = 66.8 \text{ ft}$$

In triangle APB, angle $A = 45°$.

$$\begin{aligned}
PB^2 &= AP^2 + AB^2 - 2(AP)(AB)\cos 45° \\
&= (60.5)^2 + (90)^2 - 2(60.5)(90)\cos 45° \\
PB &\approx 63.7 \text{ ft}
\end{aligned}$$

Since triangles APB and APD are congruent, $PB = PD = 63.7$ ft.

The distance to both first and third base is 63.7 feet, and the distance to second base is 66.8 feet.

Chapter 15

1. By counting, there are 7 vertices and 7 edges.

3. There are 10 vertices and 9 edges.

5. There are 6 vertices and 9 edges.

7. Two vertices have degree 3, three have degree 2, and two have degree 1. The sum of the degrees is
$3 + 3 + 2 + 2 + 2 + 1 + 1 = 14$.

9. Six vertices have degree 1. Four vertices have degree 3. The sum of the degrees is
$1 + 1 + 1 + 1 + 1 + 1 + 3 + 3 + 3 + 3 = 18$.
This is twice the number of edges, which is 9.

11. No, the two graphs are not isomorphic. There is one more vertex in the (b) graph.

13. Yes, the graphs are isomorphic.

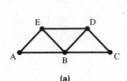

    (a)   (b)

15. No, the two graphs are not isomorphic. Two vertices in graph (b) have degree 1, while only one vertex in graph (a) has a degree of 1.

17. The graph is connected with only 1 component.

19. The graph is disconnected with 3 components.

21. The graph is disconnected with 2 components.

23. Since the sum of the degrees is given by $5 \times 4 = 20$, there are $\frac{1}{2} \times 20 = 10$ edges.

25. Since the sum of the degrees is given by $1 + 1 + 1 + 2 + 3 = 8$, there are $\frac{1}{2} \times 8 = 4$ edges.

27. (a) Yes, A→B→C is a walk.

    (b) No, B→A→D is not a walk, since there is no edge from A to D.

    (c) No, E→F→A→E is not a walk, since there is no edge from A to E.

    (d) Yes, B→D→F→B→D is a walk.

    (e) Yes, D→E is a walk.

    (f) Yes, C→B→C→B is a walk.

29. (a) No, A→B→C→D→E→F is not a circuit, since the path does not return to the starting vertex.

    (b) Yes, A→B→D→E→F→A is a circuit.

    (c) No, C→F→E→D→C is not a circuit, since there is no edge form C to F.

    (d) No, G→F→D→E→F is not a circuit, since the path does not return to the starting vertex.

    (e) No, F→D→F→E→D→F is not a circuit, since the edge from F to D is used more than once.

31. (a) No, A→B→C is not a path, since there is no edge from B to C.

    (b) No, J→G→I→G→F is not a path, since the edge from I to G is used more than once.

    (c) No, D→E→I→G→F is not a path, since it is not a walk (there is no edge from E to I).

    (d) Yes, C→A is a path.

    (e) Yes, C→A→D→E is a path.

    (f) No, C→A→D→E→D→A→B is not a path, since the edges A to D and D to E are used more than once.

33. A→B→C is a walk and also a path (no edges are repeated) but not a circuit (since the path does not return to the starting vertex).

35. A→B→A→C→D→A is a walk, not a path (since the edge A to B is used twice), and hence, not a circuit.

37. C→A→B→C→D→A→E is a walk and a path but not a circuit (since it doesn't end at C).

39. No, this is not a complete graph, since there is no edge going from A to C (nor B to D).

41. No, this is not a complete graph, since there is no edge going from A to F, for example.

43. Yes, this is a complete graph, since there is exactly one edge going from each vertex to each other vertex in the graph.

45.

Counting the edges, there are 7 games to be played in this competition.

47. One can draw a graph and count the edges or one may reason that each of the members of one team must shake the hand of each of the 6 members of the second team. Thus, there are $6 \times 6 = 36$ handshakes in total.

49. Each of the 8 people represents a vertex. Two of these people had 4 conversations. Thus, the degree for each of these vertices is 4. Another person had 3 conversations. Thus, the degree for this vertex is 3. Four people had 2 conversations each. Therefore, each of these 4 vertices have degree 2. One person had 1 conversation, so the degree of this vertex is 1. Thus, the sum of the degrees is
$4 + 4 + 3 + 2 + 2 + 2 + 2 + 1 = 20.$
Since there are one half as many edges as the sum of the degrees, we arrive at 10 edges, or telephone conversations.

51.

The circuit forms a triangle.

53. Writing exercise; answers will vary.

55. Writing exercise; answers will vary.

57. Writing exercise; answers will vary.

59.

 2

61. 3

63. 2

65. (a)

2

(b)

3

(c)

2

(d)

3

(e) A cycle with an odd number of vertices has chromatic number 3. A cycle with an even number of vertices has chromatic number 2.

67.

3

69.

3

71.

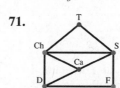

The chromatic number is 3. Thus, there must be at least 3 meeting times. The following organizations should meet at the same time: Choir and Forensics, Service Club and Dance Club, and Theater and Caribbean.

73.

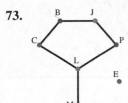

the chromatic number is 3. Thus, there must be at least 3 gatherings: The colleagues that should be invited to each are Brad, Phil, and Mary, Joe and Lindsay, and Caitlin and Eva.

75.

3 colors

77.

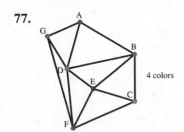

4 colors

79.

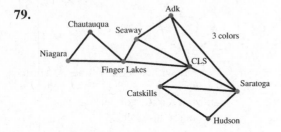

3 colors

In Exercise 81, there are many different ways to draw the map.

81.

83. By the four color theorem, this is not possible.

15.2 Exercises

1. (a) No, the sequence of vertices, A→B→C→D→A→B→C→D→A, does not represent an Euler circuit, since it is not a path. Remember that a path can traverse an edge only once.

(b) Yes, the sequence of vertices, C→B→A→D→C, is an Euler circuit, since it is a circuit and each edge is traversed only once.

(c) No, the sequence, A→C→D→B→A, is not an Euler circuit, since it is not a path.

(d) No, the sequence, A→B→C→D, is not an Euler circuit, since it is not a circuit.

3. (a) No, the sequence of vertices, A→B→C→D→E→F→A, does not represent an Euler circuit, since it doesn't use all of the edges (e.g. B→F).

(b) Yes, the sequence of vertices, A→B→C→D→E→G→C→E→F→G →B→F→A, is an Euler circuit since it is a circuit, and each edge is traversed only once.

(c) No, the sequence of vertices, A→B→C→D→E→G→C→E→F→G →B→F→A, is not an Euler circuit since it is not a path.

(d) Yes, the sequence of vertices, A→B→G→E→D→C→G→F→B→C →E→F→A, is an Euler circuit, since it is a circuit and each edge is traversed only once.

5. Yes, the graph will have an Euler circuit, since all vertices have even degree.

7. No, the graph will not have an Euler circuit, since some vertices (e.g., G) have odd degree.

9. If we assume that a vertex occurs at each intersection of curves and check the degree of each such vertex, we see that all vertices have an even degree. Thus, by Euler's theorem, all edges are traversed exactly one time—that is, we can find an Euler circuit. The answer is <u>yes</u> since this is equivalent to tracing the pattern without lifting your pencil nor going over any line more than one time.

11. Since all vertices have even degree, the graph has an Euler circuit. Because one must pass through vertex I more than once to complete any circuit, no circuit will visit each vertex exactly once.

13. Since all vertices have even degree, the graph has an Euler circuit. The sequence, A→B→H→C→G→D→F→E→A, visits each vertex exactly once.

15. Some vertices (e.g., B) have odd degree. Therefore, no Euler circuit exists. No circuit exists that will pass through each vertex exactly once.

Exercises 17–19 correspond to deciding if an Euler circuit exists when we assume that each intersection of line segments forms a vertex.

17. The upper left (and lower right) corner is of odd degree. Hence, there is no Euler circuit (or continuous path to apply the grout).

19. Since all vertices are of even degree, there exists an Euler circuit, and thus, a continuous path to apply the grout without retracing.

21. There are no cut edges, since no break along an edge will disconnect the graph.

23. The student has a choice of B→E or B→D. Choosing B→F, for example, is not allowed, since it is a cut edge.

25. The student has a choice of B→C or B→H. Choosing B→G, for example, is not allowed since it is a cut edge.

There are many different correct answers for the following exercises.

27. Beginning at A, one Euler circuit that results is A→C→B→F→E→D→C→F→D→A.

29. Since each vertex is of even degree, there is an Euler circuit. Beginning at A, one such circuit is A→G→H→J→I→L→J→K→I →H→F→G→E→F→D→E→C→D→B →C→A→E→B→A.

31. Some vertices (e.g., C) have odd degree. Therefore, no Euler circuit exists.

33. Such a route exists. Since all vertices are of even degree, an Euler circuit exists for the graph. Using Fleury's algorithm, one such route is given by A→D→B→C→A→H→D →E→B→H→G→E→F→H→J→L→C→M →A→K→M→L→K→J→A.

35. Draw a map such as the following.

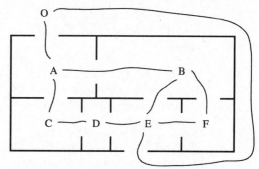

Since several vertices have odd degree (e.g., A), an Euler circuit does not exist. Thus, it is not possible to walk through each door exactly once and end up back outside.

37. There is no Euler path, since there are more than two vertices with odd degree.

39. Yes, there is an Euler path, since there are exactly two vertices with odd degree. The path must begin or end at B or G, the two vertices with odd degree.

Exercises 41–43 are solved by thinking of the corresponding graphs for Exercises 34–36. Remember that the rooms represent vertices and the doors are edges. To answer each question in the affirmative, an Euler path must exist.

41. Yes, it is possible because there are exactly two rooms with an odd number of doors (odd degree).

43. No, there is no Euler path, due to the fact that all rooms have an even number of doors (or degree).

45. One possible circuit is A→B→D→C→A. There are 4 edges in any such circuit.

47. One possible circuit is A→B→C→D→E →F→G→H→C→A→H→I→A. There are 12 edges in any such circuit.

49. Only those *complete graphs* for which the number of vertices is an odd number greater than or equal to 3 have an Euler circuit. In a complete graph with *n* vertices, the degree of each vertex is $n - 1$. And $n - 1$ is even if, and only if, *n* is odd.

15.3 Exercises

1. (a) No, A→E→C→D→E→B→A, is not a Hamilton circuit, since it visits vertex E twice.

    (b) Yes, A→E→C→D→B→A, is a Hamilton circuit, since it visits all vertices (except the first) only once.

    (c) No, D→B→E→A→B, is not a Hamilton circuit, since it does not visit C.

    (d) No, E→D→C→B→E, is not a Hamilton circuit, since it does not visit A.

3. (a) The path, A→B→C→D→E→A, is not a circuit (since BC is not an edge) and, hence, is not an Euler circuit nor a Hamilton circuit.

    (b) The path, B→E→C→D→A→B, is a circuit, is an Euler circuit (travels each edge only once), and is a Hamilton circuit (travels through each vertex only once).

    (c) The path, E→B→A→D→A→D→C→E, is not a circuit, is not an Euler circuit (travels edge AD twice), and is not a Hamilton circuit (travels through several vertices more than once).

5. This graph has a Hamilton circuit. One example is A→B→D→E→F→C→A.

7. This graph has a Hamilton circuit. One example is G→H→J→I→G.

9. This graph has a Hamilton circuit. One example is X→T→U→W→V→X.

11.
    A→B→C→D→A is a Hamilton circuit. There is no Euler circuit, since at least one of the vertices has odd degree. (In fact, all have odd degree.)

13.
    A→B→C→D→E→F→A is both a Hamilton and an Euler circuit.

15. This exercise could be solved by a Hamilton circuit, since each vertex represents a bandstand.

17. This exercise could be solved by an Euler circuit, since each path corresponds to an edge.

19. This exercise could be solved by a Hamilton circuit, since each vertex represents a country.

21. $4! = 4 \cdot 3 \cdot 2 \cdot 1 = 24$

23. $9! = 9 \cdot 8 \cdot 7 \cdot 6 \cdot 5 \cdot 4 \cdot 3 \cdot 2 \cdot 1 = 362,880$

25. There are $(10 - 1)! = 9!$ Hamilton circuits.

27. There are $(18 - 1)! = 17!$ Hamilton circuits.

29. Since this is a complete graph with 4 vertices, there are $(4 - 1)! = 3! = 3 \cdot 2 \cdot 1 = 6$ Hamilton circuits. Choosing P as a beginning vertex, they are:
    P→Q→R→S→P; P→Q→S→R→P;
    P→R→Q→S→P; P→R→S→Q→P;
    P→S→Q→R→P; P→S→R→Q→P.
    (Note: A tree diagram might be helpful here. P would represent the first node; Q, R and S, the second node; the third node would represent each of the remaining two vertices.)

31. Hamilton circuits starting with E→H→I would include: E→H→I→F→G→E and E→H→I→G→F→E.

33. Hamilton circuits starting with E→F would include:
    E→F→G→H→I→E; E→F→G→I→H→E;
    E→F→H→G→I→E; E→F→H→I→G→E;
    E→F→I→G→H→E; E→F→I→H→G→E.

35. Hamilton circuits starting with E→G would include:
E→G→F→H→I→E; E→G→F→I→H→E;
E→G→H→F→I→E; E→G→H→I→F→E;
E→G→I→F→H→E; E→G→I→H→F→E.

37. Hamilton circuits starting with A would include:
A→B→C→D→E→A; A→C→B→D→E→A; A→B→C→E→D→A; A→C→B→E→D→A;
A→B→D→C→E→A; A→C→D→B→E→A; A→B→D→E→C→A; A→C→D→E→B→A;
A→B→E→C→D→A; A→C→E→B→D→A; A→B→E→D→C→A; A→C→E→D→B→A;
A→D→B→C→E→A; A→E→B→C→D→A; A→D→B→E→C→A; A→E→B→D→C→A;
A→D→C→B→E→A; A→E→C→B→D→A; A→D→C→E→B→A; A→E→C→D→B→A;
A→D→E→B→C→A; A→E→D→B→C→A; A→D→E→C→B→A; A→E→D→C→B→A.

39. Using the Brute Force Algorithm:

Circuit	Total weight of circuit:
1. P→Q→R→S→P	550 + 640 + 500 + 510 = 2200
2. P→Q→S→R→P	550 + 790 + 500 + 600 = 2440
3. P→S→Q→R→P	510 + 790 + 640 + 600 = 2540
4. P→R→S→Q→P (opposite of 2)	2440
5. P→S→R→Q→P (opposite of 1)	2200
6. P→R→Q→S→P (opposite of 3)	2540

Thus, the Minimum Hamilton circuit is P→Q→R→S→P and the weight is 2200.

41. Observe that for a complete graph with 5 vertices we will have $(5 - 1) = 4! = 24$ Hamilton circuits.

Circuit:	Total weight of circuit:
1. C→D→E→F→G→C	12 + 10 + 17 + 15 + 10 = 64
2. C→D→E→G→F→C	12 + 10 + 13 + 15 + 15 = 65
3. C→F→D→E→G→C	15 + 21 + 10 + 13 + 10 = 69
4. C→D→F→G→E→C	12 + 21 + 15 + 13 + 15 = 76
5. C→E→D→F→G→C	15 + 10 + 21 + 15 + 10 = 71
6. C→D→G→E→F→C	12 + 13 + 13 + 17 + 15 = 70
7. C→E→F→D→G→C	15 + 17 + 21 + 13 + 10 = 76
8. C→E→D→G→F→C	15 + 10 + 13 + 15 + 15 = 68
9. C→F→E→D→G→C	15 + 17 + 10 + 13 + 10 = 65
10. C→E→G→D→F→C	15 + 13 + 13 + 21 + 15 = 77
11. C→D→F→E→G→C	12 + 21 + 17 + 13 + 10 = 73
12. C→D→G→F→E→C	12 + 13 + 15 + 17 + 15 = 72
13–24. (opposites of above)	

Thus, the Minimum Hamilton circuit is C→D→E→F→G→C and the weight is 64.

43. (a) Using the nearest neighbor algorithm, choose the first edge with the minimum weight, $A \xrightarrow{1} C.$ Keep track of weight by noting its value over the arrow. Choose the second edge, $C \xrightarrow{2} E,$ which has the minimum weight. Choose the third edge, $E \xrightarrow{7} D,$ which has the minimum weight. Choose the fourth edge, $D \xrightarrow{6} B,$ which has the minimum weight. Finally, choose the remaining edge, $B \xrightarrow{4} A.$ Note that this is your only choice for the last remaining edge. The resulting approximate minimum Hamilton circuit is, therefore, A→C→E→D→B→A. The circuit has a total weight of $1 + 2 + 7 + 6 + 4 = 20.$

(b) Using the nearest neighbor algorithm, choose the edge which has the minimum weight, $C \xrightarrow{1} A$. Keep track of weight by noting its value over the arrow. Choose the second edge with corresponding minimum weight, $A \xrightarrow{4} B$. Choose the third edge, which has the minimum weight, $B \xrightarrow{6} D$. Choose the fourth edge, which has the minimum weight, $D \xrightarrow{7} E$. Finally, choose the remaining edge, $E \xrightarrow{2} C$. The resulting approximate minimum Hamilton circuit is, therefore, $C \rightarrow A \rightarrow B \rightarrow D \rightarrow E \rightarrow C$. The circuit has a total weight of $1 + 4 + 6 + 7 + 2 = 20$.

(c) Using the nearest neighbor algorithm, choose the edge which has the minimum weight, $D \xrightarrow{3} C$. Choose the second edge with corresponding minimum weight, $C \xrightarrow{1} A$. Choose the third edge, which has the minimum weight, $A \xrightarrow{4} B$. Choose the fourth edge, which has the minimum weight, $B \xrightarrow{8} E$. Finally, choose the remaining edge, $E \xrightarrow{7} D$. The resulting approximate minimum Hamilton circuit is, therefore, $D \rightarrow C \rightarrow A \rightarrow B \rightarrow E \rightarrow D$. The circuit has a total weight of $3 + 1 + 4 + 8 + 7 = 23$.

(d) Using the nearest neighbor algorithm, choose the edge which has the minimum weight, $E \xrightarrow{2} C$. Choose the second edge with corresponding minimum weight, $C \xrightarrow{1} A$. Choose the third edge, which has the minimum weight, $A \xrightarrow{4} B$. Choose the fourth edge, which has the minimum weight, $B \xrightarrow{6} D$. Finally, choose the remaining edge, $D \xrightarrow{7} E$. The resulting

approximate minimum Hamilton circuit is, therefore, $E \rightarrow C \rightarrow A \rightarrow B \rightarrow D \rightarrow E$. The circuit has a total weight of $2 + 1 + 4 + 6 + 7 = 20$.

45. (a) Beginning with vertex A, choose the edge with the minimum weight, $A \xrightarrow{5} C$. Choose the second edge, $C \xrightarrow{7} D$. Choose the third edge, $D \xrightarrow{10} E$. Choose the fourth edge, $E \xrightarrow{50} B$. Finally, to the last vertex, A, we must choose $B \xrightarrow{15} A$. The resulting approximate minimum Hamilton circuit is $A \rightarrow C \rightarrow D \rightarrow E \rightarrow B \rightarrow A$. The circuit has a total weight of $5 + 7 + 10 + 50 + 15 = 87$.
Beginning with vertex B, choose the edge with the minimum weight, $B \xrightarrow{11} C$. Choose the second edge, $C \xrightarrow{5} A$. Choose the third edge, $A \xrightarrow{9} E$. Choose the fourth edge, $E \xrightarrow{10} D$. Finally, to the last vertex, B, we must choose $D \xrightarrow{60} B$. The resulting approximate minimum Hamilton circuit is $B \rightarrow C \rightarrow A \rightarrow E \rightarrow D \rightarrow B$. The circuit has a total weight of $11 + 5 + 9 + 10 + 60 = 95$.
Beginning with vertex C, choose the edge with the minimum weight, $C \xrightarrow{5} A$. Choose the second edge, $A \xrightarrow{9} E$. Choose the third edge, $E \xrightarrow{10} D$. Choose the fourth edge, $D \xrightarrow{60} B$. Finally, to the last vertex, C, we must choose $B \xrightarrow{11} C$. The resulting approximate minimum Hamilton circuit is $C \rightarrow A \rightarrow E \rightarrow D \rightarrow B \rightarrow C$. The circuit has a total weight of $5 + 9 + 10 + 60 + 11 = 95$.
Beginning with vertex D, choose the edge with the minimum weight, $D \xrightarrow{7} C$. Choose the second edge, $C \xrightarrow{5} A$. Choose the third edge, $A \xrightarrow{9} E$.

Choose the fourth edge, $\overset{50}{E \rightarrow B}$. Finally, to the last vertex, D, we must choose $\overset{60}{B \rightarrow D}$. The resulting approximate minimum Hamilton circuit is D→C→A→E→B→D. The circuit has a total weight of $7 + 5 + 9 + 50 + 60 = 131$. Beginning with vertex E, choose the edge with the minimum weight, $\overset{8}{E \rightarrow C}$. Choose the second edge, $\overset{5}{C \rightarrow A}$. Choose the third edge, $\overset{14}{A \rightarrow D}$. Choose the fourth edge, $\overset{60}{D \rightarrow B}$. Finally, to the last vertex, E, we must choose $\overset{50}{B \rightarrow E}$. The resulting approximate minimum Hamilton circuit is E→C→A→D→B→E. The circuit has a total weight of $8 + 5 + 14 + 60 + 50 = 137$.

(b) The best solution is the circuit A→C→D→E→B→A, since it's total weight, 87, is the smallest.

(c) One example would be A→B→C→D→E→A, with a total weight of $15 + 11 + 7 + 10 + 9 = 52$.

47. All Hamilton circuits include:
A→B→C→D→E→F→A;
A→B→C→F→E→D→A;
A→B→E→D→C→F→A;
A→B→E→F→C→D→A;
A→D→E→F→C→B→A;
A→D→E→B→C→F→A;
A→D→C→B→E→F→A;
A→D→C→F→E→B→A;
A→F→E→B→C→D→A;
A→F→E→D→C→B→A;
A→F→C→D→E→B→A;
A→F→C→B→E→D→A.
A tree diagram such as below, can be a help in creating all of the Hamilton circuits.

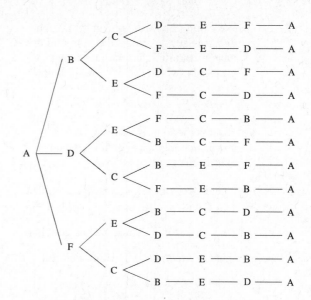

49. All Hamilton circuits include:
A→B→C→D→E→F→A;
A→B→C→E→D→F→A;
A→B→C→E→F→D→A;
A→B→C→F→E→D→A;
A→D→E→F→C→B→A;
A→D→F→E→C→B→A;
A→F→D→E→C→B→A;
A→F→E→D→C→B→A.

51. (a) For graph (1), $\frac{n}{2}$: $\frac{6}{2} = 3$. Condition not satisfied. For example, vertex A has degree 2.

For graph (2), $\frac{n}{2}$: $\frac{6}{2} = 3$. Condition is satisfied, since all vertices are 3 or larger in degree.

For graph (3), $\frac{n}{2}$: $\frac{5}{2} = 2.5$. Condition not satisfied. For example, vertex G has degree 2.

For graph (4), $\frac{n}{2}$: $\frac{5}{2} = 2.5$. Condition is satisfied, since all vertices are 3 or larger in degree.

For graph (5), $\frac{n}{2}$: $\frac{7}{2} = 3.5$. Condition not satisfied. For example, vertex B has degree 3.

(b) By Dirac's theorem, graphs (2) and (4) are predicted to have Hamilton circuits.

(c) We cannot be sure that a graph which doesn't satisfy Dirac's theorem will not have a Hamilton circuit. For example, graph (1), which doesn't satisfy Dirac's theorem, has a Hamilton circuit A→B→C→D→E→F→A.

(d) Dirac's theorem is not true for $n < 3$ since such a graph will not have any Hamilton circuit at all.

(e) The degree of each vertex in a complete graph with n vertices is $(n-1)$. If $n \geq 3$, then $(n-1) \geq \dfrac{n}{2}$. Thus, we can conclude that the graph has a Hamilton circuit.

Note that there are different possible answers for question 53.

53. A→F→G→R→S→T→U→Q→P→N→M →L→K→J→I→H→B→C→D→E→A is a Hamilton circuit.

55. Writing exercise; answers will vary.

15.4 Exercises

1. The graph is a tree, since it is connected and has no circuits.

3. The graph is not a tree, since it is not connected.

5. The graph is a tree, since it is connected and has no circuits.

7. Such a graph is not a tree, since it has a circuit. Remember that Euler's theorem tells us that all connected graphs with all vertices of even degree will have an Euler circuit.

9. It is not possible to form a tree by adding an extra edge, since the graph has a circuit.

11. Yes, such a graph is a tree, since it is connected and has no circuits.

13. No, such a graph would not necessarily be a tree since circuits may be formed.

15. The statement "Every connected graph n which each edge is a cut edge is a tree" is true. Since all edges are cut edges, there can be no circuits.

17. The statement "Every graph in which each edge is a cut edge is a tree" is false. For example, disconnected graphs as the following

are not trees.

There are many different correct answers for 19.

19.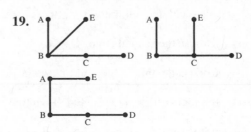

21.

23.

25. There are two circuits in this graph. We can drop any one of 4 edges from the first circuit and any one of 5 edges from the second circuit to form a spanning tree. thus, there would be a total of $4 \times 5 = 20$ different spanning trees.

27. If a connected graph has circuits, none of which have common edges, the number of spanning trees for the graph is the product of the number of edges in each circuit.

29. Choose the edge with minimum weight GF (2). Choose the next edge with minimum weight, FC (3). The next edge with minimum weight is FD (8). The next edge of choice would be DE (10). Note here that, although DC represents a remaining edge with minimum weight (9), if it were chosen, we would then have a circuit, FDC, which is not allowed. Choose next edge with minimum weight FB (12) followed by BA (16). (Note BG (15) is not allowed because the circuit, BGF, is then formed.) We now have the following minimum spanning tree.

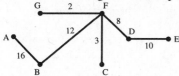

The total weight is $2 + 3 + 8 + 10 + 12 + 16 = 51$.

31. Choose initial minimum edge AD (6), followed by BG (8), then EF (9), then DC (10), and then AB (13). The last remaining minimum edge to choose, without completing a circuit, is AE (20). Thus, the minimum spanning tree is:

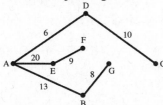

The total weight is $6 + 8 + 9 + 10 + 13 + 20 = 66$.

33. Choose the initial minimum edge, BF (23), followed by either AE (25) or EF (25). Then choose the other edge with weight 25. Finally, choose AD (32) followed by DC (35).

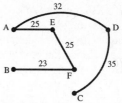

The total length of the minimum pathway is $23 + 25 + 25 + 32 + 35 = 140$ ft.

35. The tree with 34 vertices will have one less, or 33 edges.

37. A spanning tree for a complete graph with 63 vertices will have one less, or 62 edges.

39. Different spanning trees must have the same number of edges, since the number of vertices in the tree is the number of vertices in the original graph, and the number of edges has to be one less than the number of vertices.

41. Consider a tree with 10 vertices.

   (a) There will be $10 - 1 = 9$ edges.

   (b) The sum of the degrees for all vertices will be twice the number of edges, or $2 \times 9 = 18$.

   (c) The smallest number of vertices of degree four on this graph will be 0. For example, a tree may be drawn with the first and last vertex of degree 1, and the remaining 8 vertices of degree 2.

   (d) The largest number of vertices of degree four in this graph is 2. Let us use the strategy of trial and error. If we use 4 vertices with degree 4, this will contribute 16 to the total degree and would leave $18 - 16 = 2$ as the degree sum of the remaining 6 vertices. But this means that some of those vertices would have no edges joined to them, so that our graph would not be connected and would not be a tree. Similarly, using 3 vertices with degree 4 would leave $18 - 12 = 6$ as the sum of (the remaining vertex) degrees. Thus, of the last 7 vertices, at least one would have no degree (or connecting edge). This would not be a problem if we choose two vertices with degree 4. The following tree, for example, would satisfy our conditions.

43. Treating each of the 23 employee's computers as a vertex and the network as a tree, there will be one less edge than the number of vertices. Hence, there will be 22 cables that will have to be run between the computers.

45. It is possible to draw in the same number of vertices as edges. The graph must be a tree, since we still have one less edge than vertex.

47. It is possible to draw in more edges than vertices. But the graph cannot be a tree, since we would now have at least as many edges as vertices.

49. Using Cayley's theorem with $n = 3$, "A complete graph with n vertices has n^{n-2} spanning trees," we arrive at $3^{3-2} = 3^1 = 3$ spanning trees. A complete graph followed by the three spanning trees are as follows.

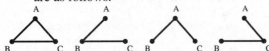

51. For a graph with 5 vertices, we will have $5^{5-2} = 5^3 = 125$ spanning trees.

53. There are just 3 non-isomorphic trees with 5 vertices. They are as follows.

55. There are 11 non-isomorphic trees with 7 vertices. They are as follows.

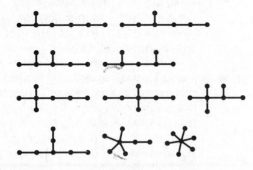

57. Writing exercise; answers will vary.

EXTENSION: ROUTE PLANNING

1. In the graph shown, not all vertices have even degree. For example, the vertex marked X has degree 5. So this graph still does not have an Euler circuit.

In Exercise 3, there are other ways of inserting the additional edges.

3.

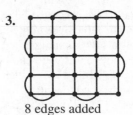

8 edges added

Chapter 15 Test

1. By counting, there are 7 vertices.

2. The sum of the degrees of the vertices are: $4 + 2 + 2 + 2 + 6 + 2 + 2 = 20$.

3. By counting, there are 10 edges.

4. **(a)** No, B→A→C→E→B→A is not a path, since edge AB is used twice.

  (b) Yes, A→B→E→A is a path.

  (c) No, A→C→D→E is not a path, since there is no edge from C to D.

5. **(a)** Yes, A→B→E→D→A is a circuit.

  (b) No, A→B→C→D→E→F→G→A is not a circuit, since, for example, there is no edge from B to C.

  (c) Yes, A→B→E→F→G→E→D→A →E→C→A is a circuit.

6. A graph with 2 components, for example, is:

7. The sum of degrees of the vertices is:
    4 + 4 + 4 + 2 + 2 + 2 + 2 + 2 + 2 + 2 = 26.
    Thus, there are 26 ÷ 2 = 13 edges.

8. The graphs are isomorphic.

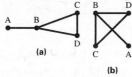

9. The graph is as follows.

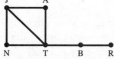

    The graph is connected. Tina knows the
    largest number of other guests.

10. Let each of the 8 contestants represent a
    vertex of a complete graph (graph with each
    vertex connected by an edge to all other
    vertices). The degree of each vertex is 7.
    Thus, the sum of the degrees is 8 · 7 = 56.
    Since the sum of the degrees is twice the
    number of edges, there must be
    56 ÷ 2 = 28 edges. Since each edge
    represents a game to be played, there are
    28 games in the competition.

11. Yes, the graph is a complete graph, since
    there is an edge from each vertex to each of
    the remaining 6 vertices.

12.

    Chromatic number: 3

13.

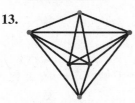

    Chromatic number: 5

14.

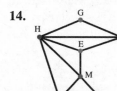

    Three exam times are needed. The exams
    that can be given at the same time are
    history and chemistry, mathematics and
    psychology, and English, biology, and
    geography.

15. (a) No, A→B→E→D→A is not an Euler
        circuit, since it does not use all of the
        edges.

    (b) No, A→B→C→D→E→F→G→A is
        not a circuit because, for example, there
        is no edge from B to C.

    (c) Yes, A→B→E→F→G→E→D→A→E
        →C→A is an Euler circuit.

16. No, the graph will not have an Euler circuit,
    since some of the vertices have odd degree.

17. Yes, the graph will have an Euler circuit,
    since all vertices have even degree.

18. No, since two of the rooms have an odd
    number of doors. Note that we are
    considering each room to be a vertex and
    asking the question "Can an Euler circuit be
    formed?"

19. A resulting Euler circuit is
    F→B→E→D→B→C→D→K→B→A→H
    →G→F→A→G→J→F. Note, BF is the
    only cut edge after F→B, thus you may
    choose any vertex in the right subgraph after
    B.

20. (a) No, A→B→E→D→A, is not a
        Hamilton circuit since it does not visit
        all vertices.

    (b) No, A→B→C→D→E→F→G→A is
        not a circuit because, for example, there
        is no edge from B to C.

    (c) No, A→B→E→F→G→E→D→A→E
        →C→A is not a Hamilton circuit since
        it visits some vertices twice before
        returning to starting vertex.

21. F→G→H→I→E→F; F→G→H→E→I→F;
    F→G→I→H→E→F; F→G→I→E→H→F;
    F→G→E→H→I→F; F→G→E→I→H→F.
    There are 6 such Hamilton circuits.

22. Using the Brute Force Algorithm and P as
    the starting vertex, we get the following
    circuits. Use a tree diagram as an aid.

Circuit:	Total weight of circuit:
1. P→Q→R→S→P	7 + 8 + 7 + 7 = 29
2. P→Q→S→R→P	7 + 9 + 7 + 4 = 27
3. P→R→Q→S→P	4 + 8 + 9 + 7 = 28
4. P→R→S→Q→P (opposite of 2)	27
5. P→S→Q→R→P (opposite of 3)	28
6. P→S→R→Q→P (opposite of 1)	29

Thus, P→Q→S→R→P is the minimum Hamilton circuit with a weight of 27.

23. Using the nearest neighbor algorithm, choose the first edge with the minimum weight, A $\xrightarrow{1.6}$ E. Keep track of weight by noting its value over the arrow. Choose the second edge, E $\xrightarrow{1.95}$ D, which has the minimum weight. Choose the third edge, D $\xrightarrow{1.8}$ C, which has the minimum weight. Choose C $\xrightarrow{2.3}$ F. Choose F $\xrightarrow{1.7}$ B. Finally, choose B $\xrightarrow{2.5}$ A. Note that this is your only choice for the last remaining edge. The resulting approximate minimum Hamilton circuit is, therefore,
A→E→D→C→F→B→A.
The circuit has a total weight of
1.6 + 1.95 + 1.8 + 2.3 + 1.7 + 2.5 = 11.85.

24. For a complete graph with 25 vertices, there will be (25 − 1)! = 24! Hamilton circuits.

25. This problem calls for a Hamilton circuit, since the band wants to visit each city (vertex) only once.

26. Any three of the following, for example, would satisfy the stated conditions.

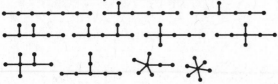

27. The statement "Every tree has a Hamilton circuit" is false, since in many trees, you will have to visit the same vertex more than once.

28. The statement "In a tree each edge is a cut edge" is true.

29. The statement "Every tree is connected" is true, since one can always move from each vertex of the graph along edges to every other vertex.

30. The following represent the different spanning trees for the accompanying graph.

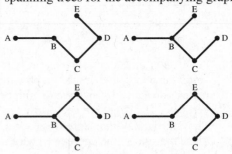

31. Using Kruskal's algorithm, choose the edge with minimum weight, B $\xrightarrow{3}$ E. Choose the next edge with minimum weight, E $\xrightarrow{5}$ D. Continuing in this fashion, we choose C $\xrightarrow{7}$ D followed by D $\xrightarrow{9}$ A.

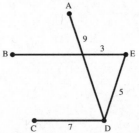

The weight is 3 + 5 + 7 + 9 = 24.

32. The number of edges in a tree is one less than the number of vertices. Thus, there are 50 − 1 = 49 edges.

Chapter 16

Many of the exercises in this chapter are done most efficiently by using a spreadsheet.

16.1 Exercises

1. **(a)** Because there are four breeds, the number of ways that a staff member can complete his or her ballot is
$$4! = 4 \cdot 3 \cdot 2 \cdot 1 = 24.$$

   (b) The voter profile is

Votes	Ranking	Voters
3	b > c > a > d	1, 2, 9
2	a > d > c > b	3, 11
1	c > d > b > a	4
1	d > c > b > a	5
1	d > a > b > c	6
2	c > a > d > b	7, 8
1	a > c > d > b	10
2	a > b > c > d	12, 13

   (c) By the plurality method, the breed with the most votes wins. The Australian shepherd wins.

Breed	1st place votes
a	5
b	3
c	3
d	2

3. The voter profile from Exercise 1(b) is

Votes	Ranking	Voter	Row
3	b > c > a > d	1, 2, 9	1
2	a > d > c > b	3, 11	2
1	c > d > b > a	4	3
1	d > c > b > a	5	4
1	d > a > b > c	6	5
2	c > a > d > b	7, 8	6
1	a > c > d > b	10	7
2	a > b > c > d	12, 13	8

(a) Using the pairwise comparison method, first make a table to compute the votes in comparison.

Compare	Votes	Rows
a > b	2 + 1 + 2 + 1 + 2 = 8	2, 5, 6, 7, 8
b > a	3 + 1 + 1 = 5	1, 3, 4
a > c	2 + 1 + 1 + 2 = 6	2, 5, 7, 8
c > a	3 + 1 + 1 + 2 = 7	1, 3, 4, 6
a > d	3 + 2 + 2 + 1 + 2 = 10	1, 2, 6, 7, 8
d > a	1 + 1 + 1 = 3	3, 4, 5
b > c	3 + 1 + 2 = 6	1, 5, 8
c > b	2 + 1 + 1 + 2 + 1 = 7	2, 3, 4, 6, 7
b > d	3 + 2 = 5	1, 8
d > b	2 + 1 + 1 + 1 + 2 + 1 = 8	2, 3, 4, 5, 6, 7
c > d	3 + 1 + 2 + 1 + 2 = 9	1, 3, 6, 7, 8
d > c	2 + 1 + 1 = 4	2, 4, 5

Make another table to award pairwise points.

Pairs	Votes	Pairwise Points
a : b	8 : 5	a, 1
a : c	6 : 7	c, 1
a : d	10 : 3	a, 1
b : c	6 : 7	c, 1
b : d	5 : 8	d, 1
c : d	9 : 4	c, 1

Using the pairwise comparison method, the winner is c with 3 pairwise points. Breed a received 2 points; breed d received 1 point; breed b received no points.

(b) Use the information from part (a) to make a table to examine the points in the Borda method.

Votes	Ranking	Points			
		3	2	1	0
3	b > c > a > d	b	c	a	d
2	a > d > c > b	a	d	c	b
1	c > d > b > a	c	d	b	a
1	d > c > b > a	d	c	b	a
1	d > a > b > c	d	a	b	c
2	c > a > d > b	c	a	d	b
1	a > c > d > b	a	c	d	b
2	a > b > c > d	a	b	c	d

Examine the last four columns, and compute the weighted sum for each issue.

Breed	Borda Points
a	$3 \cdot 1 + 2 \cdot 3 + 1 \cdot 0 + 1 \cdot 0 + 1 \cdot 2 + 2 \cdot 2 + 1 \cdot 3 + 2 \cdot 3 = 24$
b	$3 \cdot 3 + 2 \cdot 0 + 1 \cdot 1 + 1 \cdot 1 + 1 \cdot 1 + 2 \cdot 0 + 1 \cdot 0 + 2 \cdot 2 = 16$
c	$3 \cdot 2 + 2 \cdot 1 + 1 \cdot 3 + 1 \cdot 2 + 1 \cdot 0 + 2 \cdot 3 + 1 \cdot 2 + 2 \cdot 1 = 23$
d	$3 \cdot 0 + 2 \cdot 2 + 1 \cdot 2 + 1 \cdot 3 + 1 \cdot 3 + 2 \cdot 1 + 1 \cdot 1 + 2 \cdot 0 = 15$

The winner is breed a with 24 Borda points.

(c) Calculate how many points each breed obtained from the thirteen first-place votes from the voter profile.

Votes	Ranking	Voter	Row
3	b > c > a > d	1, 2, 9	1
2	a > d > c > b	3, 11	2
1	c > d > b > a	4	3
1	d > c > b > a	5	4
1	d > a > b > c	6	5
2	c > a > d > b	7, 8	6
1	a > c > d > b	10	7
2	a > b > c > d	12, 13	8

Breed	1st-place votes
a	5
b	3
c	3
d	2

None of the dog breeds received a majority, 7, of the votes. Therefore, eliminate breed d, the breed that received the least number of votes. Compare the other three.

Votes	Ranking	Voter	Row
3	b > c > a	1, 2, 9	1
2	a > c > b	3, 11	2
1	c > b > a	4	3
1	c > b > a	5	4
1	a > b > c	6	5
2	c > a > b	7, 8	6
1	a > c > b	10	7
2	a > b > c	12, 13	8

Breed a: $2 + 1 + 1 + 2 = 6$ votes
Breed b: 3 votes
Breed c: $1 + 1 + 2 = 4$ votes
None of the breeds received a majority; therefore, eliminate b and compare first-place points again for a and c.

Votes	Ranking	Voter	Row
3	c > a	1, 2, 9	1
2	a > c	3, 11	2
1	c > a	4	3
1	c > a	5	4
1	a > c	6	5
2	c > a	7, 8	6
1	a > c	10	7
2	a > c	12, 13	8

Breed a: $2 + 1 + 1 + 2 = 6$
Breed c: $3 + 1 + 1 + 2 = 7$
Breed c now has a majority and wins.

5. For $n = 5$, the number of rankings is
$$5! = 5 \cdot 4 \cdot 3 \cdot 2 \cdot 1 = 120.$$
For $n = 7$, the number of rankings is
$$7! = 7 \cdot 6 \cdot 5 \cdot 4 \cdot 3 \cdot 2 \cdot 1 = 5040.$$

7. Writing exercise; answers will vary.

9. The number of pairwise comparisons needed to learn the outcome of an election involving $n = 6$ candidates is the number of combinations of 6, taken 2 at a time. Mathematically this means
$$_6C_2 = \frac{6!}{2!(6-2)!} = \frac{6 \cdot 5 \cdot 4 \cdot 3 \cdot 2 \cdot 1}{2 \cdot 1 \cdot 4 \cdot 3 \cdot 2 \cdot 1} = 15.$$
For $n = 8$ candidates, find the combinations of 8, taken 2 at a time.
$$_8C_2 = \frac{8!}{2!(8-2)!} = \frac{8 \cdot 7 \cdot 6 \cdot 5 \cdot 4 \cdot 3 \cdot 2 \cdot 1}{2 \cdot 1 \cdot 6 \cdot 5 \cdot 4 \cdot 3 \cdot 2 \cdot 1} = 28.$$

11. Writing exercise; answers will vary.

13. Use the voter profile to answer the questions.

Votes	Ranking	Row
3	$a > c > b$	1
4	$c > b > a$	2
2	$b > a > c$	3
4	$b > c > a$	4

(a) Using the plurality method to determine the chairperson, the winner is candidate b with 6 first-place votes in rows 3 and 4. Notice from the table that candidate a received 3 votes in row 1, and candidate c received 4 votes in row 2.

(b) First, the number of comparisons is
$$_3C_2 = \frac{3!}{2!(3-2)!} = \frac{3 \cdot 2 \cdot 1}{2 \cdot 1 \cdot 1} = 3.$$
Make a table to compute the votes in comparisons.

Comparison	Votes	Rows
$a > b$	$3 = 3$	1
$b > a$	$4 + 2 + 4 = 10$	2, 3, 4
$a > c$	$3 + 2 = 5$	1, 3
$c > a$	$4 + 4 = 8$	2, 4
$b > c$	$2 + 4 = 6$	3, 4
$c > b$	$3 + 4 = 7$	1, 2

Make another table to award pairwise points.

Pairs	Votes	Pairwise Points
$a : b$	$3 : 10$	b, 1
$a : c$	$5 : 8$	c, 1
$b : c$	$6 : 7$	c, 1

Using the pairwise comparison method, the winner is candidate c with 2 pairwise points. Candidate b received only 1 point, and candidate a received 0 points.

(c) Use the information from part(a) to make a table to examine the points in the Borda method.

Votes	Ranking	Points		
		2	1	0
3	$a > c > b$	a	c	b
4	$c > b > a$	c	b	a
2	$b > a > c$	b	a	c
4	$b > c > a$	b	c	a

Examine the last three columns and compute the weighted sum for each candidate.

Candidate	Borda Points
a	$3 \cdot 2 + 4 \cdot 0 + 2 \cdot 1 + 4 \cdot 0 = 8$
b	$3 \cdot 0 + 4 \cdot 1 + 2 \cdot 2 + 4 \cdot 2 = 16$
c	$3 \cdot 1 + 4 \cdot 2 + 2 \cdot 0 + 4 \cdot 1 = 15$

The winner is candidate b with 16 Borda points.

(d) There are 13 voters, so a candidate must receive a majority of the votes or 7 votes to win by the Hare method. No candidate received a majority; candidates a, b, and c received 3, 6, and 4 votes respectively. (See part a.) By this method, candidate a is eliminated, and another vote is taken. Now compare only candidates b and c.

Comparison	Votes	Rows
$b > c$	$2 + 4 = 6$	3, 4
$c > b$	$3 + 4 = 7$	1, 2

Candidate c now receives a majority and wins.

15. Use the voter profile to answer the questions.

Votes	Ranking	Row
6	a > b > c	1
1	b > c > a	2
3	b > a > c	3
3	c > a > b	4

(a) Using the plurality method to determine the logo, the winner is logo a with 6 first-place votes in row 1. Notice from the table that logo b received a total of 4 votes in rows 2 and 3, and logo c received 3 votes in row 4.

(b) First, the number of comparisons is

$$_3C_2 = \frac{3!}{2!(3-2)!} = \frac{3\cdot2\cdot1}{2\cdot1\cdot1} = 3.$$

Make a table to compute the votes in comparisons.

Comparison	Votes	Rows
a > b	6 + 3 = 9	1, 4
b > a	1 + 3 = 4	2, 3
a > c	6 + 3 = 9	1, 3
c > a	1 + 3 = 4	2, 4
b > c	6 + 1 + 3 = 10	1, 2, 3
c > b	3 = 3	4

Make another table to award pairwise points.

Pairs	Votes	Pairwise Points
a : b	9 : 4	a, 1
a : c	9 : 4	a, 1
b : c	10 : 3	b, 1

Using the pairwise comparison method, the winner is logo a with 2 pairwise points. Logo b received only 1 point, and logo c received 0 points.

(c) Use the information from part (a) to make a table to examine the points in the Borda method.

Votes	Ranking	Points 2	1	0
6	a > b > c	a	b	c
1	b > c > a	b	c	a
3	b > a > c	b	a	c
3	c > a > b	c	a	b

Examine the last three columns and compute the weighted sum for each logo.

Logo	Borda Points
a	6·2 + 1·0 + 3·1 + 3·1 = 18
b	6·1 + 1·2 + 3·2 + 3·0 = 14
c	6·0 + 1·1 + 3·0 + 3·2 = 7

The winner is logo a with 18 Borda points.

(d) There are 13 voters, so a logo must receive a majority of the votes or at least 7 votes to win by the Hare method. No logo received a majority. Logo a received 6 votes; log b received 4 votes; logo c received 3 votes. Eliminate log c and compare the remaining logos.

Votes	Ranking	Row
6	a > b	1
1	b > a	2
3	b > a	3
3	a > b	4

Now logo a receives 6 + 3 = 9 votes; logo b receives 1 + 3 = 4 votes. Logo a now receives a majority and wins. Logo a wins this competition, as it is selected by all methods.

17. Use the voter profile to answer the questions.

Votes	Ranking	Row
6	h > j > g > e	1
5	e > g > j > h	2
4	g > j > h > e	3
3	j > h > g > e	4
3	e > j > h > g	5

(a) Using the plurality method to determine the highest priority issue, the winner is issue e with a total of 8 first-place votes in rows 2 and 5. Notice from the table that the remaining issues received the following number of first-place votes: h, 6 votes; g, 4 votes; j, 3 votes.

(b) First, the number of comparisons is
$$_4C_2 = \frac{4!}{2!(4-2)!} = \frac{4\cdot3\cdot2\cdot1}{2\cdot1\cdot2\cdot1} = 6.$$
Make a table to compute the votes in comparisons.

Comparison	Votes	Rows
e > h	5 + 3 = 8	2, 5
h > e	6 + 4 + 3 = 13	1, 3, 4
e > g	5 + 3 = 8	2, 5
g > e	6 + 4 + 3 = 13	1, 3, 4
e > j	5 + 3 = 8	2, 5
j > e	6 + 4 + 3 = 13	1, 3, 4
h > g	6 + 3 + 3 = 12	1, 4, 5
g > h	5 + 4 = 9	2, 3
h > j	6 = 6	1
j > h	5 + 4 + 3 + 3 = 15	2, 3, 4, 5
j > g	6 + 3 + 3 = 12	1, 4, 5
g > j	5 + 4 = 9	2, 3

Make another table to award pairwise points.

Pairs	Votes	Pairwise Points
e : h	8 : 13	h, 1
e : g	8 : 13	g, 1
e : j	8 : 13	j, 1
h : g	12 : 9	h, 1
h : j	6 : 15	j, 1
j : g	12 : 9	j, 1

Using the pairwise comparison method, the winner is issue j with 3 pairwise points. Issue h received 2 points; issue g received 1 point; issue e received no points.

(c) Use the information from part (a) to make a table to examine the points in the Borda method.

Votes	Ranking	Points 3	2	1	0
6	h > j > g > e	h	j	g	e
5	e > g > j > h	e	g	j	h
4	g > j > h > e	g	j	h	e
3	j > h > g > e	j	h	g	e
3	e > j > h > g	e	j	h	g

Examine the last four columns, and compute the weighted sum for each issue.

Issue	Borda Points
h	$6\cdot3 + 5\cdot0 + 4\cdot1 + 3\cdot2 + 3\cdot1 = 31$
j	$6\cdot2 + 5\cdot1 + 4\cdot2 + 3\cdot3 + 3\cdot2 = 40$
g	$6\cdot1 + 5\cdot2 + 4\cdot3 + 3\cdot1 + 3\cdot0 = 31$
e	$6\cdot0 + 5\cdot3 + 4\cdot0 + 3\cdot0 + 3\cdot3 = 24$

The winner is issue j with 40 Borda points.

(d) There are 21 voters, so an issue must receive a majority of the votes or at least 11 votes to win by the Hare method. No issue received a majority. See part (a) that issue e received 8 first-place votes; issue h received 6 votes; issue g received 4 votes; and issue j received 3 votes. Eliminate issue j because it has the least number of votes, and compare only h, g, and e.

Votes	Ranking	Row
6	h > g > e	1
5	e > g > h	2
4	g > h > e	3
3	h > g > e	4
3	e > h > g	5

Issue h receives 6 + 3 = 9 votes.
Issue g receives 4 votes.
Issue e receives 5 + 3 = 8 votes.
No issue has received a majority, so another vote is taken with issue g eliminated.

Votes	Ranking	Row
6	h > e	1
5	e > h	2
4	h > e	3
3	h > e	4
3	e > h	5

Issue h receives 6 + 4 + 3 = 13 votes.
Issue e receives 5 + 3 = 8 votes.
Now issue h has a majority and wins.

19. The voter profile is:

Votes	Ranking	Row
18	t > m > h > k > c	1
12	c > h > m > k > t	2
10	k > c > h > m > t	3
9	m > k > h > c > t	4
4	h > c > m > k > t	5
2	h > k > m > c > t	6

(a) If the plurality method is used, activity t is selected with 18 first-place votes.

(b) Using the pairwise comparison method, first make a table to compute the votes in comparison.

Compare	Votes	Rows
t > m	18 = 18	1
m > t	12 + 10 + 9 + 4 + 2 = 37	2, 3, 4, 5, 6
t > h	18 = 18	1
h > t	12 + 10 + 9 + 4 + 2 = 37	2, 3, 4, 5, 6
t > k	18 = 18	1
k > t	12 + 10 + 9 + 4 + 2 = 37	2, 3, 4, 5, 6
t > c	18 = 18	1
c > t	12 + 10 + 9 + 4 + 2 = 37	2, 3, 4, 5, 6
m > h	18 + 9 = 27	1, 4
h > m	12 + 10 + 4 + 2 = 28	2, 3, 5, 6
m > k	18 + 12 + 9 + 4 = 43	1, 2, 4, 5
k > m	10 + 2 = 12	3, 6
m > c	18 + 9 + 2 = 29	1, 4, 6
c > m	12 + 10 + 4 = 26	2, 3, 5
h > k	18 + 12 + 4 + 2 = 36	1, 2, 5, 6
k > h	10 + 9 = 19	3, 4
h > c	18 + 9 + 4 + 2 = 33	1, 4, 5, 6
c > h	12 + 10 = 22	2, 3
k > c	18 + 10 + 9 + 2 = 39	1, 3, 4, 6
c > k	12 + 4 = 16	2, 5

Make another table to award pairwise points.

Pairs	Votes	Pairwise Points
t : m	18 : 37	m, 1
t : h	18 : 37	h, 1
t : k	18 : 37	k, 1
t : c	18 : 37	c, 1
m : h	27 : 28	h, 1
m : k	43 : 12	m, 1
m : c	29 : 26	m, 1
h : k	36 : 19	h, 1
h : c	33 : 22	h, 1
k : c	39 : 16	k, 1

Using the pairwise comparison method, the winner is h with 4 pairwise points. Activity m received 3 points; activity k received 2 points; activity c received 1 point; activity t received no points.

(c) Use the information from part (a) to make a table to examine the points in the Borda method.

Votes	Ranking	Points				
		4	3	2	1	0
18	t > m > h > k > c	t	m	h	k	c
12	c > h > m > k > t	c	h	m	k	t
10	k > c > h > m > t	k	e	h	m	t
9	m > k > h > c > t	m	k	h	c	t
4	h > c > m > k > t	h	c	m	k	t
2	h > k > m > c > t	h	k	m	c	t

The winner is m with 136 Borda points.

(d) Calculate how many points each activity obtained from the 55 first-place votes from the voter profile

Votes	Ranking	Row
18	t > m > h > k > c	1
12	c > h > m > k > t	2
10	k > c > h > m > t	3
9	m > k > h > c > t	4
4	h > c > m > k > t	5
2	h > k > m > c > t	6

Activity	1st-place votes
t	18
c	12
k	10
m	9
h	6

None of the activities received a majority, 28, of the votes. Therefore, eliminate activity h, the activity that received the least number of votes. Compare the first-place votes of others.

Votes	Ranking	Row
18	t > m > k > c	1
12	c > m > k > t	2
10	k > c > m > t	3
9	m > k > c > t	4
4	c > m > k > t	5
2	k > m > c > t	6

Activity t: 18 votes
Activity c: 12 + 4 = 16 votes
Activity k: 10 + 2 = 12 votes
Activity m: 9 votes
There is no activity that has received a majority; therefore, eliminate m with only 9 votes, and compare the first-place votes of the others.

Votes	Ranking	Row
18	t > k > c	1
12	c > k > t	2
10	k > c > t	3
9	k > c > t	4
4	c > k > t	5
2	k > c > t	6

Activity t: 18 votes
Activity c: 12 + 4 = 16 votes
Activity k: 10 + 9 + 2 = 21 votes
Again there is not a majority; therefore, eliminate c and compare the first-place votes of t and k.

Votes	Ranking	Row
18	t > k	1
12	k > t	2
10	k > t	3
9	k > t	4
4	k > t	5
2	k > t	6

Activity t: 18 votes
Activity k:
12 + 10 + 9 + 4 + 2 = 37 votes
Finally k wins with 37 votes.

21. The voter profile is as follows.

Votes	Ranking	Row
6	h > j > g > e	1
5	e > g > j > h	2
4	g > j > h > e	3
3	j > h > g > e	4
3	e > j > h > g	5

The issues with the most votes are h and e. Eliminate the remaining candidates to examine a second runoff election between these two candidates.

Votes	Ranking	Row
6	h > e	1
5	e > h	2
4	h > e	3
3	h > e	4
3	e > h	5

Issue h: 6 + 4 + 3 = 13
Issue e: 5 + 3 = 8
Issue h beats e.

23. The voter profile is as follows.

Votes	Ranking	Row
18	t > m > h > k > c	1
12	c > h > m > k > t	2
10	k > c > h > m > t	3
9	m > k > h > c > t	4
4	h > c > m > k > t	5
2	h > k > m > c > t	6

The activities with the most votes are t and c. Eliminate the remaining candidates to examine a second runoff election between these two candidates.

Votes	Ranking	Row
18	t > c	1
12	c > t	2
10	c > t	3
9	c > t	4
4	c > t	5
2	c > t	6

Activity t has 18 votes.
Activity c: 12 + 10 + 9 + 4 + 2 = 37
Activity c beats t.

25. The voter profile is as follows.

Votes	Ranking	Row
18	t > m > h > k > c	1
12	c > h > m > k > t	2
10	k > c > h > m > t	3
9	m > k > h > c > t	4
4	h > c > m > k > t	5
2	h > k > m > c > t	6

The activities that rank second and third in first-place votes are c and k, respectively. Here is a runoff election between these two activities:

Votes	Ranking	Row
18	k > c	1
12	c > k	2
10	k > c	3
9	k > c	4
4	c > k	5
2	k > c	6

Activity k: 18 + 10 + 9 + 2 = 39
Activity c: 12 + 4 = 16
Now k faces the first-place activity, t.

Votes	Ranking	Row
18	t > k	1
12	k > t	2
10	k > t	3
9	k > t	4
4	k > t	5
2	k > t	6

Activity t has 18 votes.
Activity k: 12 + 10 + 9 + 4 + 2 = 37
Activity k wins with 37 votes.

27. The voter profile is as follows.

Votes	Ranking	Row
6	a > b > c	1
1	b > c > a	2
3	b > a > c	3
3	c > a > b	4

The logos that rank second and third in first-place votes are b and c, respectively. Here is a run off election between these two logos.

Votes	Ranking	Row
6	b > c	1
1	b > c	2
3	b > c	3
3	c > b	4

Logo b: 6 + 1 + 3 = 10
Logo c: 3
Now b faces the first-place logo, a.

Votes	Ranking	Row
6	a > b	1
1	b > a	2
3	b > a	3
3	a > b	4

Logo a: 6 + 3 = 9
Logo b: 1 + 3 = 4
Logo a wins.

29. **(a)** For $n = 7$ candidates, find the combinations of 7, taken 2 at a time:

$$_7C_2 = \frac{7!}{2!(7-2)!}$$
$$= \frac{7 \cdot 6 \cdot 5 \cdot 4 \cdot 3 \cdot 2 \cdot 1}{2 \cdot 1 \cdot 5 \cdot 4 \cdot 3 \cdot 2 \cdot 1}$$
$$= 21$$

The sum of the number of comparisons listed in the table in the text is:
3 + 5 + 7 + 1 + 2 + 1 = 19.
Only two comparisons remain, so f wins two points.

(b) Examine the table in the text to see that c wins with 7 pairwise points.

31. **(a)** For $n = 8$ candidates, find the combinations of 8, taken 2 at a time.

$$_8C_2 = \frac{8!}{2!(8-2)!}$$
$$= \frac{8 \cdot 7 \cdot 6 \cdot 5 \cdot 4 \cdot 3 \cdot 2 \cdot 1}{2 \cdot 1 \cdot 6 \cdot 5 \cdot 4 \cdot 3 \cdot 2 \cdot 1}$$
$$= 28$$

The sum of the number of comparisons listed in the table in the text is:
2 + 6 + 3 + 4 + 2 + 2 + 2 = 21.
28 − 21 = 7 comparisons remain, so e wins seven points.

(b) Candidate e wins with 7 pairwise points.

33. **(a)** Notice that in the Borda method, the sum of all the points must equal the product of the number of voters and the number of possible points for each voter's selection. For example, in this exercise there are 15 voters choosing among 3 candidates. Each voter has a total of 3 points to assign: 2 points for first place and 1 point for second place. Therefore, the total number of points is 15 · 3 = 45.
If Candidate a receives 15 points and Candidate b receives 14 points, then Candidate c must receive
45 − (15 + 14) = 16 points.

(b) Candidate c wins the Borda election.

35. **(a)** Notice that in the Borda method, the sum of all the points must equal the product of the number of voters and the number of possible points for each voter's selection. For example, in this exercise there are 20 voters choosing among 5 candidates. Each voter has a total of 10 points to assign: 4 points for first place, 3 points for second place, 2 points for third place, and 1 point for fourth place. therefore, the total number of points is 20 · 10 = 200.
Candidate c must receive the difference between 200 and the sum of the points of the other candidates.
200 − (35 + 40 + 40 + 30) = 55 points

(b) Candidate c wins the Borda election with 55 points.

37. The voter profile is as follows.

Votes	Ranking	Row
6	h > j > g > e	1
5	e > g > j > h	2
4	g > j > h > e	3
3	j > h > g > e	4
3	e > j > h > g	5

Using the Coombs method, eliminate the issue with the most last place votes, which is e, with 13 last place votes. Compare the remaining issues.

Votes	Ranking	Row
6	h > j > g	1
5	g > j > h	2
4	g > j > h	3
3	j > h > g	4
3	j > h > g	5

There is no majority, so again compare totals of last-place votes.
Issue g: $6 + 3 + 3 = 12$
Issue h: $5 + 4 = 9$
Eliminate g because it has the most last-place votes. Now h and j remain.

Votes	Ranking	Row
6	h > j	1
5	j > h	2
4	j > h	3
3	j > h	4
3	j > h	5

Issue j has a majority, so it wins.

39. Writing exercise; answers will vary.

41. The least possible number of rounds of voting required in a Hare method election is 1. A candidate could receive a majority of first-round votes, and thus win the election.

43. Answers will vary; here is one possible arrangement.

Votes	Ranking	Row
2	a > b > d > c	1
4	b > c > d > a	2
5	c > d > a > b	3
7	d > a > b > c	4
3	a > d > c > b	5

Plurality method
Candidate a receives $2 + 3 = 5$ votes.
Candidate b receives 4 votes.
Candidate c receives 5 votes.
Candidate d receives 7 votes.

Pairwise Comparison method
First make a table to compute the votes in comparison.

Compare	Votes	Rows
a > b	$2 + 5 + 7 + 3 = 17$	1, 3, 4, 5
b > a	$4 = 4$	2
a > c	$2 + 7 + 3 = 12$	1, 4, 5
c > a	$4 + 5 = 9$	2, 3
a > d	$2 + 3 = 5$	1, 5
d > a	$4 + 5 + 7 = 16$	2, 3, 4
b > c	$2 + 4 + 7 = 13$	1, 2, 4
c > b	$5 + 3 = 8$	3, 5
b > d	$2 + 4 = 6$	1, 2
d > b	$5 + 7 + 3 = 15$	3, 4, 5
c > d	$4 + 5 = 9$	2, 3
d > c	$2 + 7 + 3 = 12$	1, 4, 5

Make another table to award pairwise points.

Pairs	Votes	Pairwise Points
a : b	17 : 4	a, 1
a : c	12 : 9	a, 1
a : d	5 : 16	d, 1
b : c	13 : 8	b, 1
b : d	6 : 15	d, 1
c : d	9 : 12	d, 1

Using the pairwise comparison method, the winner is d with 3 pairwise points. Candidate a received 2 points; candidate b received 1 point; candidate c received no points.

Borda method
Make a table to examine the points in the Borda method.

Votes	Ranking	Points			
		3	2	1	0
2	a > b > d > c	a	b	d	c
4	b > c > d > a	b	c	d	a
5	c > d > a > b	c	d	a	b
7	d > a > b > c	d	a	b	c
3	a > d > c > b	a	d	c	b

Examine the last four columns, and compute the weighted sum for each candidate.

Issue	Borda Points
a	$2 \cdot 3 + 4 \cdot 0 + 5 \cdot 1 + 7 \cdot 2 + 3 \cdot 3 = 34$
b	$2 \cdot 2 + 4 \cdot 3 + 5 \cdot 0 + 7 \cdot 1 + 3 \cdot 0 = 23$
c	$2 \cdot 0 + 4 \cdot 2 + 5 \cdot 3 + 7 \cdot 0 + 3 \cdot 1 = 26$
d	$2 \cdot 1 + 4 \cdot 1 + 5 \cdot 2 + 7 \cdot 3 + 3 \cdot 2 = 43$

The winner is d with 43 Borda points.

Votes	Ranking	Row
2	a > d > c	1
4	c > d > a	2
5	c > d > a	3
7	d > a > c	4
3	a > d > c	5

Candidate a receives 2 + 3 = 5 votes.
Candidate c receives 4 + 5 = 9 votes.
Candidate d receives 7 votes.
Again, no candidate has received a majority, so another vote is taken with a eliminated.

Votes	Ranking	Row
2	d > c	1
4	c > d	2
5	c > d	3
7	d > c	4
3	d > c	5

Candidate c receives 4 + 5 = 9 votes.
Candidate d receives 2 + 7 + 3 = 12 votes.
Finally, d has a majority and wins.

16.2 Exercises

1. (a) Read the table in the text to see that a has the majority of first-place votes. Because there are 11 voters, 6 votes constitutes a majority.

   (b) Examine the following table to calculate the Borda points.

Votes	Ranking	Points		
		2	1	0
6	a > b > c	a	b	c
3	b > c > a	b	c	a
2	c > b > a	c	b	a

   Alternative a: $6 \cdot 2 + 3 \cdot 0 + 2 \cdot 0 = 12$
   Alternative b: $6 \cdot 1 + 3 \cdot 2 + 2 \cdot 1 = 14$
   Alternative c: $6 \cdot 0 + 3 \cdot 1 + 2 \cdot 2 = 7$
   Alternative b wins with 14 Borda points.

   (c) Yes, the Borda method violates the majority criterion because it fails to select the majority candidate.

3. (a) Read the table in the text to see that a has the majority of first-place votes. Because there are 36 voters, at least 19 votes constitutes a majority. Alternative a received 20 votes.

   (b) Use the information from the text to make a table to examine the points in the Borda method.

Votes	Ranking	Points			
		3	2	1	0
20	a > b > c > d	a	b	c	d
6	b > c > d > a	b	c	d	a
5	c > b > d > a	c	b	d	a
5	d > b > a > c	d	b	a	c

   Examine the last four columns, and compute the weighted sum for each alternative.

Alternative	Borda Points
a	$20 \cdot 3 + 6 \cdot 0 + 5 \cdot 0 + 5 \cdot 1 = 65$
b	$20 \cdot 2 + 6 \cdot 3 + 5 \cdot 2 + 5 \cdot 2 = 78$
c	$20 \cdot 1 + 6 \cdot 2 + 5 \cdot 3 + 5 \cdot 0 = 47$
d	$20 \cdot 0 + 6 \cdot 1 + 5 \cdot 1 + 5 \cdot 3 = 26$

The winner is alternative b with 78 Borda points.

(c) Yes, the Borda method violates the majority criterion because it fails to select the majority candidate.

5. (a) Read the table in the text to see that a has the majority of first-place votes. Because there are 30 voters, at least 16 votes constitutes a majority. Alternative a received 16 votes.

(b) Use the information from the text to make a table to examine the points in the Borda method.

		Points				
Votes	Ranking	4	3	2	1	0
16	$a > b > c > d > e$	a	b	c	d	e
3	$b > c > d > e > a$	b	c	d	e	a
5	$c > d > b > e > a$	c	d	b	e	a
3	$d > b > c > a > e$	d	b	c	a	e
3	$e > c > d > a > b$	e	c	d	a	b

Examine the last five columns, and compute the weighted sum for each alternative.

Alternative	Borda Points
a	$16 \cdot 4 + 3 \cdot 0 + 5 \cdot 0 + 3 \cdot 1 + 3 \cdot 1 = 70$
b	$16 \cdot 3 + 3 \cdot 4 + 5 \cdot 2 + 3 \cdot 3 + 3 \cdot 0 = 79$
c	$16 \cdot 2 + 3 \cdot 3 + 5 \cdot 4 + 3 \cdot 2 + 3 \cdot 3 = 76$
d	$16 \cdot 1 + 3 \cdot 2 + 5 \cdot 3 + 3 \cdot 4 + 3 \cdot 2 = 55$
e	$16 \cdot 0 + 3 \cdot 1 + 5 \cdot 1 + 3 \cdot 0 + 3 \cdot 4 = 20$

The winner is alternative b with 79 Borda points.

(c) Yes, the Borda method violates the majority criterion because it fails to select the majority candidate.

7. (a) Make a table to compute the votes in comparisons.

Comparison	Votes	Rows
$a > b$	$4 + 3 = 7$	1, 4
$b > a$	$2 + 4 = 6$	2, 3
$a > c$	$4 + 4 = 8$	1, 3
$c > a$	$2 + 3 = 5$	2, 4
$b > c$	$4 + 2 + 4 = 10$	1, 2, 3
$c > b$	$3 = 3$	4

Make another table to award pairwise points.

Pairs	Votes	Pairwise Points
a : b	7 : 6	a, 1
a : c	8 : 5	a, 1
b : c	10 : 3	b, 1

Using the pairwise comparison method, the winner is candidate a with 2 pairwise points and is the Condorcet candidate.

(b) Examine the table in the text to see that a receives 4 first-place votes; c receives 3 first-place votes; b receives $2 + 4 = 6$ first-place votes to win by the plurality method.

(c) Use the information from the text to make a table to examine the points in the Borda method.

		Points		
Votes	Ranking	2	1	0
4	$a > b > c$	a	b	c
2	$b > c > a$	b	c	a
4	$b > a > c$	b	a	c
3	$c > a > b$	c	a	b

Examine the last three columns, and compute the weighted sum for each alternative.

Alternative	Borda Points
a	$4 \cdot 2 + 2 \cdot 0 + 4 \cdot 1 + 3 \cdot 1 = 15$
b	$4 \cdot 1 + 2 \cdot 2 + 4 \cdot 2 + 3 \cdot 0 = 16$
c	$4 \cdot 0 + 2 \cdot 1 + 4 \cdot 0 + 3 \cdot 2 = 8$

Candidate b wins with 16 Borda points.

(d) There are 13 votes, which makes 7 votes a majority. No candidate wins a majority on this vote. Candidate c has the fewest votes, so compare only a and b.

Comparison	Votes	Rows
a > b	$4 + 3 = 7$	1, 4
b > a	$2 + 4 = 6$	2, 3

Candidate a now has a majority and is selected by the Hare method.

(e) The plurality and the Borda methods violate the Condorcet criterion because neither method selects candidate a, the Condorcet candidate. The Hare method does not violate the criterion because the Condorcet candidate wins.

9. (a) Make a table to compute the votes in comparisons.

Comparison	Votes	Rows
e > h	$3 + 5 + 4 + 3 = 15$	1, 3, 4, 5
h > e	$6 = 6$	2
e > g	$3 + 6 + 3 = 12$	1, 2, 5
g > e	$5 + 4 = 9$	3, 4
e > j	$3 + 6 + 4 = 13$	1, 2, 4
j > e	$5 + 3 = 8$	3, 5
h > g	$3 + 6 + 3 = 12$	1, 2, 5
g > h	$5 + 4 = 9$	3, 4
h > j	$3 + 6 + 4 = 13$	1, 2, 4
j > h	$5 + 3 = 8$	3, 5
j > g	$5 + 3 = 8$	3, 5
g > j	$3 + 6 + 4 = 13$	1, 2, 4

Make another table to award pairwise points.

Pairs	Votes	Pairwise Points
e : h	15 : 6	e, 1
e : g	12 : 9	e, 1
e : j	13 : 8	e, 1
h : g	12 : 9	h, 1
h : j	13 : 8	h, 1
j : g	8 : 13	g, 1

Using the pairwise comparison method, the winner is candidate e with 3 pairwise points and is the Condorcet candidate.

(b) Using the plurality method to determine the highest priority issue, the winner is issue j with a total of 8 first-place votes in rows 3 and 5.

(c) Make a table to examine the points in the Borda method.

Votes	Ranking	Points 3	2	1	0
3	e > h > g > j	e	h	g	j
6	h > e > g > j	h	e	g	j
5	j > g > e > h	j	g	e	h
4	g > e > h > j	g	e	h	j
3	j > e > h > g	j	e	h	g

Examine the last four columns and compute the weighted sum for each alternative.

Alternative	Borda Points
e	$3 \cdot 3 + 6 \cdot 2 + 5 \cdot 1 + 4 \cdot 2 + 3 \cdot 2 = 40$
h	$3 \cdot 2 + 6 \cdot 3 + 5 \cdot 0 + 4 \cdot 1 + 3 \cdot 1 = 31$
g	$3 \cdot 1 + 6 \cdot 1 + 5 \cdot 2 + 4 \cdot 3 + 3 \cdot 0 = 31$
j	$3 \cdot 0 + 6 \cdot 0 + 5 \cdot 3 + 4 \cdot 0 + 3 \cdot 3 = 24$

The winner is issue e with 40 Borda points.

(d) There are 21 votes, which makes 11 votes a majority. No candidate wins a majority on this vote. Eliminate issue e because it has the least number of votes, and compare only h, g, and j.

Votes	Ranking	Row
3	h > g > j	1
6	h > g > j	2
5	j > g > h	3
4	g > h > j	4
3	j > h > g	5

Issue h receives 3 + 6 = 9 votes.
Issue g receives 4 votes.
Issue j receives 5 + 3 = 8 votes.
Again, no issue has received a majority. Therefore, eliminate g with the least number of votes, and compare only h and j.

Votes	Ranking	Row
3	h > j	1
6	h > j	2
5	j > h	3
4	h > j	4
3	j > h	5

Issue h receives 3 + 6 + 4 = 13 votes
Issue j receives 5 + 3 = 8 votes.
Finally, issue h is the winner.

(e) The plurality and Hare methods violate the Condorcet criterion because issue e, the Condorcet candidate, does not win. The Borda method does not violate the criterion.

11. (a) Using the pairwise comparison method, first make a table to compute the votes in comparison.

Compare	Votes	Rows
t > k	18 = 18	1
k > t	12 + 10 + 9 + 4 + 2 = 37	2, 3, 4, 5, 6
t > h	18 = 18	1
h > t	12 + 10 + 9 + 4 + 2 = 37	2, 3, 4, 5, 6
t > m	18 = 18	1

Compare	Votes	Rows
m > t	12 + 10 + 9 + 4 + 2 = 37	2, 3, 4, 5, 6
t > c	18 = 18	1
c > t	12 + 10 + 9 + 4 + 2 = 37	2, 3, 4, 5, 6
k > h	18 + 9 = 27	1, 4
h > k	12 + 10 + 4 + 2 = 28	2, 3, 5, 6
k > m	18 + 12 + 9 + 4 = 43	1, 2, 4, 5
m > k	10 + 2 = 12	3, 6
k > c	18 + 9 + 2 = 29	1, 4, 6
c > k	12 + 10 + 4 = 26	2, 3, 5
h > m	18 + 12 + 4 + 2 = 36	1, 2, 5, 6
b > h	10 + 9 = 19	3, 4
h > c	18 + 9 + 4 + 2 = 33	1, 4, 5, 6
c > h	12 + 10 = 22	22, 3
m > c	18 + 10 + 9 + 2 = 39	1, 3, 4, 6
c > m	12 + 4 = 16	2, 5

Make another table to award pairwise points.

Pairs	Votes	Pairwise Points
t : k	18 : 37	k, 1
t : h	18 : 37	h, 1
t : m	18 : 37	m, 1
t : c	18 : 37	c, 1
k : h	27 : 28	h, 1
k : m	43 : 12	k, 1
k : c	29 : 26	k, 1
k : m	36 : 19	h, 1
h : c	33 : 22	h, 1
m : c	39 : 16	m, 1

Using the pairwise comparison method, the winner is h with 4 pairwise points. This is the Condorcet candidate.

(b) If the plurality method is used, activity t is selected with 18 first-place votes.

(c) Make a table to examine the points in the Borda method.

Votes	Ranking	Points				
		4	3	2	1	0
18	t > k > h > m > c	t	k	h	m	c
12	c > h > k > m > t	c	h	k	m	t
10	m > c > h > k > t	m	c	h	k	t
9	k > m > h > c > t	k	m	h	c	t
4	h > c > k > m > t	h	c	k	m	t
2	h > m > k > c > t	h	m	k	c	t

Examine the last five columns and compute the weighted sum for each alternative.

Activity	Borda Points
t	$18 \cdot 4 + 12 \cdot 0 + 10 \cdot 0$ $+ 9 \cdot 0 + 4 \cdot 0 + 2 \cdot 0 = 72$
k	$18 \cdot 3 + 12 \cdot 2 + 10 \cdot 1 + 9 \cdot 4$ $+ 4 \cdot 2 + 2 \cdot 2 = 136$
h	$18 \cdot 2 + 12 \cdot 3 + 10 \cdot 2 + 9 \cdot 2$ $+ 4 \cdot 4 + 2 \cdot 4 = 134$
m	$18 \cdot 1 + 12 \cdot 1 + 10 \cdot 4 + 9 \cdot 3$ $+ 4 \cdot 1 + 2 \cdot 3 = 107$
c	$18 \cdot 0 + 12 \cdot 4 + 10 \cdot 3 + 9 \cdot 1$ $+ 4 \cdot 3 + 2 \cdot 1 = 101$

The winner is k with 136 Borda points.

(d) Calculate how many points each activity obtained from the 55 first-place votes from the voter profile

Votes	Ranking	Row
18	t > k > h > m > c	1
12	c > h > k > m > t	2
10	m > c > h > k > t	3
9	k > m > h > c > t	4
4	h > c > k > m > t	5
2	h > m > k > c > t	6

Activity	1st-place votes
t	18
c	12
m	10
k	9
h	6

None of the activities received a majority, 28, of the votes. Therefore, eliminate activity h, the activity that received the least number of votes. Compare the first-place votes of the others.

Votes	Ranking	Row
18	t > k > m > c	1
12	c > k > m > t	2
10	m > c > k > t	3
9	k > m > c > t	4
4	c > k > m > t	5
2	m > k > c > t	6

Activity t: 18 votes
Activity c: 12 + 4 = 16 votes
Activity m: 10 + 2 = 12 votes
Activity k: 9 votes
There is no activity that has received a majority; therefore, eliminate k with only 9 votes, and compare the first-place votes of the others.

Votes	Ranking	Row
18	t > m > c	1
12	c > m > t	2
10	m > c > t	3
9	m > c > t	4
4	c > m > t	5
2	m > c > t	6

Activity t: 18 votes
Activity c: 12 + 4 = 16 votes
Activity m: 10 + 9 + 2 = 21 votes
Again there is not a majority; therefore, eliminate c and compare the first-place votes of t and m.

Votes	Ranking	Row
18	t > m	1
12	m > t	2
10	m > t	3
9	m > t	4
4	m > t	5
2	m > t	6

Activity t: 18 votes
Activity m: 12 + 10 + 9 + 4 + 2
                = 37 votes
Finally, m wins with 37 votes in the
Hare method.

(e)   All three methods violate the Condorcet
      criterion because none of them select
      the Condorcet candidate.

13. (a)   Using the pairwise comparison method,
          first make a table to compute the votes
          in comparison.

Compare	Votes	Rows
m > c	5 + 3 = 8	1, 3
c > m	4 + 2 = 6	2, 4
m > s	5 + 2 = 7	1, 4
s > m	4 + 3 = 7	2, 3
m > b	5 + 2 = 7	1, 4
b > m	4 + 3 = 7	2, 3
c > s	5 + 2 = 7	1, 4
s > c	4 + 3 = 7	2, 3
c > b	5 + 2 = 7	1, 4
b > c	4 + 3 = 7	2, 3
s > b	5 + 2 = 7	1, 4
b > s	4 + 3 = 7	2, 3

Make another table to award pairwise
points.

Pairs	Votes	Pairwise Points
m : c	8 : 6	m, 1
m : s	7 : 7	m, $\frac{1}{2}$; s, $\frac{1}{2}$
m : b	7 : 7	m, $\frac{1}{2}$; b, $\frac{1}{2}$
c : s	7 : 7	c, $\frac{1}{2}$; s, $\frac{1}{2}$
c : b	7 : 7	c, $\frac{1}{2}$; b, $\frac{1}{2}$
s : b	7 : 7	s, $\frac{1}{2}$; b, $\frac{1}{2}$

Using the pairwise comparison method,
here are the total points each city has
received.

m: $1 + \frac{1}{2} + \frac{1}{2} = 2$

s: $\frac{1}{2} + \frac{1}{2} + \frac{1}{2} = 1\frac{1}{2}$

b: $\frac{1}{2} + \frac{1}{2} + \frac{1}{2} = 1\frac{1}{2}$

c: $\frac{1}{2} + \frac{1}{2} = 1$

Montreal, m, is selected.

(b)   Make a table to compute the votes in
      comparison.

Compare	Votes	Rows
m > c	5 + 3 + 2 = 10	1, 3, 4
c > m	4 = 4	1, 4
m > s	5 + 2 = 7	1, 4
s > m	4 + 3 = 7	2, 3
m > b	5 + 2 = 7	1, 4
b > m	4 + 3 = 7	2, 3
c > s	5 + 2 = 7	1, 4
s > c	4 + 3 = 7	2, 3
c > b	5 = 5	1
b > c	4 + 3 + 2 = 9	2, 3, 4
s > b	5 = 5	1
b > s	4 + 3 + 2 = 9	2, 3, 4

Make another table to award pairwise
points.

Pairs	Votes	Pairwise Points
m : c	10 : 4	m, 1
m : s	7 : 7	m, $\frac{1}{2}$; s, $\frac{1}{2}$
m : b	7 : 7	m, $\frac{1}{2}$; b, $\frac{1}{2}$
c : s	7 : 7	c, $\frac{1}{2}$; s, $\frac{1}{2}$
c : b	5 : 9	b, 1
s : b	5 : 9	b, 1

Using the pairwise comparison method, here is the total points each city has received for this vote.

m: $1+\frac{1}{2}+\frac{1}{2}=2$

s: $\frac{1}{2}+\frac{1}{2}=1$

b: $\frac{1}{2}+1+1=2\frac{1}{2}$

c: $\frac{1}{2}$

Boston, b, is selected.

(c) The Monotonicity criterion is violated because city m does not win the second election.

15. (a) Calculate how many votes each city obtained from the 17 first-place votes from the voter profile.

City	1st-place votes
a	6
c	4 + 2 = 6
b	5
d	0

None of the cities received a majority, 9, of the votes. Therefore, eliminate city d, the city that received the least number of votes. Compare the first-place votes of the others.

Votes	Ranking	Row
6	a > c > b	1
5	b > a > c	2
4	c > b > a	3
2	c > a > b	4

a: 6 = 6
b: 5 = 5
c: 4 + 2 = 6
Again, no city has received a majority; therefore, eliminate b with only 5 points and compare the first-place votes of a and c.

Votes	Ranking	Row
6	a > c	1
5	a > c	2
4	c > a	3
2	c > a	4

a: 6 + 5 = 11
c: 4 + 2 = 6
Finally, a wins with 11 votes in the Hare method.

(b) Calculate how many votes each city obtained from the 17 first-place votes from the voter profile.

City	1st-place votes
a	6 + 2 = 8
b	5
c	4
d	0

None of the cities received a majority, 9, of the votes. Therefore, eliminate city d, the city that received the least number of votes. Compare the first-place votes of the others.

Votes	Ranking	Row
6	a > c > b	1
5	b > a > c	2
4	c > b > a	3
2	a > c > b	4

a: 6 + 2 = 8
b: 5 = 5
c: 4 = 4
Again, no city has received a majority; therefore, eliminate c with only 4 votes and compare the first-place votes of a and b.

Votes	Ranking	Row
6	a > b	1
5	b > a	2
4	b > a	3
2	a > b	4

a: 6 + 2 = 8
b: 5 + 4 = 9
Finally, b wins with 9 votes in the Hare method.

(c) Yes, the rearranging voters moved a, the winner of the non-binding election, to the top of their ranking, but b wins the official selection process.

17. (a) Calculate how many votes each candidate obtained from the 175 first-place votes from the voter profile.

City	1st-place votes
a	75
b	30
c	50
d	20

Candidate a receives the most votes, 75, and wins by the plurality method.

(b) Compare the rankings if candidate b drops out.

Votes	Ranking	Row
75	a > c > d	1
50	c > a > d	2
30	c > d > a	3
20	d > c > a	4

a: 75 = 75
c: 50 + 30 = 80
d: 20 = 20
Now, c wins with 80 votes.

(c) Yes, the Independence of Irrelevant Alternatives criterion is violated because candidate a does not win the second election.

19. In Example 4 percussionist x wins the most comparisons in the pairwise selection. If w had dropped out instead of y, here are the comparisons.

Pairs	Pairwise Points
v : x	v, 1
y : v	y, 1
v : z	v, 1
x : y	x, 1
x : z	x, 1
z : y	z, 1

Using the pairwise comparison method, here are the total points each percussionist has received.
v: 1 + 1 = 2
x: 1 + 1 = 2
y: 1 = 1
z: 1 = 1
No, the Independence of Irrelevant Alternatives criterion is not violated; this second pairwise vote has resulted in a tie between v and x.

21. (a) Calculate how many votes each candidate obtained from the 34 first-place votes from the voter profile.

City	1st-place votes
a	12
b	10
c	8 + 4 = 12

None of the candidates received a majority, 18, of the votes. Therefore, eliminate candidate b, the candidate that received the least number of votes. Compare the first-place votes of a and c.

Votes	Ranking	Row
12	a > c	1
10	a > c	2
8	c > a	3
4	c > a	4

a: 12 + 10 = 22
c: 8 + 4 = 12
Now a wins with 22 votes.

(b) If candidate c drops out.

Votes	Ranking	Row
12	a > b	1
10	b > a	2
8	b > a	3
4	a > b	4

a: 12 + 4 = 16
b: 10 + 8 = 18
Now b wins with 18 votes.

(c) Yes, according to the Independence of Irrelevant Alternatives criterion, candidate a should win the second election; therefore, the Hare method violates the criterion.

23. Writing exercise; answers will vary.

25. Answers will vary. One possibility is as follows.

Votes	Ranking
10	a > b > c > d > e > f
9	b > f > e > c > d > a

Candidate a wins the majority of the 19 votes. If a pairwise comparison is made, candidate a beats each of the others in turn to earn 5 points, from the 10 votes in the top row. From the information in the second row, notice that b beats f, e, c, and d and will receive 4 points in the pairwise comparison.

27. Answers will vary. One possible profile is from Exercise 2 and Exercise 4 of Section 16.1.

Votes	Ranking
3	a > b > c > d
1	b > c > a > d
2	d > b > c > a
1	c > b > a > d
2	a > d > c > b
1	b > c > d > a
1	c > b > d > a
2	a > b > d > c

Candidate a is the Condorcet candidate because a is preferred over each of the other candidates.

Borda method
Here is the table from Exercise 4 of Section 16.1 to examine the points in the Borda method.

Votes	Ranking	Points 3	2	1	0
3	a > b > c > d	a	b	c	d
1	b > c > a > d	b	c	a	d
2	d > b > c > a	d	b	c	a
1	c > b > a > d	c	b	a	d
2	a > d > c > b	a	d	c	b
1	b > c > d > a	b	c	d	a
1	c > b > d > a	c	b	d	a
2	a > b > d > c	a	b	d	c

Examine the last four columns, and compute the weighted sum for each.

Candidate	Borda Points
a	$3 \cdot 3 + 1 \cdot 1 + 2 \cdot 0 + 1 \cdot 1 + 2 \cdot 3 + 1 \cdot 0 + 1 \cdot 0 + 2 \cdot 3$ $= 23$
b	$3 \cdot 2 + 1 \cdot 3 + 2 \cdot 2 + 1 \cdot 2 + 2 \cdot 0 + 1 \cdot 3 + 1 \cdot 2 + 2 \cdot 2$ $= 24$
c	$3 \cdot 1 + 1 \cdot 2 + 2 \cdot 1 + 1 \cdot 3 + 2 \cdot 1 + 1 \cdot 2 + 1 \cdot 3 + 2 \cdot 0$ $= 17$
d	$3 \cdot 0 + 1 \cdot 0 + 2 \cdot 3 + 1 \cdot 0 + 2 \cdot 2 + 1 \cdot 1 + 1 \cdot 1 + 2 \cdot 1$ $= 14$

The winner is b with 24 Borda points. The Condorcet candidate, a, has not won.

Hare method
Here is the table of 1st-place votes from Exercise 4 in 16.1

Breed	1st-place votes
a	7
b	2
c	2
d	2

Use the Hare method, breed a has a majority of the votes and wins.

Plurality method
From the table above, it can be seen that a has the most first-place votes to win.

29. (a) Using the pairwise comparison method, first make a table to compute the votes in comparison.

Compare	Votes	Rows
$a > x$	$6 + 4 = 10$	1, 3
$x > a$	$5 + 3 = 8$	2, 4
$a > y$	$6 + 3 = 9$	1, 4
$y > a$	$5 + 4 = 9$	2, 3
$a > z$	$6 + 3 = 9$	1, 4
$z > a$	$5 + 4 = 9$	2, 3
$x > y$	$6 + 3 = 9$	1, 4
$y > x$	$5 + 4 = 9$	2, 3
$x > z$	$6 + 3 = 9$	1, 4
$z > x$	$5 + 4 = 9$	2, 3
$y > z$	$6 + 3 = 9$	1, 4
$z > y$	$5 + 4 = 9$	2, 3

Make another table to award pairwise points.

Pairs	Votes	Pairwise Points
$a : x$	$10 : 8$	$a, 1$
$a : y$	$9 : 9$	$a, \frac{1}{2}$; $y, \frac{1}{2}$
$a : z$	$9 : 9$	$a, \frac{1}{2}$; $z, \frac{1}{2}$
$x : y$	$9 : 9$	$x, \frac{1}{2}$; $y, \frac{1}{2}$
$x : z$	$9 : 9$	$x, \frac{1}{2}$; $z, \frac{1}{2}$
$y : z$	$9 : 9$	$y, \frac{1}{2}$; $z, \frac{1}{2}$

Using the pairwise comparison method, here are the total points each has received.

$$a: 1 + \frac{1}{2} + \frac{1}{2} = 2$$

$$x: \frac{1}{2} + \frac{1}{2} = 1$$

$$y: \frac{1}{2} + \frac{1}{2} + \frac{1}{2} = 1\frac{1}{2}$$

$$z: \frac{1}{2} + \frac{1}{2} + \frac{1}{2} = 1\frac{1}{2}$$

Candidate a wins with 2 pairwise points.

(b) Answers will vary. A possible new ranking is

Voters	Ranking
6	$a > x > y > z$
5	$z > y > x > a$
4	$z > y > a > x$
3	$a > z > x > y$

Now make a table to compute the votes in comparison.

Compare	Votes	Rows
$a > x$	$6 + 4 + 3 = 13$	1, 3, 4
$x > a$	$5 = 5$	2
$a > y$	$6 + 3 = 9$	1, 4
$y > a$	$5 + 4 = 9$	2, 3
$a > z$	$6 + 3 = 9$	1, 4
$z > a$	$5 + 4 = 9$	2, 3
$x > y$	$6 + 3 = 9$	1, 4
$y > x$	$5 + 4 = 9$	2, 3
$x > z$	$6 = 6$	1
$z > x$	$5 + 4 + 3 = 12$	2, 3, 4
$y > z$	$6 = 6$	1
$z > y$	$5 + 4 + 3 = 12$	2, 3, 4

Make another table to award pairwise points.

Pairs	Votes	Pairwise Points
$a : x$	$13 : 5$	$a, 1$
$a : y$	$9 : 9$	$a, \frac{1}{2}$; $y, \frac{1}{2}$
$a : z$	$9 : 9$	$a, \frac{1}{2}$; $z, \frac{1}{2}$
$x : y$	$9 : 9$	$x, \frac{1}{2}$; $y, \frac{1}{2}$
$x : z$	$6 : 12$	$z, 1$
$y : z$	$6 : 12$	$z, 1$

Using the pairwise comparison method, here are the total points each has received.

a: $1 + \dfrac{1}{2} + \dfrac{1}{2} = 2$

x: $\dfrac{1}{2} = \dfrac{1}{2}$

y: $\dfrac{1}{2} + \dfrac{1}{2} = 1$

z: $\dfrac{1}{2} + 1 + 1 = 2\dfrac{1}{2}$

Candidate z wins with $2\dfrac{1}{2}$ pairwise points.

31. Answers will vary. Here is one possibility. If candidate c is deleted, the voter profile is:

Votes	Ranking
15	a > b > d
8	b > a > d
9	b > a > d
6	d > b > a

Now candidate b wins with 8 + 9 = 17 votes. Because candidate a wins in the original profile, the plurality method violates the Independence of Irrelevant Alternatives criterion in the second election.

33. Answers will vary. Here is one possible profile:

Votes	Ranking	Points 2	1	0
21	g > j > e	g	j	e
12	j > e > g	j	e	g
8	j > g > e	j	g	e

Examine the last three columns, and compute the weighted sum for each candidate.

Candidate	Borda Points
g	$21 \cdot 2 + 12 \cdot 0 + 8 \cdot 1 = 50$
j	$21 \cdot 1 + 12 \cdot 2 + 8 \cdot 2 = 61$
e	$21 \cdot 0 + 12 \cdot 1 + 8 \cdot 0 = 12$

The winner is candidate j with 61 Borda points. For the second election, candidate e drops out.

Votes	Ranking	Points 1	0
21	g > j	g	j
12	j > g	j	g
8	j > g	j	g

Examine the last two columns, and compute the weighted sum for each candidate.

Candidate	Borda Points
g	$21 \cdot 1 + 12 \cdot 0 + 8 \cdot 0 = 21$
j	$21 \cdot 0 + 12 \cdot 1 + 8 \cdot 1 = 20$

Candidate g wins with 21 Borda points, which violates the Independence of Irrelevant Alternatives criterion.

35. Writing exercise; answers will vary.

16.3 Exercises

1. (a) The population in 1790 was 3,615,920 and the number of seats was 105. The average number of people represented per seat was $\dfrac{3,615,920}{105} \approx 34,437$.

(b) There were 15 states, so the average number of House seats per state was $\dfrac{105}{15} = 7$.

3. Virginia received 19 seats rather than 18. Delaware received only 1 seat rather than 2.

5. (a)

State Park	Acres
a	1429
b	8639
c	7608
d	6660
e	5157
Total	29,493

The standard divisor, d, is found by dividing the total number of acres by the number of trees.

$$d = \dfrac{29493}{239} \approx 123.40$$

(b) Using the Hamilton method to apportion the trees, set up a table as seen below. Remember that the standard quota, Q, is found by dividing the number of acres of land by the standard divisor, d, from part (a).

Park	Acres	mQ	Rounded Q
a	1429	$\dfrac{1429}{123.4017}$ ≈ 11.580	11
b	8639	$\dfrac{8639}{123.4017}$ ≈ 70.007	70
c	7608	$\dfrac{7608}{123.4017}$ ≈ 61.652	61
d	6660	$\dfrac{6660}{123.4017}$ ≈ 53.970	53
e	5157	$\dfrac{5157}{123.4017}$ ≈ 41.790	41
Totals	29,493		236

There are 3 trees remaining to be apportioned to those parks that have the largest fractional parts of the standard quota: c, d, and e. Here are the final numbers.

Park	Rounded Q	Trees Apportioned
a	11	11
b	70	70
c	61	61 + 1 = 62
d	53	53 + 1 = 54
e	41	41 + 1 = 42
Totals	236	239

(c) Using the Jefferson method to apportion the trees, first calculate the standard divisor, md, by dividing the total number of acres by the number of trees as in part (a).

$$d = \frac{29493}{239} \approx 123.40$$

Set up a table as seen below by using the table in part (a). Remember that the standard quota, Q, is found by dividing the number of acres of land by the standard divisor, d, from part (a). If d is decreased to 123, the modified quotas add up to approximately 239.8, which is too high. An md of 122 works.

Park	Acres	mQ	Trees Apportioned
a	1429	$\dfrac{1429}{122}$ ≈ 11.713	11
b	8639	$\dfrac{8639}{122}$ ≈ 70.811	70
c	7608	$\dfrac{7608}{122}$ ≈ 62.361	62
d	6660	$\dfrac{6660}{122}$ ≈ 54.590	54
e	5157	$\dfrac{5157}{122}$ ≈ 42.270	42
Totals	29,493		239

(d)

Standard Quota	Rounded Traditionally
11.580	12
70.007	70
61.652	62
53.970	54
41.790	42
	240

This sum is greater than the number of trees to be apportioned.

(e) The value of md for the Webster method should be greater than the standard divisor, $d = 123.40$, because larger divisors create smaller quotas with a smaller total sum.

(f) Using the Webster method, again build a table using a modified divisor of 124.

Park	Acres	mQ	Trees Apportioned
a	1429	$\frac{1429}{124} \approx 11.524$	12
b	8639	$\frac{8639}{124} \approx 69.669$	70
c	7608	$\frac{7608}{124} \approx 61.355$	61
d	6660	$\frac{6660}{124} \approx 53.710$	54
e	5157	$\frac{5157}{124} \approx 41.589$	42
Totals	29,493		239

(g) The apportionment for the Hamilton and Jefferson methods are the same; the apportionment for the Webster method is different.

7. (a) The total enrollment is $56 + 35 + 78 + 100 = 269$.

The standard divisor is $d = \frac{269}{11} \approx 24.45$.

(b) Using the Hamilton method, we have the following.

Course	Enrollment	Q	Rounded Q
Fiction	56	$\frac{56}{24.45455} \approx 2.290$	2
Poetry	35	$\frac{35}{24.45455} \approx 1.431$	1
Short Story	78	$\frac{78}{24.45455} \approx 3.190$	3
Multicultural	100	$\frac{100}{24.45455} \approx 4.089$	4
Totals	269		10

There is 1 section remaining to be apportioned to the course that has the largest fractional parts of the standard quota, Poetry. Here are the final numbers.

Course	Rounded Q	Sections Apportioned
Fiction	2	2
Poetry	1	$1 + 1 = 2$
Short Story	3	3
Multicultural	4	4
Totals	10	11

(c) Using the Jefferson method to apportion, the standard divisor must be modified. Set up a table as seen below. An *md* of 20 works.

Course	Enrollment	mQ	Sections Apportioned
Fiction	56	$\frac{56}{20} = 2.8$	2
Poetry	35	$\frac{35}{20} = 1.75$	1
Short Story	78	$\frac{78}{20} = 3.9$	3
Multicultural	100	$\frac{100}{20} = 5$	5
Totals	269		11

(d)

Standard Quota	Rounded Traditionally
2.290	2
1.431	1
3.190	3
4.089	4
	10

This sum is less the number of sections to be apportioned.

(e) The value of *md* for the Webster method should be less than the standard divisor, $d = 24.45$, because lesser divisors create greater quotas with a greater total sum.

(f) Use the Webster method with an *md* of 23.

Course	Enrollment	Q	mQ
Fiction	56	$\frac{56}{23} \approx 2.435$	2
Poetry	35	$\frac{35}{23} \approx 1.522$	2
Short Story	78	$\frac{78}{23} \approx 3.391$	3
Multicultural	100	$\frac{100}{23} \approx 4.348$	4
Totals	269		11

(g) The Hamilton and Webster apportionments are the same; the Jefferson apportionment differs.

(h) A Poetry student would hope that either the Hamilton or Webster method would be used, because both apportion 2 sections of the class rather than 1. This would create a smaller class size.

(i) A Multicultural student would hope that the Jefferson method be used, because this apportions 5 sections rather than 4.

9. (a) Use the Hamilton Method to apportion the 131 seats. The standard divisor is found by dividing the entire population by the number of seats.

$$d = \frac{47841}{131} \approx 365.1985$$

State	Population	Q	Seats Apportioned
Abo	5672	$\frac{5672}{365.1985} \approx 15.531$	15
Boa	8008	$\frac{8008}{365.1985} \approx 21.928$	21
Cio	2400	$\frac{2400}{365.1985} \approx 6.572$	6
Dao	6789	$\frac{6789}{365.1985} \approx 18.590$	18
Effo	4972	$\frac{4972}{365.1985} \approx 13.615$	13
Foti	20,000	$\frac{20,000}{365.1985} \approx 54.765$	54
Totals	47,841		127

There are $131 - 127 = 4$ seats remaining to be apportioned to those states that have the largest fractional parts of the standard quota: Boa, Foti, Effo, and Dao. Here are the final numbers.

State	Rounded IQ	Seats Apportioned
Abo	15	15
Boa	21	21 + 1 = 22
Cio	6	6
Dao	18	18 + 1 = 19
Effo	13	13 + 1 = 14
Foti	54	54 + 1 = 55
Totals	127	131

(b) Use the Jefferson method with $md = 356$ (found by trial and error).

State	Population	Modified Quota mQ	Rounded-down mQ Seats Apportioned
Abo	5672	$\frac{5672}{356} \approx 15.933$	15
Boa	8008	$\frac{8008}{356} \approx 22.494$	22
Cio	2400	$\frac{2400}{356} \approx 6.742$	6
Dao	6789	$\frac{6789}{356} \approx 19.070$	19
Effo	4972	$\frac{4972}{356} \approx 13.966$	13
Foti	20,000	$\frac{20,000}{356} \approx 56.180$	56
Totals	47,841		131

(c) The Hamilton, Jefferson, and Webster apportionments are all different.

11. Use the Webster method with an *mc* of 366.99.

State	Population	*mQ*	Rounded *Q*
Abo	5672	$\frac{5672}{366.99}$ ≈ 15.455	15
Boa	8008	$\frac{8008}{366.99}$ ≈ 21.821	22
Cio	2400	$\frac{2400}{366.99}$ ≈ 6.540	7
Dao	6789	$\frac{6789}{366.99}$ ≈ 18.499	18
Effo	4972	$\frac{4972}{366.99}$ ≈ 13.548	14
Foti	20,000	$\frac{20,000}{366.99}$ ≈ 54.497	54
Totals	47,841		130

Notice that the total of the last column is 130 seats.

13. (a) The total number of beds is
137 + 237 + 337 + 455 + 555 = 1721.
The standard divisor for the apportionment of nurses is

$$d = \frac{1721}{40} = 43.025.$$

(b) Use the Hamilton method to apportion the nurses.

Hosp.	No. of Beds	*Q*	Rounded *Q*
A	137	$\frac{137}{43.025}$ ≈ 3.184	3
B	237	$\frac{237}{43.025}$ ≈ 5.508	5
C	337	$\frac{337}{43.025}$ ≈ 7.833	7
D	455	$\frac{455}{43.025}$ ≈ 10.575	10
E	555	$\frac{555}{43.025}$ ≈ 12.899	12
Totals	1721		37

There are 3 more nurses to apportion to those hospitals with the greatest fractional part remaining in the *Q* column. The final apportionment is as follows.

Hospital	*Q*	*mQ*
A	$\frac{137}{43.025} \approx 3.184$	3
B	$\frac{237}{43.025} \approx 5.508$	5
C	$\frac{337}{43.025} \approx 7.833$	7 + 1 = 8
D	$\frac{455}{43.025} \approx 10.575$	10 + 1 = 11
E	$\frac{555}{43.025} \approx 12.899$	12 + 1 = 13
Totals		40

(c) Use the Jefferson method with an *md* of 40.

Hosp.	No. of Beds	*mQ*	Rounded *mQ*
A	137	$\frac{137}{40}$ ≈ 3.425	3
B	237	$\frac{237}{40}$ ≈ 5.925	5
C	337	$\frac{337}{40}$ ≈ 8.425	8
D	455	$\frac{455}{40}$ $= 11.375$	11
E	555	$\frac{555}{40}$ ≈ 13.875	13
Totals	1721		40

(d)

Standard Quota	Rounded Traditionally
3.184	3
5.508	6
7.833	8
10.575	11
12.899	13
Total	41

The traditionally rounded values add to 41, which is greater than the number of nurses to be apportioned.

(e) The value of *md* for the Webster method should be greater than the standard divisor, $d = 43.025$, because larger divisors create lesser modified quotas with a lesser total sum.

(f) Use the Webster method with an *md* of 43.1.

Hosp.	No. of Beds	mQ	Rounded mQ
A	137	$\frac{137}{43.1}$ ≈ 3.179	3
B	237	$\frac{237}{43.1}$ ≈ 5.499	5
C	337	$\frac{337}{43.1}$ ≈ 7.819	8
D	455	$\frac{455}{43.1}$ $= 10.557$	11
E	555	$\frac{555}{43.1}$ ≈ 12.877	13
Totals	1721		40

(g) All three apportionments are the same.

15. Here is one possible ridership profile.

Route	a	b	c	d	e	Total
No. of riders	131	140	303	178	197	949

Hamilton method

The standard divisor is $d = \dfrac{949}{16} = 59.3125$.

Bus Route	Riders	Q	Rounded Q
a	131	$\frac{131}{59.3125}$ ≈ 2.209	2
b	140	$\frac{140}{59.3125}$ ≈ 2.360	2
c	303	$\frac{303}{59.3125}$ ≈ 5.109	5
d	178	$\frac{178}{59.3125}$ ≈ 3.001	3
e	197	$\frac{197}{59.3125}$ ≈ 3.321	3
Totals	198		15

The remaining bus will be assigned to route b because it has the largest fractional portion in its quotient. The final apportionment is as follows.

Bus Route	Riders	Q	Rounded Q
a	131	2.209	2
b	140	2.360	2 + 1 = 3
c	303	5.109	5
d	178	3.001	3
e	197	3.321	3
Totals	949	1	16

Jefferson method
Using a modified divisor of 50, here is the apportionment.

Bus Route	Riders	mQ	Bus Apportioned
a	131	$\dfrac{131}{50} = 2.62$	2
b	140	$\dfrac{140}{50} = 2.8$	2
c	303	$\dfrac{303}{50} = 6.06$	6
d	178	$\dfrac{178}{50} = 3.56$	3
e	197	$\dfrac{197}{50} = 3.94$	3
Totals	949		16

Webster method

Using a modified divisor of 56.2, round each value of Q according to normal rules of rounding.

Bus Route	Riders	mQ	Bus Apportioned
a	131	$\dfrac{131}{56.2} \approx 2.331$	2
b	140	$\dfrac{140}{56.2} \approx 2.491$	2
c	303	$\dfrac{303}{56.2} \approx 5.391$	5
d	178	$\dfrac{178}{56.2} \approx 3.167$	3
e	197	$\dfrac{197}{56.2} \approx 3.505$	4
Totals	949		16

17. Answers will vary. Here is one possible population profile.

State	a	b	c	d	e	Total
Pop.	50	230	280	320	120	1000

Hamilton method

The standard divisor is $d = \dfrac{1000}{100} = 10$.

State	Pop.	Q
a	50	$\dfrac{50}{10} = 5$
b	230	$\dfrac{230}{10} = 23$
c	280	$\dfrac{280}{10} = 28$
d	320	$\dfrac{320}{10} = 32$
e	120	$\dfrac{120}{10} = 12$
Totals		100

It is unnecessary to find a modified divisor for the Jefferson and Webster methods, because the value of d divides into each population evenly. No rounding is necessary. That is, the modified divisor is 10 for both methods.

EXTENSION: TWO ADDITIONAL APPORTIONMENT METHODS

1. Writing exercise: answers will vary.

3. (a) Using the Adams method with an *md* of 29

Course	Enr.	Q	mQ
Fiction	56	$\dfrac{56}{29} \approx 1.931$	2
Poetry	35	$\dfrac{35}{29} \approx 1.207$	2
Short Story	78	$\dfrac{78}{29} \approx 2.690$	3
Multicultural	100	$\dfrac{100}{29} \approx 3.448$	4
Totals	269		11

(b) The Adams apportionment is the same as the Hamilton and Webster apportionments. It is different from the Jefferson apportionment.

5. (a) Use the Adams method with an *md* of 377.3.

State	Pop.	*mQ*	Seats Apportioned
Abo	5672	$\frac{5672}{377.3}$ ≈ 15.033	16
Boa	8008	$\frac{8008}{377.3}$ ≈ 21.224	22
Cio	2400	$\frac{2400}{377.3}$ ≈ 6.361	7
Dao	6789	$\frac{6789}{377.3}$ ≈ 17.994	18
Effo	4972	$\frac{4972}{377.3}$ ≈ 13.178	14
Foti	20,000	$\frac{20,000}{377.3}$ ≈ 53.008	54
Totals	47,841		131

(b) Each method produces a different apportionment.

7. The cutoff point for rounding the modified quota of 56.498 up to 57 is calculated by finding the geometric mean of 56 (the integer part of 56.498) and 57.

$$\sqrt{56 \cdot 57} = \sqrt{3192} \approx 56.498$$

9. The cutoff point for rounding the modified quota of 32.497 up to 33 is calculated by finding the geometric mean of 32 (the integer part of 32.497) and 33.

$$\sqrt{32 \cdot 33} = \sqrt{1056} \approx 32.496$$

11. If the sum of the traditionally rounded Q values is greater than the number of objects being apportioned, then the modified divisor is found by slowly increasing the value of d. A greater divisor produces lesser modified quotas with a lesser sum.

13. (a) Use the Huntington-Hill method with an *md* of 24.

Course	Enr.	*mQ*	Seats Apportioned
Fiction	56	$\frac{56}{24}$ $= 2.\overline{3}$	2
Poetry	35	$\frac{35}{24}$ ≈ 1.458	2
Short Story	78	$\frac{78}{24}$ $= 3.25$	3
Multicultural	100	$\frac{100}{24}$ $= 4.1\overline{6}$	4
Totals	269		11

(b) The Huntington-Hill apportionment is the same as the Hamilton, Webster, and Adams apportionments. It is different from the Jefferson apportionment.

15. (a) Using the Huntington-Hill method with an *md* of 367, each value of *mQ* is found by dividing the population figure by 367. See the explanation in the text for computation of the geometric mean.

State	Pop.	*mQ*	Geo Mean	Seats Apportioned
Abo	5672	15.4550	15.4919	15
Boa	8008	21.8202	21.4942	22
Cio	2400	6.5395	6.4807	7
Dao	6789	18.4986	18.4932	19
Effo	4972	13.5477	13.4907	14
Foti	20,000	54.4959	54.4977	54
Totals	47,841			131

(b) The Huntington-Hill apportionment is the same as the Webster apportionment. The other three are all different.

16.4 Exercises

1. The sum of the populations is
$17179 + 7500 + 49400 + 5824 = 79,903$.

The standard divisor is $d = \dfrac{79903}{132} = 605.326$.

Use the standard quota, we have the following.

State	Pop.	Q	round down/up
a	17179	$\dfrac{17179}{605.326}$ ≈ 28.380	28/29
b	7500	$\dfrac{7500}{605.326}$ ≈ 12.390	12/13
c	49400	$\dfrac{49400}{605.326}$ ≈ 81.609	**81/82**
d	5824	$\dfrac{5824}{605.326}$ ≈ 9.621	9/10
Totals	79,903		130/134

Use the Jefferson method with *md* = 595.

State	Pop.	Q	number of seats
a	17179	$\frac{17179}{595}$ ≈ 28.872	28
b	7500	$\frac{7500}{595}$ ≈ 12.605	12
c	49400	$\frac{49400}{595}$ ≈ 83.025	**83**
d	5824	$\frac{5824}{595}$ ≈ 9.788	9
Totals	79,903		132

The Jefferson method violates the Quota Rule because state c receives more seats than its apportionment from the standard quota.

3. The sum of the populations is
    $2567 + 1500 + 8045 + 950 + 1099 = 14161$.
    The standard divisor is

    $$d = \frac{14161}{290} \approx 48.8310.$$

    Use the standard quota.

State	Pop.	Q	round down/up
a	2567	$\frac{2567}{48.8310}$ ≈ 52.569	52/53
b	1500	$\frac{1500}{48.8310}$ ≈ 30.718	30/31
c	8045	$\frac{8045}{48.8310}$ ≈ 164.752	**164/165**
d	950	$\frac{950}{48.8310}$ ≈ 19.455	19/20
e	1099	$\frac{1099}{48.8310}$ ≈ 22.506	22/23
Totals	14,161		287/292

Use the Jefferson method with $md = 48.4$.

State	Pop.	mQ	Number of seats
a	2567	$\frac{2567}{48.4}$ ≈ 53.037	53
b	1500	$\frac{1500}{48.4}$ ≈ 30.992	30
c	8045	$\frac{8045}{48.4}$ ≈ 166.219	**166**
d	950	$\frac{950}{48.4}$ ≈ 19.628	19
e	1099	$\frac{1099}{48.4}$ ≈ 22.707	22
Totals	14,161		290

The Jefferson method violates the Quota Rule because state c receives more seats than its apportionment from the standard quota.

5. First calculate the apportionment with $n = 204$. The total population is
    $3462 + 7470 + 4265 + 5300 = 20,497$.
    The standard divisor for 204 seats is

    $$d = \frac{20497}{204} \approx 100.4755.$$

    Use the Hamilton method.

State	Pop.	Q	rounded Q
a	3462	$\frac{3462}{100.4755}$ ≈ 34.456	34
b	7470	$\frac{7470}{100.4755}$ ≈ 74.346	74
c	4265	$\frac{4265}{100.4755}$ ≈ 42.448	42
d	5300	$\frac{5300}{100.4755}$ ≈ 52.749	52
Totals	20497		202

The two remaining seats will be apportioned to State d and State a, because the fractional parts of their quotas are the largest. The final apportionment is as follows.

State	Pop.	Q	Number of seats
a	3462	34.456	34 + 1 = **35**
b	7470	74.346	74
c	4265	42.448	42
d	5300	52.749	52 + 1 = 53
Totals	20497		204

Now increase the number of seats to $n = 205$. The standard divisor is

$$d = \frac{20497}{205} \approx 99.9854.$$

Use the Hamilton method with the new d.

State	Pop.	Q	rounded Q
a	3462	$\frac{3462}{99.98537} \approx 34.625$	**34**
b	7470	$\frac{7470}{99.98537} \approx 74.711$	74
c	4265	$\frac{4265}{99.98537} \approx 42.656$	42
d	5300	$\frac{5300}{99.98537} \approx 53.008$	53
Totals	20497		203

The two remaining seats will be apportioned to State b and State c, because the fractional parts of their quotas are the largest. The final apportionment is as follows.

State	Q	Number of seats
a	34.623	**34**
b	74.711	74 + 1 = 75
c	42.656	42 + 1 = 43
d	53.008	53
Totals		205

State a is a victim of the Alabama Paradox, because it has lost a seat despite the fact that the overall number of seats has increased.

7. First calculate the apportionment with $n = 126$. The total population is $263 + 808 + 931 + 781 + 676 = 3459$. The standard divisor for 126 seats is

$$d = \frac{3459}{126} \approx 27.4524.$$

Use the Hamilton method.

State	Pop.	Q	rounded Q
a	263	$\frac{263}{27.4524} \approx 9.580$	9
b	808	$\frac{808}{27.4524} \approx 29.433$	29
c	931	$\frac{931}{27.4524} \approx 33.913$	33
d	781	$\frac{781}{27.4524} \approx 28.449$	28
e	676	$\frac{676}{27.4524} \approx 24.624$	24
Totals	3459		123

The remaining seats will be apportioned to States c, e, and a, because the fractional parts of their quotas is the largest. The final apportionment is

State	Q	Number of seats
a	9.580	9 + 1 = **10**
b	29.433	29
c	33.913	33 + 1 = 34
d	28.449	28
e	24.624	24 + 1 = 25
Totals		126

Now increase the number of seats to $n = 127$. The standard divisor is

$$d = \frac{3459}{127} \approx 27.2362.$$

Use the Hamilton method with the new d.

State	Pop.	Q	rounded Q
a	263	$\dfrac{263}{27.2362}$ ≈ 9.656	**9**
b	808	$\dfrac{808}{27.2362}$ ≈ 29.666	29
c	931	$\dfrac{931}{27.2362}$ ≈ 34.182	34
d	781	$\dfrac{781}{27.2362}$ ≈ 28.675	28
e	676	$\dfrac{676}{27.2362}$ ≈ 24.820	24
Totals	3459		124

The three remaining seats will be apportioned to States e, d, and b, because the fractional parts of their quotas are the largest. The final apportionment is

State	Q	Number of seats
a	9.656	**9**
b	29.666	29 + 1 = 30
c	34.182	34
d	28.675	28 + 1 = 29
e	24.820	24 + 1 = 25
Totals		127

State a is a victim of the Alabama Paradox, because it has lost a seat despite the fact that the overall number of seats has increased.

9. First calculate the apportionment for the initial populations. The total population is $55 + 125 + 190 = 370$.
The standard divisor for 11 seats is

$$d = \frac{370}{11} = 33.6364.$$

Use the Hamilton method.

State	Pop.	Q	rounded Q
a	55	$\dfrac{55}{33.6364}$ ≈ 1.635	1
b	125	$\dfrac{125}{33.6364}$ ≈ 3.716	3
c	190	$\dfrac{190}{33.6364}$ ≈ 5.649	5
Totals	370		9

The remaining seats will be apportioned to States b and c, because the fractional parts of their quotas are the largest. The final apportionment is as follows.

State	Q	Number of seats
a	1.635	1
b	3.716	3 + 1 = 4
c	5.649	5 + 1 = 6
Totals		11

Now apportion the seats for the revised populations.
$61 + 148 + 215 = 424$

The standard divisor is $d = \dfrac{424}{11} \approx 38.5455$.

Use the Hamilton method with the new d.

State	Pop.	Q	rounded Q
a	61	$\dfrac{61}{38.5455}$ ≈ 1.583	1
b	148	$\dfrac{148}{38.5455}$ ≈ 3.840	3
c	215	$\dfrac{215}{38.5455}$ ≈ 5.578	5
Totals	424		9

The two remaining seats will be apportioned to States a and b, because the fractional parts of their quotas are the largest. The final apportionment is as follows.

State	Q	Number of seats
a	1.583	1 + 1 = 2
b	3.840	3 + 1 = 4
c	5.578	5
Totals		11

Here is a final summary.

State	Old Pop.	New Pop.	% growth	Old no. of seats	New no. of seats
a	55	61	**10.91**	**1**	**2**
b	125	148	18.40	4	4
c	190	215	**13.16**	**6**	**5**
Totals	370	424		11	11

Notice from the table that there was a greater percent increase in growth for State c than for State a. Yet, State a gained a seat, and State c lost a seat. This is an example of the Population Paradox.

11. First calculate the apportionment for the initial populations. The total population is 930 + 738 + 415 = 2083.
The standard divisor for 13 seats is

$$d = \frac{2083}{13} = 160.2308.$$

Use the Hamilton method.

State	Pop.	Q	rounded Q
a	930	$\frac{930}{160.2308} \approx 5.804$	5
b	738	$\frac{738}{160.2308} \approx 4.606$	4
c	415	$\frac{415}{160.2308} \approx 2.590$	2
Totals	2083		11

The two remaining seats will be apportioned to States a and b, because the fractional parts of their quotas are the largest. The final apportionment is as follows.

State	Q	Number of seats
a	5.804	5 + 1 = 6
b	4.606	4 + 1 = 5
c	2.590	2
Totals		13

Now apportion the seats for the revised populations.
975 + 750 + 421 = 2146.
The standard divisor is

$$d = \frac{2146}{13} \approx 165.0769.$$

Use the Hamilton method with the new d.

State	Pop.	Q	rounded Q
a	975	$\frac{975}{165.0769} \approx 5.906$	5
b	750	$\frac{750}{165.0769} \approx 4.543$	4
c	421	$\frac{421}{165.0769} \approx 2.550$	2
Totals	2146		11

The two remaining seats will be apportioned to States a and c, because the fractional parts of their quotas are the largest. The final apportionment is as follows.

State	Q	Number of seats
a	5.906	5 + 1 = 6
b	4.543	4
c	2.550	2 + 1 = 3
Totals		13

Here is a final summary.

State	Old Pop.	New Pop.	% growth	Old no. of seats	New no. of seats
a	930	975	4.84	6	6
b	738	750	**1.63**	**5**	**4**
c	415	421	**1.45**	**2**	**3**
Totals	2083	2146		13	13

Notice from the table that there was a greater percent increase in growth for State b than for State c. Yet, State c gained a seat, and State b lost a seat. This is an example of the Population Paradox.

13. First calculate the apportionment for the initial populations. The total population is $3184 + 8475 = 11{,}659$.
The standard divisor for 75 seats is

$$d = \frac{11659}{75} \approx 155.4533.$$

Use the Hamilton method.

State	Pop.	Q	rounded Q
a	3184	$\frac{3184}{155.4533} \approx 20.482$	20
b	8475	$\frac{8475}{155.4533} \approx 54.518$	54
Totals	11,659		74

The one remaining seat will be apportioned to State b because the fractional parts of its quota is larger. The final apportionment is as follows.

State	Q	Number of seats
a	20.482	20
b	54.518	54 + 1 = 55
Totals		75

Now apportion the seats to include the new state. The standard quota of the new state is

$$Q = \frac{330}{155.4533} \approx 2.123.$$

Rounded down to 2, add 2 new seats to the original 75 to obtain 77 seats to be apportioned. Now the new population is $3184 + 8475 + 330 = 11{,}989$.
The standard divisor is

$$d = \frac{11989}{77} \approx 155.7013.$$

State	Pop.	Q	rounded Q
a	3184	$\frac{3184}{155.7013} \approx 20.449$	20
b	8475	$\frac{8475}{155.7013} \approx 54.431$	54
c	330	$\frac{330}{155.7013} \approx 2.119$	2
Totals	11,989		76

The one remaining seat will be apportioned to State a because the fractional part of its quota is the largest. The final apportionment is as follows.

State	Q	Number of seats
a	20.449	20 + 1 = 21
b	54.431	54
c	2.119	2
Totals		77

The New State Paradox has occurred because the addition of the new state has caused a shift in the apportionment of the original states. States a and b originally had 20 and 55 seats, respectively; now they have 21 and 54, respectively.

15. First calculate the apportionment for the initial populations. The total population is $7500 + 9560 = 17{,}060$.
The standard divisor for 83 seats is

$$d = \frac{17060}{83} \approx 205.5422.$$

Use the Hamilton method.

State	Pop.	Q	rounded Q
a	7500	$\frac{7500}{205.5422} \approx 36.489$	36
b	9560	$\frac{9560}{205.5422} \approx 46.511$	46
Totals	17,060		82

The one remaining seat will be apportioned to State b because the fractional parts of its quota is larger. The final apportionment is as follows.

State	Q	Number of seats
a	36.489	36
b	46.511	46 + 1 = 47
Totals		83

Now apportion the seats to include the new state. The standard quota of the new state is

$$Q = \frac{1500}{205.5422} \approx 7.298.$$

Rounded down to 7, add 7 new seats to the original 83 to obtain 90 seats to be apportioned. Now the new population is 7500 + 9560 + 1500 = 18,560.
The standard divisor is

$$d = \frac{18560}{90} \approx 206.2222.$$

State	Pop.	Q	rounded Q
a	7500	$\frac{7500}{206.2222} \approx 36.369$	36
b	9560	$\frac{9560}{206.2222} \approx 46.358$	46
c	1500	$\frac{1500}{206.2222} \approx 7.274$	7
Totals	18,560		89

The one remaining seat will be apportioned to State a because the fractional part of its quota is the largest. The final apportionment is as follows.

State	Q	Number of seats
a	36.369	36 + 1 = 37
b	46.358	46
c	7.274	7
Totals		90

The New State Paradox has occurred because the addition of the new state has caused a shift in the apportionment of the original states. States a and b originally had 36 and 47 seats, respectively; now they have 37 and 46, respectively.

17. The sum of the populations is
1720 + 3363 + 6960 + 24223 + 8800 = 45066.
The standard divisor is

$$d = \frac{45066}{220} \approx 204.8455.$$

State	Pop.	Q	Q rounded down
a	1720	$\frac{1720}{204.8455} \approx 8.397$	8
b	3363	$\frac{3363}{204.8455} \approx 16.417$	16
c	6960	$\frac{6960}{204.8455} \approx 33.977$	33
d	24223	$\frac{24223}{204.8455} \approx 118.250$	**118**
e	8800	$\frac{8800}{204.8455} \approx 42.959$	42
Totals	45066		217

Use the Adams method with a modified divisor of 208.

State	Pop.	Q	number of seats
a	1720	$\frac{1720}{208} \approx 8.269$	9
b	3363	$\frac{3363}{208} \approx 16.168$	17
c	6960	$\frac{6960}{208} \approx 33.462$	34
d	24223	$\frac{24223}{208} \approx 116.457$	**117**
e	8800	$\frac{8800}{208} \approx 42.308$	43
Totals	45066		220

This is a violation of the Quota Rule because State d receives only 117 seats, although it should receive at least 118.

19. Here is the apportionment from Example 4, with a population of 531 for the second subdivision. The total population is $8500 + 1671 + 531 = 10{,}702$.
The standard divisor for 105 seats is

$$d = \frac{10702}{105} \approx 101.9238.$$

Community	Pop.	Standard Quota	rounded down Q
Original	8500	$\frac{8500}{101.9238} \approx 83.396$	83
1st Annexed	1671	$\frac{1671}{101.9238} \approx 16.395$	16
2nd Annexed	531	$\frac{531}{101.9238} \approx 5.210$	5
Totals	10,702		104

The one remaining seat will be apportioned to the original community because the fractional parts of its quota is larger. The final apportionment is as follows.

Community	Q	Number of seats
a	83.396	$83 + 1 = 84$
b	16.395	16
c	5.210	5
Totals		105

Now increasing the population of the second annexed subdivision to 532, here are the calculations. The total population would be $8500 + 1671 + 532 = 10{,}703$.
The standard divisor would be

$$d = \frac{10703}{105} \approx 101.9333.$$

Community	Pop.	Standard Quota	rounded down Q
Original	8500	$\frac{8500}{101.9333} \approx 83.388$	83
1st Annexed	1671	$\frac{1671}{101.9333} \approx 16.393$	16
2nd Annexed	532	$\frac{532}{101.9333} \approx 5.219$	5
Totals	10,703		104

Now the one remaining seat would be apportioned to the 1st annexed community because the fractional part of its quota is the largest. The final apportionment is as follows.

Community	Q	Number of seats
a	83.388	83
b	16.393	$16 + 1 = 17$
c	5.219	5
Totals		105

The New State Paradox does not occur if the new population is 531, but it does occur when the population of the 2nd annexed subdivision increases to 532.

21. Writing exercise; answers will vary.

23. Writing exercise; answers will vary.

25. Writing exercise; answers will vary.

Chapter 16 Test

1. The voter profile is as follows

Votes	Ranking
5	a > b > d > c
6	b > c > a > d
5	c > d > b > a
7	d > a > b > c
4	c > a > d > b

By the plurality method the destination with the most votes wins. Cancun receives $5 + 4 = 9$ first-place votes to win.

2. Use the ranking information from Exercise 1 to make a table to examine the points in the Borda method.

Votes	Ranking	Points			
		3	2	1	0
5	a > b > d > c	a	b	d	c
6	b > c > a > d	b	c	a	d
5	c > d > b > a	c	d	b	a
7	d > a > b > c	d	a	b	c
4	c > a > d > b	c	a	d	b

Examine the last four columns, and compute the weighted sum for each candidate.

Issue	Borda Points
a	$5 \cdot 3 + 6 \cdot 1 + 5 \cdot 0 + 7 \cdot 2 + 4 \cdot 2 = 43$
b	$5 \cdot 2 + 6 \cdot 3 + 5 \cdot 1 + 7 \cdot 1 + 4 \cdot 0 = 40$
c	$5 \cdot 0 + 6 \cdot 2 + 5 \cdot 3 + 7 \cdot 0 + 4 \cdot 3 = 39$
d	$5 \cdot 1 + 6 \cdot 0 + 5 \cdot 2 + 7 \cdot 3 + 4 \cdot 1 = 40$

The winner is Aruba with 43 Borda points.

3. There are 27 sorority sisters voting, so a destination must receive a majority of the votes or at least 14 votes to win by the Hare method. No destination received a majority. See from Exercise 1 that Aruba received 5 first-place votes; the Bahamas received 6 votes; Cancun received 9 votes; and the Dominican Republic received 7 votes. Eliminate Aruba because it has the least number of votes, and compare only b, c, and d.

Votes	Ranking	Row
5	b > d > c	1
6	b > c > d	2
5	c > d > b	3
7	d > b > c	4
4	c > d > b	5

The Bahamas received $5 + 6 = 11$ votes.
Cancun receives $5 + 4 = 9$ votes.
The Dominican Republic receives 7 votes. Again, no destination has received a majority. Therefore, eliminate d with the least number of votes, and compare only b and c.

Votes	Ranking	Row
5	b > c	1
6	b > c	2
5	c > b	3
7	b > c	4
4	c > b	5

The Bahamas receives $5 + 6 + 7 = 18$ votes. Cancun receives $5 + 4 = 9$ votes. Finally, the Bahamas is the winner.

4. Using the pairwise comparison method make a table to compute the votes in comparison.

Compare	Votes	Rows
a > b	$5 + 7 + 4 = 16$	1, 4, 5
b > a	$6 + 5 = 11$	2, 3
a > c	$5 + 7 = 12$	1, 4
c > a	$6 + 5 + 4 = 15$	2, 3, 5
a > d	$5 + 6 + 4 = 15$	1, 2, 5
d > a	$5 + 7 = 12$	3, 4
b > c	$5 + 6 + 7 = 18$	1, 2, 4
c > b	$5 + 4 = 9$	3, 5
b > d	$5 + 6 = 11$	1, 2
d > b	$5 + 7 + 4 = 16$	3, 4, 5
c > d	$6 + 5 + 4 = 15$	2, 3, 5
d > c	$5 + 7 = 12$	1, 4

Make another table to award pairwise points.

Pairs	Votes	Pairwise Points
a : b	16 : 11	a, 1
a : c	12 : 15	c, 1
a : d	15 : 12	a, 1
b : c	18 : 9	b, 1
b : d	11 : 16	d, 1
c : d	15 : 12	c, 1

Using the pairwise comparison method, here are the final point totals.
a: $1 + 1 = 2$
b: $1 = 1$
c: $1 + 1 = 2$
d: $1 = 1$
Aruba and Cancun each have two pairwise points and Bahamas and Dominican Republic each have one pairwise point, so the pairwise comparison method vote results in a tie.

5. There are $7! = 5040$ different complete rankings.

6. In a pairwise comparison election with 10 alternatives, $_{10}C_2 = 45$ comparisons must be made.

7. Writing exercise; answers will vary.

8. Writing exercise; answers will vary.

9. Writing exercise; answers will vary.

10. Writing exercise; answers will vary.

11. Use the ranking information from the text to make a table to examine the points in the Borda method.

Votes	Ranking	Points 2	1	0
16	a > b > c	a	b	c
8	b > c > a	b	c	a
7	c > b > a	c	b	a

Examine the last three columns, and compute the weighted sum for each candidate.

Candidate	Borda Points
a	$16 \cdot 2 + 8 \cdot 0 + 7 \cdot 0 = 32$
b	$16 \cdot 1 + 8 \cdot 2 + 7 \cdot 1 = 39$
c	$16 \cdot 0 + 8 \cdot 1 + 7 \cdot 2 = 22$

Although a has the majority of first-place votes in the ranking, candidate b wins by the Borda method. This violates the majority criterion.

12. Make a table to compute the votes in comparisons.

Comparisons	Votes	Rows
a > b	5 = 5	1
b > a	6 + 4 + 6 = 16	2, 3, 4
a > c	5 + 4 = 9	1, 3
c > a	6 + 6 = 12	2, 4
b > c	4 + 6 = 10	3, 4
c > b	5 + 6 = 11	1, 2

Make another table to award pairwise points.

Pairs	Votes	Pairwise Points
a : b	5 : 16	b, 1
a : c	9 : 12	c, 1
b : c	10 : 11	c, 1

Using the pairwise comparison method, the

winner is c with 2 pairwise points. alternative b received only 1 point, and a received 0 points.

13. Alternative c is the Condorcet candidate.
<u>Plurality</u> method
a: 5 = 5
b: 4 + 6 = 10
c: 6 = 6
Alternative b is selected, which violates the Condorcet criterion.

<u>Borda</u> method
Use the ranking information from the text to make a table to examine the points in the Borda method.

Votes	Ranking	Points 2	1	0
5	a > c > b	a	c	b
6	c > b > a	c	b	a
4	b > a > c	b	a	c
6	b > c > a	b	c	a

Examine the last three columns, and compute the weighted sum for each candidate.

Candidate	Borda Points
a	$5 \cdot 2 + 6 \cdot 0 + 4 \cdot 1 + 6 \cdot 0 = 14$
b	$5 \cdot 0 + 6 \cdot 1 + 4 \cdot 2 + 6 \cdot 2 = 26$
c	$5 \cdot 1 + 6 \cdot 2 + 4 \cdot 0 + 6 \cdot 1 = 23$

Alternative b wins by the Borda method. This violates the Condorcet criterion.

<u>Hare</u> method
None of the candidates has a majority of first-place votes, which is at least 11. Eliminate a because it has the least number of votes at 5, and compare only b and c.

Votes	Ranking
5	c > b
6	c > b
4	b > c
6	b > c

Now, alternative c receives 5 + 6 = 11 votes and wins. This does not violate the Condorcet criterion.

14. Make a table to compute the votes in comparisons.

Comparison	Votes	Rows
c > m	8 + 6 = 14	1, 3
m > c	7 + 5 = 12	2, 4
c > b	8 + 5 = 13	1, 4
b > c	7 + 6 = 13	2, 3
c > s	8 + 5 = 13	1, 4
s > c	7 + 6 = 13	2, 3
m > b	8 + 5 = 13	1, 4
b > m	7 + 6 = 13	2, 3
m > s	8 + 5 = 13	1, 4
s > m	7 + 6 = 13	2, 3
b > s	8 + 5 = 13	1, 4
s > b	7 + 6 = 13	2, 3

Make another table to award pairwise points.

Pairs	Votes	Pairwise Points
c : m	14 : 12	c, 1
c : b	13 : 13	c, $\frac{1}{2}$; b, $\frac{1}{2}$
c : s	13 : 13	c, $\frac{1}{2}$; s, $\frac{1}{2}$
m : b	13 : 13	m, $\frac{1}{2}$; b, $\frac{1}{2}$
m : s	13 : 13	m, $\frac{1}{2}$; s, $\frac{1}{2}$
b : s	13 : 13	b, $\frac{1}{2}$; s, $\frac{1}{2}$

Using the pairwise comparison method, here are the final point totals.

c: $1 + \frac{1}{2} + \frac{1}{2} = 2$

b: $\frac{1}{2} + \frac{1}{2} + \frac{1}{2} = 1\frac{1}{2}$

m: $\frac{1}{2} + \frac{1}{2} = 1$

s: $\frac{1}{2} + \frac{1}{2} + \frac{1}{2} = 1\frac{1}{2}$

Using the pairwise comparison method, the winner is c with 2 pairwise points.
Now, change the ranking of the last 5 voters as shown in the text and make a table to compute the votes in comparisons.

Comparison	Votes	Rows
c > m	8 + 6 + 5 = 19	1, 3, 4
m > c	7 = 7	2
c > b	8 + 5 = 13	1, 4
b > c	7 + 6 = 13	2, 3
c > s	8 + 5 = 13	1, 4
s > c	7 + 6 = 13	2, 3
m > b	8 + 5 = 13	1, 4
b > m	7 + 6 = 13	2, 3
m > s	8 = 8	1
s > m	7 + 6 + 5 = 18	2, 3, 4
b > s	8 = 8	1
s > b	7 + 6 + 5 = 18	2, 3, 4

Make another table to award pairwise points.

Pairs	Votes	Pairwise Points
c : m	19 : 7	c, 1
c : b	13 : 13	c, $\frac{1}{2}$; b, $\frac{1}{2}$
c : s	13 : 13	c, $\frac{1}{2}$; s, $\frac{1}{2}$
m : b	13 : 13	m, $\frac{1}{2}$; b, $\frac{1}{2}$
m : s	8 : 18	s, 1
b : s	8 : 18	s, 1

Using the pairwise comparison method, here are the final point totals.

c: $1 + \frac{1}{2} + \frac{1}{2} = 2$

b: $\frac{1}{2} + \frac{1}{2} = 1$

m: $\frac{1}{2} = \frac{1}{2}$

s: $\frac{1}{2} + 1 + 1 = 2\frac{1}{2}$

Using the pairwise comparison method, the winner is s with $2\frac{1}{2}$ pairwise points.

Although c was moved to the top of the ranking when the 5 voters changed their votes, c did not win. This shows that the pairwise comparison method can violate the monotonicity criterion.

15. Count the number of first-place votes to see that choice a has 9; choice b has 8, choice c has 12, and choice d has 0. None of the candidates has a majority of first-place votes, which is at least 15. The elimination of d does not change the number of first-place votes of a, b, and c. Therefore, eliminate b and compare only a and c.

Votes	Ranking
9	a > c
8	a > c
7	c > a
5	c > a

Now a receives 9 + 8 = 17 votes and wins. Now change the ranking of the last 5 voters as shown in the text and apply the Hare method again. Count the number of first-place votes to see that choice a now has 14, choice b still has 8, and choice c now has 7. Because none of the candidates has a majority, eliminate c and compare only a and b.

Votes	Ranking
9	a > b
8	b > a
7	b > a
5	a > b

This time b received 8 + 7 = 15 votes and wins. Although a was moved to the top of the ranking when the 5 voters changed their votes, a did not win. This shows that the Hare method can violate the monotonicity criterion.

16. Examine the voter profile in the text to see that candidate a receives the most votes at 10 and is the winner by the plurality method. If alternative c is dropped from the selection process the rankings are as follows.

Votes	Ranking
10	a > b
7	b > a
5	b > a

Now alternative b has 12 votes and wins. Although losing alternative c is dropped,

candidate a does not win the second election.

The two outcomes show that the plurality method can violate the irrelevant alternatives criterion. A losing alternative was dropped from the selection process, but the original preferred alternative is not selected a second time.

17. Use the ranking information from the text to make a table to examine the points in the Borda method.

Votes	Ranking	Points 3	2	1	0
7	a > b > c > d	a	b	c	d
5	b > c > d > a	b	c	d	a
4	d > c > a > b	d	c	a	b

Examine the last four columns, and compute the weighted sum for each candidate.

Issue	Borda Points
a	$7 \cdot 3 + 5 \cdot 0 + 4 \cdot 1 = 25$
b	$7 \cdot 2 + 5 \cdot 3 + 4 \cdot 0 = 29$
c	$7 \cdot 1 + 5 \cdot 2 + 4 \cdot 2 = 25$
d	$7 \cdot 0 + 5 \cdot 1 + 4 \cdot 3 = 17$

Alternative b wins by the Borda method. Now repeat the Borda method after eliminating alternative d. Make a table to examine the points in the Borda method.

Votes	Ranking	Points 2	1	0
7	a > b > c	a	b	c
5	b > c > a	b	c	a
4	c > a > b	c	a	b

Examine the last three columns, and compute the weighted sum for each candidate.

Issue	Borda Points
a	$7 \cdot 2 + 5 \cdot 0 + 4 \cdot 1 = 18$
b	$7 \cdot 1 + 5 \cdot 2 + 4 \cdot 0 = 17$
c	$7 \cdot 0 + 5 \cdot 1 + 4 \cdot 2 = 13$

This time alternative a wins by the Borda method.

The two outcomes show that the Borda method can violate the irrelevant alternatives criterion. A losing alternative was dropped from the selection process, but the original preferred alternative is not selected a second time.

18. Count the number of first-place votes to see that choice a has 7; choice b has 5, choice d has 4, and choice c has 0. None of the candidates has a majority of first-place votes, which is at least 9. The elimination of c does not change the number of first-place votes of a, b, and d. Therefore, eliminate d and compare only a and b.

Votes	Ranking
7	a > b
5	b > a
4	a > b

Now a receives 7 + 4 = 11 votes and wins. Now eliminate candidate b, and apply the Hare method again. Here is a new voter profile.

Votes	Ranking
7	a > c > d
5	c > d > a
4	d > c > a

Count the number of first-place votes to see that choice a now has 7, choice c has 5, and choice d now has 4.

Votes	Ranking
7	a > c
5	c > a
4	c > a

This time c received 5 + 4 = 9 votes and wins. Although b was dropped from the ranking, the originally preferred candidate a did not win. This is a violation of the Independence of Irrelevant Alternative criterion.

The two outcomes show that the Hare method can violate the irrelevant alternatives criterion. A losing alternative was dropped from the selection process, but the original preferred alternative is not selected a second time.

19. Writing exercise; answers will vary.

20. Using the Hamilton method, the total population is
    1429 + 8639 + 7608 + 6660 + 1671
    = 26007.
    The standard divisor is
    $$d = \frac{26007}{195} \approx 133.3692.$$

Ward	Pop.	Q	Rounded Q
1st	1429	$\frac{1429}{133.369}$ ≈ 10.715	10
2nd	8639	$\frac{8639}{133.369}$ ≈ 64.775	64
3rd	7608	$\frac{7608}{133.369}$ ≈ 57.045	57
4th	6660	$\frac{6660}{133.369}$ ≈ 49.937	49
5th	1671	$\frac{1671}{133.369}$ ≈ 12.529	12
Totals	26007		192

The three remaining seats will be apportioned to the 1st 2nd and 4th wards, because the fractional parts of their quotas are the largest. The final apportionment is as follows.

Ward	Q	Number of seats
1st	10.715	10 + 1 = 11
2nd	64.775	64 + 1 = 65
3rd	57.045	57
4th	49.937	49 + 1 = 50
5th	12.529	12
Totals		195

21. Use the Jefferson method with $md = 131$.

Ward	Pop.	mQ	No. of seats
1st	1429	$\frac{1429}{131}$ ≈ 10.908	10
2nd	8639	$\frac{8639}{131}$ ≈ 65.947	65
3rd	7608	$\frac{7608}{131}$ ≈ 58.076	58
4th	6660	$\frac{6660}{131}$ ≈ 50.840	50
5th	1671	$\frac{1671}{131}$ ≈ 12.756	12
Totals	26007		195

22. Use the Webster method with $md = 133.7$.

Ward	Pop.	mQ	No. of seats
1st	1429	$\frac{1429}{133.7}$ ≈ 10.688	11
2nd	8639	$\frac{8639}{133.7}$ ≈ 64.615	65
3rd	7608	$\frac{7608}{133.7}$ ≈ 56.904	57
4th	6660	$\frac{6660}{133.7}$ ≈ 49.813	50
5th	1671	$\frac{1671}{133.7}$ ≈ 12.498	12
Totals	26007		195

23. Writing exercise; answers may vary.

24. Writing exercise; answers may vary.

25. Writing exercise; answers may vary.

26. Writing exercise; answers may vary.

27. The sum of the populations is
$2354 + 4500 + 5598 + 23000 = 35452$.

The standard divisor is $d = \dfrac{35452}{100} = 354.52$.

Using the standard quota, we have the following.

State	Pop.	Q	No. of seats (Q rounded down)
a	2354	$\frac{2354}{354.52}$ ≈ 6.640	6
b	4500	$\frac{4500}{354.52}$ ≈ 12.693	12
c	5598	$\frac{5598}{354.52}$ ≈ 15.790	15
d	23000	$\frac{23000}{354.52}$ ≈ 64.876	**64**
Totals	35,452		97

Use the Jefferson method with $md = 347$.

State	Pop.	mQ	No. of seats (mQ rounded down)
a	2354	$\frac{2354}{347}$ ≈ 6.784	6
b	4500	$\frac{4500}{347}$ ≈ 12.968	12
c	5598	$\frac{5598}{347}$ ≈ 16.133	16
d	23000	$\frac{23000}{347}$ ≈ 66.282	**66**
Totals	35,452		100

This Jefferson method violates the Quota Rule because State d receives two more seats than its apportionment from the standard quota.

28. First, calculate the apportionment with $n = 126$. The total population is $263 + 809 + 931 + 781 + 676 = 3460$. The standard divisor for 126 seats is

$$d = \frac{3460}{126} \approx 27.4603.$$

Use the Hamilton Method.

State	Pop.	Q	rounded Q
a	263	$\frac{263}{27.4603}$ ≈ 9.577	9
b	809	$\frac{809}{27.4603}$ ≈ 29.461	29
c	931	$\frac{931}{27.4603}$ ≈ 33.903	33
d	781	$\frac{781}{27.4603}$ ≈ 28.441	28
e	676	$\frac{676}{27.4603}$ ≈ 24.617	24
Totals	3460		123

The three remaining seats will be apportioned to States c, e, and a, because the fractional parts of their quotas are the largest. The final apportionment is as follows.

State	Q	Number of seats
a	9.577	$9 + 1 = 10$
b	29.461	29
c	33.903	$33 + 1 = 34$
d	28.441	28
e	24.617	$24 + 1 = 25$
Totals		126

Now increase the number of seats to 127. The standard divisor is

$$d = \frac{3460}{127} \approx 27.2441.$$

Use the Hamilton method with the new d.

State	Pop.	Q	rounded Q
a	263	$\frac{263}{27.2441}$ ≈ 9.653	9
b	809	$\frac{809}{27.2441}$ ≈ 29.695	29
c	931	$\frac{931}{27.2441}$ ≈ 34.173	34
d	781	$\frac{781}{27.2441}$ ≈ 28.667	28
e	676	$\frac{676}{27.2441}$ ≈ 24.813	24
Totals	3460		124

The three remaining seats will be apportioned to States e, b, and d, because the fractional parts of their quotas are the largest. The final apportionment is as follows.

State	Q	Number of seats
a	9.653	**9**
b	29.695	$29 + 1 = 30$
c	34.173	34
d	28.667	$28 + 1 = 29$
e	24.813	$24 + 1 = 25$
Totals		127

State a is a victim of the Alabama paradox, because it has lost a seat despite the fact that the overall number of seats has increased.

29. First, calculate the apportionment for the initial populations. The total population is $55 + 125 + 190 = 370$. The standard divisor for 11 seats is

$$d = \frac{370}{11} = 33.6364.$$

Use the Hamilton Method.

State	Pop.	Q	rounded Q
a	55	$\frac{55}{33.6364}$ ≈ 1.635	1
b	125	$\frac{125}{33.6364}$ ≈ 3.716	3
c	190	$\frac{190}{33.6364}$ ≈ 5.649	5
Totals	370		9

The two remaining seats will be apportioned to States b and c, because the fractional parts of their quotas are the largest. The final apportionment is as follows.

State	Q	Number of seats
a	1.635	1
b	3.716	3 + 1 = 4
c	5.649	5 + 1 = 6
Totals		11

Now apportion the seats for the revised populations.
63 + 150 + 220 = 433
The new standard divisor is

$$d = \frac{433}{11} \approx 39.3636.$$

Use the Hamilton method with the new values.

State	Pop.	Q	rounded Q
a	63	$\frac{55}{33.3636}$ ≈ 1.600	1
b	150	$\frac{150}{33.3636}$ ≈ 3.811	3
c	220	$\frac{220}{39.3636}$ ≈ 5.589	5
Totals	433		9

The two remaining seats will be apportioned to States a and b, because the fractional parts of their quotas are the largest. The final apportionment is as follows.

State	Q	Number of seats
a	1.600	1 + 1 = 2
b	3.811	3 + 1 = 4
c	5.589	5
Totals		11

Here is a final summary.

State	Old Pop.	New Pop.	% growth	Old no. of seats	New no. of seats
a	55	63	**14.55**	**1**	**2**
b	125	150	20.00	4	4
c	190	220	**15.79**	**6**	**5**
Totals				11	11

Notice from the table that there was a greater percent increase in growth for State c than for State a. Yet, State a gained a seat, and State c lost a seat. This is an example of the Population paradox.

30. First, calculate the apportionment for the initial populations. The total population is 49 + 160 = 209.
The standard divisor for 100 seats is

$$d = \frac{209}{100} = 2.09.$$

Use the Hamilton Method.

State	Pop.	Q	rounded Q
a	49	$\frac{49}{2.09}$ ≈ 23.445	23
b	160	$\frac{160}{2.09}$ ≈ 76.555	76
Totals	209		99

The one remaining seat will be apportioned to State b because the fractional parts of its quota is larger. The final apportionment, for the initial number of seats, follows.

State	Q	Number of seats
a	23.445	23
b	76.555	76 + 1 = 77
Totals		100

Now apportion the seats to include the new state. The standard quota of the new state is:

$$Q = \frac{32}{2.09} \approx 15.311.$$

Rounded down 15, add 15 new seats to the original 100 to obtain 115 seats to be apportioned.
Now the new population is
49 + 160 + 32 = 241.
The new standard divisor is

$$d = \frac{241}{115} \approx 2.0957.$$

State	Pop.	Q	rounded Q
a	49	$\frac{49}{2.0957}$ ≈ 23.381	23
b	160	$\frac{160}{2.0957}$ ≈ 76.347	76
c	32	$\frac{32}{2.0957}$ ≈ 15.269	15
Totals	241		114

The one remaining seat will be apportioned to State a because the fractional part of its quota is the largest.
The final apportionment, for the revised number of seats, is as follows.

State	Q	Number of seats
a	23.381	23 + 1 = 24
b	76.347	76
c	15.269	15
Totals		115

The new states paradox has occurred because the addition of the new state has caused a shift in the apportionment of the original states. States a and b originally had 23 and 77 seats, respectively; now they have 24 and 76, respectively.

The Metric System

1. $\dfrac{8 \text{ m}}{1} \cdot \dfrac{1000 \text{ mm}}{1 \text{ m}} = 8000 \text{ mm}$

2. $\dfrac{14.76 \text{ m}}{1} \cdot \dfrac{100 \text{ cm}}{1 \text{ m}} = 1476 \text{ cm}$

3. $\dfrac{8500 \text{ cm}}{1} \cdot \dfrac{1 \text{ m}}{100 \text{ cm}} = 85 \text{ m}$

4. $\dfrac{250 \text{ mm}}{1} \cdot \dfrac{1 \text{ m}}{1000 \text{ mm}} = 0.25 \text{ m}$

5. $\dfrac{68.9 \text{ cm}}{1} \cdot \dfrac{10 \text{ mm}}{1 \text{ cm}} = 689 \text{ mm}$

6. $\dfrac{3.25 \text{ cm}}{1} \cdot \dfrac{10 \text{ mm}}{1 \text{ cm}} = 32.5 \text{ mm}$

7. $\dfrac{59.8 \text{ mm}}{1} \cdot \dfrac{1 \text{ cm}}{10 \text{ mm}} = 5.98 \text{ cm}$

8. $\dfrac{3.542 \text{ mm}}{1} \cdot \dfrac{1 \text{ cm}}{10 \text{ mm}} = 0.3542 \text{ cm}$

9. $\dfrac{5.3 \text{ km}}{1} \cdot \dfrac{1000 \text{ m}}{1 \text{ km}} = 5300 \text{ m}$

10. $\dfrac{9.24 \text{ km}}{1} \cdot \dfrac{1000 \text{ m}}{1 \text{ km}} = 9240 \text{ m}$

11. $\dfrac{27,500 \text{ m}}{1} \cdot \dfrac{1 \text{ km}}{1000 \text{ m}} = 27.5 \text{ km}$

12. $\dfrac{14,592 \text{ m}}{1} \cdot \dfrac{1 \text{ km}}{1000 \text{ m}} = 14.592 \text{ km}$

13. 2.54 cm; 25.4 mm

14. 3.3 cm; 33 mm

15. 5 cm; 50 mm

16. 2.54 cm; 25.4 mm

17. $\dfrac{6 \text{ L}}{1} \cdot \dfrac{10^2 \text{ cl}}{1 \text{ L}} = 6 \times 100 = 600 \text{ cl}$

18. $\dfrac{4.1 \text{ L}}{1} \cdot \dfrac{10^3 \text{ ml}}{1 \text{ L}} = 4.1 \times 1000 = 4100 \text{ ml}$

19. $\dfrac{8.7 \text{ L}}{1} \cdot \dfrac{10^3 \text{ ml}}{1 \text{ L}} = 8.7 \times 1000 = 8700 \text{ ml}$

20. $\dfrac{12.5 \text{ L}}{1} \cdot \dfrac{10^2 \text{ cl}}{1 \text{ L}} = 12.5 \times 100 = 1250 \text{ cl}$

21. $\dfrac{925 \text{ cl}}{1} \cdot \dfrac{1 \text{ L}}{10^2 \text{ cl}} = \dfrac{925}{100} = 9.25 \text{ L}$

22. $\dfrac{412 \text{ ml}}{1} \cdot \dfrac{1 \text{ L}}{10^3 \text{ ml}} = \dfrac{412}{1000} = 0.412 \text{ L}$

23. $\dfrac{8974 \text{ ml}}{1} \cdot \dfrac{1 \text{ L}}{10^3 \text{ ml}} = \dfrac{8974}{1000} = 8.974 \text{ L}$

24. $\dfrac{5639 \text{ cl}}{1} \cdot \dfrac{1 \text{ L}}{10^2 \text{ cl}} = \dfrac{5639}{100} = 56.39 \text{ L}$

25. $\dfrac{8000 \text{ g}}{1} \cdot \dfrac{1 \text{ kg}}{10^3 \text{ g}} = \dfrac{8000}{1000} = 8 \text{ kg}$

26. $\dfrac{25000 \text{ g}}{1} \cdot \dfrac{1 \text{ kg}}{10^3 \text{ g}} = \dfrac{25000}{1000} = 25 \text{ kg}$

27. $\dfrac{5.2 \text{ kg}}{1} \cdot \dfrac{10^3 \text{ g}}{1 \text{ kg}} = 5.2 \times 1000 = 5200 \text{ g}$

28. $\dfrac{12.42 \text{ kg}}{1} \cdot \dfrac{10^3 \text{ g}}{1 \text{ kg}} = 12.42 \times 1000 = 12,420 \text{ g}$

29. $\dfrac{4.2 \text{ g}}{1} \cdot \dfrac{10^3 \text{ mg}}{1 \text{ g}} = 4.2 \times 1000 = 4200 \text{ mg}$

30. $\dfrac{3.89 \text{ g}}{1} \cdot \dfrac{10^2 \text{ cg}}{1 \text{ g}} = 3.89 \times 100 = 389 \text{ cg}$

31. $\dfrac{598 \text{ mg}}{1} \cdot \dfrac{1 \text{ g}}{10^3 \text{ mg}} = \dfrac{598}{1000} = 0.598 \text{ g}$

32. $\dfrac{7634 \text{ cg}}{1} \cdot \dfrac{1 \text{ g}}{10^2 \text{ cg}} = \dfrac{7634}{100} = 76.34 \text{ g}$

33. $C = \dfrac{5}{9}(F - 32) = \dfrac{5}{9}(86 - 32) = \dfrac{5}{9}(54) = 30°$

34.
$$C = \frac{5}{9}(F-32)$$
$$= \frac{5}{9}(536-32)$$
$$= \frac{5}{9}(504)$$
$$= 280°$$

35.
$$C = \frac{5}{9}(F-32)$$
$$= \frac{5}{9}(-114-32)$$
$$= \frac{5}{9}(-146)$$
$$\approx -81°$$

36.
$$C = \frac{5}{9}(F-32)$$
$$= \frac{5}{9}(-40-32)$$
$$= \frac{5}{9}(-72)$$
$$= -40°$$

37. $F = \frac{9}{5}C + 32 = \frac{9}{5} \cdot 10 + 32 = 18 + 32 = 50°$

38. $F = \frac{9}{5}C + 32 = \frac{9}{5} \cdot 25 + 32 = 45 + 32 = 77°$

39.
$$F = \frac{9}{5}C + 32$$
$$= \frac{9}{5} \cdot -40 + 32$$
$$= -72 + 32$$
$$= -40°$$

40. $F = \frac{9}{5}C + 32 = \frac{9}{5} \cdot -15 + 32 = -27 + 32 = 5°$

41. $\frac{1\text{ kg}}{1} \cdot \frac{1000\text{ g}}{1\text{ kg}} \cdot \frac{1\text{ nickel}}{5\text{ g}} = 200\text{ nickels}$

42. $\frac{1\text{ L}}{1} \cdot \frac{1000\text{ ml}}{1\text{ L}} \cdot \frac{3.5\text{ g}}{1000\text{ ml}} = 3.5\text{ g}$

43. $\frac{1\text{ L}}{1} \cdot \frac{1000\text{ ml}}{1\text{ L}} \cdot \frac{0.0002\text{ g}}{1\text{ ml}} = 0.2\text{ g}$

44. $\frac{1\text{ ml}}{1} \cdot \frac{1\text{ L}}{1000\text{ ml}} \cdot \frac{1500\text{ g}}{1\text{ L}} = 1.5\text{ g}$

45. $\frac{7\text{ strips}}{1} \cdot \frac{67\text{ cm}}{1\text{ strip}} \cdot \frac{1\text{ m}}{100\text{ cm}} \cdot \frac{\$8.74}{1\text{ m}} \approx \$40.99$

46. $\frac{15\text{ pieces}}{1} \cdot \frac{384\text{ mm}}{1\text{ piece}} \cdot \frac{1\text{ m}}{1000\text{ mm}} \cdot \frac{\$54.20}{1\text{ m}}$
$\approx \$312.19$

47. $A = 128\text{ cm} \cdot 174\text{ cm} = 22,272\text{ cm}^2$
$$\frac{22,272\text{ cm}^2}{1} \cdot \frac{(1\text{ m})^2}{(100\text{ cm})^2} \cdot \frac{\$174.20}{1\text{ m}^2}$$
$\approx \$387.98$

48. $A = 9\text{ cm} \cdot 14\text{ cm} = 126\text{ cm}^2$
$$\frac{80\text{ pieces}}{1} \cdot \frac{126\text{ cm}^2}{1\text{ piece}} \cdot \frac{(1\text{ m})^2}{(100\text{ cm})^2} \cdot \frac{\$63.79}{1\text{ m}^2}$$
$\approx \$64.30$

49. $\frac{82\text{ cm}}{1} \cdot \frac{1\text{ m}}{100\text{ cm}} = 0.82\text{ m}$
$V = 0.82\text{ m} \cdot 1.1\text{ m} \cdot 1.2\text{ m} = 1.0824\text{ m}^3$
$$\frac{1.0824\text{ m}^3}{1} \cdot \frac{(100\text{ cm})^3}{(1\text{ m})^3} = 1,082,400\text{ cm}^3$$

50. $\frac{1.5\text{ m}}{1} \cdot \frac{100\text{ cm}}{1\text{ m}} = 150\text{ cm}$
$V = 150\text{ cm} \cdot 74\text{ cm} \cdot 97\text{ cm} = 1,076,700\text{ cm}^3$
$$\frac{1,076,700\text{ cm}^3}{1} \cdot \frac{(1\text{ m})^3}{(100\text{ cm})^3} = 1.0767\text{ m}^3$$

51. $\frac{160\text{ L}}{1} \cdot \frac{1000\text{ ml}}{1\text{ L}} \cdot \frac{1\text{ bottle}}{800\text{ ml}} = 200\text{ bottles}$

52. $\frac{80\text{ people}}{1} \cdot \frac{400\text{ ml}}{1\text{ person}} \cdot \frac{1\text{ L}}{1000\text{ ml}} \cdot \frac{1\text{ bottle}}{2\text{ L}}$
$= 16\text{ bottles}$

53. $\frac{982\text{ yd}}{1} \cdot \frac{0.9144\text{ m}}{1\text{ yd}} \approx 897.9\text{ m}$

54. $\frac{12.2\text{ km}}{1} \cdot \frac{0.6214\text{ mi}}{1\text{ km}} \approx 7.581\text{ mi}$

55. $\frac{125\text{ mi}}{1} \cdot \frac{1.609\text{ km}}{1\text{ mi}} \approx 201.1\text{ km}$

56. $\frac{1000\text{ mi}}{1} \cdot \frac{1.609\text{ km}}{1\text{ mi}} = 1609\text{ km}$

57. $\dfrac{1816 \text{ g}}{1} \cdot \dfrac{0.0022 \text{ lb}}{1 \text{ g}} \approx 3.995 \text{ lb}$

58. $\dfrac{1.42 \text{ lb}}{1} \cdot \dfrac{454 \text{ g}}{1 \text{ lb}} = 644.68 \text{ g}$

59. $\dfrac{47.2 \text{ lb}}{1} \cdot \dfrac{454 \text{ g}}{1 \text{ lb}} = 21,428.8 \text{ g}$

60. $\dfrac{7.68 \text{ kg}}{1} \cdot \dfrac{2.2 \text{ lb}}{1 \text{ kg}} \approx 16.90 \text{ lb}$

61. $\dfrac{28.6 \text{ L}}{1} \cdot \dfrac{1.0567 \text{ qt}}{1 \text{ L}} \approx 30.22 \text{ qt}$

62. $\dfrac{59.4 \text{ L}}{1} \cdot \dfrac{1.0567 \text{ qt}}{\text{L}} \approx 62.77 \text{ qt}$

63. $\dfrac{28.2 \text{ gal}}{1} \cdot \dfrac{3.785 \text{ L}}{1 \text{ gal}} \approx 106.7 \text{ L}$

64. $\dfrac{16 \text{ qt}}{1} \cdot \dfrac{0.9464 \text{ L}}{1 \text{ qt}} \approx 15.14 \text{ L}$

65. Unreasonable; $\dfrac{2 \text{ kg}}{1} \cdot \dfrac{2.2 \text{ lb}}{1 \text{ kg}} = 4.4 \text{ lb}$

66. Unreasonable; $\dfrac{4 \text{ L}}{1} \cdot \dfrac{0.2642 \text{ gal}}{1 \text{ L}} \approx 1.1 \text{ gal}$

67. Reasonable;
$\dfrac{25 \text{ ml}}{1} \cdot \dfrac{1 \text{ L}}{1000 \text{ ml}} \cdot \dfrac{1.0567 \text{ qt}}{1 \text{ L}} \cdot \dfrac{32 \text{ oz}}{1 \text{ qt}} \cdot \dfrac{2 \text{ T}}{1 \text{ oz}}$
$\approx 1.7 \text{ T}$

68. Reasonable; $\dfrac{6 \text{ L}}{1} \cdot \dfrac{0.2642 \text{ gal}}{1 \text{ L}} \approx 1.6 \text{ gal}$

69. Unreasonable; $\dfrac{0.5 \text{ L}}{1} \cdot \dfrac{1.0567 \text{ qt}}{1 \text{ L}} \approx 0.5 \text{ qt}$

70. Unreasonable;
$\dfrac{40 \text{ g}}{1} \cdot \dfrac{0.0022 \text{ lb}}{1 \text{ g}} \cdot \dfrac{16 \text{ oz}}{1 \text{ lb}} \approx 1.4 \text{ oz}$

71. B; $\dfrac{3 \text{ m}}{1} \cdot \dfrac{3.2808 \text{ ft}}{1 \text{ m}} \approx 9.8 \text{ ft}$

72. C; $\dfrac{5 \text{ m}}{1} \cdot \dfrac{3.2808 \text{ ft}}{1 \text{ m}} \approx 16 \text{ ft}$

73. B; $\dfrac{5000 \text{ km}}{1} \cdot \dfrac{0.62124 \text{ mi}}{1 \text{ km}} \approx 3100 \text{ mi}$

74. A; $\dfrac{3 \text{ cm}}{1} \cdot \dfrac{1 \text{ in.}}{2.54 \text{ cm}} \approx 1.2 \text{ in.}$

75. C; $\dfrac{193 \text{ mm}}{1} \cdot \dfrac{1 \text{ cm}}{10 \text{ mm}} \cdot \dfrac{1 \text{ in.}}{2.54 \text{ cm}} \approx 7.6 \text{ in.}$

76. A; $\dfrac{1 \text{ kg}}{1} \cdot \dfrac{2.20 \text{ lb}}{1 \text{ kg}} = 2.2 \text{ lb}$

77. A; $\dfrac{1300 \text{ kg}}{1} \cdot \dfrac{2.20 \text{ lb}}{1 \text{ kg}} \approx 2900 \text{ lb}$

78. B; $\dfrac{355 \text{ ml}}{1} \cdot \dfrac{1 \text{ L}}{1000 \text{ ml}} \cdot \dfrac{1.0567 \text{ qt}}{1 \text{ L}} \cdot \dfrac{32 \text{ oz}}{1 \text{ qt}}$
$\approx 12.0 \text{ oz}$

79. A; $\dfrac{180 \text{ cm}}{1} \cdot \dfrac{1 \text{ m}}{100 \text{ cm}} \cdot \dfrac{3.2808 \text{ ft}}{1 \text{ m}} \approx 5.9 \text{ ft}$

80. C; $\dfrac{13,000 \text{ km}}{1} \cdot \dfrac{0.6214 \text{ mi}}{1 \text{ km}} \approx 8100 \text{ mi}$

81. C; $\dfrac{800 \text{ m}}{1} \cdot \dfrac{1 \text{ km}}{1000 \text{ m}} \cdot \dfrac{0.6214 \text{ mi}}{1 \text{ km}} \approx 0.5 \text{ mi}$

82. A; $\dfrac{1 \text{ L}}{1} \cdot \dfrac{1.0567 \text{ qt}}{1 \text{ L}} \approx 1.1 \text{ qt}$

83. A; $\dfrac{70 \text{ cm}}{1} \cdot \dfrac{1 \text{ m}}{100 \text{ cm}} \cdot \dfrac{39.37 \text{ in.}}{1 \text{ m}} \approx 28 \text{ in.}$

84. C; $\dfrac{70 \text{ kg}}{1} \cdot \dfrac{2.20 \text{ lb}}{1 \text{ kg}} = 154 \text{ lb}$

85. B; $\dfrac{50 \text{ cm}}{1} \cdot \dfrac{1 \text{ m}}{100 \text{ cm}} \cdot \dfrac{39.37 \text{ in.}}{1 \text{ m}} \approx 20 \text{ in.}$

86. A; $\dfrac{1 \text{ m}}{1} \cdot \dfrac{3.2808 \text{ ft}}{1 \text{ m}} \approx 3.3 \text{ ft}$

87. B; $\dfrac{9 \text{ mm}}{1} \cdot \dfrac{1 \text{ in.}}{25.4 \text{ mm}} \approx 0.35 \text{ in.}$

88. A; $\dfrac{300 \text{ mm}}{1} \cdot \dfrac{1 \text{ in.}}{25.4 \text{ mm}} \approx 12 \text{ in.}$

89. A; this is the freezing temperature of water in Celsius.

90. B; $F = \dfrac{9}{5} \cdot 40 + 32 = 72 + 32 = 104°F$

91. B; $F = \dfrac{9}{5} \cdot 60 + 32 = 108 + 32 = 140°F$

92. A; this is the boiling temperature of water in Celsius.

93. C; $F = \dfrac{9}{5} \cdot 10 + 32 = 18 + 32 = 50°F$

94. A; $F = \dfrac{9}{5} \cdot 30 + 32 = 54 + 32 = 86°F$

95. B; $F = \dfrac{9}{5} \cdot 170 = 32 = 306 + 32 \approx 340°F$

96. A; $F = \dfrac{9}{5} \cdot 35 + 32 = 63 + 32 = 95°F$